Desktop Encyclopedia of
Voice and Data Networking

Desktop Encyclopedia of Voice and Data Networking

Nathan J. Muller

McGraw-Hill

New York • San Francisco • Washington, D.C. • Auckland
Bogotá • Caracas • Lisbon • London • Madrid
Mexico City • MilanMontreal • New Delhi • San Juan
Singapore • Sydney • Tokyo • Toronto

Library of Congress Cataloging-in-Publication Data

Muller, Nathan J.
  Desktop encyclopedia of voice and data networking / Nathan J.
Muller.
    ISBN 0-07-134711-9
    1. Telecommunication systems Encyclopedias.   2. Data transmission
systems Encyclopedias.   I. Title.
TK5102.M853   2000
621.382—dc21   99-32794
                                                         CIP

McGraw-Hill

*A Division of The **McGraw·Hill** Companies*

1 2 3 4 5 6 7 8 9 0   DOC/DOC   9 0 4 3 2 1 0 9

ISBN 0-07-134711-9

The sponsoring editor for this book was Steve Chapman, the editing supervisor was Kellie Hagan,
and the production supervisor was Sherri Souffrance. It was set in Vendome ICG by Wanda Ditch
through the services of Barry E. Brown (Broker—Editing, Design and Production).

Printed and bound by R.R. Donnelley & Sons Company.

McGraw-Hill books are available at special quantity discounts to use as premiums and sales promo-
tions, or for use in corporate training programs. For more information, please write to the Director
of Special Sales, McGraw-Hill, 11 West 19th Street, New York, NY 10011. Or contact your local book-
store.

 This book is printed on recycled, acid-free paper containing a minimum of 50% recycled, de-inked fiber.

This book is dedicated to the person I love and admire the most:

CONTENTS

■ ■ ■ PREFACE

The telecommunications industry is changing rapidly in its quest to converge voice and data on high-performance multiservice networks that are designed to meet virtually any business and consumer need. These next-generation networks are being driven as much by the revenue-generating potential of new integrated applications as they are by the promise of more efficient, low-cost, easier-to-manage networks that will provide operators and users significant competitive advantage in the global economy.

The focus of the telecommunications industry is on packet-switching networks—the kind that make the Internet possible—whose performance doubles every year. By contrast, conventional circuit-switching networks—the kind that carry ordinary phone calls—take at least four years to achieve the same improvement. With traffic doubling on the Internet every six months, only packet switches have a chance of keeping pace. In fact, several major regional and long-distance telephone companies have announced that they are beefing up their investments in packet-switching technologies and cutting back on spending for circuit-switching technology.

An emerging generation of routing switches handles massive amounts of IP packets—for voice as well as data—at a performance and cost dramatically different from that of the past. In 1998 alone, prices for data network equipment have dropped by a factor of 10, while performance has gone up by about the same amount. Packetized voice allows many conversations to take place over the same bandwidth pipe simultaneously. And what takes 64 Kbps on a traditional voice network can fit into 8 Kbps and less on a packet-switched network without a perceptible loss of quality. Carriers can offer multiple services more efficiently and cheaply by moving them onto a single, improved packet-based infrastructure.

Of course, this is only one of the many trends influencing the direction of the telecommunications industry. Accordingly, this book straddles both the circuit-switched and packet-switched environments at several levels: voice-data, wireline- wireless, and business-consumer. This comprehensive volume gives you a painless way to fill any knowledge gaps you might have, while providing new insights about the operational aspects of today's increasingly complex technologies, networks, and applications. It also contains management concepts that help put many other topics into perspective. Of course, the book makes an excellent reference for those outside the industry who want to better understand how these and other developments are advancing and changing our everyday lives.

This book is the third in a series of desktop encyclopedias, which attempts to chronicle the sweeping changes in technologies and applications. For those of you who want an Internet-oriented perspective, there is the *Desktop Encyclopedia of the Internet*, published by Artech House in Norwood, MA. If you want a voice-oriented perspective, there is the *Desktop Encyclopedia of Telecommunications*, published by McGraw-Hill. All three encyclopedias can be conveniently purchased on the Web through Amazon.com or BarnesandNoble.com. The information contained in this book, especially as it relates to specific vendors and products, is accurate at the time it was written and is, of course, subject to change with continued advancements in technology and shifts in market forces. Mention of specific products and services is for illustration purposes only and does not constitute an endorsement of any kind by either the author or the publisher.

ACRONYMS

AAL   ATM Adaptation Layer
ABATS   Automated Bit Access Test System
ABM   Accunet Bandwidth Manager (AT&T)
ABR   Available Bit Rate
AC   Access Control
AC   Address Copied
AC   Alternating Current
AC   Authentication Center
ACD   Automatic Call Distributor
ACELP   Algebraic Code Excited Linear Predictive
ACK   Acknowledge
ACL   Asynchronous Connectionless
ACP   Access Control Point
ACPI   Advanced Configuration and Power Interface
ADA   Americans with Disabilities Act
ADC   Analog to Digital Converter
ADCR   Alternate Destination Call Routing (AT&T)
ADM   Add-Drop Multiplexer
ADN   Advanced Digital Network (Pacific Bell)
ADPCM   Adaptive Differential Pulse Code Modulation
ADSI   Active Directory Services Interface
ADSL   Asymmetrical Digital Subscriber Line
AFLC   Adaptive Frame Loss Concealment
AFP   Apple File Protocol
AGRAS   Air-Ground Radiotelephone Automated Service
AIOD   Automatic Identification of Outward Dialed calls
AIN   Advanced Intelligent Network
AJBM   Automatic Jitter Buffer Management
ALI   Automatic Location Information
AM   Amplitude Modulation
AMI   Alternate Mark Inversion
AMPS   Advanced Mobile Phone Service
ANI   Automatic Number Identification
ANR   Automatic Network Routing (IBM Corp.)
ANSI   American National Standards Institute

ANT   ADSL Network Terminator
AO/DI   Always On/Dynamic ISDN
AOL   America Online
AP   Access Point
APC   Access Protection Capability (AT&T)
APD   Avalanche Photo Diode
API   Application Programming Interface
APPC   Advanced Program-to-Program Communications (IBM Corp.)
APPN   Advanced Peer-to-Peer Network (IBM Corp.)
APC   Automatic Protection Switching
ARCnet   Attached Resource Computer Network (Datapoint Corp.)
ARB   Adaptive Rate Based (IBM Corp.)
ARP   Address Resolution Protocol
ARPA   Advanced Research Projects Agency
ARQ   Automatic Repeat Request
ARS   Action Request System (Remedy Systems Inc.)
AS   Autonomous System
ASCII   American Standard Code for Information Interchange
ASIC   Application-Specific Integrated Circuit
ASN.1   Abstract Syntax Notation 1
ASTN   Alternate Signaling Transport Network (AT&T)
AT&T   American Telephone & Telegraph
ATE   Advanced Television Enhancement
ATIS   Alliance for Telecommunications Industry Solutions (formerly, ECSA)
ATM   Asynchronous Transfer Mode
ATSC   Advanced Television Systems Committee
ATVEF   Advanced Television Enhancement Forum
AUI   Attachment Unit Interface
AWG   American Wire Gauge

B   Byte
B8ZS   Binary Eight Zero Substitution
BACP   Bandwidth Allocation Control Protocol
BBS   Bulletin Board System
BCC   Block Check Character
BCCH   Broadcast Control Channel

BDCS   Broadband Digital Cross-connect System
BECN   Backward Explicit Congestion Notification
Bellcore   Bell Communications Research, Inc.
BER   Bit Error Rate
BERT   Bit Error Rate Tester
BGP   Border Gateway Protocol
BHCA   Busy Hour Call Attempts
BIB   Backward Indicator Bit
BIOS   Basic Input-Output System
BMC   Block Multiplexer Channel (IBM Corp.)
BMS-E   Bandwidth Management Service-Extended (AT&T)
BNC   Bayonet Nut Connector
BOC   Bell Operating Company
BONDING   Bandwidth on Demand Interoperability Group
BootP   Boot Protocol
BPDU   Bridge Protocol Data Unit
BPS   Bits Per Second
BPV   Bipolar Violation
BRI   Basic Rate Interface (ISDN)
BSA   Basis Serving Arrangement
BSA   Business Software Alliance
BSC   Base Station Controller
BSC   Binary Synchronous Communication (IBM Corp.)
BSD   Berkeley Software Distribution
BSE   Basic Service Element
BSN   Backward Sequence Number
BSSC   Base Station System Controller
BTS   Base Transceiver Station
BUS   Broadcast/Unknown Server

CA   Communications Assistant
CAD   Computer Aided Design
CAM   Computer Aided Manufacturing
CAN   Campus Area Network
CAP   Carrierless Amplitude/Phase (modulation)
CAP   Competitive Access Provider
CARP   Cache Array Routing Protocol
CARS   Cable Antenna Relay Services
CASE   Computer Aided Software Engineering
CATV   Community Antenna Television
CB   Citizens Band
CBR   Constant Bit Rate

CBSC   Centralized Base Site Controller
CCC   Clear Channel C
CCCH   Common Control Channel
CCITT   Consultative Committee for International Telegraphy and Telephony
CCR   Customer Controlled Reconfiguration
CCS   Common Channel Signaling
CCSNC   Common Channel Signaling Network Controller
CCSS 6   Common Channel Signaling Systems 6
CCTV   Closed Circuit Television
CCU   Central Control Unit
CD   Compact Disk
CDCS   Continuous Dynamic Channel Selection
CDG   CDMA Development Group
CD-R   Compact Disk-Recordable
CD-ROM   Compact Disk-Read Only Memory
CDMA   Code Division Multiple Access
CDO   Community Dial Office
CDPD   Cellular Digital Packet Data
CDR   Call Detail Recording
CEI   Comparably Efficient Interconnection
CEMA   Consumer Electronics Manufacturers Association
CENTREX   Central Office Exchange
CGI   Common Gateway Interface
CGSA   Cellular Geographic Servicing Areas
CHAP   Challenge Handshake Authentication Protocol
CICS   Customer Information Control System (IBM Corp.)
CIF   Common Intermediate Format
CIO   Chief Operations Officer
CIR   Committed Information Rate
CIX   Commercial Internet Exchange
CLASS   Custom Local Area Signaling Services
CLEC   Competitive Local Exchange Carrier
CLI   Calling Line Identification
CLP   Cell Loss Priority
CMI   Cable Microcell Integrator
CMIP   Common Management Information Protocol
CMIS   Common Management Information Services
CMRS   Commercial Mobile Radio Service

CNR   Customer Network Reconfiguration
CNS   Complementary Network Service
CO   Central Office
CON   Concentrator
COPS   Common Open Policy Service
CORBA   Common Object Request Broker Architecture
CoS   Class of Service
COT   Central Office Terminal
CP   Communications Processor
CP   Connection Point
CP   Coordination Processor
CPE   Customer Premises Equipment
CPID   Calling Party Identification
CPS   cycles per second (Hertz)
CPU   Central Processing Unit
CRC   Cyclic Redundancy Check
CRIMP   Connectivity Routing and Infrastructure Modeling Program (Cablesoft, Inc.)
CSA   Carrier Serving Area
CSM   Communications Services Management
CSMA/CD   Carrier Sense Multiple Access with Collision Detection
CSU   Channel Service Unit
CT   Cordless Telecommunications
CTI   Computer-Telephony Integration
CUG   Closed User Group
CVSD   Continuously Variable Slope Delta (modulation)

D-AMPS   Digital Advanced Mobile Phone Service
DA   Destination Address
DAC   Data Acquisition and Control
DACS   Digital Access and Cross-connect System (AT&T)
DAC   Digital to Analog Converter
DAP   Demand Access Protocol
DAP   Directory Access Protocol
DAS   Dual Attached Station
DASD   Direct Access Storage Device (IBM Corp.)
DAT   Digital Audio Tape
dB   decibel
DBA   Dynamic Bandwidth Allocation
DBMS   Data Base Management System
DBS   Direct Broadcast Satellite
DBU   Dial Backup Unit
DCCH   Digital Control Channel

DCE   Data Communications Equipment
DCE   Distributed Computing Environment
DCF   Data Communication Function
DCOM   Distributed Component Object Model (Microsoft Corp.)
DCS   Digital Cross-connect System
DDS   Digital Data Services
DDS/SC   Digital Data Service with Secondary Channel
D/E   Debt/Equity (ratio)
DECT   Digital Enhanced (formerly, European) Cordless Telecommunication
DES   Data Encryption Standard
DFB   Distributed Feedback
DFSMS   Data Facility Storage Management Subsystem (IBM Corp.)
DHCP   Dynamic Host Configuration Protocol
DID   Direct Inward Dialing
DIF   Digital Interface Frame
Diff-Serv   Differentiated Services
DDL   Data Link Layer
DLCI   Data Link Connection Identifier
DLCS   Digital Loop Carrier System
DLL   Dynamic Link Library
DLSw   Data Link Switching (IBM Corp.)
DLU   Digital Line Unit
DM   Distributed Management
DME   Distributed Management Environment
DMI   Desktop Management Interface
DMT   Discrete Multi Tone
DMTF   Desktop Management Task Force
DNS   Domain Name Service
DOCSIS   Data Over Cable Service Interface Specification
DoD   Department of Defense
DOD   Direct Outward Dialing
DOM   Document Object Model
DOS   Disk Operating System
DOV   Data Over Voice
DPL   Digital Power Line
DQDB   Distributed Queue Dual Bus
DQPSK   Differential Quadrature Phase-Shift Keying
DSO   Digital Signal-Level 0 (64 Kbps)
DS1   Digital Signal-Level 1 (1.544 Mbps)
DS1C   Digital Signal-Level 1C (3.152 Mbps)
DS2   Digital Signal-Level 2 (6.312 Mbps)
DS3   Digital Signal-Level 3 (44.736 Mbps)
DS4   Digital Signal-Level 4 (274.176 Mbps)

DSA Directory System Agent
DSI Digital Speech Interpolation
DSL Digital Subscriber Line
DSN Defense Switched Network
DSP Digital Signal Processor
DSP Directory System Protocol
DSS Decision Support System
DSSS Direct Sequence Spread Spectrum
DSU Data Service Unit
DSX1 Digital Systems Cross-connect 1
DTE Data Terminal Equipment
DTMF Dual Tone Multi Frequency
DTR Dedicated Token Ring
DTU Data Transfer Unit
DTV Digital Television
DUA Directory User Agent
DWDM Dense Wavelength Division Multiplexer
DWMT Discrete Wavelet Multi Tone
DXI Data Exchange Interface

E&M Ear and Mouth (a signaling method)
E-Mail Electronic Mail
E-TDMA Expanded Time Division Multiple Access
EB Exabyte (1,000,000,000,000,000,000 bytes)
EBCDIC Extended Binary Code Decimal Interchange Code (IBM Corp.)
ECMA European Computer Manufacturers Association
ECSA Exchange Carriers Standards Association
ED Ending Delimiter
EDFA Erbium-Doped Fiber Amplifier
EDI Electronic Data Interchange
EDIFACT Electronic Data Interchange For Administration and Transport
EDRO Enhanced Diversity Routing Option (AT&T)
EEROM Electronically Erasable Read-Only Memory
EFF Electronic Frontier Foundation
EFRC Enhanced Full Rate Codec
EFS Encrypting File System (Microsoft Corp.)
EFT Electronic Funds Transfer
EGP External Gateway Protocol
EHF Extremely High Frequency (more than 30 GHz)
EIA Electronic Industries Association

EIR Equipment Identity Register
EISA Extended Industry Standard Architecture
EMI Electromechanical Inteference
EMS Element Management System
EOC Embedded Overhead Channel
EOT End of Transmission
EP Extension Point
ERP Enterprise Resource Planning
ESCON Enterprise System Connection (IBM Corp.)
ESD Electronic Software Distribution
ESF Extended Super Frame
ESMA Enterprise Storage Management Architecture (Legato Systems)
ESMR Enhanced Specialized Mobile Radio
ESMTP Extended Simple Mail Transfer Protocol
ESN Electronic Serial Number
ETC Exempt Telecommunication Companies
ETSI European Telecommunication Standards Institute
EVRC Enhanced Variable Rate Encoder

4GL Fourth-Generation Language
FACCH Fast Associated Control Channel
FASB Financial Accounting Standards Board
FASC Fraud Analysis and Surveillance Center (AT&T)
FASTAR Fast Automatic Restoral (AT&T)
FAT File Allocation Table
FC Frame Control
FC Fibre Channel
FC-0 Fibre Channel—Layer 0
FC-1 Fibre Channel—Layer 1
FC-2 Fibre Channel—Layer 2
FC-3 Fiber Channel—Layer 3
FC-4 Fibre Channel—Layer 4
FC-AL Fibre Channel-Arbitrated Loop
FCC Federal Communications Commission
FCS Frame Check Sequence
FDDI Fiber Distributed Data Interface
FDIC Federal Deposit Insurance Corporation
FDL Facilities Data Link
FEC Forward Error Correction
FECN Forward Explicit Congestion Notification

FEP   Front-End Processor
FHSS   Frequency Hopping Spread Spectrum
FIB   Forward Indicator Bit
FIB   Forwarding Information Base
FIFO   First-In/First-Out
FIRST   Flexible Integrated Radio Systems Technology
FITL   Fiber-In-The-Loop
FM   Frequency Modulation
FOCC   Forward Control Channel
FOD   Fax on Demand
FPF   Fraud Protection Feature
FRAD   Frame Relay Access Device
FRF   Frame Relay Forum
FRS   Family Radio Service
FS   Frame Status
FSN   Forward Sequence Number
FTAM   File Transfer, Access, and Management
FT1   Fractional T1
FTP   File Transfer Protocol
FTS   Federal Telecommunications System
FTTB   Fiber To The Building
FTTC   Fiber To The Curb
FTTH   Fiber To The Home
FWA   Fixed Wireless Access
FWT   Fixed Wireless Terminal
FX   Foreign Exchange (line)
FXO   Foreign Exchange Office
FXS   Foreign Exchange Station

GATT   General Agreement on Tariffs and Trade
GB   Gigabyte (1,000,000,000 bytes)
GDS   Generic Digital Services
GEO   Geostationary-earth-orbit
GFC   Generic Flow Control
GFR   Guaranteed Frame Rate
GHz   Gigahertz (billions of cycles per second)
GIS   Geographic Information Systems
GloBanD   Global Bandwidth on Demand
GMRS   General Mobile Radio Service
GPI   General Purpose Intel
GPS   Global Positioning System
GSA   General Services Administration
GSM   Global System for Mobile (GSM) telecommunications (formerly, Groupe Spéciale Mobile)
GSN   Gigabit System Network
GUI   Graphical User Interface

H0   High-capacity ISDN channel operating at 384 Kbps
H11   High-capacity ISDN channel operating at 1.536 Mbps
HDLC   High-level Data Link Control
HDML   Handheld Device Markup Language
HDSL   High-bit-rate Digital Subscriber Line
HDTV   High Definition Television
HEC   Header Error Check
HF   High Frequency (3 MHz to 30 MHz)
HFC   Hybrid Fiber/Coax
HIC   Head-end Interface Converter
HIPPI   High Performance Parallel Interface
HLR   Home Location Register
HNF   High-performance Networking Forum
HPNA   Home Phoneline Networking Alliance
HIPPI   High Performance Parallel Interface
HomePNA   Home Phoneline Networking Alliance
HPR   High Performance Routing (IBM Corp.)
HSCSD   High Speed Circuit Switched Data
HSM   Hierarchical Storage Management
HST   Helical Scan Tape
HSTR   High Speed Token Ring
HSTRA   High Speed Token Ring Alliance
HTML   HyperText Markup Language
HTTP   HyperText Transfer Protocol
HVAC   Heating, Ventilation, and Air Conditioning
Hz   Hertz (cycles per second)

I/O   Input/Output
IAB   Internet Architecture Board
IANA   Internet Assigned Numbers Authority
IAPP   Inter Access Point Protocol
ICI   Interexchange Carrier Interface
ICMP   Internet Control Message Protocol
ICP   Internet Cache Protocol
ICR   Intelligent Call Routing
ICS   Intelligent Calling System
ICSA   International Computer Security Association
ID   Identification
IDDD   International Direct Dialing Designator
IDPR   Inter-Domain Policy Routing
IEC   International Electrotechnical Commission

IEEE Institute of Electrical and Electronic Engineers

IESG Internet Engineering Steering Group

IETF Internet Engineering Task Force

IF Intermediate Frequence

IGMP Internet Group Management Protocol

IGP Interior Gateway Protocol

IIS Internet Information Server (Microsoft Corp.)

ILEC Incumbent Local Exchange Carrier

IMAP Internet Mail Access Protocol

IMEI International Mobile Equipment Identity

IMS/VS Information Management System/Virtual Storage (IBM Corp.)

IMSI International Mobile Subscriber Identity

IMT Intelligent Multimode Terminals

IMTS Improved Mobile Telephone Service

IN Intelligent Network

INMARSAT International Maritime Satellite Organization

INMS Integrated Network Management System

IOC Inter Office Channel

IP Internet Protoco

IPH Integrated Packet Handler

IPI Intelligent Peripheral Interface

IPN Intelligent Peripheral Node

IPX Internetwork Packet Exchange

IR Infrared

IRC Internet Relay Chat

IrDA Infrared Data Association

IrLAN Infrared LAN

IrLAP Infrared Link Access Protocol

IrLMP Infrared Link Management Protocol

IrPL Infrared Physical Layer

IRQ Interrupt Request

IrTTP Infrared Transport Protocol

IS Information System

IS Industry Standard

IS-IS Intra-autonomous System to Intra-autonomous System

ISA Industry Standard Architecture

ISAPI Internet Server Applications Programming Interface

ISDL ISDN Subscriber Digital Line

ISDN Integrated Services Digital Network

ISM Industrial, Scientific, and Medical (frequency bands)

ISO International Organization for Standardization

ISOC Internet Society

ISP Internet Service Provider

ISR Intermediate Session Routing (IBM Corp.)

ISSI Inter-Switching Systems Interface

IT Information Technology

ITFS Instructional Television Fixed Service

ITG Internet Telephony Gateway

ITR Intelligent Text Retrieval

ITU International Telecommunication Union

IVR Interaction Voice Response

IXC Interexchange Carrier

JCE Java Cryptography Extensions

JIT Just In Time

JEPI Joint Electronic Payments Initiative

JPEG Joint Photographic Experts Group

JRE Java Runtime Environment

JTAPI Java Telephony Application Programming Interface

JTC Joint Technical Committee

JVM Java Virtual Machine

K (Kilo) One Thousand (e.g., Kbps)

KB Kilobyte (1,000 bytes)

KHz kilohertz (thousands of cycles per second)

KSU Key Service Unit

KTS Key Telephone System

KVM Keyboard, Video, Mouse (a type of switch)

L2F Layer 2 Forwarding

L2TP Layer 2 Tunneling Protocol

LAN Local Area Network

LANCES LAN Resource Extension and Services (IBM Corp.)

LANE Local Area Network Emulation

LAPB Link Access Procedure-Balanced

LAT Local Area Transport (Digital Equipment Corp.)

LATA Local Access and Transport Area

LBO Line Build Out

LCD Liquid Crystal Display

LCN   Local Channel Number
LCP   Link Control Protocol
LD   Laser Diode
LEC   Local Exchange Carrier
LED   Light Emitting Diode
LEO   Low-earth-orbit
LF   Low Frequency (30 KHz to 300KHz)
LI   Length Indicator
LIPS   Lightweight Internet Person Schema
LLC   Logical Link Control
LMDS   Local Multipoint Distribution
System
LMS   Location and Monitoring Service
LNP   Local Number Portability
LOM   LAN on Motherboard
LSAPI   Licensing Service Application
Programming Interface
LSI   Large Scale Integration
LTG   Line Trunk Group
LU   Logical Unit (IBM Corp.)
LVD   Low Voltage Differential

M   (Mega) One Million (e.g., Mbps)
ma   milliamp
MAC   Media Access Control
MAC   Moves, Adds, Changes
MAE   Metropolitan Area Exchange
MAN   Metropolitan Area Network
MAPI   Messaging Applications
Programming Interface (Microsft Corp.)
MAS   Multiple Address System
MATV   Metropolitan Antenna Television
MAU   Multistation Access Unit
MB   Megabyte (1,000,000 bytes)
MCA   Micro Channel Architecture (IBM
Corp.)
MCU   Multipoint Control Unit
MD   Mediation Device
MDF   Main Distribution Frame
MDI   Medium Dependent Interface
MDS   Multipoint Distribution Service
MEO   Middle-earth-orbit
MF   Mediation Function
MES   Master Earth Station
MF   Medium Frequency (300 KHz to 3
MHz)
MHz   megahertz (millions of cycles per
second)
MIB   Management Information Base
MIC   Management Integration Consortium

MIF   Management Information Format
MII   Media Independent Interface
MIME   Multipurpose Internet Mail
Extensions
MIN   Mobile Identification Number
MIPS   Millions of Instructions Per
Second
MIS   Management Information Services
MISR   Multiprotocol Integrated Switch-
Routing
MJU   Multipoint Junction Unit
MM   Mobility Manager
MMDS   Multichannel, Multipoint
Distribution Service
MMITS   Modular Multifunction
Information Transfer System
MO   Magneto-Optical
Modem   Modulation/demodulation
MOSPF   Multicast Open Shortest Path
First
MPEG   Moving Pictures Experts Group
MPOA   Multi-Protocol Over ATM
MPPP   Multilink Point to Point Protocol
MRI   Magnetic Resonance Imaging
ms   millisecond (thousandths of a second)
MS   Mobile Station
MSC   Mobile Switching Center
MSF   Multiservice Switching Forum
MSN   Microsoft Network
MSRN   Mobile Station Roaming Number
MSS   Mobile Satellite Service
MTA   Message Transfer Agent
MTA   Multimedia Terminal Adapter
MTBF   Mean Time Between Failure
MTBSO   Mean Time Between Service
Outages
MTP   Message Transfer Part
MTSO   Mobile Telephone Switching Office
MTTR   Mean Time To Restore
MVC   Multicast Virtual Circuit
MVDS   Microwave Video Distribution
System
MVPRP   Multi-Vendor Problem
Resolution Process
mW   milliwatt

N-AMPS   Narrowband Advanced Mobile
Phone Service
NAC   Network Applications Consortium
NAK   Negative Acknowledge
NAM   Numeric Assignment Module

NAP   Network Access Point

NAU   Network Addressable Unit or Network Accessible Unit (IBM Corp.)

NAUN   Nearest Active Upstream Neighbor

NC   Network Computer

NCP   Network Control Program (IBM Corp.)

NCP   Network Control Point

NCSA   National Center for Supercomputer Applications

NDIS   Network Driver Interface Specification

NDS   Network Directory Service

NDS   Novell Directory Services (Novell, Inc.)

NE   Network Element

NEBS   New Equipment Building Specifications

NECA   National Exchange Carrier Association

NetBIOS   Network Basic Input/Output System

NEF   Network Element Function

NFS   Network File System (or Server)

NIC   Network Interface Card

NiCd   Nickel Cadmium

NIF   Network Interconnection Facilities

NiMH   Nickel-Metal Hydride

NIST   National Institute of Standards and Technology

NLM   NetWare Loadable Module (Novell Inc.)

nm   nanometer

NM   Network Manager

NMS   NetWare Management System (Novell, Inc.)

NMS   Network Management System

NNM   Network Node Manager (Hewlett-Packard Co.)

NNTP   Network News Transfer Protocol

NOC   Network Operation Center

NOS   Network Operating System

NPC   Network Protection Capability (AT&T)

NPV   Net Present Value

NRC   Network Reliability Council

NRIC   Network Reliability and Interoperability Council

ns   nanosecond

NSA   National Security Agency

NSF   National Science Foundation

NTSA   Networking Technical Support Alliance

NTSC   National Television Standards Committee

OAM   Operations, Administration, Management

OAM&P   Operations, Administration, Maintenance and Provisioning

OC   Optical Carrier

OC-1   Optical Carrier Signal-Level 1 (51.84 Mbps)

OC-3   Optical Carrier Signal-Level 3 (155.52 Mbps)

OC-9   Optical Carrier Signal-Level 9 (466.56 Mbps)

OC-12   Optical Carrier Signal-Level 12 (622.08 Mbps)

OC-18   Optical Carrier Signal-Level 18 (933.12 Mbps)

OC-24   Optical Carrier Signal-Level 24 (1.244 Gbps)

OC-36   Optical Carrier Signal-Level 36 (1.866 Gbps)

OC-48   Optical Carrier Signal-Level 48 (2.488 Gbps)

OC-96   Optical Carrier Signal-Level 96 (4.976 Gbps)

OC-192   Optical Carrier Signal-Level 192 (9.952 Gbps)

OC-256   Optical Carrier Signal-Level 256 (13.271 Gbps)

OCR   Optical Character Recognition

OCUDP   Office Channel Unit Data Port

ODBC   Open Data Base Connectivity (Microsoft Corp.)

ODI   Open Datalink Interface

ODS   Operational Data Store

OEM   Original Equipment Manufacturer

OFX   Open Financial Exchange

OLAP   On-Line Analytical Processing

OLE   Object Linking and Embedding

OMA   Object Management Architecture

OMAP   Operations, Maintenance, Administration & Provisioning

OMC   Operations and Maintenance Center

OMF   Object Management Framework

OMG   Object Management Group

OOP   Object Oriented Programming

OPX   Off Premises Extension
ORB   Object Request Broker
OS   Operating System
OS/2   Operating System/2 (IBM Corp.)
OSF   Open Software Foundation
OSF   Operations Systems Function
OSI   Open Systems Interconnection
OSS   Operations Support Systems
OTDR   Optical Time Domain Reflectometry

PA   Preamble
PACS   Personal Access Communications System
PAD   Packet Assembler-Disassembler
PAL   Phase Alternating by Line
PAN   Personal Area Network
PAP   Password Authentication Protocol
PB   Petabyte (1,000,000,000,000,000 bytes)
PBX   Private Branch Exchange
PC   Personal Computer
PCB   Printed Circuit Board
PCH   Paging Channel
PCI   Peripheral Component Interface
PCM   Pulse Code Modulation
PCN   Personal Communications Networks
PCS   Personal Communications Services
PCT   Private Communication Technology
PDA   Personal Digital Assistant
PDF   Portable Document Format (Adobe Systems)
PDN   Packet Data Network
PDU   Payload Data Unit
PEM   Privacy Enhanced Mail
PERL   Practical Extraction and Reporting Language
PGP   Pretty Good Privacy
PHS   Personal Handyphone System
PHY   Physical Layer
PIM   Personal Information Manager
PIN   Personal Identification Number
PIN   Positive-Intrinsic-Negative
PIP   Picture-in-a-Picture
PKE   Public Key Encryption
PLMRS   Private Land Mobile Radio Services
PMA   Physical Medium Attachment
PMD   Physical Media Dependent
PnP   Plug and Play
PoS   Packet over SONET
POP   Point of Presence

POP   Post Office Protocol
POS   Point of Sale
POTS   Plain Old Telephone Service
PPN   Policy Powered Network (3Com Corp.)
PPP   Point to Point Protocol
PPS   Packets Per Second
PPTP   Point-to-Point Tunneling Protocol
PQ   Priority Queuing
PRI   Primary Rate Interface (ISDN)
PSAP   Public Safety Answering Point
PSN   Packet Switched Network
PSTN   Public Switched Telephone Network
PT   Payload Type
PTT   Post Telephone & Telegraph
PU   Physical Unit (IBM Corp.)
PUC   Public Utility Commission
PUK   Personal Unblocking Key
PVC   Permanent Virtual Circuit
PWT   Personal Wireless Telecommunications
PXE   Preboot Execution Environment

QA   Quality Assurance
QAM   Quadrature Amplitude Modulation
QCIF   Quarter Common Intermediate Format
QIC   Quarter Inch Cartridge
QoS   Quality of Service
QPSK   Quadrature Phase Shift Keying

RAC   Remote Access Concentrator
RACH   Random Access Channel
RAD   Remote Antenna Driver
RADIUS   Remote Authentication Dial-In Service
RAID   Redundant Array of Inexpensive Disks
RAM   Random Access Memory
RAS   Remote Access Server
RASDL   Rate Adaptive Digital Subscriber Line
RASP   Remote Antenna Signal Processor
RBES   Rule Based Expert Systems
RCU   Remote Control Unit
RDBMS   Relational Data Base Management System
RDSS   Radio Determination Satellite Service
RED   Random Early Detection
RECC   Reverse Control Channel

RFC   Request for Comment
RF   Radio Frequency
RF   Routing Field
RFI   Radio Frequency Interference
RFI   Request for Information
RFP   Request For Proposal
RFQ   Request for Quotation
RG/U   Radio Guide/Utility
RI/RO   Ring In/Ring Out
RIP   Routing Information Protocol
RISC   Reduced Instruction Set Computing
RJE   Remote Job Entry
RJU   Residential Junction Unit
RLL   Run Length Limited
RMON   Remote Monitoring
ROI   Return on Investment
ROM   Read Only Memory
RPC   Remote Procedure Call
RSCN   Registered State Change
Notification
RSS   Residential Service System
RSVP   resource ReSerVation Protocol
RT   Remote Terminal
RTNR   Real Time Network Routing
(AT&T)
RTP   Rapid Transfer Protocol (IBM Corp.)
RTP   Real-time Transfer Protocol
RX   Receive

SA   Source Address
SACCH   Slow Associated Control Channel
SAFER   Split Access Flexible Egress Routing
(AT&T)
SAN   Storage Area Network
SAP   Second Audio Program
SAP   Service Access Point
SAP   Service Advertising Protocol
SAS   Single Attached Station
SATAN   Security Administrator Tool for
Analyzing Networks
SBCCS   Single Byte Command Code Set
(IBM Corp.)
SC   Subscriber Channel
SC   System Controller
SCC   Standards Coordinating Committees
(IEEE)
SCO   Synchronous Connection Oriented
SCP   Service Control Point
SCSI   Small Computer Systems Interface
SD   Starting Delimiter
SDCCH   Standalone Dedicated Control
Channel

SDH   Synchronous Digital Hierarchy
SDK   Software Development Kit
SDLC   Synchronous Data Link Control
(IBM Corp.)
SDM   Subrate Data Multiplexing
SDN   Software Defined Network(AT&T)
SDP   Service Delivery Point
SDR   Software Defined Radio
SDSL   Symmetric Digital Subscriber Line
SET   Secure Electronic Transaction
SFD   Start Frame Delimiter
SGML   Standard Generalized Markup
Language
SHF   Super High Frequency (3 GHz to 30
GHz)
SHTTP   Secure HyperText Transfer
Protocol
SIF   Signaling Information Field
SIF   SONET Interoperability Forum
SIM   Subscriber Identity Module
SIMA   Simple Integrated Media Access
SIP   SMDS Interface Protocol
SLA   Service Level Agreement
SLD   Service Level Definition
SLIC   Serial Line Interface Coupler (IBM
Corp.)
SLIP   Serial Line Internet Protocol
SLP   Service Location Protocol
SMDI   Station Message Desk Interface
SMDR   Station Message Detail Recording
SMDS   Switched Multimegabit Data
Services
SMP   Symmetric Multi-Processor
SMR   Specialized Mobile Radio
SMS   Service Management System
SMS   Short Message Service
SMT   Station Management
SMTP   Simple Mail Transfer Protocol
SN   Switching Network
SNA   Systems Network Architecture (IBM
Corp.)
snagas   SNA Gateway Access Server
SNI   Subscriber Network Interface
SNMP   Simple Network Management
Protocol
SNS   Simple Name Service
SOHO   Small Office Home Office
SONET   Synchronous Optical Network
SPA   Software Publishers Association
SPC   Stored Program Control

SPI   Service Provider Interface
SPX   Synchronous Packet Exchange (Novell, Inc.)
SQE   Signal Quality Error
SQL   Structured Query Language
SRB   Source Route Bridging
SS   Switching System
SS7   Signaling System No. 7
SSA   Serial Storage Architecture (IBM Corp.)
SSCP   System Services Control Point (IBM Corp.)
SSCP/PU   System Services Control Point/Physical Unit (IBM Corp.)
SSG   Service Selection Gateway
SSL   Secure Sockets Layer
SSP   Service Switching Point
SSP   Switch-to-Switch Protocol
ST   Somple Twist
STDM   Statistical Time Division Multiplexing
STP   Shielded Twisted Pair
STP   Signal Transfer Point
STP   Spanning Tree Protocol
STS   Shared Telecommunications Services
STS   Synchronous Transport Signal
STX   Start of Transmission
SUBT   Subscriber Terminal
SVC   Switched Virtual Circuit
SWC   Serving Wire Center
SYNTRAN   Synchronous Transmission
SYSGEN   System Generation
SWAP   Shared Wireless Access Protocol

3G   Third Generation
T1   Transmission service at the DS1 rate of 1.544 Mbps
T3   Transmission service at the DS3 rate of 44.736 Mbps
TA   Technical Advisor
TA   Technical Advisory
TAG   Technical Advisory Group
TAPI   Telephony Application Programming Interface (Microsoft Corp.)
TASI   Time Assigned Speech Interpolation
TB   Terabyte (1,000,000,000,000 bytes)
TBOP   Transparent Bit Oriented Protocol
Tbps   Terabit-per-second
TCAP   Transaction Capabilities Applications Part
TCL   Tool Command Language
TCO   Total Cost of Ownership

TCP   Transmission Control Protocol
TDD   Time Division Duplex
TDM   Time Division Multiplexer
TDMA   Time Division Multiple Access
TDMA/TDD   Time Division Multiple Access with Time Division Duplexing
TDR   Time Domain Reflectometry
TFTP   Trivial File Transfer Protocol
TIA   Telecommunications Industry Association
TIB   Tag Information Base
TIMS   Transmission Impairment Measurement Set
TL1   Transaction Language 1
TLS   Transparent LAN Service
TMN   Telecommunications Management Network
ToS   Type of Service
TRS   Telecommunications Relay Services
TSAPI   Telephony Services Application Programming Interface (Novell Inc.)
TSI   Time Slot Interchange
TSR   Terminal Stay Resident
TTRT   Target Token Rotation Time
TTY   Text Telephone
TV   Television
TWX   Teletypewriter Exchange (also known as Telex)
TX   Transmit

UART   Universal Asynchronous Receiver/Transmitter
UAWG   Universal ADSL Working Group
UBR   Unspecified Bit Rate
UDP   User Datagram Protocol
UHF   Ultra High Frequency (300 MHz to 3 GHz)
UMTS   Universal Mobile Telecommunication System
UN   United Nations
UNI   User-Network Interface
UPS   Uninterruptible Power Supply
USB   Universal Serial Bus
USDLA   United States Distance Learning Association
USNC   US. National Committee
UTP   Unshielded Twisted-Pair

VAR   Value Added Reseller
VBNS   Very High-Speed Backbone Network Service
VBR   Variable Bit Rate

VBR-rt   Variable Bit Rate real-time

VC   Virtual Circuit

VCI   Virtual Channel Identifier

VCR   Video Cassette Recorder

VCSEL   Vertical-Cavity Surface-Emitting Laser

VDSL   Very high-speed Digital Subscriber Line

VFN   Vendor Feature Node

VG   Voiced Grade

VHF   Very High Frequency (30 MHz to 300 MHz)

VLF   Very Low Frequency (less than 30 KHz)

VLR   Visitor Location Register

VLSI   Very Large Scale Integration

VM   Virtual Machine

VMS   Virtual Machine System (Digital Equipment Corp.)

VoD   Video on Demand

VoFR   Voice over Frame Relay

VoIP   Voice over Internet Protocol

VPI   Virtual Path Identifier

VP   Virtual Path

VPN   Virtual Private Network

VSAT   Very Small Aperture Terminal

VT   Virtual Terminal

VT   Virtual Tributary

VTAM   Virtual Telecommunications Access Method (IBM Corp.)

W3C   World Wide Web Consortium

W-CDMA   Wideband Code Division Multiple Access

WACS   Wireless Access Communications System

WAN   Wide Area Network

WAP   Wireless Access Protocol

WATS   Wide Area Telecommunications Service

WBM   Web Based Management

WCS   Wireless Communications Service

WDCS   Wideband Digital Cross-connect System

WDM   Wave Division Multiplexer

WEP   Wired Equivalent Privacy

WfM   Wired for Management

WFQ   Weighted Fair Queuing

WGS   Worldwide Geodetic System

WIMS   Wireless Multimedia and Messaging Services

WINS   Windows Internet Name Service (Microsoft Corp.)

WLAN   Wireless Local Area Network

WLL   Wireless Local Loop

WORM   Write Once Read Many

WRED   Weighted Random Earl Detection

WTO   World Trade Organization

WWW   World Wide Web

WWW3   World Wide Web consortium

XNS   Xerox Network System (Xerox Corp.)

Y2K   Year 2000

YAG   Yttrium-Aluminum-Garnet

YB   Yottabyte (1,000,000,000,000,000,000,000,000 bytes)

ZB   Zettabyte (1,000,000,000,000,000,000,000 bytes)

Advanced Television Enhancement

Despite the impressive growth in the number of households with multiple PCs and the meteoric rise of the Internet, one in five Americans do not want to own a computer. These households constitute the potential market for advanced television enhancement (ATE) services, which integrate telephone, television, and Internet services without requiring a PC platform. The first ATE service was tested in mid-1999, when U S WEST started offering its @TV service in selected areas within its 14-state western region.

In partnership with Network Computer, Inc. (NCI), the U S WEST service allows customers to send and receive e-mail, place and answer telephone calls, view caller ID, and alternately surf TV channels and the Web—or even surf the Web and TV channels at the same time, using high-speed DSL or dial-up connections with their televisions.

Subscribers use a television set-top box equipped with a speaker phone and NCI software to receive and make telephone calls as well as to access Internet-based features. Some of these features include programming guides, electronic commerce, news, and electronic mail. In addition to access over conventional connections, U S WEST provides support for high-speed DSL technologies that use existing copper-wire networks, offering data transmission speeds up to 200 times faster than dial-up connections.

On the client side, NCI TV Navigator software supports integrated telephony and enhanced television, allowing U S WEST customers to display Internet content and applications while watching television, and also make and receive telephone calls (Figures 1 and 2). The platform uses NCI Connect ISP Suite server software to manage the set-top boxes, provide security, and administer the network.

NCI's IQView technology formats standard Internet content for the television automatically, ensuring high-quality graphics and crisp on-screen text. Color correction is performed and real-time flicker is reduced to further enhance the image quality. Intelligent graphic scaling adapts Web sites and eliminates the need for horizontal scrolling.

Standards

A vendor group called the Advanced Television Enhancement Forum (ATVEF) has defined protocols for television programming enhanced with data, such as Internet content. The goal is to allow content creators to design enhanced programming that can be delivered over any form of transport—analog or digital TV, cable, or satellite—to any type of broad-cast receiver that complies with the specification.

The broad capabilities of Internet authoring standards on which the ATVEF specification was developed allow content producers and adver-tisers to create interactive content once without having to customize it for viewing on every type of receiver. Among the industry standards sup-ported in ATVEF's specification are:

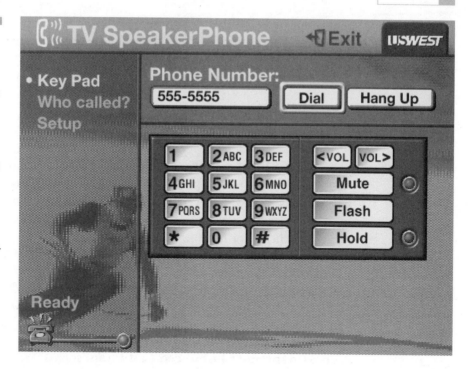

Figure 2
Upon selecting Phone, the @TV service pushes a TV SpeakerPhone interface to the television screen, which displays a telephone pad. To place calls and implement features, the user points and clicks with the remote device.

HYPERTEXT MARKUP LANGUAGE (HTML) This is the same language used to create Web pages, including image maps, which are a popular way of making graphic objects on an HTML-based page "clickable."

EXTENSIBLE ARCHITECTURE ATVEF specifies the familiar extensibility scheme of "plug-ins," which are used by most Web browsers to accommodate new capabilities.

JAVA ATVEF currently supports the European Computer Manufacturers Association's ECMAScript, a general-purpose, cross-platform programming language. The provision for hooks into Java are likely to be added to the ATVEF specification.

DOCUMENT OBJECT MODEL (DOM) This platform- and language-neutral interface allows programs and scripts to dynamically access and update the content, structure, and style of documents. The document can be further processed and the results of that processing can be incorporated back into the presented page.

JAVASCRIPT ECMAScript plus DOM is equivalent to JavaScript 1.1. Although the ATVEF's goal is to enhance a broad spectrum of differentiated services and receivers from which consumers can make selections, consumers will be able to choose whether to turn on the enhanced elements or not. Once they turn them on, there can be two levels of interactivity: content only as an enhancement, or content with links to additional information for two-way interactivity.

SUMMARY

Advanced Television Enhanced services offers new entertainment and commerce opportunities, such as sports events with complementary information on players and teams, news with additional details and related stories, and advertisements that allow consumers to order merchandise with the click of a button. A common specification, such as that developed by the ATVEF, promises to accelerate the creation and distribution of enhanced television programs and allow consumers to access such programs cost-effectively and conveniently, regardless of which transport or broadcast receiver they use. The specification also allows content providers and distributors to choose from a variety of enhanced television business models and delivery methods.
See Also **Internet-Enhanced Cell Phones**

All-Optical Networks

Optical-network technologies have reached an enormous capacity in recent years, yet carriers still cannot keep up with bandwidth demand. For example, while the demand for high-speed data services has driven interconnect vendors to introduce high-capacity interfaces (OC-12, OC-48) for backbone routers, now there is growing demand for DS-3 and OC-3 service payloads. In turn, this has created the need for more capacity at network access points that will accommodate OC-12/OC-48 trunks capable of supporting these large payloads.

Handling such emerging requirements—and enable carriers to position themselves for more demanding ones in the future—requires a new paradigm for optical-network service deployment in the next century. This new paradigm will give carriers the means to provide wavelength-allocated bandwidth, wavelength routing, and wavelength-translation capabilities.

Wavelength-allocated bandwidth will provide the capacity necessary for large users, who will need ubiquitous access throughout the country via optical virtual private networks. Wavelength routing will enable diverse point-to-point and point-to-multipoint optical payload transport for users in regional networks. Wavelength translation will enable optical services to traverse multiple carriers, regardless of vendor technologies. It enhances optical-network topology designs by allowing a wavelength frequency to be selected at one port and switched over to another frequency for acceptance by another port in the network.

Architecture

The key elements of the all-optical network architecture are dense wavelength division multiplexers (DWDM), optical add-drop multiplexers (ADM), and optical cross-connects.

DWDM DWDM systems are already in operation today, particularly for long-distance and so-called next-generation networks. These systems multiplex channels on a single fiber to vastly increase their capacity for point-to-point applications. Current DWDM systems multiplex up to 16 SONET 2.5-Gbps signals into one 40-Gbps composite multiwavelength signal. Carriers are optimizing the best attributes of SONET ring protection and survivability with DWDM virtual capacity. This allows them to maintain existing SONET rings by deploying DWDM as fiber expansion instead of using DWDM for network replacement. With DWDM systems now in deployment, the next step is to internetwork them with a variety of tributary interfaces such as asynchronous transfer mode (ATM), Internet Protocol (IP), and Gigabit Ethernet. After that, more diverse wide-area network topologies must be developed, such as mesh and ring.

ADM Optical ADMs are being integrated into DWDM-based networks. This type of multiplexer enables carriers to access wavelength-based services and create route diversity for network topologies, and also provide a migration path to optical ADM rings. In conjunction with ADMs, rings provide a fail-safe means of surviving the most severe disasters. ADMs work together to take traffic off the affected ring segment and place it onto another ring with spare capacity.

CROSS-CONNECTS Optical cross-connect systems give carriers more options in building network topologies consistent with today's telecommunications infrastructure, including mesh, ring, and star. Mesh designs, for example, make it possible to build network topologies with enhanced survivability. Essentially, optical cross-connects let carriers establish mesh designs at the optical layer, regardless of payload or service application. Optical cross-connects will complement optical ADM by centralizing all network topologies.

Transparency

The new paradigm for optical-network services also includes the concept of transparency—the use of the light path itself as the transmission medium, which eliminates the need for optical-to-electrical conversions in the network.

Existing transmission networks—which consist of fiber, DWDM multiplication products, and SONET transmission equipment—are subject to potential congestion because they require optical signals to be converted to electrical signals and then converted back into optical. Congestion can be relieved by eliminating the need for these conversions and by using the light path as the transport medium rather than the fiber.

Sycamore Networks Inc., for example, envisions a network that will at first coexist with SONET and DWDM equipment, before phasing into an all-optical network. The first phase will be optical-networking products aimed at increasing fiber capacity to relieve congestion.

To fulfill its vision, Sycamore's first products in 1999 will be for distances of less than 500 km, for such applications as corporate Internet access and virtual private networks. Sycamore's plan is for the carrier networks to evolve into first-generation optical networks and what the company calls Lambda, or light-path networking, where services are mapped to light paths, optical/electrical conversions are eliminated, and wide-area network services are delivered at local-area network speeds. In future rollouts, Sycamore will introduce products for distances of 1,000 km and then 10,000 km.

SUMMARY

All-optical networks are essential to provide the bandwidths required for the future. Improved technology in optical components—such as lasers, amplifiers, and filters—will enable information to be sent over longer distances and at lower cost. Being able to use light waves that can flow freely without having to be converted into electrical energy and then back again will enable carriers to offer OC-3 service (155 Mbps) at the price of T3 (45 Mbps) service. This transparency also provides additional advantages, such as reduced delay through the elimination of optical-electrical conversion, reduced cost due to a need for fewer system components, and added flexibility carriers will have in service provisioning and management.
See Also **Next Generation Networks, Synchronous Optical Network**

AppleTalk

AppleTalk is Apple Computer's LAN protocol. It is built into every Macintosh computer and facilitates communications between a variety of

Apple and non-Apple products linked on LANs. AppleTalk provides access to print and file servers, e-mail applications, and other network services. The AppleTalk network itself can be configured in either a bus, star, or ring topology.

When AppleTalk is used among Macintosh computers in a small workgroup LAN, the physical connection is called LocalTalk. It lets Macintosh computers in any network configuration communicate among themselves or with a printer at 230.4 Kbps over unshielded twisted pair (UTP) wiring or a RJ-11 phone cable (i.e., PhoneNet). Although LocalTalk networks are relatively slow, they are popular because they are easy and inexpensive to install and maintain. Up to 32 nodes are supported on a LocalTalk network.

AppleTalk can also be run over other types of higher-speed networks. When AppleTalk runs over Ethernet, token ring, and FDDI, the networks are referred to as EtherTalk, TokenTalk, and FDDITalk, respectively. AppleTalk networks can be interconnected through routers. An AppleTalk internet can consist of a mix of LocalTalk, TokenTalk, EtherTalk, and FDDITalk networks, or it can consist of more than one network of a single type, such as several LocalTalk networks.

AppleTalk can also be carried within different protocols through a process known as *encapsulation* or *tunneling*. This would allow two Macintosh computers to communicate through a TCP/IP-based Virtual Private Network (VPN), for example, using various tunneling protocols.

Other Connectivity Options

Additional Apple software may be required to connect computers to a network or implement special features, such as:

APPLETALK REMOTE ACCESS A family of products that allows individuals or workgroups to access information and services remotely over a variety of connections, ranging from telephone lines to wireless links over cellular networks.

APPLE INTERNET ROUTER Allows users to increase the size, performance, and manageability of their AppleTalk networks. Features include the isolation of local traffic, built-in data compression for maximizing throughput, Macintosh-based administration and configuration, and administrator password options to lock out unauthorized access.

APPLESEARCH Turns a collection of documents stored on a server into an online library where users can research a variety of information. Apple-

Search recognizes and processes natural-language search queries, as well as the Boolean queries. In addition, it uses a relevance-ranking algorithm to present the most useful references first. The documents are not altered; instead, they are stored in their native format and translated into searchable text by Apple's XTND technology.

APPLESHARE Provides file and print services, centralized storage, and administration capabilities to users in a workgroup or department.

APPLESHARE IP Combines traditional file sharing using TCP/IP and AppleTalk, with File Transfer Protocol (FTP) capabilities, a POP3/SMTP mail server, an HTTP Web server, and an AppleTalk print server to provide users with a one-stop productivity solution for shared information and resources.

MACX.400 Turns a Macintosh computer into a Message Transfer Agent (MTA) that supports the X.400 messaging standards. It allows e-mail and other electronic documents to be sent from a Macintosh computer to any number of users who work on other computer platforms and use other e-mail systems. This can be accomplished over the Internet and through a variety of network environments.

MACX High-performance software that enables users to seamlessly run both network-based X applications and Mac OS applications on one Macintosh computer. It delivers the rich application and development resources of network-based UNIX and VMS systems with the productivity and ease of use of the Mac OS.

MACX25 MacX25 server software allows users to link Macintosh computers to X.25 packet data networks. It can also be used with the Apple Internet Router to connect remote AppleTalk networks over X.25 links. Using this server software, a Macintosh can be set up as a gateway to an X.25 network. The MacX25 server distributes access to host computers and end-user services to Macintosh computers over the AppleTalk network system.

In addition, a network of Macintosh computers can be administered with Apple Network Assistant, an all-in-one solution for remote assistance, system profiling and configuration, and software distribution. With this software, administrators can provide assistance to end users, get system profiles, reconfigure system settings, and distribute applications across hundreds of computers—all from a central location, over both TCP/IP and AppleTalk networks.

Figure 3
LANsurveyor is
Neon Software's
Mac OS network
mapping,
monitoring, and
reporting
application.

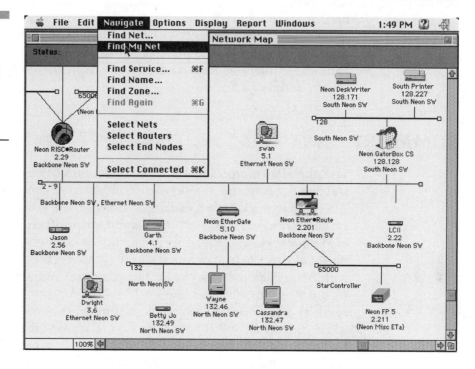

Management

AppleTalk networks can be managed with the SNMP Administrator program provided by Apple Computer. This program lets administrators set up and monitor groups of computers on the network using the Simple Network Administration Protocol (SNMP). When SNMP is not enough, there are third-party network management products that run on the Macintosh. One of these products is Neon Software's LANsurveyor.

LANsurveyor graphically draws the network, which displays network objects—computers, printers, routers, and other nodes—and their relationships (Figure 3), provides real-time troubleshooting capability, and uses SNMP to profile and monitor network devices. Network objects are queried in real time for details including SNMP data, responder information, zone information, and AppleTalk services. If the devices on the network change, the map can easily be updated to show any changes.

LANsurveyor tests the network for responsiveness, alerts managers to problems, and shows how problems relate to other parts of the map. Device profiling, reporting, and traffic monitoring features provide the data needed to configure the network for efficient performance and to determine resource requirements. Other capabilities of LANsurveyor include the collection of asset and resource management data, software inventory and license metering, electronic software distribution, remote administration, and task scheduling.

SUMMARY

The AppleTalk protocol makes it easy to set up small to midsize networks, and does not consume as much bandwidth as some other protocols, such as IPX (Internetwork Packet Exchange). However, the features that make AppleTalk easy to use, such as computers announcing their presence to the rest of the network whenever they "wake up," tend to bog down the performance of larger networks. Moving to IP eliminates the problems associated with AppleTalk's network chatter. File, print, and Web services, among others, are all seamlessly accessible via IP on the Macintosh. Mac-to-NT connectivity is now easy to achieve with third-party software, and Apple has added support for SNMP to the MacOS and is working with other vendors to improve Mac management tools in cross-platform environments.
***See Also* Linux, NetWare, Windows NT, UNIX**

Applications Hosting

Applications hosting is a type of service that allows a company's employees to access enterprise applications via an online subscription with a third-party firm. This arrangement eliminates the cost and management hassles associated with acquiring, deploying, and maintaining corporate-wide applications—or developing them in-house. Instead, the different software options are managed, monitored, and maintained by a service provider that hosts the applications in its own data center.

The service provider enters into a relationship with software vendors, who agree to license their products for distribution on a subscription basis. Virtually any kind of enterprise software a company wants can be made available through an applications hosting service on a monthly-fee basis, including:

- Financial management
- Human resources
- Enterprise relationship management
- Data warehousing

- Electronic commerce
- Messaging
- Web site management
- Popular software suites

The applications are supported on dedicated servers that sit behind fire-walls and are continuously monitored by the service provider's staff, and they are accessed by customers through the Internet. The service provider uses dedicated high-speed access connections at each of its data centers from different Internet service providers (ISPs) to offer its customers fast, robust access. Depending on the service provider, customers may have the option of using encryption to safeguard their data. To further protect customer data, the service provider mirrors it on at least one other data center. If there is a failure at the customer's primary data center, the data will be available from the second data center.

Subscriptions range in price from $25,000 to more than $100,000 per month, depending upon the size of the account and the services requested. Subscription to multiple services can qualify a company for a discount on the monthly charge.

SUMMARY

Applications hosting is a relatively new type of service. It suits IT managers who are looking for ways to cut costs associated with administrating and updating diverse applications, especially those in midsize companies that do not have the internal resources to keep up with the technical demands for new and innovative applications. With a focus on business needs and objectives, this new delivery model results in immediate productivity benefits. *See Also* Applications Metering, Outsourcing

Applications Metering

In today's business environment it has become necessary to track computer assets—both hardware and software—to determine the cost of ownership and depreciation, and aid departmental budgeting and theft deterrence. A key aspect of asset management is applications metering, which entails monitoring how many copies of a particular software program are in use at any given time. Not only is software usually the most expensive asset, but it often determines how much an organization spends in other areas, such as hardware, staff training, and support. Several compelling reasons why small and large organizations should monitor software usage is that is allows them to:

- Minimize software costs by not overspending for software licenses and upgrades.
- Eliminate exposure to litigation and financial penalties by ensuring compliance with vendor licensing agreements.
- Identify patterns of usage that can aid in capacity planning, training, and budgeting.
- Limit the amount of time personnel are engaged in using nonessential applications.
- Minimize exposure to viruses and other hostile code that may be lurking in bootleg copies.

Applications metering tools are available as stand-alone products, such as CentaMeter from Tally Systems Corp., or they may be embedded in a larger systems management package, such as Intel's LANDesk or Microsoft's SMS. All provide scanning capabilities and management reports of varying degrees of flexibility and detail. Other related functions may be included in these tools such as automated software distribution and installation.

Operation

Monitoring software allows IT administrators to control the number of people concurrently using stand-alone applications as well as those in a software suite. In the case of a software suite, the metering software tracks each application in the suite individually to automatically allow the correct number of concurrent users, per the vendor's license agreement. In addition, information is provided on which applications have been used, for how long, and by which users.

The metering software automatically tracks which applications are being used on each system on the network by scanning local hard drives and file servers for all installed software. The executable files of each application are examined to determine the product name and the publisher. Files that cannot be readily identified are listed as found, but flagged as unidentified. Once the file is eventually identified, the administrator can supply the missing information.

Each application in the software inventory is automatically assigned a unique name or tag, which is used for metering. When a user launches a metered application, the tag is checked to determine if there are available copies (licenses). When a user starts an application, one less copy is available to run. Likewise, when the application is closed, the copy becomes available again. Before users are granted access to an application, the software inventory is checked to determine whether or not there are copies available. If all copies are being used, a status message is issued to the requesting user, indicating that there are no copies available. The user can then decide to wait in queue for the next available copy or try again later.

Functionality

The IT administrator can choose to be notified when users are denied access to particular applications because all available copies are in use. If this happens frequently, it may help justify the need to purchase additional copies of the software or pay an additional license charge to the vendor so more users can access the application.

Some software metering packages allow application usage to be tracked by department, project, workgroup, and individual for charge-back purposes. Charges can be assigned on the basis of general network use, such as time spent logged on to the network or disk space consumed. Reports and graphs of workgroup or department charges can be printed out or exported to other programs, such as an accounting application. Even if workgroups and departments are not actually required to pay for application or network usage, the charge-back feature can still be a valuable tool for identifying operations costs and for budget planning.

Metering software can also be used for capacity planning. For example, if a company has accounting software running on six different servers, it might want to consolidate applications on fewer servers. To do this properly, the company needs to know which servers are being accessed the most and from which locations. With this information, the company can decide which servers can safely handle the redistributed load without imposing undue performance penalties on frequent users.

Another function performed by some metering products is automated software distribution. After a software inventory is compiled, this information can be used to create a software distribution list that includes workstation addresses. When an upgrade or patch becomes available from the vendor, the IT administrator can send it to all the workstations appearing on the distribution list.

Some metering products not only automate software distribution, but they also facilitate remote installation by allowing administrators to add conditional logic to customize mass installations across a network. Ordinarily, it would take considerable effort for an administrator to develop a script for automating software installations because he or she would first have to anticipate all the changes an application makes when it is installed. With the right installation package, however, the administrator can concentrate on adding the customization logic and let the installation com-

Application	License File	Total	Used	Inactive	Free	Queued
WordPerfect 6.0a	WP	19	19	4	0	0
Microsoft Office 4.2	MSOFFICE	5	3	1	2	0
Microsoft Excel for Windows 5.0a	EXCEL	5	3	1	2	0
WordPerfect 5.1+	WPDOS	3	3	1	0	1
Lotus SmartSuite	SMART	5	2	1	3	0
Microsoft Project for Windows 4.0	WINPROJ	3	1	1	2	0
Corel CorelDRAW 5.0	CORELDRW	5	5	2	0	0
Borland Paradox 4.5	PDOXDOS	2	1	0	1	0
Fifth Generation FastBack Plus 6.0	FB	1	1	0	0	1
Computer Select 3.0	COMPSEL	2	0	0	2	0

ponent of the metering software figure out the hundreds of files that are changed by an individual installation.

License Tracking

In monitoring the usage status of all software on the network, each application can be identified by total licenses, licenses used, licenses inactive, licenses free, and users queued for the next available copy (Figure 4). This and other information is used to monitor software license compliance and optimize software usage for both network and locally installed software packages. Some products, including CentaMeter from Tally Systems, even provide an analysis of software usage by time of day (Figure 5).

Microsoft's SMS 2.0 is among the systems management products that features dynamic license sharing across servers and domains. SMS also has extensive license metering capabilities, letting users with laptop computers check out licenses, providing pools of licenses or static licensing, and supporting accounting for charge-back on application usage. SMS can also search for unsupported applications on the network and report their existence to the administrator, or even disable them so they cannot be run. It also has tamper-resistant features that report when a malicious user attempts to disable client software or rename applications.

Features

Some software metering products issue a warning if the limit on the number of legal copies in use has been exceeded. Depending on the prod-

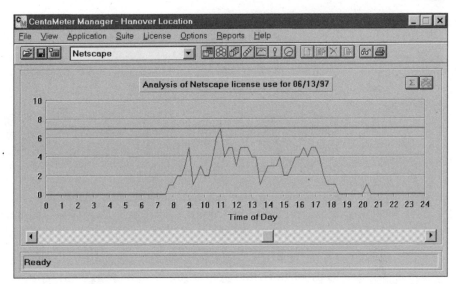

Figure 5
Tally Systems'
CentaMeter
Manager provides
a visual indication
of software usage
by time of day. In
this analysis of
Netscape browser
usage, for
example, the peak
hour was 11:00
A.M., when seven
of the ten browser
licenses were in
use at the same
time.

uct, the tracking program may even specify the directories where these files were found. This saves the IT administrator from having to track down any illegal copies and delete them. Some of the advanced features of today's software metering products include:

CUSTOM SUITE METERING WITH OPTIMIZATION CAPABILITY Allows the administrator to monitor the distribution of software suites in compliance with license agreements. This ensures that a suite will never be broken up illegally. The metering software automatically monitors the usage of individual components by end users and switches whole suite licenses to users working with more than one suite application at a time, leaving stand-alone licenses available whenever possible. With the optimization capability, single and suite licenses are used efficiently and legally—automatically.

INTERACTIVITY TRACKING AND REMINDERS Allows the administrator to track the amount of time open applications are inactive and reminds users to close inactive applications to make those resources available to others.

LICENSE ALLOCATION Allows the administrator to allocate licenses to an individual, group, machine, or any combination of these. Access to applications is given on a priority basis to users who need it most. Overflow pool licenses can be created for common access.

PRIORITY QUEUING Gives users the option of joining a queue when all eligible licenses for an application are in use. Different queuing arrangements and access limits can be implemented for each application. High-priority users can jump ahead in the queue, while VIPs can get immediate access to an application regardless of license count restrictions.

Applications Metering

POINT-OF-EXECUTION METERING Allows tracking of applications, regardless of where they are executed—even on a local disk drive where users think they have privacy and can get away with ignoring company policy. Unauthorized programs can be shut down wherever they reside—even for VIP users.

LICENSE SHARING ACROSS LOCALLY CONNECTED SERVERS The metering software can be installed on any server for tracking applications across multiple servers. Access to licenses for a product installed on more than one server can be pooled together.

ENTERPRISE MANAGEMENT CAPABILITIES Allows licenses for applications to be transferred to remote locations across WAN connections, facilitating configuration changes and organizational moves.

LOCAL APPLICATION METERING Tracks software usage and restricts access to unauthorized applications installed on local hard drives.

UNREGISTERED APPLICATIONS LOGGING An IT administrator can log all executions of local applications that do not have defined license profiles. For example, this feature can be helpful for learning what downloads are being used and who may be using unregistered software on a network, further preventing software piracy and the spread of harmful viruses.

ENHANCED APPLICATION IDENTIFICATION Allows the administrator to use a variety of categories to identify an application for metering, including filename, size, date, drive and path, or any combination of these.

DYNAMIC REALLOCATION OF LICENSES Allows the administrator to transfer licenses between groups and users to accommodate special access needs.

FILES-TRACKED METERING This is a relatively new feature, offered by Tally Systems, in which storage hierarchy decisions are made based on which files are frequently and infrequently used. An infrequently used file might be archived to tape, for example, while a frequently used file would stay on a local hard disk. Files can be tracked by any extension and reports are generated on usage. This feature can save money by minimizing the need for more disk drives or other types of online storage.

Reports

Application use can be metered on a global basis or selectively, according to such parameters as users, workgroups, workstations, hardware configuration, and networks. The information gathered by the monitoring software can be reported in a variety of ways, including:

GRAPHS A graph is created daily for each application, showing peak usage over a 24-hour day. Administrators can also view usage by group or user, as well as queuing patterns over time.

ERROR REPORT An error report describes users who have been denied access to an application, attempted unauthorized access, or restarted their applications in mid-operation.

COLOR-CODED STATUS SCREEN A single screen displays the ongoing status of each metered application, and administrators may select different colors to indicate that a license limit has been reached or that users are in the queue.

COLOR-CODED USER SCREEN Displays which users are active and which are inactive on any application.

E-MAIL ALERTS Messages can be sent via any mail application to a designated address when an unauthorized user attempts to access an application, when a user has been denied access to an application, or when license limits have been exceeded.

PAGER ALERTS Messages that are deemed crucial by the administrator can be sent to a pager so immediate action can be taken. For example, a pager alert can be sent whenever unauthorized software has been installed on a workstation. Since this can cause problems on the network, as well as unwanted legal trouble, immediate notification by pager allows the offending user to be tracked down as soon as possible.

Vendors provide different kinds of reports. The standard usage report calculates the number of users that have started applications during a specified time period. Another type of report shows peak usage over a 24-hour period. There is even a report that shows all unmetered applications that are running on the network. Examples of unmetered applications are freeware and beta programs, demo software, and internally developed applications.

A license report offers an enterprise-wide view of license usage by calculating usage for each application by location in order to determine whether more or less licenses should be allocated. This also helps determine how software-license load should be balanced among servers. There are also reports that provide historical data on such events as total number of executions, total time the application has been in use, and the total number of reject/queue occurrences.

Some vendors include a built-in report writer in their monitoring products to help the administrator summarize historical data. Seagate Software, for example, provides trending reports through its Crystal Reports engine, an application that is included as part of its Desktop Management Suite.

More than 15 types of reports can be generated, including one that shows compliance with Microsoft's Select Agreement. Hewlett-Packard includes a copy of the Crystal Reports engine in its Desktop Administrator.

Global Licensing

A global license is created by defining the total number of purchased licenses for any given product. The monitoring software keeps track of user demands for each application and will automatically move unused licenses from a server with excess counts to a server in need of a license. When a global license is defined, one server is designated as the dispatcher for that license.

In the case of Elron Software's SofTrack, for example, messages between servers keep each other informed of user demand for various applications. If a server on the network runs out of licenses for a particular application, it will request one from the dispatch server. The dispatch server, in turn, will borrow a license from the server with the most available licenses. The communication continues so the server in need of a license borrows it, and the maximum concurrent user counts for that application are automatically updated on each server. From the user's perspective, it looks as if he or she is just waiting in queue for the next available license.

SofTrack is also useful for organizations that are in the process of migrating their networks from NetWare to Windows NT. In addition to running over TCP/IP and IPX nets, it provides concurrent global license sharing between the two types of servers under central administration. This gives organizations the ability to share licenses between different server platforms to ensure license compliance and maximize license availability, while allowing them to migrate at their own pace.

Another product that spans multiple operating environments and also works over TCP/IP networks is KeyServer from Sassafras Software. The license metering tool can run on a Windows NT/Windows 95 computer or a Macintosh/Power Macintosh computer. Regardless of which platform and operating system is actually running KeyServer, its license management services can support thousands of clients on both platforms with negligible traffic overhead. Using a common console from the Windows or Macintosh platform, all of KeyServer's features can be configured locally or remotely, enabling global administration via a corporate intranet or the Internet (Figure 6).

A key advantage of products that work over WANs is that the IT administrator can distribute and meter software to remote IP servers that are not part of the local network. Depending on the specific product,

Figure 6
Using the Key
Details window in
Sassafras Software'
KeyServer, the IT
administrator can
set up permissions
for multiple
groups of users
via TCP/IP nets.

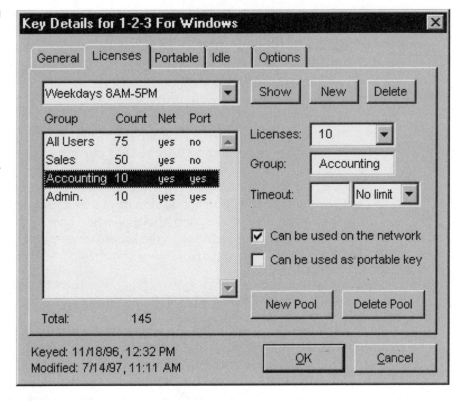

software usage can be monitored and controlled across file servers for Windows, OS/2, UNIX, and Macintosh clients.

The IT administrator can tag applications or suites for metering and can set several properties, such as maximum concurrent users, rules for license-borrowing between servers, program files to meter, and the amount of time a user waits in a queue. With the ability to monitor all the metered servers in the enterprise and print comprehensive reports, the IT administrator is saved the time-consuming chore of collating reports from many different locations.

License Registration Standard

Most legitimate computer software use is regulated by an explicit license. The license typically states who may use the software and under what conditions. There are many different types of licenses, each of which reflects the intended use of the software.

Until a few years ago, software use licenses were often nothing more than a printed license statement included in the product's packaging.

Software vendors relied on the integrity of their customers to not violate the license; in many cases, this was sufficient to protect the vendor's investment in developing the software. However, in an attempt to further reduce losses that result from illegal software distribution and use, a group of software companies cooperated in developing the Licensing Service Application Programming Interface (LSAPI).

The LSAPI specification was developed in 1994 by a consortium of vendors including Sassafras, Microsoft, Novell, Apple, and DEC to provide a common programming interface to licensing services while hiding the underlying details. LSAPI lets supporting software products monitor and control the number of application instances concurrently in use. Through the LSAPI, an application can register itself with a license server. When run on a workstation, the application asks the license server if the license agreement recorded in the license server permits another instance of the application to run.

Legal Ramifications

Another use of metering software is for auditing purposes. It lets companies know if more copies of programs are being used than were paid for under each vendor's licensing agreement. An audit can also spot software that has been installed by individual users without company authorization, but for which the company is ultimately responsible.

The ability to identify and locate unauthorized software is important because it is a felony under U S federal law to copy and use (or sell) software. Companies found guilty of copyright infringements face financial penalties of up to $100,000 per violation.

The two leading trade associations of the software industry are the Software Publisher's Association (SPA) and the Business Software Alliance (BSA). Both organizations run antipiracy campaigns and provide opportunities for whistleblowers to report software piracy through toll-free numbers and forms posted on their Web pages.

Software piracy is actually commonplace among businesses. In a 1997 study of global software piracy jointly sponsored by the SPA and BSA, it was estimated that of the 523 million new business software applications used globally during 1996, 225 million units—nearly half—were pirated. This figure represented a 20-percent increase in the number of units pirated in 1995. Although revenue losses in the worldwide software industry due to piracy were estimated at $11.2 billion in 1996, a 16- percent decrease over the estimated losses of $13.3 billion in 1995, the decline was attributed to lower software prices rather than a decrease in piracy.

The 1200-member SPA represents companies that develop and publish software applications and tools for use on the desktop, client/server networks, and the Internet. The SPA claims to receive about 40 calls a day from whistleblowers over its toll-free hotline. It sponsors an average of 250 lawsuits a year against companies suspected of software copyright violations. Since 1988, every case the SPA has been involved in has been settled successfully.

The BSA is another organization that fights software piracy. Its membership includes leading publishers of software for personal computers, such as Adobe Systems, Lotus Development, Microsoft, Novell, and Symantec. According to the BSA, software piracy cost U S businesses $6 million in fines and legal fees in 1997. In the U S alone, software piracy cost 130,000 jobs in 1996, and piracy is expected to cost another 300,000 jobs by 2005.

Both the SPA and BSA offer free audit tools that help IS managers find licensed and unlicensed software installed on computer systems. BSA offers Softscan for Windows and MacScan for Macintosh. SPA offers SPAudit by McAfee (now part of Network Associates, Inc.) and KeyAudit by Sassafras Software. SPAudit does hardware and software audits on both network and stand-alone Windows NT and NetWare machines using a catalog to identify over 4,400 software applications. KeyAudit works on Macintosh machines.

One of the most comprehensive and accurate audit tools available is GASP from Attest Systems. This is the tool that SPA, BSA, Microsoft, and other software companies use for conducting enforcement audits worldwide. Its database of applications exceeds 7,000 and is user-expandable. Most audits are completed in less than two minutes. GASP comes in two versions: GASP Net is compatible with Windows NT, SMS, NetWare, Vines, and other network operating systems, and GASP Audit is compatible with Windows NT, Windows 95, Windows 3.*x*, DOS, and Macintosh.

SUMMARY

When software metering arrived on the scene, it was primarily used to prevent the unauthorized use of software and manage concurrent-use software licenses. Today, information—not control—is the primary use of software license metering programs. Knowing what users are doing at their desktops promotes effective asset management. Specifically, the focus is on reporting actual usage, trends in usage site to site, what software packages are being used effectively, and what ones are not being used at all.

Gathering all license usage information in one place allows companies to see the big picture and use the information for decision support. Among other things, this can save money on software, support, and training. It can also aid capacity planning and budget development. The use of asset management information as decision support criteria is a trend that will likely continue.

See Also Asset Management

Applications Monitoring

Applications monitoring involves the continuous surveillance of data networks and their components, and the reporting of potential problems so they can be resolved quickly. Organizations can monitor applications with standard network management tools and resolve problems with in-house technical staff. Alternatively, the organization can outsource the applications monitoring function to a third-party network management firm or carrier.

Among the types of applications monitoring offerings that are available are WAN circuit management and remote network monitoring services. WAN circuit management enables organizations to easily and cost-effectively manage the performance of wide-area networks by providing real-time information about data traffic flow. The service provides customers with instant access to reports and information about their WANs. Remote network monitoring services provide complete and continuous surveillance of data networks and their components, 24 hours a day and seven days a week. Customers are quickly alerted to potential problems. Both services lower a customer's operating and capital costs by eliminating the need to build expensive in-house systems or hire and train staff.

This approach to network management empowers organizations with the information they need—anytime, anywhere, via the Web—to plan for future network demand. Access to this information offers customers flexibility, greatly improving their overall efficiency and productivity.

Bell Atlantic is among the first carriers to address the growing challenge among its customers of how to optimize their networks to support crucial business functions in this era of IT skills shortage. Bell Atlantic's Network Integration (BANI) unit offers a portfolio of services that enable customers to better leverage their networks to achieve their financial and competitive objectives.

The services are packaged to support specific applications, including Internet and corporate applications. Bell Atlantic engineers oversee the day-to-day management and operations of a customer's data network, leaving companies free to run their business and serve the needs of their own customers. Companies can view information on network performance, as well as analyses and recommendations from the engineers, via secure Web access.

For example, a company might receive feedback on how a new application affects network performance or how peaks in activity affect communications with customers. A company could use this information to avoid potential problems and improve network performance, or get details on what parts of the financial accounting software are being used. In knowing what part of the financials are getting the most traffic, business decision-makers can take steps to optimize their network to facilitate access to the application and improve response time by such means as increasing server capacity, adding a network cache, or segmenting the network.

Among those that are watched are SAP, Baan, PeopleSoft, and other Enterprise Resource Planning (ERP) applications. The service provides customers with immediate Web access to performance statistics and reports for planning and troubleshooting purposes. This relieves the strain on IT departments that are strapped for staff, while giving executives decision-making information.

BANI also gives corporate managers a real-time view of their networks, which BANI also installs and maintains. BANI delivers its analysis of customer networks and recommends improvements based on current and projected network usage. Customers can accept recommended upgrades via the Web.

Bell Atlantic is not the only carrier that monitors customer applications. AT&T also offers a server monitoring service as part of its Managed Network Solutions. Initially, these offerings have focused largely on configuration and monitoring of routers and frame relay access devices. Now, technicians at AT&T's Global Client Support Center monitor corporate application servers across dozens or even hundreds of sites. AT&T watches for such things as excessive CPU usage, disk space exhaustion, performance degradation, and other problems that can affect the performance of applications. AT&T also assumes responsibility for distributing software electronically and tracking hardware and software inventories.

SUMMARY

Many companies monitor applications through third-party tools that are integrated into the major network management platforms. A growing trend among large companies is to outsource applications monitoring because there is a serious shortage of skilled IT staff, which is expected to continue through 2003. By outsourcing this task, companies can focus on core business issues without worrying about attracting and keeping qualified technical staff.

See Also **Managed Applications Services, Outsourcing, Service Level Agreements, Web-Based Management Tools**

Applications Recovery

Applications recovery is the ability of systems to recover transactions after an application has crashed. When an application goes down, a recovery program checks back-end databases to find which transactions were actually completed and compares that with its records of which transactions were being processed when the crash occurred. It then flags those that were not completed, issues an alarm to the management station, and proceeds to restart them (Figure 7).

Figure 7
BMC Software's PATROL Knowledge Module (KM) for Microsoft Exchange Server monitors more than 275 performance objects to ensure they are running, and can be configured to automatically restart those that are not. Knowledge modules are available for other environments as well, such as Baan, IBM, Informix, Novell, Oracle, and PeopleSoft.

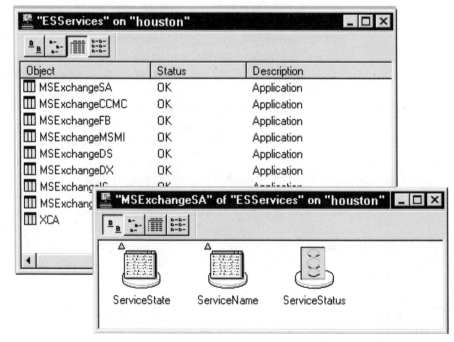

Such proactive monitoring tools recover data in specific applications at the particular point in time that the business process took place, without having to recover the entire database. The benefit is a reduction in application downtime because the software recovers specific database objects used by the application without having to bring down the database for whole-system data recovery. Using recovery software to do batch jobs on a DB2 mainframe, for example, enables the jobs to run as much as 50 percent faster by reducing application downtime.

The network plays a key role as the application delivery mechanism and can significantly affect overall application performance and availability. By understanding the relationship between the network and applications, IT organizations can better pinpoint and diagnose existing or potential application and network problems before users experience application service degradation.

Accordingly, some vendors have extended the monitoring and recovery processes to operate over networks by providing a comprehensive view of how applications perform on the network and how the network affects application performance. The correlated management view between the application and network enables IT organizations to manage application response time and bandwidth issues, thus improving the service delivery of business-critical applications to users.

The monitoring and recovery tools for networked applications integrate and correlate management information on applications, databases, operating systems, middleware, Web servers, and networks for comprehensive detailed analysis and fault isolation. With crucial network coverage, these tools monitor and manage application traffic and events, including clients accessing servers from different subnets.

A software agent, capable of working in switched network environments, collects performance data without interfering with network resources, enabling IT managers to quickly pinpoint the cause of a problem and rapidly address it before there is a significant end-user impact. Therefore, it can be accurately determined if network, application, or database teams need to respond to problems. Given the complexity of today's business applications and their tight integration across various types of networks, such capabilities can go a long way in enhancing a company's competitive position.

SUMMARY

Applications monitoring and recovery tools offer IT organizations greater end-to-end application management and service delivery support. They help manage the service level by enabling IT managers to understand how the network affects applications and how applications consume network resources. They generate application consumer reports, allowing IT managers to understand who is generating the most traffic (client, subnet, or server), which helps determine bandwidth allocation needs. They identify the impact of application rollouts and deployments on the network and other applications. They also facilitate cooperation between IT teams by helping to quickly pinpoint application performance problems to the server or network to resolve problems.

See Also **Event Monitoring and Reporting, Managed Applications Services**

ARCnet

ARCnet (Attached Resource Computer Network) is a network technology that was introduced in 1977 by Datapoint Corp., the originator of local-area networks. It became an ANSI standard in 1982.

ARCnet can be configured in a simple star or bus topology, or in a combined star and bus topology. ARCnet can use unshielded twisted pair (UTP) wiring, coaxial cable, or optical fiber. The primary characteristics of ARCnet include:

- 2.5 Mbps transmission
- Support for up to 255 nodes
- Coax cable lengths up to 2000 feet (600 m)
- Twisted pair cable lengths up to 400 feet (120 m)
- Fiber cable lengths up to 8000 feet (2.5 km)

ARCnet uses a token-bus scheme for managing line sharing among the stations and other devices connected to the LAN. The LAN server continuously circulates empty message frames on a bus. When a device wants to send a message, it attaches a "token" (i.e., sets the token bit to 1) to an empty frame, into which it then inserts the message. When the destination device or LAN server reads the message, it resets the token to 0 so the frame can be reused by another device. The scheme is very efficient under a high traffic load, since all devices are given the same opportunity to use the shared network.

A key component of ARCnet networks is the active hub, which is used when more than four devices are connected in a single network or when there are distances greater than 200 feet between the network components. Active hubs may be interconnected via coaxial cable to provide up to 255 ports. The active hub also provides signal detection and repeating, which eliminates noise caused by line reflections. Since the hub functions as the central connector of all network components, it can help network administrators identify and isolate problems on the network.

SUMMARY

The success of ARCnet technology initially played an important role in the growth of local-area networks. It has been eclipsed by other networks, however, namely Ethernet and token ring. Ethernet and token ring provide upgrade paths to 100 Mbps and 1000 Mbps speeds, whereas ARCnet is stuck at 2.5 Mbps, relegating it to niche applications. It is used in embedded design to communicate between controllers and manage commercial applications such as weighing scales, temperature controllers, scoring systems for bowling alleys, security systems, amusement-park rides, and other low-bandwidth applications. A plan in 1992 to increase the speed of ARCnet to 20 Mbps (ARCnet Plus) failed to generate significant interest. At this writing, a proposal is being circulated by the ARCnet Trade Association that would revise the current ANSI/ATA 878.1 specification to allow for optional speeds between 19.2 Kbps and 10 Mbps.

See Also **Ethernet, StarLAN, Token Ring**

Asynchronous Communication

Asynchronous communication describes how a computer uses a modem to connect with other computers over phone lines. With this method of data communication, the transmitter and receiver do not explicitly coordinate each transmission. The transmitter can wait arbitrarily long periods between transmissions, which is typical when a user is typing a message from the keyboard.

Years ago, when teletypes and dumb terminals were prevalent, data was sent synchronously—that is, sending and receiving devices relied on a shared timer that marked the boundary of the transmission. However, this method of data communication is not very efficient for moving large blocks of data over analog phone lines because noise disrupts the timing between the sending and receiving devices. So vendors began building modems that dispensed with a required timing mechanism to mark the start of transmission. Instead, they used a start bit, stop bit, and optional error-checking parity bit to indicate the boundary of each character.

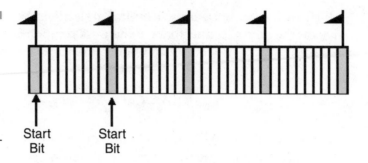

Figure 8
In asynchronous
communication,
start and stop bits
define the
boundary of each
eight-bit character
or byte.

Start
Bit

Start
Bit

Start Bit

Legacy modems alternate between two tones, one representing 0 and the other representing 1. Together, 0 and 1 constitute the binary language. The absence of 0s or 1s is not an option; to a modem, silence means the call has been disconnected. So when a modem has no user data to send, it continuously transmits 1s to indicate to the other modem that it is in an idle state. When there is more data to send, the modem sends a single 0 bit to indicate that there is user data on the way. The next eight bits the remote modem receives is the real data. Then the line goes back to idle for at least one more bit. The 0 just before the user data is called the *start bit*. The 1 at the end is the *stop bit* (Figure 8). These control bits are often called *flags*.

Parity Bit

Since legacy modems transmit data one character at a time, an error check is applied to each character. The usual way to do this is by appending an extra bit—the parity bit—to the end of each character before the stop bit. This bit is set to 0 or 1 based on the value of the previous data bits. Today's modems gather data into packets and send a larger two- or four-byte error-check (i.e., checksum) value to validate data in the entire packet. This eliminates the need for a parity bit.

Cyclical Redundancy Check

The CRC is a mathematical technique used to check for errors when sending data by modem. Because many phone lines are noisy, data can become corrupted during transmission. If the CRC adds up, the receiving modem sends back an ACK, or acknowledgement. However, if the CRC does not add up, the receiving modem sends back a NAK, or negative acknowledgement, which acts as a request to send the data again.

SUMMARY ▪ ▪ ▪ ▪ ▪ ▪ ▪ ▪ ▪ ▪

With the advent of high-quality digital lines, synchronous communication has become much more reliable than asynchronous communication. However, the latter is still very much in use today—typically when used to access a local bulletin board system (BBS) or directly connect two computers over a telephone line. A BBS is a text-oriented service that offers files and software for download, e-mail, job postings, news, and chat rooms. BBSs were popular in the 1980s, but have largely given way to the more graphic-oriented Web sites, which have global reach through the Internet. Asynchronous communication also comes into play when modem connections are established between two computers. In this case, one computer is set up as the host and the other as the terminal. When the terminal calls the host and the host grants access, the terminal can access its files.

See Also **Isochronous Communication, Bisynchronous Communication, Synchronous Communication**

Asynchronous Transfer Mode

Asynchronous Transfer Mode (ATM) is a protocol-independent, cell-switching technology that offers high speed and low latency for the support of data, voice, and video traffic. ATM provides for the automatic and guaranteed assignment of bandwidth to meet the specific needs of applications, making it ideally suited to supporting multimedia applications. ATM is also highly scalable, making it equally suited for interconnecting local-area networks and building wide-area networks. ATM-based networks may be accessed through a variety of standard interfaces, including frame relay.

Applications

ATM provides users with high bandwidth, low latency, and guaranteed Quality of Service (QoS) for multimedia applications, making it the transport medium of choice for real-time delivery of multimedia. In addition, ATM is an ideal transport medium for applications such as software on demand, interactive multimedia, bulk data transfer, and collaborative/clustering applications.

ATM serves a broad range of applications very efficiently by allowing an appropriate QoS to be specified for each application. These categories of QoS and their requirements are summarized in the table on page 30:

Quality of Service Requirements

Category	Application	Bandwidth Guarantee	Delay Variation Guarantee	Throughput Guarantee	Congestion Feedback
Constant Bit Rate (CBR)	Provides a fixed virtual circuit for applications that require a steady supply of bandwidth, such as voice, video, and multimedia traffic	Yes	Yes	Yes	No
Variable Bit Rate (VBR)	Provides enough bandwidth for bursty traffic such as transaction processing and LAN interconnection, as long as rates do not exceed a specified average	Yes	Yes	Yes	No
Unspecified Bit Rate (UBR)	Makes use of any available bandwidth for routine communications between computers, but does not guarantee when or if data will arrive at its destination	No	No	No	No
Available Bit Rate (ABR)	Makes use of available bandwidth and minimizes data loss through congestion notification. Applications include e-mail and file transfers	Yes	No	Yes	Yes
Guaranteed Frame Rate (GFR)	Provides a minimum rate guarantee to virtual circuits at the frame/packet level and allows for the fair usage of any extra network bandwidth	Yes	Yes	Yes	Yes

The ATM Forum, a vendor association whose charter is to speed the development and deployment of ATM products and services, guides standards development as well. Because ATM can be implemented across nearly every aspect of communications, from within homes and businesses to central offices and onto WANs, the ATM Forum continues to work on the wide variety of specifications associated with these platforms. Among the Forum's most recent accomplishments is a ratified specification for voice and telephony over ATM to the desktop. The ATM Forum is also working on setting industry standards that will enable Java programmers to make efficient use of ATM, and speed deployment of Java/ATM applications. These and other efforts help to assure interoperability and make it easier to migrate to ATM.

Operation

QoS allows ATM to admit a constant bite rate (CBR) voice connection, while protecting a variable bite rate (VBR) connection for a transaction processing application and allowing an available bit rate (ABR) or unspecified bit rate (UBR) data transfer to proceed over the same network. Each virtual circuit has its own QoS contract, which is established at the time of connection setup at the user-to-network interface (UNI). The network will not allow any new QoS contracts to be established if they adversely affect its ability to meet existing contracts. In such cases, the application cannot get on the network until the network is fully capable of meeting the new contract.

When the QoS is negotiated with the network, there are performance guarantees that go along with it: maximum cell rate, available cell rate, cell transfer delay, and cell loss ratio. The network reserves the resources needed to meet the performance guarantees and the user is required to honor the contract by not exceeding the negotiated parameters. Several methods are available to enforce the contract. Among them is traffic policing and traffic shaping.

Traffic policing is a management function performed by switches or routers on the ATM network. To police traffic, the switches or routers use a buffering technique referred to as a *leaky bucket*. This technique entails traffic flowing (leaking) out of the buffer (bucket) at a constant rate (the negotiated rate), regardless of how fast it flows into the buffer. If the traffic flows into the buffer too fast, the cells will be allowed onto the network only if enough capacity is available. If there is not enough capacity, the cells are discarded and must be retransmitted by the sending device.

Traffic shaping is a management function performed at the UNI of the ATM network. It ensures that traffic matches the contract negotiated between the user and network during connection setup. Traffic shaping helps guard against cell loss in the network. If too many cells are sent at

Figure 9
ATM cell structure.

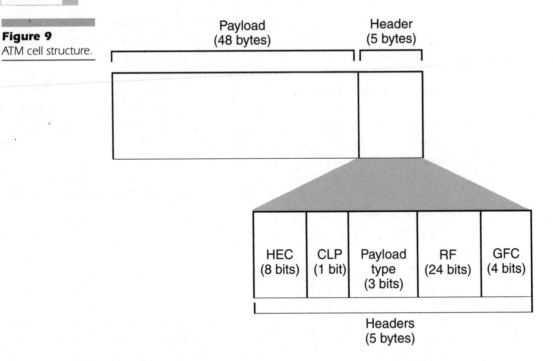

once, cell discards can result, which will disrupt time-sensitive applications. Because traffic shaping regulates the data transfer rate by evenly spacing the cells, discards are prevented.

Cell Structure

Voice, video, and data traffic is usually comprised of bytes, packets, or frames. When the traffic reaches an ATM switch, it is segmented into small, fixed-length cells of 53 bytes: 5 bytes for the header and 48 bytes for the data payload (Figure 9).

The cell header contains the information needed to route the data payload through the ATM network. The header supports five functions:

GENERIC FLOW CONTROL (GFC) This four-bit field has only local significance; it enables customer premises equipment at the UNI to regulate the flow of traffic for different grades of service.

ROUTING FIELD (RF) This 24-bit field contains a virtual path identifier/ virtual channel identifier (VPI/VCI) combination to route the cell through the network. The number of bits available for VPI and VCI subfields is negotiated at subscription time to the network.

PAYLOAD TYPE (PT) This three-bit field indicates whether the cell contains user information or connection management information. This field also provides for network congestion notification.

CELL LOSS PRIORITY (CLP) This one-bit field, when set to a 1, indicates that the cell may be discarded in the event of congestion.

HEADER ERROR CHECK (HEC) This eight-bit field is used by the physical layer to detect and correct bit errors in the cell header. The header carries its own error check to validate the VPIs and VCIs and prevent misdelivery of cells to the wrong UNI at the remote end. Cells received with header errors are discarded. Higher-layer protocols are responsible for initiating lost cell recovery procedures.

Despite the need to break larger variable-rate frames into fixed-size cells when interconnecting LANs, the latency of ATM is orders of magnitude less than frame relay alone. For example, on a five-node network spanning 700 miles, ATM exhibits a 0.3 millisecond latency versus a 60 millisecond latency for frame relay at T1 speeds. At T3, the latency of ATM is only 0.15 milliseconds. Thus, ATM makes for fast, reliable switching and eliminates the potential congestion problems of frame relay networks.

There has been some concern about the high overhead of cell relay, with its ratio of 5 header bytes to 48 data bytes. However, with recent innovations in Wave Division Multiplexing (WDM) to increase fiber's already high capacity, ATM's overhead is no longer a serious issue. Instead, the focus is on ATM's unique ability to provide a quality of service in support of all applications on the network.

ATM Layers

Like other technologies, ATM uses a layered protocol model. It has only three layers: the physical layer, the ATM layer, and the adaptation layer:

PHYSICAL LAYER (MEDIUM DEPENDENT) Defines several transport systems, including SONET, T3, optical fiber, and twisted pair. The Synchronous Optical Network (SONET) provides the primary transmission infrastructure for implementing public ATM networks, offering service at OC-1 (51.84 Mbps) to OC-12 (622.08 Mbps). Current definitions of SONET go up to OC-256 (13.271 Gbps). SONET facilities, however, are of limited availability for many users. Therefore, UNI outlines the use of DS-3 and a physical layer definition similar to Fiber Distributed Data Interface (FDDI) to provide a 100-Mbps private ATM network interface.

ATM LAYER Provides segmentation and reassembly operations for data services that may use protocol data units (PDUs) different from those of an ATM cell. This layer is then responsible for relaying and routing—as well as multiplexing—the traffic through an ATM network.

ATM ADAPTATION LAYER (AAL) Residing between the ATM Layer and the higher-layer protocols, this layer provides the necessary services that are not part of the ATM Layer in order to support the higher-layer protocols.

SUMMARY

A solid base of standards now exists to allow equipment vendors, service providers, and end users to implement a wide range of applications via ATM. The standards will continue to evolve as new applications emerge. The rapid growth of the Internet is one area where ATM can have a significant impact. With the Internet forced to handle a growing number of multimedia applications—telephony, video conferencing, faxes, and collaborative computing, to name a few—congestion and delays are becoming ever more frequent and prolonged. High-speed ATM backbones will play a key role in alleviating these conditions, enabling the Internet to be used to its full potential. *See Also* **Frame Relay, Synchronous Optical Network**

Audioconferencing Systems

Audioconferencing refers to the capability of three or more people to communicate in voice-only mode over a network. The key benefit of audioconferencing is that it eliminates travel, while enhancing communications by allowing many people to share information directly and simultaneously. It can be implemented through a PC, telephone, or corporate PBX/key system and over a variety of lines and services, including private leased lines, ISDN, and the Internet.

For a basic telephone conference involving a limited number of participants, a telephone set with either a three-way calling feature on the line or a conferencing feature supported by the PBX or key system is required. For an audioconference with more than three parties, the attendant console operator can establish the connections through the PBX and add more participants to the call than can be accommodated from a normal telephone.

As another option, companies can initiate a conference through the telephone company conference operator. The operator sets up a conference by calling each person until all participants are online together. A conference can be established at a prearranged time or organized so the participants can phone in to a preassigned number at a designated time.

System Components

While individuals use their telephone to participate in a conference, sometimes several people at a given location participate in the conference as a group. In such cases, specialized equipment is required, such as an integrated table-top audioconferencing system. This type of system consists of the following components:

BASE UNIT Contains the electronics for echo-free system operation in full-duplex mode for interactive conversations.

KEYPAD Used for dialing telephone numbers of conference participants and implementing various system features such as microphone mute and last-number redial.

OMNIDIRECTIONAL MICROPHONE Provides 360-degree room coverage with a pick-up range of 100 to 600 square feet.

SPEAKER Picks up conversations from the remote conference participants. The table-top audioconferencing system is easy to set up. It plugs into a power outlet and phone jack—just like a cordless telephone (Figure 10).

Role of Bridges

When more than three locations must be linked together, the most economical way to audioconference is to call the various participants on separate telephone lines, then join the phone lines with a bridge. Bridges can link several hundred participants in a single conference call, with multi-

Figure 10
Table-top audioconferencing system set-up.

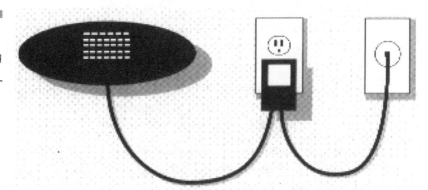

ple operators per conference. The bridge can operate as a stand-alone device or connect through ports of a PBX or Centrex switch—using either line-side ports or trunk-side ports.

A bridged conference can be implemented in several ways. In one method, each participant is called and then transferred to one of the PBX ports assigned to support the conference. Another method entails participants dialing a designated phone number at a preset time and automatically being placed on the bridge. Still another method of access requires participants to dial a main number and then be manually transferred onto the bridge by a live operator. In any case, whenever a new person joins the party, a tone indicates his or her presence. The more conversations brought into the conference, the more free extensions are required.

An optional moderator phone can be used to set up bridged conferences. The moderator phone allows the operator to initiate conferences, actively participate in conversations, and terminate connections, and also conduct isolated conversations with one party and then either admit or readmit the call to the conference or disconnect the call.

Ports and Interfaces

Bridges can accommodate a number of different ports and interfaces. Depending on vendor and the particular model selected, these ports and interfaces may be standard or optional:

ANALOG INTERFACE An analog interface board provides up to 24 channels over 4 wires. This interface is used with various analog I/O such as operators, music, record/playback, and link lines for connecting together two or more bridges.

T1 INTERFACE The bridge may have slots for one or more T1 interface boards, and each board may accommodate one or more T1 channels. The T1 connector is typically RJ45.

ADMINISTRATION PORT This is usually the COM1 port to which an ASCII terminal is connected. Although mostly used for administration and maintenance, this terminal can be used by an operator to set up conferences.

REMOTE MAINTENANCE PORT Via an internal or external modem, this port provides the means to access the bridge to perform maintenance from a remote location over a dial-up line.

PRINTER PORT A parallel printer port provides the means for printing call detail reports, dialing lists, logs, configurations, alarm indications, and network statistics.

SERIAL I/O INTERFACE A multiport I/O card implements local and remote operator displays. ASCII terminals interface via RS-232 connectors.

EXTERNAL ALARM INTERFACE Via a DB15 connector, the external alarm interface provides several contact closures in the event of an alarm condition. These include major audible, major visual, minor audible, and minor visual.

LAN CONNECTION Multiple bridges can be linked via a local-area network for the purpose of transferring files such as dialing lists and digitally recorded conferences. A LAN connection can also support automatic, real-time transfer of call detail reports to the server.

Conference Modes

Depending on the vendor, there are a number of conference modes to choose from that are implemented by the bridge:

OPERATOR DIAL-OUT Participants are brought into the conference by an operator using a method such as manual or abbreviated dialing.

ORIGINATOR DIAL-OUT Additional participants are added to the conference by the conference moderator, who accesses available lines using a touchtone telephone.

PREARRANGED Conferences are dialed automatically or by a user dialing a predefined code from a touchtone phone. In either case, the information needed to set up the conference is stored in a scheduler.

MEET-ME Participants call into a bridge at a specified time to begin the conference. If audioconferencing is used frequently, a dedicated 800 number can be justified for this purpose.

SECURITY CODE ACCESS Participants enter a conference code and are automatically routed to the appropriate conference. If an invalid password is entered, the call is routed to an attendant station where an operator screens the caller and offers assistance.

AUTOMATIC NUMBER IDENTIFICATION (ANI) Automatically processes incoming calls based on the phone number of the caller. This includes call branding, call routing, and conference identification. Custom greetings may be designed for each incoming call.

System Features

The various features of a audioconferencing system can be grouped into five categories: participant, moderator, conference, administration, and maintenance.

PARTICIPANT FEATURES   Individual participants in any audioconference need only one piece of equipment—a touchtone telephone. Some bridges facilitate touchtone interaction, increasing the level of participation and end-user control dramatically. The conference administrator can configure dual tone multifrequency (DTMF) detection for one or two digits. This allows conference participants to implement various features by pressing one or two buttons on their touchtone phone. Among the features a participant can implement are:

HELP   If the participant needs help, pressing 0 or *0 signals the conference operator for assistance.

MUTE   If a participant wants to mute the line, possibly to talk without having remarks conveyed to the other conferees, pressing 6 or *6 (M for mute) will place the line in listen-only mode.

POLLING   Conferees can participate in voting sessions. By pressing one or two digits—to indicate yes or no, for example—each participant's preference can be recorded by the bridge. The results can be read immediately or stored on disk or printed out via the parallel printer port.

QUESTION AND ANSWER   Conference participants can enter a question queue to signal the moderator that a question is waiting. At the appropriate time, the moderator can address each participant's question.

MODERATOR FEATURES   Moderators are given special privileges that provide a higher level of conference control than that afforded to participants. These privileges can be activated using their touchtone phone:

SECURITY   A moderator can secure a conference by pressing one or two buttons on the touchtone keypad.

CONFERENCE GAIN   To level all signals in the conference, the moderator can implement the gain control feature of the bridge, if it is not already set for automatic gain control.

LECTURE   The moderator can initiate a lecture in which all lines, except those designated as moderators, are muted. This allows the presenter to convey information without interruption from conference participants.

CONFERENCE FEATURES   Some bridges provide an extensive array of features, all of which contribute to a productive and successful audioconference. Depending on vendor, these features may include:

POLLING The moderator can poll participants using one of the following methods: yes/no, true/false, multiple choice, or assigned ranges. Participants use their touchtone phones to make their selection. Results are compiled immediately and can be printed or saved to disk.

QUESTION AND ANSWER Participants can indicate that they have questions by pressing one or two buttons on their touchtone phones. When the lecture is over, the moderator can take each question randomly or in the order received. Participants can also remove their own line from the queue.

SPEED DIAL Allows an operator to quickly initiate outbound calls from an attendant console. The operator accesses a stored list of phone numbers and highlights the individual to be called into the conference. By simply pressing the Enter key, the operator can dial the number.

AUTO DIAL Similar to speed dial, except it allows the operator to highlight all the phone numbers of individuals who must join in a conference. By pressing the Enter key, the operator dials all the numbers simultaneously. Each conferee is greeted by a recorded announcement that provides further instructions, such as a prompt for a security code.

LECTURE Automatically mutes all lines in the conference, except that of the moderator, for uninterrupted sessions.

SECURITY Provides confidential conferencing, prohibiting unauthorized individuals from entering. Secured conferences lock out the operator and cannot be recorded. By pressing the same buttons on the touchtone keypad, the moderator can remove security from the conference.

MUTE Places a specific line in listen-only mode. Also, lines are automatically muted by the system when certain features are implemented, such as lecture, polling, and Q&A.

MUSIC Via an external device attached to the bridge, music is provided to entertain participants until the conference begins. Music also provides assurance to waiting participants that the connection is still alive and that they should continue to wait.

RECORD/PLAYBACK Conferences can be recorded and played back. Also, recorded material can be played into a conference with an external system. All lines are muted during playback. Some systems can be configured to allow the moderator to be heard during the playback.

HELP Allows conferees to signal an operator for assistance. Help can be set on a per-line basis whereby the operator removes the conferee from the conference to provide help. Help can also be set on a conference-wide basis whereby the operator responds by entering the conference to address the entire group.

CONFERENCE ID Provides for the identification of conferences for report generation. The operator can assign the conference ID, or the system can be configured to assign IDs automatically.

CONFERENCE NOTE The operator can record notations during a conference that will appear on the conference report.

CONFERENCE SCAN An automatic audio scan of the entire conference can be performed at assigned intervals. Secured conferences are not scanned.

OPERATOR CHAT Two or more operators can send electronic messages to each other without disrupting active conferences.

LISTEN MODE For quality control purposes, operators can listen to individual lines or a range of lines without affecting conference activity.

DISCONNECT NOTIFICATION The system can be configured to notify the operator of a disconnect during the conference.

OPERATOR ALARMS The system can be configured to provide the operator with audible and visual signals upon disconnect, help request, or queue activity.

ADMINISTRATION FEATURES From an audioconferencing product's system administration menu, an administrator can:

- Modify system configurations
- Perform supervision of the system during operation
- Configure the system for auto-dial
- Configure operator functions
- Configure channels
- Perform file management
- Implement disk utilities
- Configure the conference scheduler

MAINTENANCE FEATURES From a system's maintenance menu, a maintenance person can access all conference and administrative functions, plus such special maintenance features as:

- Power-up diagnostics
- Online diagnostics
- Remote diagnostics
- Warm boot
- Maintenance reports
- Alarms

SUMMARY

Audioconferencing is an effective, economical way for business people to collaborate on projects or stay in contact with colleagues without resorting to expensive and time-consuming travel. While the same can be said about videoconferencing, audioconferencing continues to be the more available, economical, and simpler solution for the majority of business needs. In the case of IP networks, companies can fashion their own internal telephone networks by leveraging existing intranets and conduct audioconferences (and videoconferences) at virtually no extra charge. Alternatively, subscription services are emerging that offer audioconferencing services over private IP networks that can be accessed from a corporate intranet or the Internet at greatly reduced rates.

See Also **Videoconferencing Systems**

Automated Intrusion Detection

Companies of all types and sizes are connecting their internal networks to the public Internet and their own TCP/IP-based intranets for a variety of reasons, including productivity improvement, electronic commerce, customer service, and collaboration. At the same time, these organizations realize the importance of protecting themselves from outside attacks against vital systems on their LAN. The problem has been a lack of staff to watch all systems and then physically reconfigure them to respond to the latest security threats.

A solution to this problem comes in the form of real-time, automated intrusion detection that can stop unauthorized activities immediately, even if IT managers are not around to intervene. With such tools, network intrusions are detected and appropriate responses are taken automatically before they can cause serious damage.

Many companies now offer automated intrusion detection tools, but one in particular is notable for its ease of use. AXENT Technologies, Inc. offers 200 so-called "drop and detect" security scenarios in its Omni-Guard/Intruder Alert. These preconfigured scenarios let organizations install Intruder Alert to instantly protect systems against hundreds of the most common and dangerous security threats to Windows NT and other key enterprise systems.

Intruder Alert uses a real-time, manager/agent architecture to monitor the audit trails of distributed systems for "footprints" that signal suspicious or unauthorized activity on all major operating systems, Web servers, firewalls, routers, applications, databases, and SNMP traps from other network devices. Unlike other intrusion detection tools, which typically report suspicious activity hours or even days after it occurs, Intruder Alert instantly takes action to alert IT managers, shut systems down, terminate offending sessions, execute commands, and take other actions to stop intrusions before they damage crucial systems.

As new security threats emerge, IT managers can quickly protect their systems by loading new drop-and-detect scenarios researched and developed by AXENT's Information Security SWAT Team, a group of computer security professionals focused on hacking techniques and the latest computer security threats. These new scenarios, which can be downloaded from the SWAT Team Web site and installed enterprise-wide, make it easy for IT manager to keep systems safe from evolving threats.

From a single management workstation, IT managers can quickly drag new security policies and attack scenarios to different enterprise domains, implementing additional protection for hundreds or thousands of systems in a matter of minutes. The enterprise console also provides a correlated, graphical view of security trends, letting IT managers view graphs that illustrate real-time security trends and drill-down to additional details on activity.

Intruder Alert 3.0 can detect and respond to hundreds of Windows NT, UNIX, and NetWare security threats and attacks, including:

TROJAN HORSE ATTACKS   Hackers often attack Windows NT with Trojan horses, convincing a systems administrator or user into running a particular program, ActiveX, or Java applet. The Trojan horse code appears to run normally, but works behind the scenes to change system files or steal data. For example, one Trojan horse ActiveX applet changes the Windows NT log-in program (msgina.dll) to a hacker version. The hacker version appears to run normally and lets users log in, but then records user names and passwords and e-mails them back to the hacker. Intruder Alert contains a drop-and-detect scenario that automatically monitors the Windows NT log-in program for changes and immediately notifies the management console. It can also page IT professionals and execute an action, such as shutting down network access to the Windows NT server.

SYN FLOOD/DENIAL OF SERVICE ATTACKS Hackers often flood system ports with communications requests, requiring NT servers to keep processing the requests and preventing them from doing their primary work, such as hosting a Web page or supporting an e-mail service. This attack is particularly dangerous for Internet service providers (ISPs), who can easily lose customers to busy signals. Intruder Alert actively checks NT and other operating systems for SYN Flooding activity and instantly alert administrators.

DISABLED AUDIT LOGS To cover their tracks, hackers will often stop, disable, or attempt to edit audit trails that track their activities. Among other actions, Intruder Alert can detect audit changes, issue alerts, automatically restart auditing, and page the administrator.

NT ACCOUNT LOCKOUTS After receiving a set number of failed log-in attempts, Windows NT automatically "locks out" that username. Even after the account is locked out, hackers can still log into that system under another username or move to a different system and try the same name again. Intruder Alert detects single and multiple account lockout attacks, monitoring for a rapid succession of lockouts that signal attempts to gain unauthorized access.

Hackers can also use a variety of tools to guess passwords, making just a few log-on attempts on a single system or camouflaging their attempts by distributing them across several platforms to avoid triggering an account lockout. The Administrator account and Guest account are popular Windows NT username targets, since they are created by default when the NT operating system is installed. In addition, the default Guest account does not require a password until it has been configured to require one. Intruder Alert closely monitors failed log-ins across all systems, alerting administrators to disturbing trends and failed attempts to gain Administrator, Guest, and user privileges.

SUMMARY

Taking effective security measures to protect a corporate network often requires several products from different vendors because no single tool is capable of addressing every potential threat. Of particular value are security tools that continually monitor the network for possible break-ins.

Automated Intrusion Detection

This can be done with automated intrusion detection tools that identify break-in attempts in real-time and take protective measures to stop intruders from going any further. The magnitude of the security problem corporations face today easily justifies the investment in such tools. In fact, security is such a concern that companies are spending an estimated $6 billion annually on security hardware, software, and services.

See Also **Firewalls, LAN Security, Security Risk Assessment**

B

Bisynchronous Communications

Bisynchronous communications, or binary synchronous communications (also known as bisync or BSC), is one of the two commonly used methods of encoding data for transmission between devices in IBM mainframe environments.

As in synchronous communication, bisync communication requires that both sending and receiving devices be synchronized before transmission of data begins. Data characters are gathered in a package called a *frame*, which is marked by two synchronization bits, hence the term *bisynchronous*.

Each frame contains leading and trailing characters that allow the computers to synchronize their clocks. The structure of a bisynchronous communications frame starts with initial synchronizing characters, followed by optional header characters, then the data message, which is preceded by a start-of-transmission (STX) character and followed by an end-of-transmission (ETX) character. The ETX is followed by a block check character (BCC), which verifies the accuracy of the transmission. The BCC is a one- or two-character result of a transmission verification algorithm performed on the block of data being transmitted.

SUMMARY

While bisynchronous communication is still used extensively in IBM environments, the more modern protocol is Synchronous Data Link Control (SDLC), the data link layer of IBM's Systems Network Architecture (SNA). SDLC is a more efficient method than the older bisync protocol when it comes to packaging data for transmission between computers. Packets of data are sent over the line without the overhead created by synchronization and other padding bits.

See Also **Synchronous Data Link Control**

■ ■ ■ Bit Error Rate Testing

Bit error rate testing is used to determine whether data is being passed reliably over a digital communications link. Bit Error Rate Testers (BERTs) send and receive various bit patterns and data characters so what is transmitted and what is received can be compared. Any difference between the two is displayed as an error. The bit error rate is calculated as a ratio: the total number of bit errors divided by the total number of bits received.

Some low-end BERTs may indicate only that an error occurred, not how many errors or what kind. High-end BERTs display a real-time cumulative total of bit errors, as well as a real-time calculation of the bit error rate itself. In addition, high-end BERTs can include information on sync losses, sync loss seconds, errored seconds, error-free seconds, time unavailable, elapsed time, frame errors, and parity errors.

Interfaces

The BERT must be able to physically and electrically accommodate the interfaces on the devices under test. Some BERTs can be configured to accommodate a variety of interfaces through plug-in modules that attach to the base unit. Modules are available for the following common interfaces:

- RS-232-C
- V.35
- RS-449-B
- EIA-530
- X.21
- MIL 188-114
- DDS
- DS1/T1
- T1/Fractional T1
- T1/FT1 Data/Voice Frequency Drop and Insert
- E1/Fractional E1
- G.703 64 Kbps Co-Directional
- G.703 2.048 Mbps (E1)
- G.703 8.448 Mbps (E2)
- Universal Lab Interface

Test Patterns

Most BERTs offer various bit and byte test patterns, including continuous 1s or 0s, alternate 1s and 0s, and pseudo-random patterns such as 63, 511, 2047, and 4095. These are typically used for testing synchronous devices. Testing several different types of patterns is useful to determine the degree of performance degradation of digital lines or equipment. For example, when testing a particular device, the 511 pattern may "run clear" and the all ones (111) test may fail.

Some BERTs even allow customization of test patterns, which can extend the range of stress testing. The idea is to "stress" the device under test in a variety of ways in an effort to identify weak components on the network so they can be replaced before causing serious problems.

Testing asynchronous devices, such as asynchronous statistical multiplexers, with random bit patterns is not feasible because of their use of start and stop bits. However, there are BERTs that can send asynchronous byte test patterns.

Some BERTs are also referred to as BLERTs (block error rate testing) because they calculate block error rates in addition to bit error rates. By transmitting in blocks, BLERTs may more accurately represent the flow of user data in a network.

Types of Products

BERTs can be stand-alone devices or portable pocket units. Alternatively, BERT functions can be implemented in software for installation on a laptop computer and accessed through a graphical interface, such as Windows 95 (Figure 11).

SUMMARY ■ ■ ■ ■ ■ ■ ■ ■ ■

The bit error rate is a fundamental performance indicator of a digital line. On many types of networks, errors force the retransmission of affected data, which slows throughput and response time. Being able to measure the bit error rate enables technicians to keep corporate networks running at peak performance.

See Also **Network Monitoring, Network Troubleshooting**

Figure 11
Reach
Technologies
offers BERT
software that can
be accessed
through a
Windows 95
interface.

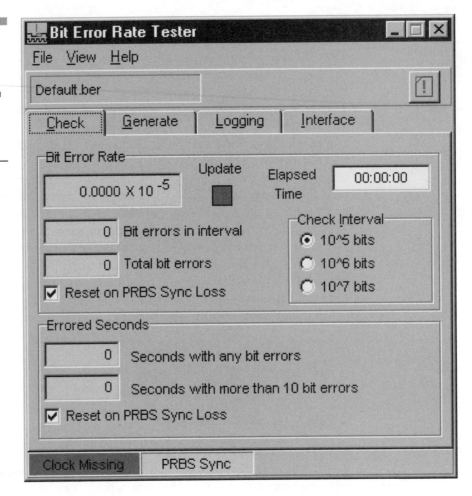

Bluetooth

Bluetooth is the code name for the technology specification of short-range radio links between mobile PCs, mobile phones, and other portable devices. As a potential global standard for wireless connectivity, Bluetooth aims to eliminate the tangle of cables currently required to connect peripherals to computers and cellular phones.

In one application scenario, a Bluetooth-equipped cellular phone synchronizes the numbers in its memory with the most frequently accessed numbers in a person's PDA. Messages arriving on a pager are routinely logged into the to-do list on the laptop and PDA.

Cable Replacement

Bluetooth technology provides one universal short-range radio link that allows for the replacement of the many proprietary cables that connect one device to another. For instance, Bluetooth radio technology built into both a cellular telephone and a laptop computer could replace the cumbersome cable used today to connect a laptop to a cellular telephone. Printers, desktops, fax machines, keyboards, joysticks, and virtually any other digital device can be part of a Bluetooth system.

But beyond untethering devices by replacing the cables, Bluetooth radio technology provides a universal bridge to existing data networks, a peripheral interface, and a mechanism to form small private ad-hoc groups of connected devices away from fixed network infrastructures. Bluetooth security accommodates both private and public devices, and uses streaming encryption with up to 128-bit keys.

Radio Link

Bluetooth radios operate in the unlicensed ISM band at 2.4 GHz to provide a gross data rate of 1 Mbps. A Time-Division Duplex (TDD) scheme is used for full-duplex transmission, and up to 80 devices can operate in a network. Designed to operate in a noisy radio frequency environment, Bluetooth's radio technology uses a fast acknowledgment and frequency hopping scheme to make the link robust.

Bluetooth radio modules avoid interference from other signals by hopping to a new frequency after transmitting or receiving a packet. Compared with other systems operating in the same frequency band, a Bluetooth radio typically hops faster and uses shorter packets, which makes it more robust than other systems. Short packets and fast hopping also limit the impact of domestic and professional microwave ovens, wireless LANs, and metropolitan-area wireless telecommunications systems such as Metricom—all of which use the same 2.4-GHz frequency. Use of Forward Error Correction (FEC) limits the impact of random noise on the links. The encoding is optimized for an uncoordinated environment.

Channels

The Bluetooth baseband protocol is a combination of circuit and packet switching. Slots can be reserved for synchronous packets, and each packet is transmitted in a different hop frequency. A packet nominally covers a single slot, but can be extended to cover up to five slots. Bluetooth can support an asynchronous data channel, up to three simultaneous synchronous voice channels, or a channel that simultaneously supports asynchronous data and synchronous voice. Each voice channel supports a 64-Kbps synchronous (voice) link. The asynchronous channel can support an asymmetric link at 721 Kbps in either direction while permitting 57.6 Kbps in the return direction, or a 432.6-Kbps symmetric link.

SUMMARY

Bluetooth (named for Denmark's first Christian king) is a specification first advanced by Ericsson as an extension to the Global System for Mobile (GSM) telecommunications—a world standard for cellular communications. It is a combination of software, communications protocols, and a tiny radio transceiver on a chip that will let devices communicate with one another over short distances. Bluetooth can handle phone-quality voice as well as data.

See Also **Personal Area Networks, Wireless LANs**

Bridges

Bridges are used to minimize the impact of workgroup traffic on overall network performance. Bridges can connect networks running at different speeds with different topologies or communication protocols. Any type of network can benefit from bridges, including Ethernet, token ring, and FDDI.

Through filtering, bridges keep local traffic within the workgroup LAN and forward only the traffic with a destination elsewhere on the backbone. The filtering rate is usually much faster than the forwarding rate. For example, the DECbridge 90 filters at 29,694 packets per second (pps) and forwards at 14,847 pps.

Bridges monitor all traffic on the subnets they link, and they read both the source and destination addresses of all the packets sent through them. If a bridge encounters a source address not already contained in its address table, it assumes that a new device has been added to the local network. The bridge then adds the new address to its table.

For each received frame, the bridge stores the address in its table, in the frame's source-address field together with the port from which the frame arrived. The table also stores the time that an entry is recorded. The bridge deletes an address in the table after a period of time in which no frames are received with that address as the source address. This self-learning capability permits bridges to keep up with changes on the network without requiring that their tables be manually updated.

While a bridge supports an unlimited number of nodes on the backbone side, it also supports a limited number of addresses on the workgroup side. The DECbridge 90, for example, supports 200 such addresses.

Bridges isolate traffic by examining the destination address of each packet. If the destination address matches any of the source addresses in its table, the packet is not allowed to pass over the bridge because the traffic is local. If the destination address does not match any of the source addresses in the table, the packet is allowed to pass onto the adjacent network. This filtering process is repeated at each bridge on the internetwork until the packet eventually reaches its destination. Not only does this process prevent unnecessary traffic from leaking onto the backbone network, but it acts as a simple security mechanism that can screen unauthorized packets from accessing various corporate resources.

Bridges can also interconnect LANs that use different media, such as twisted pair, coaxial, and fiber-optic cabling, and various types of wireless links. An Ethernet backbone connection, for example, can be made using unshielded twisted pair (UTP) or AUI transceiver cable. The AUI port supports various transmission media, such as coaxial and fiber-optic cables. LED indicators show the status of backbone and workgroup activity and integrity, as well as configuration and power status.

In office environments that use wireless communications technologies, such as spread spectrum and infrared, bridges can function as an access point to wired LANs. On the WAN, bridges can even switch traffic to a secondary port if the primary port fails. For example, a full-time wireless bridging system can establish a dial-up connection on the public network if the primary wireline or wireless link is lost due to environmental interference.

In reference to the OSI model, a bridge connects LANs at the Media Access Control (MAC) sublayer of the data link layer. It routes by means of the Logical Link Control (LLC), the upper sublayer of the data link layer (Figure 12).

SUMMARY

Because bridges connect LANs at a relatively low level, throughput hovers around 30,000 packets per second (pps). Traditional multiprotocol routers

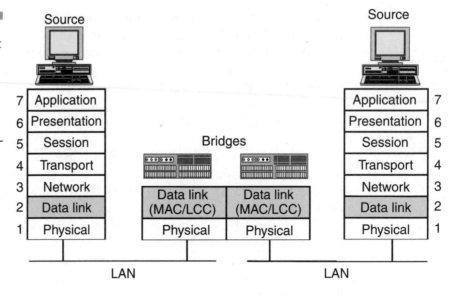

Figure 12
Bridges operate at
the data link layer
(layer 2) of the
Open Systems
Interconnection
(OSI) reference
model.

and gateways, which can also be used for LAN interconnection, operate at higher levels of the OSI model. In performing more protocol conversions, routers and gateways are usually slower than bridges.

***See Also* Gateways, Repeaters, Routers**

Broadband Information Gateways

A broadband information gateway is a relatively new class of network product devised by Motorola to provide household members with shared access to the Internet. The gateway integrates a cable modem with the "no new wires" approach to in-home data networking. The device allows families with multiple PCs to simultaneously surf the Internet, play interactive games, and subscribe to emerging low-cost IP telephony services.

The gateway is intended to extend the benefits of broadband Internet services to all members of a household, who would otherwise have to take turns. Family members can now access the Internet from any room in their home wirelessly or over existing telephone or electrical wiring using a single broadband information gateway. These in-home networks will enable data, voice, and video services throughout homes without the need for new cabling.

This "no new wires" approach to home networking is a key element in the strategies of Motorola and other vendors for addressing the burgeoning field of home networking, which seeks to connect PCs, IP phones, and other Internet-aware devices in the home to broadband cable networks operated by such cable-modem access providers as @Home and Road Runner, which together control about 90 percent of the market. @Home is controlled by cable operator Tele-Communications Incorporated (TCI), which is part of AT&T, and Road Runner, which is a joint venture between Time Warner and MediaOne.

SUMMARY

The architecture of broadband information gateways allows consumers to use IP telephony services with a look and feel identical to today's telephone services. The gateways provide standard telephones and Internet appliances with access to the integrated voice and data networks currently being built by cable companies and telecommunications carriers.

See Also **Cable Modems, Phoneline Networking, Powerline Networking, Wireless Home Networking**

Bytes

A byte is a group of eight binary digits, or bits. A bit can be a zero or a one. There are 256 different combinations of zeros and ones that can be made with one byte, from 00000000 to 11111111.

A byte is the standard unit by which data is stored; it is the standard increment of data that computers work with to process data. Hard-disk capacity, for example, is cited in terms of megabytes (MB) or gigabytes (GB), and is priced accordingly.

Term	Approximate Capacity
Byte (B)	8 bits
Kilobyte (KB)	1,000 bytes
Megabyte (MB)	1,000,000 bytes
Gigabyte (GB)	1,000,000,000 bytes
Terabyte (TB)	1,000,000,000,000 bytes
Petabyte (PB)	1,000,000,000,000,000 bytes
Exabyte (EB)	1,000,000,000,000,000,000 bytes
Zettabyte (ZB)	1,000,000,000,000,000,000,000 bytes
Yottabyte (YB)	1,000,000,000,000,000,000,000,000 bytes

Interestingly, the speed of data transmission is almost always cited in terms of bits per second (bps). A T1 line, for example, moves data at 1.544 megabits per second (Mbps or Mb/s), while the speed available over a SONET OC-256 fiber-optic line is 13.271 gigabits per second (Gbps or Gb/s). However, if the speed of a data transmission link must be expressed in terms of bytes, it is done with a capital B, as in KBps or MBps, for kilobytes per second and megabytes per second, respectively. These terms can also be written as KB/s or MB/s.

SUMMARY

The distinction between *bits* and *bytes*, therefore, is not just technical jargon; it is the standard way vendors and service providers distinguish their offerings and set pricing.

Cable Management

A key aspect of network maintenance and support involves cabling and rewiring. Although not the most glamorous aspect of maintenance and support, it is probably one of the most planning-intensive. When a new system or network is installed, invariably some wiring already exists and there is usually strong economic incentive for its continued use. When new wiring is installed, the desire to make it as useful as possible in subsequent feature and capacity expansions must be balanced against the initial installation cost.

Elements of Premises Wiring

There are typically several hierarchical layers of wiring that merit attention. With PBX installations, for example, these layers are:

- The telephone carrier's distribution frame or demarcation point, which is the termination of the carrier circuits on the premises and also the termination of carrier responsibility for the wiring. Often, this is simply a series of terminal blocks through which user CPE is attached to carrier circuits.

- The incoming circuit distribution frame, a place where cross-connection between carrier lines and user CPE can be made. Cut-throughs or PBX bypass lines used to answer or originate calls when a PBX fails may be attached here, in front of the PBX. This is normally placed with the carrier's demarcation point and may be omitted if there is only one CPE destination for all lines.

- The user's telephone equipment, linked to the incoming circuit distribution frame on one side and the private network or intermediate distribution frame on the other. The station pairs from the telephone system exit this equipment.

- The intermediate private-network wiring distribution frame, where the telephone system station pairs are connected to terminal blocks or panels for matching with the building wiring.

- Riser cables, which terminate in the intermediate distribution frame and link it to, for example, horizontal distribution panels on floors.

- Horizontal distribution or wiring-closet panels that take riser connections and distribute them to the actual station wiring.

- Outside wiring, serving the combined functions of riser cables and horizontal distribution cables where the run must exit the premises and transit an outdoor space.

- Station wiring, which links the horizontal distribution panel with the instruments.

The wiring process is designed to achieve an important goal—permit restructuring of the system without actually stringing new wiring or performing extensive rewiring to accommodate moves, adds, and changes. Virtually the same considerations apply to hub-based LANs.

Cable Plant as Asset

In a properly designed wiring system, each cable pair should be viewed as a manageable asset, which can be manipulated to satisfy any user requirement. Individual cable pairs should be color-coded for easy identification, and cables should be identified at both ends by a permanent tag with serial numbers. This allows the individual pairs to be selected at either end with a high degree of reliability and connected to a patch panel or punch-down block as appropriate to implement moves, adds, or changes.

Patch panels and punch-down blocks should be designed to segregate different functions or circuit types to expedite the easy location of pairs. Data and voice connections, for example, can be terminated in different areas of the panel or block to avoid confusion. Many panels and blocks are precolored or have color-tagging capabilities, which can assist in locating pairs later.

The cabling used in wiring a telephone system or LAN depends on the requirements of the devices being used and the formal distribution plan provided by the telephone system and/or computer vendor. Formal plans, such as IBM's Cabling System or AT&T's Premises Distribution System, compete with similar plans by nearly every major vendor. All of these plans have the common goal of establishing a wiring strategy that will support present needs and future growth. In-house technicians should be familiar with these cabling schemes.

A number of asset management applications are available that keep track of the wiring associated with connectors, patch panels, and wiring hubs. These cable management products offer color maps and floor plans that illustrate the cabling infrastructure of one or more offices, floors, and buildings. Managers can create both logical and physical views of their facilities, and even view a complete data path simply by clicking on a connection.

Some products provide complete descriptions of the cabling and connections, showing views of cross-connect cabling, network diagrams by floor, and patch panels and racks. Work orders can be generated for moving equipment or rewiring, and can include automatically generated instructions describing exactly what work must be done (Figure 13). With this information, network administrators know where the equipment should go, what needs to be disconnected, and what should be reconnected. Then technicians can take this job description to the location and make the necessary changes.

Other products provide a CAD interface, enabling equipment locations and cable runs to be tracked through punch-down blocks, multiconductor cables, and cable trays. In addition, bill-of-materials reports can be produced for new and existing cable installations.

Cable management applications can be run as stand-alone systems or may be integrated with help-desk products, hub management systems, and network management platforms such as IBM's NetView, Hewlett-Packard's OpenView, and Sun's Solstice SunNet Manager. Some cable management products integrate well with all of these management platforms.

Figure 13
With the ETEC cable management system from Exan Technologies, every activity can be recorded and issued as a work order. A work order is created automatically, including the instructions, then users can add comments to amplify the instructions.

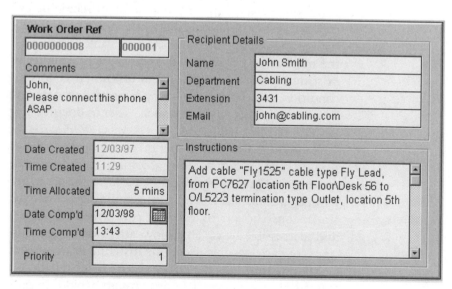

When coupled with a hub management system and help desk, a high degree of automation can be introduced to the problem resolution process. When the hub management system detects a media failure, the actual cable run can be extracted from the cable management application and submitted along with a trouble ticket generated by the help desk. And when the hub management system is integrated with a network management platform such as OpenView, all this activity can be monitored from a single management console, which expedites problem resolution.

Cable Planning

To determine if existing wiring can support new system installations requires a complete wiring plan, preferably prepared before the start of system installation. The plan acts not only as a guide to the installation process but as a check on the capacity and planning that can be carried out before installers appear and begin working. In the case of a PBX, for example, the plan should associate each instrument with a complete path back to the station wiring on the PBX, through all panels, horizontal feeds, and risers. During this process, the capacity of each trunk/riser should be checked one last time, and additional cabling run as needed. Spare pairs on each cable can be identified for possible future use.

As with station cabling, there are many factors to consider in deciding whether to purchase new or continue to use existing house cabling. The high cost of large-paired distribution cabling (300, 400, 600, or 900 pair) and its installation make the lease or purchase of existing cabling cost-effective. As with station cabling, the labor costs for identifying, reterminating, and documenting the existing cable remains. This cost increases as the number of pairs increase, as does the probability of an error. Because existing cabling often introduces unknown factors, such as pair counts, condition, and destination, many interconnect companies prefer to install new cabling. This cabling is much easier to document, install, and cutover because it is not being used.

A building under construction is the ideal environment for cable planning because it can be done according to the company's needs without concern for the requirements of existing cable plant. Moreover, factors that hamper installation in an existing building are not present in a building under construction (i.e., cosmetic concerns, disruption of office personnel, and inaccessible areas). This allows attention to be focused on meeting the needs of each possible telephone instrument or terminal location.

Many companies specialize in cable installation, and may be contracted on a per-project basis through a bidding process. Some cabling contracting companies only install cabling; others both install and maintain cable networks. Because of the complexity of a cable network, it is convenient to deal with only a single company—one that endeavors to become familiar with the organization's current and future cabling requirements.

SUMMARY

If information systems and communications networks constitute a strategic resource to a company, so must its wiring. The proper planning for each wiring and rewiring of a facility can preserve these resources intact for the support of future applications. Improper wiring and poor record-keeping often leave no recourse but to fully rewire a facility—a task that is often far more expensive than new installation.

***See Also* Cable Testing, Physical Network Management**

Cable Modems

Cable modems link computers to the Internet through local cable television (i.e., CATV) networks (Figure 14). The CATV network may provide unidirectional or bidirectional access to the Internet.

With older unidirectional service, two modems are used, one in the computer and the external cable modem (Figure 15). Bidirectional service means the CATV network is used for both uploading and downloading. In this case, only a cable modem is required. About 20 percent of American homes have cable service that supports two-way communications, and most cable operators are upgrading their networks for bidirectional communication.

With both uni- and bidirectional services, information can be downloaded at speeds of 10 Mbps and potentially 28 Mbps in the future. At 10 Mbps, a cable modem could transmit all 857 pages of Melville's *Moby Dick* in about two seconds, while a 28.8 Kbps modem—which is still the most common way of accessing the Internet—would not get past the third page within the same period of time.

Modulation Schemes

The most common digital modulation scheme used by cable modems for downstream data transmission is 64-state quadrature amplitude modula-

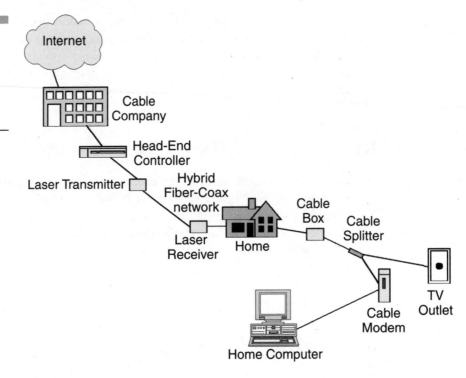

Internet

Cable
Company

Head-End
Controller

Laser Transmitter

Hybrid
Fiber-Coax
network

Laser
Receiver

Home

Cable
Box

Cable
Splitter

TV
Outlet

Cable
Modem

Home Computer

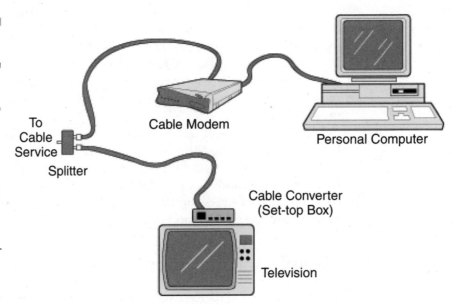

To
Cable
Service

Splitter

Cable Modem

Personal Computer

Cable Converter
(Set-top Box)

Television

tion (64 QAM). For upstream transmission, the most common digital modulation scheme is differential quadrature phase shift keying (DQPSK).

The 64 QAM digital frequency modulation technique is used primarily to send data downstream over a CATV network. It is very efficient, supporting up to 28 Mbps peak transfer rates over a single 6-MHz channel. But 64 QAM's susceptibility to interferance makes it ill suited for noisy upstream transmissions (from the cable subscriber to the Internet). For the upstream transmissions, DQPSK is used.

DQPSK is a digital modulation technique commonly used with cellular systems. It is highly efficient, allowing for a greater number of narrow 600-KHz channels. When the modem is first installed, DQPSK picks out the least noisy channel for sending data upstream. However, since the modem uses an ordinary phone line, its peak upstream transmission rate is limited to 768 Kbps.

Connection and Setup

A cable modem connects directly to the same coaxial cable that delivers television programming via standard gauge RG-59 coaxial cable with an F-type connector. With a splitter attached to the main cable, one segment of coaxial cable goes to the modem and the other segment goes to the television's set-top box. (Another piece of cable links the set-top box to the television.) This arrangement allows simultaneous television viewing and Internet access. Since different frequencies are used for each, there is no signal interference.

If the cable operator offers unidirectional Internet access, a phone line must also be connected to the cable modem. The cable modem contains an internal analog modem to establish the initial Internet connection via a stored telephone number, and the phone line cannot be used for telephone calls while the Internet connection is in use.

Next, the cable modem must be connected to the PC's 10BaseT Ethernet card using standard four-pair category 5 modular PVC cable. Each end of this cable plugs into the RJ45 jacks of the cable modem and Ethernet card (Figure 16). Of course, this assumes that the Ethernet card is already installed, the appropriate drivers have been loaded, and TCP/IP has been enabled.

Several computers can access the Internet simultaneously through a single cable modem. This can be achieved by connecting all the computers together through an Ethernet hub and then connecting the hub, via its uplink port, to the cable modem. The firmware of the cable modem may have to be updated to enable multiple computer access. This can usually be done through an online configuration process.

Figure 16
A basic
bidirectional
Internet
connection via a
cable modem.
Source: Motorola

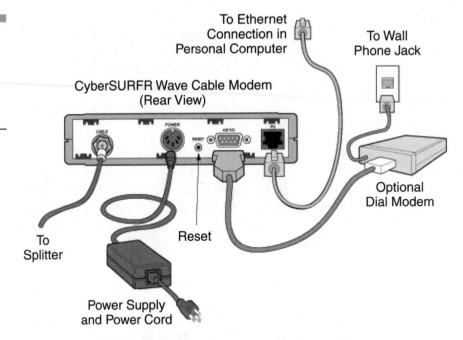

Figure 16
A basic
bidirectional
Internet
connection via a
cable modem.
Source: Motorola

Depending on the specific hub, two types of cable are used to connect the cable modem and hub: straight-through (normal) and cross-over. The right selection depends on the designation found on the hub's uplink port. A straight line indicates that straight-through or normal cable should be used, whereas an X indicates that a special cross-over cable should be used. With this type of cable, the send/receive pairs within the cable are crossed. If no designation appears on the uplink port, normal straight-through cable is used.

With multiple computers, each system must have its own IP address for purposes of communicating with the cable modem. The cable modem has it own IP address, which is presented to the CATV network. The assignment of IP addresses may look like this:

Cable modem	=	192.168.100.1
Computer #1	=	192.168.100.2
Computer #2	=	192.168.100.3
Computer #3	=	192.168.100.4
Computer #4	=	192.168.100.5

Since the cable modem is always on, the network connection is always open. Getting on the Internet is as simple as opening the Web browser (Figure 17). Unlike dial-up Internet access, there is no waiting through a slow log-on procedure.

Figure 17
With a cable modem, the initial connection to the Internet is simply a matter of opening the Web browser and clicking on the Connect button. The network connection stays open until the user clicks on the Disconnect button.

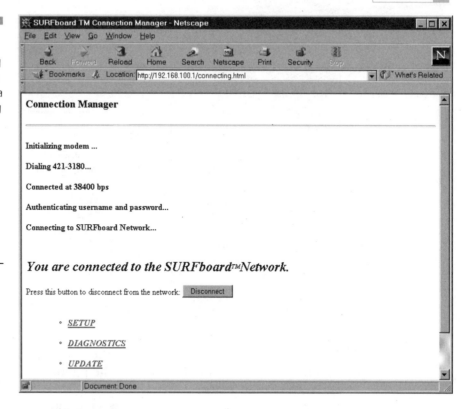

IP Telephony

Cable operators offering Internet access now offer IP telephony services as well. Although this has been possible for a long time, a new wrinkle is the support of data and voice through the same cable modem. Previously, to put voice and data on the cable network required a modem and a separate voice gateway, which is more complex and expensive.

This new type of cable modem is an integrated multimedia terminal adapter (MTA). Motorola's MTA, for example, has a telephone port and a four-port Ethernet hub built into it. Data speeds on the cable can be as high as 10 Mbps, depending on how many other users are on the same shared cable subnet to exchange information. The MTA uses proprietary traffic prioritization techniques to ensure that there is enough bandwidth to prevent voice packets from getting delayed.

The MTA sits on the customer premises (Figure 18), where it converts phone and PC traffic into packets for transport over a cable network. It then sends the packets to a router or packet switch on the cable network, where voice and data are sorted out. Data is routed to the cable operator's packet network, while voice is routed to an IP/PSTN gateway. The gateway

Figure 18
Placing a
telephone call
over a cable
network.

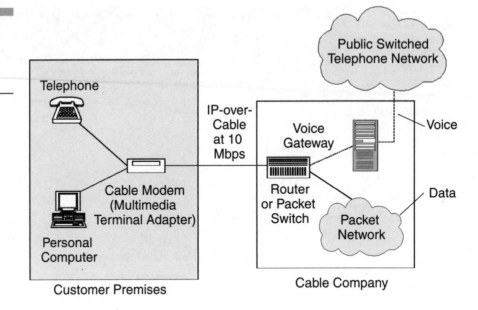

decompresses the packets and returns them to analog form so voice can be received by an ordinary phone on the public network or a corporate phone in back of a PBX.

With many corporations devising ways to internally implement voice over IP, the cable option allows them to extend it to telecommuters. Cable modems already support data at the Ethernet speed of 10 Mbps, which gives telecommuters a viable way to retrieve information from the corporate LAN via an Internet access connection.

Although ISDN is more flexible in terms of bandwidth allocation and call handling features, making it a good choice for many telecommuters, its usage-based pricing makes it far more expensive than cable when used extensively for Internet access.

Two online services, America Online and MindSpring, have studied possible ways to get faster Internet connections to customers. Their conclusion, as reported to the Federal Communications Commission (FCC), is that cable modems will be the preferred method over the next three to five years. The main challenger from phone companies, Digital Subscriber Line (DSL) technologies, offer as much bandwidth as cable, but like ISDN they are far more expensive. DSL technologies are also not yet widely available and will likely be overtaken by cable modems before they have a chance to establish significant market share among consumers.

Standards

CableLabs, on behalf of the cable television industry, is pursuing interoperability aggressively in several high-level projects—DOCSIS (Data over Cable Service Interface Specification), OpenCable, and PacketCable. This interoperability will enable the cable industry to choose equipment from a variety of suppliers building similar products. It will also provide cable customers with the economic benefits of price competition and product innovation, as well as the opportunity to purchase interoperable products at retail stores in the near future.

DOCSIS is aimed at attaining interoperable high-speed cable modems, OpenCable is seeking to advance interoperable digital set-top boxes, and PacketCable is focusing on interoperable Internet protocol-based services that will be distributed over hybrid fiber/coax systems. One of the initial services on which PacketCable is concentrating is IP telephony, which uses the Internet or other packet networks to deliver telephone calls at competitive prices.

At this writing, a common media-access control scheme for sending data over cable is under development by the IEEE 802.14 committee, which will also lead to interoperable products.

SUMMARY

Cable modems provide economical, high-speed access to the Internet via local CATV networks. The devices cost about $150 and can usually be purchased or leased through the cable operator, who decides what vendor it is going to use. About 75 percent of cable modems in use today are from Motorola.

***See Also* CATV Networks, Hubs, Routers**

Call Accounting

Call accounting systems capture detailed information about telephone calls and store them on a PC or other collection device. The collected data is then processed into a variety of cost and usage reports. Such systems are offered in the form of stand-alone products, or they may be one of several modules within a suite of telemanagement applications that include one or more of the following functions:

Call Accounting

- Call center management
- Move, add, and change administration
- Cable management
- Inventory management
- Invoice management
- Lease management
- Trouble reporting and tracking
- Work order management
- Traffic analysis
- Network design and optimization

Benefits

Call accounting systems represent a powerful management tool that can monitor complete telephone system usage by every employee—including phone, fax, modem, and dial-up Internet connection time. Such systems provide organizations with many benefits:

REDUCED TOLL FRAUD Online, real-time detection can notify telecom managers of fraud as it is occurring so it can be stopped it before it gets out of hand.

REDUCED UNAUTHORIZED USE If employees know that telephone usage is monitored, including incoming calls and local calls, the number of personal calls decreases.

CHARGE ALLOCATION Call charges can be billed back to the actual extensions, projects, or departments that incurred the cost.

COST ALLOCATION Administrative charges can be allocated to departments to help pay for the telephone system and its management.

TENANT BILLING If the organization leases office space to other companies or individuals, they can be billed directly for their phone usage.

INCREASED PRODUCTIVITY Since telephone usage can be monitored, unnecessary calls are decreased, which increases productivity; if the job description requires phone contact, being able to monitor the number of calls is an indicator of productivity.

Call Detail Records

Call accounting systems produce reports from Station Message Detail Recording (SMDR) devices, which capture call detail records generated by PBX, Centrex, hybrid, or key telephone systems, or from tip-and-ring line scanners. Most call accounting systems compute costs for each incoming or outgoing call, whether local or long distance. Call detail records usually contain the following basic information:

- Date of the call
- Duration of the call
- Extension number
- Number dialed
- Trunk group used
- Account number (optional)

This information can be printed in detail or summarized by categories such as individual station, department, or project. Other summary categories might also include the most frequently dialed numbers, longest-duration calls, and highest-cost calls.

Standard call accounting reports can be used to locate toll fraud by identifying short and frequent calls, long-duration calls, unusual calling patterns, and unusual activity on 800 and 900 numbers. In addition, the reports can identify calls made after hours or on weekends or holidays. Some vendors have introduced special toll fraud detection packages. These programs alert managers to changes in calling patterns or breached overflow thresholds. Several types of alarms indicating suspicious activity can be generated and sent to a printer, a local PC, remote PC, or pager.

Call Costing

A key feature of call accounting systems is call costing. For each call record, the cost per call can be figured in a variety of ways to suit various organizational needs (Figure 19), including:

- Actual route cost
- Comparison route cost
- User-defined least cost route
- What-if recosting for comparison
- Equipment charge assignment

Call Accounting

- User-definable pricing by route group
- Tariff table pricing
- Usage-sensitive pricing
- Flat rate with percentage surcharge
- Minimum charge
- Evening and night discounts
- Operator-assisted charge

The ability to price calls according to various parameters gives organizations flexibility in allocating costs and meeting budgetary targets.

Cost Allocation

Another key feature of call accounting systems is cost allocation, which distributes call costs to the appropriate internal departments, projects and clients, workgroups and subsidiaries, or external customers and individuals. With this feature, costs can be applied to calls that also include the cost for equipment, trunks, lines, maintenance, or other administrative charges. Some systems can also depreciate equipment according to organizational

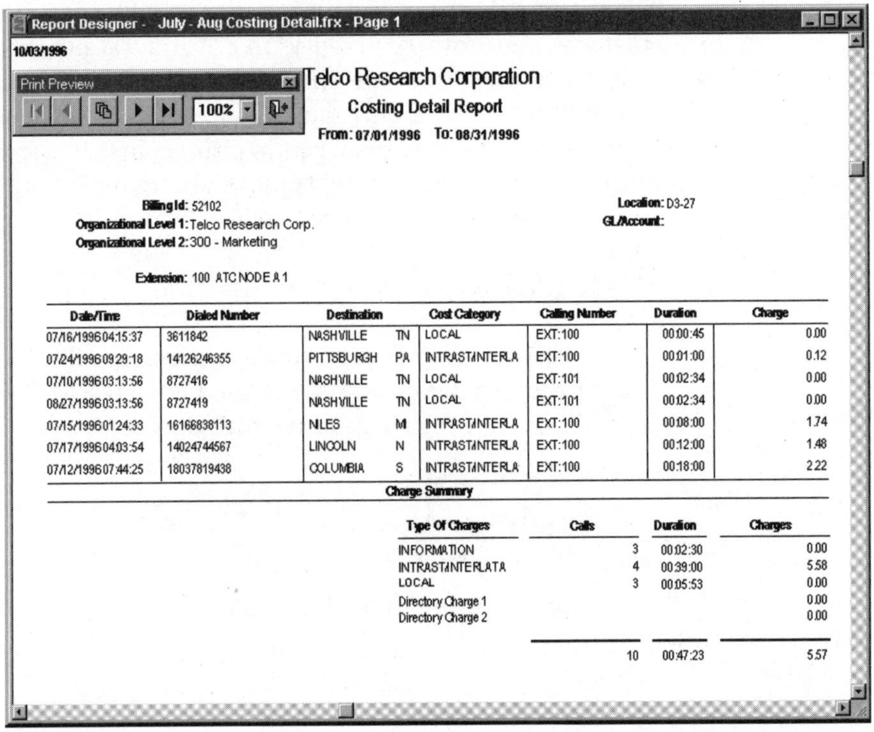

Figure 19
A typical call-costing detail report generated by Telco Research's TRU Call Accountant, a Windows 95 application.

Report Designer - July - Aug Costing Detail.frx - Page 1

10/03/1996

Print Preview | 100%

Telco Research Corporation
Costing Detail Report
From: 07/01/1996 To: 08/31/1996

Billing Id: 52102   Location: D3-27
Organizational Level 1: Telco Research Corp.   GL/Account:
Organizational Level 2: 300 - Marketing

Extension: 100 ATC NODE A 1

Date/Time	Dialed Number	Destination		Cost Category	Calling Number	Duration	Charge
07/16/1996 04:15:37	3611842	NASHVILLE	TN	LOCAL	EXT:100	00:00:45	0.00
07/24/1996 09:29:18	14126246355	PITTSBURGH	PA	INTRAST/INTERLA	EXT:100	00:01:00	0.12
07/10/1996 03:13:56	8727416	NASHVILLE	TN	LOCAL	EXT:101	00:02:34	0.00
08/27/1996 03:13:56	8727419	NASHVILLE	TN	LOCAL	EXT:101	00:02:34	0.00
07/15/1996 01:24:33	16166838113	NILES	MI	INTRAST/INTERLA	EXT:100	00:08:00	1.74
07/17/1996 04:03:54	14024744567	LINCOLN	N	INTRAST/INTERLA	EXT:100	00:12:00	1.48
07/12/1996 07:44:25	18037819438	COLUMBIA	S	INTRAST/INTERLA	EXT:100	00:18:00	2.22

Charge Summary

Type Of Charges	Calls	Duration	Charges
INFORMATION	3	00:02:30	0.00
INTRAST/INTERLATA	4	00:39:00	5.58
LOCAL	3	00:05:53	0.00
Directory Charge 1			0.00
Directory Charge 2			0.00
	10	00:47:23	5.57

depreciation schedules. Most cost allocation applications also provide an interface to the organization's general ledger.

Polling

Call records are collected from one or more locations via a polling process, which is usually done by a PC. Multiple sites can feed their data simultaneously into a single recording PC via a dial-up or dedicated line. During the polling session, Cyclic Redundancy Checking (CRC) is used for error correction. Compression can be added to minimize collection time and save on long-distance call charges. User-defined filters can be applied to exclude certain types of call records from being captured. Once the relevant data is collected, it can be processed into the appropriate report formats.

At each location, there is typically a solid-state recording unit that attaches to the PBX. These devices are available with various memory capacities, from 256 KB to 8 MB. They include a battery backup that preserves data integrity for 30 to 60 days in case of a local power failure. They also have a graphical interface that can be used for management. Through this interface, the system settings can be specified, including the system and polling parameters and the alarm and callback schedules.

Several polling methods are available. Polling can be triggered when the SMDR device reaches 80-percent full to ensure that buffers are not overwritten before the call data is retrieved. Polling sessions can occur on a scheduled basis for each location. Polling can also be initiated manually. The telecom manager determines the number of automatic retries (if the line is busy) and restarts (if transmission is aborted) during a polling session.

Some polling applications can collect call detail records over TCP/IP networks, including the Internet, at a much lower cost than traditional long-distance services (Figure 20). The TRU Network Poller offered by Telco Research, for example, works in conjunction with a serial server and acts as a protocol converter that can translate RS-232 serial ASCII data into packets. Commands can automatically be sent from a PC over the network through a serial server to the solid-state recorder's buffer box to begin polling the call detail records. The records move via the solid-state recorder's serial port into the serial server where the data is packetized and sent across the TCP/IP network to the PC where the call accounting system resides (Figure 21). There, the call detail records are processed into various reports.

Outsourcing

An alternative to in-house call detail record collection and processing is a third-party service bureau. This type of service bureau collects call records

Figure 20
A typical
configuration
using Telco
Research's TRU
Network Poller. A
backup modem
can be used to poll
over the Public
Switched
Telephone
Network (PSTN) if
the TCP/IP network
is unavailable.

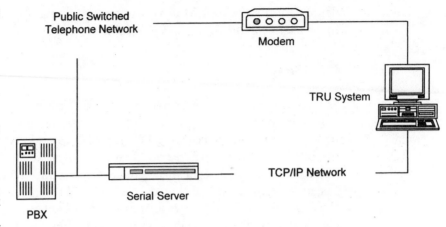

Figure 21
Detailed view of
how the TRU
Network Poller is
configured.

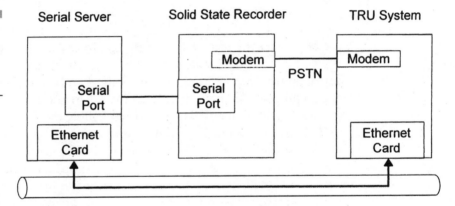

from a PC or other type of recording device at a customer's premises via a dial-up connection or dedicated line on a daily, weekly, or monthly basis—depending on the call record volume. The call detail records are processed into reports for the customer. The client may choose from among a set of standard reports or have the data processed into custom reports.

Some third-party processing firms offer carrier billing verification services. Audits verify that carrier invoices accurately reflect charges for telephone services (voice and data) and equipment actually contracted for and used, assuring that appropriate rates, taxes, and surcharges have been properly applied. Refund requests are prepared and submitted to ensure that customers receive the appropriate reimbursement for incorrect prior bills, as well as ensuring that future bills are accurate. Refunds of historical overcharges have produced ongoing savings to customers generally ranging from 5 percent to as much as 15 percent of basic monthly service charges.

SUMMARY

A new trend in call accounting is the consolidation of enterprise-wide data and voice call detail records into a single database that can be accessed by a Java-enabled Web browser. Telecom or IT managers can gather telephone and Internet call record information from network collection devices such as PBXs, Centrex systems, routers, firewalls, remote access servers, and network access servers. This data is rolled into management applications (such as Telco Research's TRU Call Accountant). Department managers can be restricted to viewing only information related to their respective departments. Executives can be given access to view telephone and Internet usage statistics and costs for everyone in the organization.
See Also **Application Metering, Cable Management, Inventory Management**

Campus-Area Networks

Campus-area networks (CANs) provide high-speed connections between several hundred to several thousand users located within a group of buildings or floors and connected to various local-area networks (LANs). Typically, optical fiber is used for the connections. To ensure high availability for mission-critical applications, a Fiber Distributed Data Interface (FDDI) is the preferred type of network for the campus environment because of its self-healing, dual ring architecture (Figure 22). It is also capable of handling Ethernet and token ring traffic reliably. A drawback of FDDI is that there are no plans to upgrade it to gigabit speeds.

Gigabit Ethernet can also be used to build campus-area networks. This is a good choice if the organization has already standardized on Ethernet/Fast Ethernet because Gigabit Ethernet switches can simply be dropped into the existing environment to interconnect LANs in different buildings. If the organization has standardized using token ring, then a High-Speed Token Ring (HSTR) backbone would be the best choice. HSTR also provides an upgrade path to Gigabit Token Ring.

If different types of LANs must be interconnected, ATM is a better choice for the campus backbone. Layer 3 switches are most often used when ATM provides the connectivity for the campus-area network. A network layer switching standard known as Multi-Protocol over ATM (MPOA) defines a way for IP traffic to run in ATM cells within a campus network. Edge devices connect traditional LANs to an MPOA-capable network. In this manner, Ethernet, Fast Ethernet, token ring, FDDI, and other LAN traffic enter and exit the ATM network. Another benefit of

Figure 22
A campus-area
network based on
FDDI.

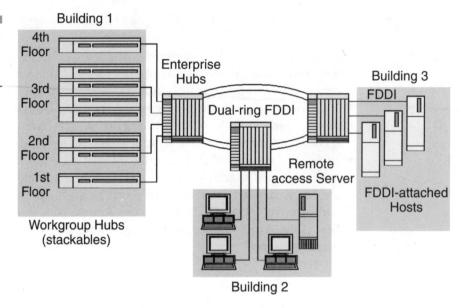

Building 1

4th Floor

3rd Floor

2nd Floor

1st Floor

Workgroup Hubs
(stackables)

Enterprise
Hubs

Dual-ring FDDI

Remote
access Server

Building 3

FDDI

FDDI-attached
Hosts

Building 2

ATM is that it can handle constant bit rate (CBR) services such as voice and interactive video.

SUMMARY

There are several viable technologies for implementing campus-area networks. The choice will hinge mostly on what types of networks must be interconnected and the types of applications that need to be supported. Other considerations include scalability and upgradability to higher speeds.
See Also **Enterprise Networks, Metropolitan-Area Networks, Wide-Area Networks**

Carrier-Based WAN Restoral

There was a time when most organizations relied on their long-distance carrier for maintaining acceptable network performance. More often than not, the carriers were not up to the task. This led to the emergence of private networks in the mid-1980s and their tremendous growth, which continued to the mid-1990s. Private networks allowed companies to exercise close control of leased lines with an inhouse staff of network managers and technicians. Metrics such as network availability, throughput, and delay were manageable.

In their eagerness to recapture lost market share, the local exchange carriers (LECs) and interexchange carriers (IXCs) have made great strides in improving overall performance as well as their response to network congestion and outages. Today, many carriers offer a range of wide-area network protection and restoral solutions. When problems occur, automated processes perform functions such as raising alarms, rerouting traffic, activating redundant systems, performing diagnostics, isolating the causes of problems, generating trouble tickets and work orders, and dispatching repair technicians. The result of these processes is the return of primary facilities and systems to their original service configuration in a timely fashion.

Many carriers now provide customers with service level agreements (SLAs) that specify levels of circuit reliability and availability, as well as financial penalties for nonperformance. In some cases, the carrier even provides management tools to allow customers to independently verify the performance of their circuits. Today, most companies are once again comfortable in relying on the carriers for maintaining acceptable network performance.

Local Restoral Services

Local exchange carriers offer protection and restoral services intended to provide customers with failsafe communications throughout their operating territory. There are several types of protection and restoral services to choose from, depending on the specific needs of the organization.

DEDICATED CIRCUITS FROM AN ALTERNATIVE CENTRAL OFFICE This lets users place outgoing calls and receive incoming switched voice calls even if an outage occurs at the primary serving office.

AUTOMATIC REROUTING A similar service for private lines, automatic rerouting is implemented with a digital cross-connect system (DCS). Normally, DCS units are used to set up semipermanent, high-speed routes between central offices on a demand basis. However, when configured with its own processors and software, a DCS unit becomes intelligent enough to automatically redirect network traffic to and from various central offices.

FIBER RINGS Many LECs, especially those in major metropolitan areas, offer disaster recovery services to their customers via self-healing SONET (synchronous optical network)-compliant fiber-optic rings that are configured in a dual counter-rotating ring topology. This topology makes use of self-healing mechanisms in SONET-compliant equipment (i.e., the Add-Drop Multiplexers, or ADMs) to ensure the highest degree of network availability and reliability. In the event of a break in the line, traffic is automatically switched from one ring to the other in a matter of milliseconds, thus maintaining the integrity of the network. When the break is fixed, the network automatically returns to its original state. SONET's embedded management channels allow carriers (and users) to continuously monitor the links and take preemptive corrective action in response to impending trouble conditions.

ADVANCED INTELLIGENT NETWORK ACCESS Some LECs offer customers direct access to their advanced intelligent network (AIN) through an on-premises management workstation. Among other things, this enables them to add circuits, increase bandwidth, and reroute circuits in case of disaster without carrier involvement.

CAP Restoral Services

In areas where the services of a competitive access provider (CAP) are available, large corporate users can obtain protection and restoral services. Often, the services of a CAP are cheaper and more reliable than those of the incumbent LEC. Some users switch to the CAP solely for diverse routing, while others do so for the improved responsiveness CAPs give to trouble situations.

Local exchange carriers and CAPs sometimes cooperate in disaster prevention by interconnecting their networks. In broadening the interconnection of their networks, users are provided with more choice for disaster recovery services. With passage of the Telecommunications Act of 1996, LECs, CAPs, and long-distance carriers are free to compete with each other in the provision of local services. This will offer customers more opportunities for network diversification and, consequently, network protection.

Options and Limitations

LECs and CAPs also offer optional services to meet individual requirements. These services can include planning and building a complete network with diverse routing, spare bandwidth capacity, and redundant equipment to meet specific reliability guarantees. In many cases, the LEC will commit to specific network availability and reliability parameters via a service level agreement.

Currently, the restoral services offered by LECs are available only on a regional basis. Companies with WANs, which span multiple regions and must rely on local carriers for whatever restoral services might be available at each end of the circuits within the region, implement their own restoral mechanisms via customer premises equipment (CPE) or opt for an outsourcing agreement with a long-distance carrier to handle end-to-end restoral.

Fast Automatic Restoration System

The interexchange carriers offer the same protection and restoral services as the LECs. Using the same equipment and network management tools, many of these services can be extended to include the access portion of the long-haul circuits as well.

One of the most comprehensive services is AT&T's Fast Automatic Restoration System (FASTAR), which provides automated facilities restoration for all types of services (special services and switched) traveling over its fiber-optic transmission systems. FASTAR is designed to restore a substantial portion of traffic following a fiber-optic facilities problem, such as a cable cut, in less than five minutes. Moving into the future, the goal is to continually reduce that time frame to eventually make it negligible. A number of FASTAR options are available from AT&T:

SPLIT ACCESS FLEXIBLE EGRESS ROUTING (SAFER) For users of MEGACOM 800 and Software Defined Network service, delivery of traffic from the network (egress) is of primary importance. The use of SAFER provides routes from two separate switches to the customer's location. In the event of a network disruption, SAFER automatically directs calls to the working switch.

ALTERNATE DESTINATION CALL ROUTING (ADCR) This is also an 800 service reliability feature. With ADCR, if traffic is blocked to a customer's location for any reason (problems with the customer's PBX or local service, etc.), calls are automatically sent to another of the customer's locations.

NETWORK PROTECTION CAPABILITY (NPC) This reliability option for AT&T's digital service customers provides a geographically diverse backup facility and will usually switch traffic to this backup route within milliseconds when there is a service interruption.

ENHANCED DIVERSITY ROUTING OPTION (EDRO) This offering provides AT&T customers with a documented physical and electrical circuit diversity program.

ACCESS PROTECTION CAPABILITY (APC) This offering protects the access portion of a customer's circuit. APC provides immediate recovery of access circuits from certain network failures by automatically transferring service to a dedicated, separately routed access circuit.

CUSTOMER CONTROLLED RECONFIGURATION (CCR) Available in conjunction with AT&T's Digital Access and Cross-Connect System (DACS), this offers a means to route around failed facilities. The DACS is not a switch to set up calls or perform alternate routing; rather, it is a routing device. Originally designed to automate the process of circuit provisioning to avoid having a carrier's technician manually patch the customer's derived DS0 channels to designated long-haul transport facilities, the DACS allows CCR subscribers to organize and manage their own circuits from an on-premises terminal.

BANDWIDTH MANAGER AND BANDWIDTH MANAGEMENT SERVICE-EXTENDED (BMS-E) These services provide customer flexibility in configuration, reconfiguration, fault management, and restoration capability.

RESERVED T1 SERVICE This service supports applications requiring 1.544-Mbps (T1) speeds. AT&T brings a dedicated T1 facility online only after customers verbally request one over the phone. This restoral solution requires a subscription to the service and access facilities at each end of the link.

Network Management Capabilities

Interexchange carriers have network management centers that oversee their long-distance networks. Among other things, these management centers monitor the performance of networks, reroute traffic to smooth out peak-hour flows, and isolate problems until repair crews can be dispatched.

AT&T has three network management centers. The Network Operations Center in Bedminster, N.J., is responsible for the bulk movement of traffic across the AT&T Worldwide Intelligent Network. The network management centers in Denver, Col., and Conyers, Ga., (near Atlanta) are responsible for managing traffic coming onto or leaving the AT&T network from the local exchange carriers.

The network management centers use real-time information for managing the AT&T network and can implement controls directly from their centers to manage traffic flow. Personnel use specific network techniques and have skills for managing network situations during earthquakes, hurricanes, and other natural disasters, so a high degree of call completion can be maintained during a crisis situation.

SUMMARY

Despite the migration from private to public WANs in recent years, telecom managers are still accountable for overall network performance. This means working closely with the various carriers to manage service quality. Many carriers are now making near real-time performance information available to their customers through their Web sites, enabling them to track various performance metrics without having to buy expensive network monitoring systems of their own. By working closely with carriers, telecom managers can increase network availability and performance, reduce the need for recurring support, and ensure that business needs are met at the lowest possible cost.

See Also **CPE-Based WAN Restoral, Service Level Agreements**

Cellular Telephone Networks

Cellular networks provide wireless voice and data services to mobile phone users. Through their interconnection with the public switched telephone network (PSTN), cellular networks allow users to originate or receive communications with more portability and nearly the same degree of functionality as conventional wireline telephones.

Participating Technologies

Cellular networks combine a variety of technologies into a hybrid system. Radio technology provides the link between the mobile user and the nearest base station, serving a small geographical area called a *cell* (Figure 23). The base stations are connected via wirelines, fiber, or microwave to a mobile telephone switching office (MTSO). Traditional telephone switching technology provides the interconnection between the MTSO and the wireline public switched network. Finally, computer technology continually monitors the location of the mobile party and arranges for the signal's handoff to another base station as the user moves from cell to cell.

The mobile telephone service that preceded cellular service was known as Improved Mobile Telephone Service (IMTS), but it suffered from limited coverage areas, high cost, and an overall lack of functionality. These problems were overcome with the introduction of Advanced Mobile Phone Service (AMPS). Although several advanced digital technologies provide a class of service known as Personal Communications Services (PCS), the most prevalent version of cellular service used today is still AMPS, which is analog.

A digital version of AMPS—referred to as D-AMPS—provides increased capacity and a greater range of services. Both AMPS and D-AMPS operate in the 800-MHz band and can coexist with one another. D-AMPS can be implemented with time division multiple access (TDMA) as the underlying technology. TDMA provides 10 to 15 times more channel capacity than

Figure 23
A typical cellular
network.

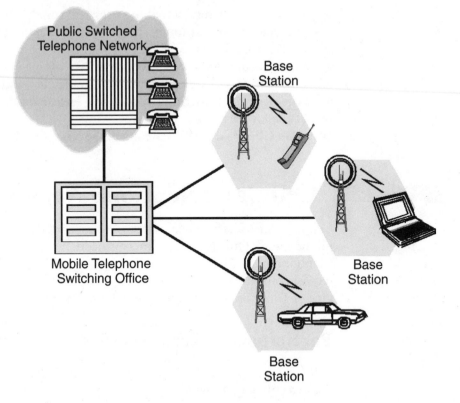

AMPS networks and allows the introduction of new feature-rich services such as data communications, voice mail, call waiting, call diversion, voice encryption, and calling line identification. A digital control channel supports such advanced features as a sleep mode, which increases battery life on newer cellular phones by as much as ten times over the current battery capabilities of analog phones. D-AMPS can also be implemented with code division multiple access (CDMA) technology to increase channel capacity by as much as 20 times and provide a comparable range of services and features. Unlike TDMA, which can be overlayed onto existing AMPS networks, CDMA requires an entirely new network infrastructure.

Operation

The system is engineered so the cell sites are located in close enough proximity to one another to provide seamless networking capability. The coverage areas for adjacent cells actually overlap in order to allow continuous coverage for a user in motion across the network, as well as to allow for some load balancing of network traffic. Three hundred and twelve radio channels are available for use by each carrier for voice communications between telephones and the cell site. The channels used by one cell can be reused by other nonadjacent cells, since the transmitted power levels are relatively low.

The radio frequencies used for communication between mobile users and the cell site are in the nominal range of 825 to 890 megahertz (MHz). Separate channels are used to transmit and receive voice communications, and the telephone equipment allows transmit and receive channels to be used simultaneously so parties communicating with each other experience a full-duplex conversation not unlike that of a wired telephone.

Additional radio communication between a telephone and cell site takes place over control channels. The special channels exchange data between mobile phone and cellular network, reporting the active phones operating within a particular service area. These control channels also provide functions crucial to the establishment of calls and the management of the voice communications channels. From the moment the telephone is turned on, even when idle, communication periodically takes place between the phone and the nearest cell site. The phone and the cellular network repeatedly exchange information via control channel protocols as to the location and status of the phone and the relative strength of the radio signal between them. This allows the network to find the optimal cell site through which it should route incoming calls to the mobile phone, to determine when the network should "hand off" an established connection from one cell site to another in order to maintain a strong radio connection, and to allow the phone and the network to synchronize their dynamic use of the available communications frequencies.

A mobile phone operating outside its local service area is considered to be "roaming." The user's account is established with a local provider, but other providers will allow visitors to their network to use the service. Billing is through the local provider. A local provider's coverage area might be statewide or might represent a particular area code. Billing to the user represents all on-air use or airtime, whether for outgoing or incoming calls, plus any long-distance charges.

Most carriers offer an arrangement whereby basic airtime charges, on a per-minute basis, are the only usage cost for an extended calling area. Calls to locations that might incur toll charges within the carrier's service area if made by conventional phone might not incur those charges for a cellular call, but they will be billed based on a flat airtime basis, usually in the area of 20 to 45 cents per minute. Calls made while roaming outside this area are at a higher per-minute rate and/or with additional per-call surcharges.

Types of Providers

Cellular network service providers have been licensed by the Federal Communications Commission (FCC) to operate based on limited competition in each service area. One provider is usually the local telephone company (also known as the "wireline" provider due to its traditional operation of the wired telephone network), and the other licensee is a competitor to the local telephone company, also known as the "nonwireline" carrier. Due to this limited competition, carriers can feel confident that their investment in developing a network will be rewarded with a significant enough portion of the subscriber base to support continued operation. Without this arrangement, it is unlikely that the cellular networks would have evolved in such a rapid manner.

Each carrier, wireline and nonwireline, has been assigned separate radio frequencies under which their license permits them to operate. This allows the competitors to coexist within the same physical operating area without interfering with each other's systems. Cellular telephones are manufactured with the inherent capability to operate on either carrier's network, since they have the capability to transmit and receive on either group of frequencies or channels.

Performance Issues

Since a mobile phone is so dependent on a radio link to establish and maintain communications, most of the factors that affect their operation are related to aspects of radio technology. Some of these factors are outside the control of the end user and are specific to the engineering of the carrier's network. The location of cell sites, proximity of adjacent cells, transmitter power, receiver sensitivity, antenna location, and performance can all affect the quality of communications. This can vary from one service provider to another, and better service might be obtained by registering with an alternate service. In many locations, however, performance between providers is virtually indistinguishable. It is quite likely that each service provider has its own strengths and weaknesses, especially pertaining to signal coverage in any specific location.

Service providers are not always able to place their cell sites and antennas in the ideal locations, but they do continually test and tune their network to attempt to provide the best level of service possible. An additional factor somewhat beyond the user's control is that of network traffic loading. Service can suffer even on the best of networks merely due to the network congestion that occurs from too many users attempting to access too few transmission facilities. Newer cellular network technologies enabling a greater number of channels to be derived from existing frequencies (i.e., frequency reuse), the addition of additional frequencies, and the implementation of smaller cell coverage areas or microcells provide relief from this problem, which can be especially troublesome in busy metropolitan areas.

Performance issues also come into play with the selection of a mobile phone, especially issues of power, selectivity (the ability of the receiver to discern one signal from another, unwanted signal), and sensitivity, since the manufacturer's design and production of the equipment have a direct impact on these characteristics. To a large degree, the performance of the equipment is a case of "you get what you pay for," since equipment with high power levels and an appropriately sensitive and selective receiver section is likely to cost more than a unit with lesser performance.

The performance of the best equipment, however, can be significantly diminished by a substandard installation in a motor vehicle or selection of a mediocre antenna. Installation of any equipment in a motor vehicle creates the possibility of ignition and alternator system noise being introduced into the audio circuits of the system. Electrical and antenna connections are subject to continual vibration, temperature changes, moisture, and corrosion. Antenna installation is especially significant, since the output power of the system and the received signal are more dependent on the antenna and connecting cables than any other components. These happen to be the components over which the manufacturer of the equipment has the least control, and they are system components that are required to withstand significant physical ill treatment.

The location chosen for the antenna installation is important, since the antenna radiates an electrical signal pattern that is dependent on the characteristics of the mounting location. The metal surfaces of the vehicle act as a "ground" and contribute to the effective performance of the antenna system. The ideal location on a motor vehicle for any antenna is in the center of the roof. This provides the antenna a high point as well as an effectively omnidirectional signal radiation pattern relative to the metal surfaces of the vehicle. Many people find this an undesirable location since they are then faced with the choice of having a hole drilled through the roof to route the antenna cable connection, or to opt for a magnetic mount antenna that uses a temporary-looking cable installation running across the roof via a door seam.

Trunk center-mounted antennas bring some of the same potential issues, but are often preferable to the rooftop location. Fender or trunk lip-mount units can be an acceptable alternative, and although the radiation pattern that results may be more directional in nature than that of a roof mount, performance of the system is usually adequate. The popular alternative to these mounting locations has become the "through-the-glass" option. As the antenna is usually mounted to a car's rear window, this option allows the user to mount the antenna in a desirable location that is high, exhibits relatively omnidirectional signal characteristics, and avoids detracting significantly from the vehicle's appearance or creating a potential for water leaks. The antenna and cable assembly are bonded to opposing sides of the glass in a configuration that allows the signal to pass through the glass with a minimum amount of power loss.

SUMMARY

A cellular network operates by dividing a large geographical service area into cells and assigning the same channels to multiple, nonadjacent cells. This allows channels to be reused, increasing spectrum efficiency. As a subscriber travels across the service area the call is transferred (handed off) from one cell to another without noticeable interruption. All the cells in a cellular system are connected to an MTSO by landline or microwave links. The MTSO controls the switching between the PSTN and the cell site for all wireline-to-mobile and mobile-to-wireline calls. The MTSO also processes mobile unit status data received from the cell-site controllers, switches calls to other cells, processes diagnostic information, and compiles billing statistics.
See Also Mobile Phone Fraud, Mobile Phones, Software Defined Radio, Third-Generation Wireless Networks

Channel Service Units

The user end of every T1 and DDS line requires a piece of equipment called a Channel Service Unit (CSU). The CSU can be a stand-alone, rack-mounted, or router-integrated device. The CSU can also be combined with a DSU as a dual-function device. Regardless of how it is packaged, the CSU performs three main functions:

- Protection for the T1 line and the user equipment from lightening strikes and other types of electrical interference and a keep-alive signal
- Generation of a keep-alive signal
- Storage for gathered statistics
- Carrier-initiated loopback testing

The service provider requires a CSU/DSU unit in any situation where a user has purchased a high-speed service such as a T1, Fractional T1, or a DDS 56/64 Kbps line. Since digital transmission links are capable of transporting signals between data terminal equipment (DTE), nearer to their original form there is no need for complex modulation/demodulation (modem) techniques, as is the case with sending data over analog connections. Instead, CSUs are used at the front end of the digital circuit to equalize the received signal, filter the transmitted and received wave-forms, and interact with both the user's and carrier's test facilities via a su-pervisory terminal or network management system. The FCC's Part 68 registration rules require that every T-carrier circuit be terminated by a CSU. These devices can be used to set up a T1 line with a PBX, channel bank, T1 multiplexer, or any other DSX1-compliant DTE.

Line build out (LBO) is a functional requirement of all Part 68 registered T1 CSUs. LBO is an electronic simulation of a length of wire line that adjusts the signal power so that it falls within a certain decibel range at both ends of the circuit. This is determined by looping a test signal back over the re-ceive pair and measuring for signal loss. This procedure also helps reduce the potential for one T1 transmitter to cross-talk into the receiver of other services within the same cable binder. Once line loss is determined, the tele-phone company can tell the customer what setting to use on the local CSU.

All T1 CSUs provide a repeater to reconstitute signals that have been at-tenuated and distorted by the T1 span line and/or the customer's in-house cabling. This function is also part of the FCC Part 68 equipment registration requirements for CSUs.

In addition to equalizing the transmitted signal through LBO and re-generating the received line signal, the CSU ensures that the user's DTE does not send signals that could disrupt the carrier's network. For exam-ple, very long strings of zeros do not provide timing pulses for the span line repeaters to maintain synchronization. The CSU monitors the data stream from the attached DTE so the "ones density" rule (e.g., the cus-tomer's data must have at least a 12.5 percent pulse density) is not violated. This rule ensures that there are sufficient pulse transitions for the span line repeaters to maintain timing synchronization. The CSU will inject pulses if excessive zeros are being transmitted by the attached DTE.

The CSU provides the functionality to troubleshoot circuit and trans-mission problems. For proactive network management, these include LEDs that indicate both the status of the network and equipment con-nections and whether or not any alarm thresholds or error conditions have been detected. This lets technicians at either end of the circuit iso-late problems in minutes instead of hours. The CSU also contains a buffer that stores collected performance information that can be accessed by both the carrier and user for diagnostic purposes.

SUMMARY

Because the CSU interfaces user equipment to carrier facilities, it provides a window on the network, allowing both the carrier and the customer to perform testing up to the same point. CSU access to the network has prompted vendors to equip it with increasingly sophisticated diagnostic and network management features. If the CSU/DSU has more than one user port, it can function as a multiplexer that allocates the DS0 time slots between the ports in multiples of 56 Kbps or 64 Kbps. All of these capabilities go a long way toward enhancing user control and ensuring the integrity of T-carrier facilities.

See Also **Data Service Units**

Client/Server

The client/server architecture came from the need to bring computing power, information, and decision-making to users so businesses could respond faster to customer needs, competitive pressures, and market dynamics. This was a welcome alternative to the overly bureaucratic mainframe environment where departments, workgroups, and individuals typically had to put in written requests for the information they needed.

In the client/server model, an application program is broken out into two parts on the network. The client portion of the program, or front end, is run by individual users at their desktops and performs tasks such as querying a database, producing a printed report, and entering a new record. These functions are carried out through a database specification and access language, better known as Structured Query Language (SQL), which operates in conjunction with existing applications. The front end of the program executes on the user's workstation, drawing upon its random-access memory (RAM) and central processing unit (CPU).

The server portion of the program, or back end, is resident on specially equipped computers that are configured to support multiple clients, offering them shared access to numerous applications as well as printers, file storage, database management, communications, and other resources. The server not only handles simultaneous requests from multiple clients, but performs administrative tasks such as transaction management, security, event logging, database creation and updating, and concurrency management. A data dictionary standardizes terminology so database records can be maintained in a standard way across a broad base of users.

Distributed Networks

The migration from central to distributed computing has produced two types of networks involving servers: the traditional hierarchical architecture employed by mainframe vendors and the peer architecture employed by LANs.

The hierarchical approach uses layers of servers that are subordinate to a central server. In this case, PCs and workstations are connected to servers that, in turn, are connected to a remote "server of servers." This server contains extensive files of addresses of individuals, databases, and programs, as well as a corporate SQL database or file of common read-only information, such as a data dictionary. A workstation on a LAN making a request for data has its request routed to the central server. The server adds any pertinent information from its own database and sends the combined message to the end user as a unit. To the end user it appears to be a single, integrated request. Also, the local LAN server passes database updates to the central server, where the most recent files reside.

This type of server network maintains the hierarchical relationship of mainframe communication architectures, such as IBM's SNA, thus simplifying software development. An added benefit is that there is more programming expertise available here than in the distributed environment. The disadvantage is in the vulnerability of the hierarchical network to congestion or failure of the central server, unless standby links and redundant server subsystems are employed at considerable expense.

In contrast, the distributed server architecture maintains the peer-to-peer relationship employed in LANs. Each workstation on the LAN can connect to multiple specialized servers as needed, regardless of where the servers are located. There are local servers for services such as databases, user authentication, facsimile, and electronic mail. Some servers are responsible for managing connections to servers outside the LAN. The workstation merges data from the server with its own local data and presents the data as a composite whole to the requesting user.

The distributed client/server architecture is more difficult to implement than the hierarchical server architecture because the software is vastly more complex. After all, the workstations need to know where to find each necessary service, and they must be configured with appropriate access privileges and have the applications software to access the required data. When the data is finally accessed, it must be the most current data. Keeping the network and computing process transparent to end users is another challenge. And, of course, everyone has come to expect a graphically rich user interface.

The network also consists of the transmission medium and communications protocols used between clients and servers. The transmission medium used in the client/server environment is no different from that found in any other computing environment. Among the commonly used media for LANs is coaxial cable (thick or thin), twisted pair wiring (shielded or unshielded), and optical fiber (single- or multimode). However, wireless media, such as infrared and spread-spectrum radio signals, are more often found in the media mix for LANs than for mainframes. Throw in the requirement for remote access, and there may also be the need to support wireless services such as Cellular Digital Packet Data (CDPD).

A medium-access protocol grants users access to the transmission facility. For Ethernet and Token Ring networks, the medium-access protocols are Carrier Sense Multiple Access with Collision Detection (CSMA/CD) and token passing, respectively. When linking client/server computing environments over long distances, other communications protocols are used over private facilities such as point-to-point T1 links, which provide a transmission rate of up to 1.544M bps. Frame relay and Asynchronous Transfer Mode (ATM) are among the latest technologies for carrying LAN traffic over the wide-area network (WAN), while Transmission Control Protocol/Internet Protocol (TCP/IP) networks are the oldest.

Of these, more companies seem to be returning to TCP/IP-based networks to support mainstream business applications. Corporate intranets and the global Internet, in essence, are large-scale client/server networks. These networks are comprised of Web servers that run various applications and provide access to vast amounts of information. These servers are interconnected by communication lines of different types and speeds. Gateways are used to join private intranets with the public Internet. Firewalls protect corporate assets from unauthorized access, usually by filtering out the addresses of unknown sources and applying authorization schemes to prevent hackers from masquerading as corporate users.

The choice of media and protocols will hinge on price, performance, and reliability factors weighed against the requirements of the applications. More often than not, no single solution will meet all of the diverse needs of an organization. This adds to the complexity of developing client/server applications.

Assessment of Client/Server

Only a few years ago, the client/server concept was somewhat of a disappointment. Because client/server systems are distributed, costs become nearly impossible to track, and administration and management difficulties tend to multiply.

It has been estimated that the total cost of owning a client/server system is about three to six times greater than it is for a comparable mainframe system, while the software tools for managing and administering client/server cost two-and-a-half times more than mainframe tools. Some industry studies have revealed that less than 20 percent of client/server projects were on time and on budget and that only about 40 percent of companies view the client/server architecture a worthwhile investment.

Today client/server is undergoing resurgence with the growing popularity of corporate intranets and the Internet, and the advancement of the network-centric computing concept. Net-centric computing provides new opportunities for integration and synthesis between networks and applications inside the company and those outside the company. With proper security precautions, intranets can easily be set up to reach international divisions, customers, suppliers, distributors, and strategic partners. The greater Internet can be used for worldwide connectivity between distributed intranets. Companies are also partnering through so-called "extranets" to access each other's resources for a common purpose, such as research and development, joint marketing, and electronic data interchange (EDI).

The greatest potential of corporate intranets may be in making client/server truly open. The average desktop PC today has 7 to 15 applications—50 to 100 software components—each of which has different versions. With the single universal Web client, IS managers do not have to worry about configuring hundreds or thousands of desktops with appropriate drivers.

However, the Web-based client/server approach is not without problems of its own. For example, the HyperText Transfer Protocol (HTTP) and gateway interfaces are not designed to support high-performance, high-volume applications. This means that the Web is not a panacea and many companies might have to settle for blending Web and client/server applications. This can be done by moving some application logic to the client browser in the form of Java applets.

Web-based technology is certainly less costly. A business does not need to standardize desktops and operating systems for Internet technology to work. An existing client/server infrastructure can serve as the backbone of the intranet. Also, with free and low-cost application development tools available on the Internet, the creation of distributed systems and applications is much cheaper than in the client/server environment, allowing IS managers to get more done within available budgets, while preserving existing investments in legacy and client/server systems and applications. A Web browser provides a single window to all data, regardless of location. There are even gateway products that allow legacy data on mainframes to be viewed from within a Web browser with no modification to existing databases.

SUMMARY

The rise of the global economy has forced businesses to improve their operating efficiency, customer satisfaction levels, and product-to-market implementation cycle to a degree rarely seen before. This has contributed to the widespread acceptance of client/server, particularly as manifested in corporate intranets, as the most efficient and economical solution for distributing applications and sharing information. The economy of the client/server computing model may best be realized with a thin-client implementation. *See Also* **Thin Client Architectures**

Clustering

Clustering refers to the interconnection of servers in a way that makes them appear to the operating environment as a single system. As such, the cluster draws on the power of all the servers to handle the demanding processing requirements of a broad range of technical applications. It also takes advantage of parallel processing in program execution. Shared resources in a cluster may include physical hardware devices such as disk drives and network cards, TCP/IP addresses, entire applications, and databases. The cluster service is a collection of software on each node that manages all cluster-specific activity, including traffic flow and load balancing. The nodes are linked together by standard Ethernet, FDDI, ATM, or Fibre Channel connections.

Advantages

There are many advantages to clustering:

PERFORMANCE Throughput and response time are improved by using a group of machines at the same time.

AVAILABILITY If one node fails, the workload is redistributed among the other nodes for uninterrupted operation.

INCREMENTAL GROWTH Performance and availability can be enhanced by adding more nodes to the cluster.

SCALING Theoretically, there is no limit on the number of machines that can belong to the cluster.

PRICE AND PERFORMANCE The individual nodes of a cluster typically offer very good performance for their price. Because clustering does not involve the addition of expensive high-performance processors, buses, or cooling systems, the cluster retains the price/performance advantage of its individual members.

Comparison with SMP

Another form of parallel computing is the symmetric multiprocessor (SMP), which has been around since the early 1970s. An SMP computer has multiple processors, each with the same capabilities. The computer's operating system distributes the processing tasks among two or more processors. Each processor can run the operating system as well as user applications. Not only can any processor execute any job, but the jobs can be shifted from one processor to another as the load changes.

Traditionally, SMPs and clusters have been considered competitive technologies, even though they can coexist. Nevertheless, there are some differences. For example, each machine in a cluster has its own local memory, and communication with other machines is less efficient than access to a machine's own memory. In addition, each machine in a cluster has its own attached I/O. Access to another machine's I/O is less efficient than access to a machine's own I/O. By contrast, an SMP does not have multiple I/O systems or memories, and each processor has equal access to every location in the I/O system and memory.

Despite all this, there are some performance advantages to clustering. As the number of processors increases, SMP designs require that more memory and I/O bandwidth be added as well, which increases the cost. Since clusters remain intrinsically balanced in their memory and I/O capabilities—because each node comes with its own memory and I/O subsystems—performance increases are a natural result of adding more nodes.

At one time, SMPs had advantages over clusters in the areas of software and usability—specifically, central system administration, load balancing, and middleware support. With new server operating system enhancements and extensions, however, clusters offer the same capabilities and features.

Cluster Software

In comprising multiple nodes, clusters require special software. For example, there are products for batch job submission. These products typically perform at least rudimentary load balancing as well, and allow users to query the status of jobs. The degree to which the multiple-machine nature of the cluster is hidden varies among products. Some also provide checkpoint/restart facilities for more reliable operation. Others provide cluster-wide administration and accounting services.

Every major vendor of database management systems has a version of its product that operates across multiple computers. Such systems effectively merge the machines into a single database entity for application and database administration purposes.

There are also software packages available that enable applications to be reprogrammed to run in parallel. In being reorganized or "optimized" to execute simultaneously on multiple nodes in a cluster, the application is able to run faster.

Another kind of software is the operating system. Special extensions to existing operating systems such as Microsoft Windows NT enable multiple nodes running the operating system to be managed as if they were one machine, and they appear to client machines as a single system. Clients connect to the cluster, not to any single machine. The computational load is automatically balanced across the machines, and should one (or more) of the machines in the cluster fail, client systems will never know since the other nodes will automatically pick up the load without clients having to reconnect or take any other action.

UNIX operating systems provide similar capabilities and go beyond them in being able to detect software failures in addition to hardware failures. This capability is especially important, since software failures far outnumber hardware failures. The degree of software failure coverage varies among individual products, but all are capable of detecting operating system failures within the cluster.

SUMMARY

Traditionally, clusters were used in very high-end UNIX and proprietary server environments to improve application uptime and increase overall processing capacity. Today, the rapid growth of applications such as Web serving, electronic commerce and enterprise resource planning (ERP) are beginning to push high-availability, high-performance clustering technologies into the commercial mainstream. This migration has been facilitated by the increasing performance of microprocessors and decreasing prices, which makes clusters a viable way to meet the needs of server systems in a cost-effective way.

See Also **Client/Server, Disk Arrays, Network Servers, Storage-Area Networks**

Coaxial Cable

Coaxial cable is the type of line that brings television signals into homes and businesses. It has enough bandwidth to carry video, voice, and data at the same time. In fact, many cable companies now offer Internet access and IP-based telephone service, as well as television programs, over their cable infrastructure. Telephone companies sometimes use coaxial cable to link the central office to neighborhood telephone poles. Despite the increasing popularity of cheaper, easier-to-install twisted pair wire, coaxial cable still is widely used in corporate local-area networks (LANs).

Government and military applications led to the initial development of coaxial cable in 1929. Built to military specifications and classified according to Radio Guide Utility (RG/U) numbers, different cable products were developed to support high-frequency radio transmissions. Later, the RG/U numbers indicated the different impedance characteristics of the cables.[1]

Coaxial cable was first used for commercial telecommunications in 1940 on AT&T's coast-to-coast transmission system. The steady growth of the computer industry further fueled the demand for coaxial cables. Manufacturers of proprietary systems required a variety of cable designs for data communications.

Cable Construction

The cable is called "coaxial" because it includes one physical channel surrounded by another concentric physical channel, both running along the same axis and separated by a layer of insulation (Figure 24). Depending on the application, different types of conductor constructions may be found in coaxial cables. Solid copper conductors are popular on many Closed Circuit Television (CCTV) installations. Solid conductors provide less chance for distortion and line loss than copper-covered steel. However, copper is a soft material and will break if repeatedly flexed, making it better suited for use in permanent installations.

Many small strands of copper can be used to make up any gauge size. Stranded conductors provide increased flexibility over solid conductors. This design has become popular for use on pan and tilt cameras, robotics, and other applications that require repeated flexing.

In some applications, strength is a key requirement. Steel conductors covered with copper may be needed to prevent breakage in an active environment. Steel provides added strength and RF support, and is often used in Community Antenna Television (CATV) and Metropolitan Antenna Television (MATV) applications where the cable is exposed to wind.

The inner conductor of a coaxial cable is separated by an insulating material from the surrounding shield(s). This "dielectric" material is often

---

1. Impedance is a measurement of the resistance to an electrical current that a transmission medium offers. The measurement is expressed in ohms.

Figure 24
Basic construction
of a coaxial cable.

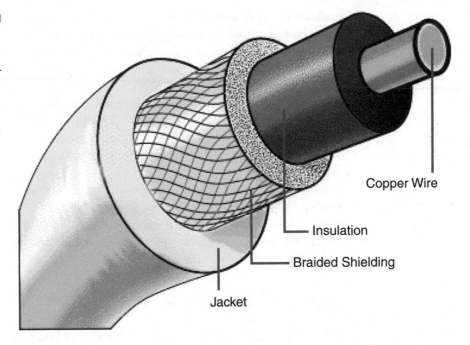

Copper Wire

Insulation

Braided Shielding

Jacket

chosen in order to maintain consistent electrical properties and minimize signal loss. The result is a clear, trouble-free transmission.

In coaxial applications, shielding is an important part of the overall composition of the cable. Shielding not only protects the loss of signal in high frequency application, but also helps to prevent EMI (electromagnetic interference) and RFI (radio frequency interference) in the circuit.

The outer sheath found on most coaxial cables is called the *jacket*. The main function of the jacket is for protection from the environment and as an additional form of insulation. The compounds used to make the jacket may have different temperature ratings. The temperature rating of a cable, along with the location rating (i.e., plenum, wet, or sunlight-resistant) determine the minimum or maximum operating temperature of the cable.

Use in LANs

Until the 10Base-T standard (unshielded twisted pair), coaxial cable was the preferred media for local-area networks. Like its counterpart used on CATV networks, this type of cable contains a conductive cylinder with insulation around a wire in the center. Coaxial cable is typically shielded to reduce interference from external sources. Coaxial cable can transmit at a much higher frequency than a wire pair, allowing more data to be transmitted in a given period. By providing a "wider" channel, coaxial cable allows multiple LAN users to share the same medium for communicating with host computers, servers, front-end processors (FEP), peripheral devices, and personal computers.

The initial Ethernet implementations used 50-ohm coaxial cable with a diameter of 10 mm, which is referred to as "thick Ethernet cable." This type of cable was heavy, rigid, and difficult to install. The maximum cable length for Ethernet is 500 meters, which is referred to as 10Base-5 transmission. This technical shorthand means "10 Mbps data rate, baseband signaling, 500 meter maximum cable segment length."

Another cable standard, 10Base-2 (meaning 10 Mbps, baseband signaling, 200 meters), uses ordinary CATV-type coaxial cable, called "thin Ethernet cable." This is now the most common type of coaxial cable used for Ethernet, although twisted pair wire has emerged as the preferred media for new LANs and optical fiber increasingly is being used for the backbone between floors and between buildings.

SUMMARY

In recent years, coaxial cables have become an essential component of the so-called "information superhighway." They are found in a wide variety of residential, commercial, and industrial installations. From broadcast, community antenna television (CATV), local-area network (LAN), closed-circuit television (CCTV), and many other applications, coaxial cable has laid the foundation for a simple, economical, and scalable communications infrastructure.

See Also **Cable Modems, CATV Networks, Fiber-Optic Technology, Microwave, Twisted Pair Cable**

Communications Processors

The widespread perception of communications processors (CPs) is that they are general-purpose devices from a family of equipment types, including feeder multiplexers, packet assembler/disassemblers (PADs), terminal servers, and protocol converters. In the typical IBM mainframe environment, however, system administrators usually take a much narrower view, limiting the definition to FEPs, establishment controllers, and network gateway controllers.

Even though most mainframe central processors include parallel circuits and clocking mechanisms to handle I/O in a quasi-parallel fashion, they still consume a significant percentage of available CPU cycles for specialized, high-overhead, front-end applications. This has motivated system designers to find ways to offload these support processes onto less expensive resources. This motivation gave rise to the introduction of communications processor technology.

These special-purpose computers are designed to manage the complexities of computer communications such as protocol processing, data format conversion, data buffering and routing, error checking and correction, and network control. Depending on the environment, vendor, or configuration, communications processors are commonly referred to as front-end processors (FEPs), local/remote concentrators or hubs, communication servers, gateway switches, intelligent routers, and controllers.

The market for communications processors is fully mature and dominated by one product, the IBM 3475 Communication Controller. Lately, the term *communications processor* has been used increasingly to refer to network add-in boards for both PC and VMEbus systems. The concept is the same—that these boards offload communications functions from the CPU.

Functions

The traffic management responsibilities of CPs include establishing, maintaining, and controlling any communications sessions between a host computer and data terminal equipment (DTE), switching devices, other hosts (peer-to-peer), and intranet and Internet activities. Several vendors have enhanced their CP offerings by adding features that include utilities for response time monitoring, event logging, terminal status indication, system administration, and diagnostic testing.

The original communications processors were relegated to mainframe systems. However, the advent of client/server computing and network topologies brought with them new, modernized devices for communications connectivity. Instead of a processor dedicated to servicing a single, central host, these newer communications servers offer front ends for any number of processors connected via a local or enterprise-wide network.

Communications servers allow any piece of data communications equipment supporting the EIA-232 standard (terminals, modems, printers, hosts, and personal computers) to attach to a network. In some configurations, several servers operate concurrently in the same network, thereby providing an extensive range of communications services.

Communications processors, on the other hand, are generally classed by their range of features, flexibility, and capabilities. These include the number and types of host interfaces and protocols supported, aggregate bandwidth, and the types of terminal equipment and other devices supported. Creating an optimum configuration generally entails customizing the CP application software according to the functional and operational requirements of the system. This process, called System Generation (SYSGEN), prompts users through a series of questions about the desired configuration and estimates of data traffic. The result of this activity is a customized, resident program and associated tables that are automatically loaded and run each time the processor is initialized.

Communications processors support a number of protocols, including OSI, TCP/IP, IBM's SNA/SDLC and 3270/3780 BSC, Ethernet, Fast Ethernet, token-ring, FDDI, ATM, X.25, frame relay, SONET, and Digital Equipment's DECnet and Local Area Transport (LAT) protocols. A number of CPs are able to operate in a multi-stack environment, supporting and translating multiple protocols concurrently.

To address the growing number of protocols, interfaces, and features, equipment vendors are now offering modular systems that use a bus concept to support a variety of specialized plug-in hardware modules. This allows the user to grow the system incrementally to satisfy future requirements. Many of these devices are capable of transparent translation (e.g., gateways) from one protocol to another for interconnecting incompatible networks.

Front-End Processors

The functionality provided by FEPs includes establishing the communications link, message framing and deframing, protocol conversion, message sequencing, error detection and correction, route selection, and session management. Some FEPs perform the packet assembly/disassembly (PAD) activities used in X.25 packet-switched networks. They also provide similar functionality for frame relay, TCP/IP, and ATM networking. In general, FEPs handle the Data Link through the Session layers of the OSI reference model (Layers 1 through 5).

FEPs offload communications functions from the host to support high-speed data transfers and low response-time interactive data traffic generated by terminal users. Depending on the vendor, the FEP also may support file transfers between hosts, and between hosts and LANs. Integrated modems support communications over analog lines, while integrated Data Service Units/Channel Service Units (DSU/CSUs) support communications over carrier-provided digital facilities (e.g., T1 and Fractional T1) and services (e.g., X.25 and frame relay).

Programmable, software-controlled FEPs isolate the host computer from changes in the network topology and buffer the network from changes in host applications. Software-controlled FEPs are more flexible than their hardware-based predecessors and, therefore, are easier to upgrade as newer networking techniques emerge.

An FEP can maintain a network's operation even when one or more of its host processors fail. In a multihost network, FEPs switch users from applications in a failed host to similar or identical applications in a backup host. In a single-host network, a functioning FEP allows service to degrade gracefully in the event of a host failure, often allowing users to terminate their current tasks before the system shuts down.

Most FEPs have a hierarchical architecture based on a parallel bus. For example, the Central Control Unit (CCU) is in a separate module (or modules), while communications scanners and host-interface channel adapters occupy their own independent modules. This design simplifies internal communications and makes adding and deleting components, such as line interface modules and line sets, a matter of adding or removing modules. The modular approach also safeguards the user's investment because system upgrades involve only inserting new components as opposed to replacing an entire FEP.

A classic example of a multifunction communications or front-end processor is the IBM 3745. Since its introduction in 1989, IBM's 3745 Communication Controller has dominated the market. The 3745 can provide three distinct functions. As a front-end processor, the 3745 offloads communications processing from the host. For example, token-ring processing is done in the 3745 instead of the host. As an intelligent switch, the 3745 uses subarea routing techniques to switch packets through an SNA network. Finally, as a concentrator, the 3745 acts as a connectivity hub for establishment controllers (a.k.a. cluster controllers) such as the IBM 3174, and other communications devices.

Line interface modules handle the interface between the line sets and the FEP. The line interface module contains a microcomputer that works with a communications scanner to provide a link between line sets and the FEP and memory. The communications scanner serializes and deserializes data, buffers characters, and transfers data to the FEP memory. Most FEPs accommodate multiple communications scanners, and each scanner handles multiple line interface modules that in turn handle multiple line sets. The number of communications lines accommodated by each line set determines the total number of lines supported by the FEP. A line set can handle up to 32 lines.

The FEP is responsible for establishing communications line connections, endpoint polling, and line disconnections. Line connection is generally a straightforward process of establishing an interface with the endpoint in accordance with the communications protocol, end-terminal codes (e.g., ASCII or EBCDIC), and transmission speed. For certain applications, the FEP must have auto-dial, auto-answer, and auto-disconnect capabilities to support dial-up operation. Transmitting messages to a remote location involves being able to "call" the location and recognize the calling device designation code, device type, transmission protocol, and data speed.

Establishment Controller

An establishment or cluster controller manages the activities of a group of printers and terminals, allowing them to communicate with the host. In a traditional SNA network, a PC can appear as one or more Logical Units (LUs) connected to a PU 2.0 device. These LUs are always referred to as dependent nodes because they cannot establish an SNA session. The host must do it for them. A 3270 terminal (known as a dumb terminal) can also be a type 2 PU.

Establishment controllers, such as IBM's 3174, interface multiple devices (e.g., terminals and printers) and handle all communications functions needed to establish and maintain a communications session with an FEP. Like the FEP, a cluster controller accepts inputs from several data transmission sources and routes them to designated end locations. It also handles message buffering, sequencing, and protocol conversion.

IBM's 3174 is the preferred replacement for older controllers that remain in a dependent display environment. It supports the attachment of 32 dependent workstations, facilitating SNA host access through remote communications. It can be directly attached to a mainframe channel or link-attached to a communications controller such as the IBM 3745. LAN connections for Ethernet and token-ring are also supported.

Using the IEEE V.35 interface, the 3174 can achieve communications speeds of 256 Kbps and more in SDLC mode. Additionally, it allows for the:

- Establishment of a logical pathway from SNA networks to token-ring and Ethernet LANs

- Availability of LAN functions over existing coaxial wiring

- Interoperation of dependent terminals and intelligent workstations using TCP/IP

- Integration on AS/400 systems, PCs, and ASCII terminals

- Links to Advanced Peer-to-Peer Networking (APPN) networks

IBM's Workstation Networking Module for its 8250 LAN hub provides 3174 functions, enabling users to attach 3270 devices directly to the hub without requiring a separate 3174 controller, making the 8250 hub a more important element in enterprise environments.

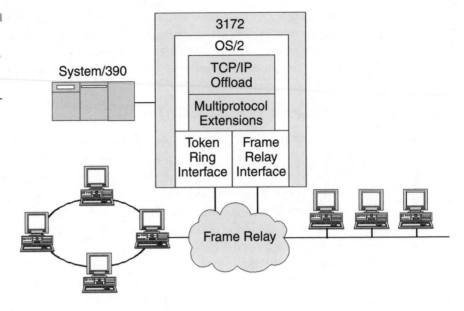

Figure 25
The IBM 3172
linking a variety of
systems to the
mainframe.

Network Gateway Processor

The network gateway processor (or interconnect controller) is a specialized communications processor introduced by IBM in 1989 to provide
SNA mainframe access for LANs based on Ethernet, token-ring, TCP/IP,
and FDDI. However, IBM's 3172 Interconnect Controller has long been
criticized by industry analysts, who contend that its position in the marketplace is somewhat fuzzy and that it conflicts with the traditional role
of the 3745. IBM, however, contends that unlike the 3745 or 3174, the IBM
3172 is its primary strategic solution for LAN-to-mainframe connectivity.

The 3172 is used primarily to connect LANs to IBM channels. It supports TCP/IP and SNA, which are the primary protocols used in IBM networks today. It also supports the most popular LAN architectures,
including FDDI, token-ring, and Ethernet (Figure 25).

SNA Network Controller

The typical mainframe networking environment is based on IBM's SNA
network, which consists of nodes and links. Nodes are network components that contain host protocol implementations; links are transmission
facilities that carry data between two SNA nodes. The term *link* also assumes the ability of the two connected nodes to operate a data link control procedure between them. Traditional SNA networking involves four
types of physical entities:

HOSTS Control all or part of a network. They provide computation, program execution, data base access, directory services, and network management.

COMMUNICATIONS CONTROLLERS Sometimes called FEPs, these devices manage the physical network, control communications links, and route data through a network. An FEP acts as an intelligent controller that is directly attached to one or more host mainframe computers. It relieves hosts from the overhead associated with establishing communications sessions and passing messages between the mainframe and the network endpoints (Figure 26).

Figure 26
Typical FEP
connections.

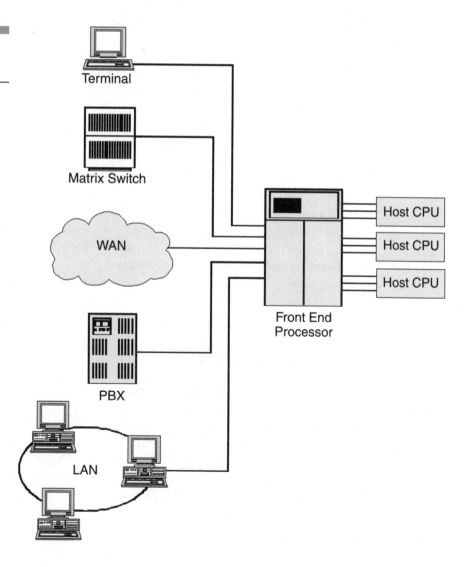

ESTABLISHMENT CONTROLLERS Sometimes called cluster controllers, these devices control the I/O operations of devices attached to them.

TERMINALS AND PRINTERS Terminals provide the user interface to the network. Printers provide hardcopies of requested information.

Remote Concentrator

The termination point for all wiring of a common access method into one area is called the concentrator. The combination of termination points for different access methods and different components is called the hub. The hub allows for Ethernet (any wiring type), token-ring (usually UTP), and FDDI (Fiber or UTP) to be housed in a single unit.

Host computers often process information created by terminals that are physically remote from the host processor. Rather than pay for separate communications links (dedicated leased lines or virtual circuits) for each terminal connection, organizations can employ remote concentrators to consolidate the number of circuits required. These devices concentrate multiple low-speed transmission lines into one or more high-speed lines.

The role of the remote concentrator depends on its position in the network hierarchy. When concentrating data from several cluster controllers, remote concentrators act as a key communications processor. If a concentrator provides local communications control of directly attached terminals, its role would be more like a terminal controller.

As a terminal controller, the remote concentrator establishes and maintains low-speed communications links between multiple terminal devices and communicates with other network nodes via high-speed trunk lines. In this type of application, a remote concentrator controls I/O operations from multiple terminal devices and consolidates the quantity of high-order communications lines needed to service connected terminals (Figure 27).

Gateway Switch

A gateway switch routes messages among the various network endpoints and provides protocol conversion between dissimilar systems, workstations, and networks. A gateway switch could be used, for example, to establish connectivity between Digital Equipment's VAX systems and IBM's SNA-oriented devices. Similarly, Ethernet-based Digital-to-IBM connectivity allows Ethernet users to implement a token-ring gateway connection to SNA hosts. This type of LAN/WAN internetworking saves the cost of adding a token-ring LAN to communicate with IBM hosts (Figure 28).

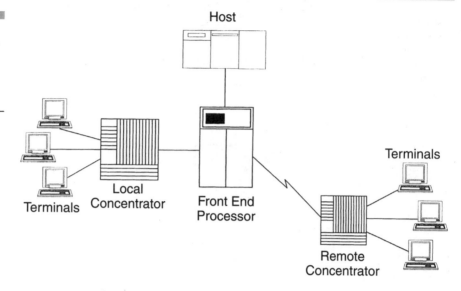

Figure 27
Local and remote concentrators functioning as communications processors.

Host

Terminals

Local Concentrator

Terminals

Front End Processor

Terminals

Remote Concentrator

SUMMARY

Communications processors offer superior network management capabilities over LAN-WAN internetworks that employ bridges, routers, and gateways. They also provide more advanced fault management, problem determination, performance measurement, and accounting functions than today's bridges and routers. However, some other types of products are incorporating the features of communications processors, rendering them unnecessary in many cases. A new breed of intelligent wiring hubs, for example, integrates many of the functions of communications processors, plus support for SNMP and hooks into IBM's NetView.
See Also Hubs

Computer-Telephony Integration

Computer-Telephony Integration (CTI) bridges the advanced call-processing capabilities of digital telephone systems and the open application environment of PCs, LANs, and the World Wide Web. CTI is used in telephone-intensive environments such as customer service and telemarketing, where customer account or product information from a host computer is immediately delivered to the desktops of appropriate call center agents so they can serve customers faster, more efficiently, and at the lowest cost per call.

CTI applications are typically built with standardized application programming interfaces (APIs): the Telephony Application Programming Interface (TAPI) developed by Intel and Microsoft or the Telephony Services

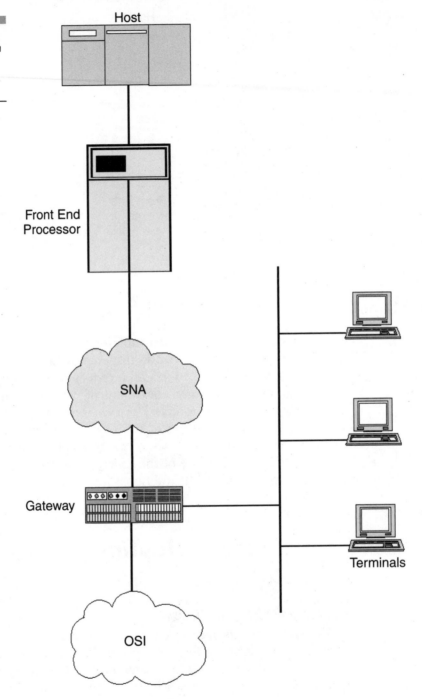

Application Programming Interface (TSAPI) developed by AT&T and Novell. The former offers a single specification for Windows application developers to use in connecting their products to the telephone network. The latter is intended to make NetWare the platform for this kind of integration. Other vendors, such as Sun Microsystems and Hewlett-Packard, also offer APIs for developing CTI applications.

In addition, the platform independence of Java is being exploited to provide computer-telephony integration. The Java Telephony Application Programming Interface (JTAPI) offers the means to build applications that run on a variety of operating systems and hardware platforms over a variety of telephony networks. It was developed by a consortium of vendors, and the specification is published by JavaSoft.

TAPI

TAPI allows custom applications to be built around inexpensive personal computers; specifically, the Windows Telephony API provides a standard development interface between PCs and the myriad telephone network APIs. TAPI is intended to insulate software developers from the underlying complexity of the telephone network. TAPI allows developers to focus entirely on the application without having to take into account the type of telephone connection: PBX, ISDN, Centrex, cellular, or plain old telephone service (POTS). They can specify the features they want to use without worrying how the hardware is ultimately linked.

APPLICATION CLASSES TAPI facilitates the development of three classes of Windows applications. The first class of applications are telephone-enabled versions of existing applications, such as word processors. TAPI creates standard access to telephone functions such as call initiation, call answering, call hold, and call transfer for Windows applications. TAPI addresses only the control of the call, not its content. However, the specification can be applied to any type of call, whether voice, data, fax, or even video.

The second class of telephone-centric applications might embrace visual call control or telephone-based conferencing and collaborative computing. Although such applications have been available for several years, they have been limited by incompatible APIs. The third class of applications enables the telephone to act as an input/output device for audio, including voice across data networks.

APPLICATION COMPONENTS An actual TAPI product implementation is comprised of three distinct components:

- TAPI-aware application
- A TAPI dynamic link library (DLL)
- One or more Windows drivers to interface to the telephone hardware

A TAPI application is any piece of software that makes use of the telephone system. An obvious example might be a personal information manager (PIM), which could dial phone numbers automatically. An application becomes TAPI-compliant by writing to the applications programming interfaces defined in the TAPI specification.

The TAPI DLL is the next major component. The application talks to the DLL using the standard APIs. The DLL translates those API calls and controls the telephone system using the device driver.

The final component is the Service Provider Interface (SPI), which is a driver that is unique to each TAPI hardware product. It is analogous to each sound card having its own driver. The TAPI specification supports more than one type of telephone adapter. In turn, the adapters can support more than one line.

TAPI-ENABLED FEATURES   TAPI facilitates the development of applications that allow the user to control the telephone from a Windows PC. A number of possible control features may become available in future applications, including:

VISUAL CALL CONTROL   Provides a Windows interface to such common PBX functions as call hold, call transfer, and call conferencing. Replacing difficult to remember dialing codes with Windows icons makes even the most complicated telephone system functions easy to implement.

CALL FILTERING   In conjunction with ANI, this function allows the user to specify the telephone numbers allowed to get through. All others are routed to an attendant, message center, or voice mailbox. Or the call can be automatically forwarded to another extension while the user is out of the office.

CUSTOMIZED MENU SYSTEMS   Allows developers to build menu systems to help callers find the right information, agent or department. Using the drag-and-drop technique, the menu system can be revised daily to suit changing business needs. The menu can be interactive, allowing the caller to respond to voice prompts by dialing different numbers. A different voice message can be associated with each response. Voice messages can be created instantly via the PC's microphone.

TSAPI

While TAPI defines the connection between a single phone and a PC, TSAPI defines the connection between a networked file server and a PBX, resulting in the integration of computer and telephone functions at the desktop using a logical connection established over the LAN.

In connecting a NetWare server to the PBX, individual PCs are given control over telephone system functions. TSAPI is implemented with NetWare Loadable Modules (NLMs) that run on Novell servers, along with another NLM containing a PBX driver. No special hardware is required at the desktop; the PBX supports its own physical connection and uses its own software. The physical link is an ISDN Basic Rate Interface card in the server that allows for the connection between the NetWare server and the PBX.

NetWare Telephony Services consists of a Telephony Server NLM, a set of dynamic link libraries (DLLs) for the client, and a sample server application (a simple point-and-click telephone listing that is integrated with directory services). Novell also offers a driver for every major PBX. Alternatively, users can obtain a driver from their PBX vendor.

The NLM's features include drag-and-drop conference calling, the ability to put voice, facsimile and electronic mail messages in one mailbox, third-party call control, and integration between telephones and computer databases. Noteworthy among these is third-party call control, which provides the ability to control a call without being a part of it. This feature would be used for setting up a conference call, for example.

Unlike TAPI, which allows only first-party call control, third-party constructs are an integral part of NetWare Telephony Services. The command Make Call, for example, has two parameters: one for addressing the originating party; the other for addressing the destination party. An application using this command would therefore allow users to designate an address different from their own as the originating party and establish a connection without becoming a participant in the call. This third-party call control also lets users set up automatic routing schemes.

JTAPI

JTAPI provides the definition for a reusable set of telephone call control objects, which enable application portability across computer platforms. The scalability of JTAPI enables it to be implemented on devices ranging from hand-held phones to desktop computers to large servers. This allows enterprises to blend together Internet and telephony technology components within a single application environment as they design and deploy new business strategies for improving customer service levels, including launching their presence on the Web.

JTAPI is composed of a set of Java language packages. Each package provides a specific piece of functionality for a certain aspect of computer-telephony applications. Implementations of telephony servers choose the packages they support, depending on the capabilities of their underlying platform and hardware. Applications may query for the packages supported by the implementation they are currently using. Additionally, application developers may concern themselves with only the supported packages the application needs to accomplish a task.

At the center of the Java Telephony API is the "core" package. The core package provides the basic framework to model telephone calls and rudimentary telephony features. These features include placing a telephone call, answering a telephone call, and disconnecting a telephone call. Simple telephony applications need to use the core to accomplish their tasks, and do not need to concern themselves with the details of other packages. For example, the core package permits applet designers to easily add telephone capabilities to a Web page.

A number of standard packages extend the JTAPI core package. Among the extension packages are those for call control, call center, media, phone, private data, and capabilities:

CALL CONTROL Extends the core package by providing more advanced call-control features such as placing calls on hold, transferring telephone calls, and conferencing telephone calls.

CALL CENTER Provides applications the ability to perform advanced features necessary for managing large call centers such as routing, automated call distribution, predictive calling, and associating application data with telephony objects.

MEDIA Provides applications access to the media streams associated with a telephone call. They are able to read and write data from these media streams. DTMF (touch-tone) and non-DTMF tone detection and generation is also provided by this package.

PHONE Permits applications to control the features of telephone sets.

CAPABILITIES Allows applications to query whether certain actions may be performed. Capabilities take two forms: static capabilities indicate whether an implementation supports a feature; dynamic capabilities indicate whether a certain action is allowed given the current state of the call model.

PRIVATE DATA Enables applications to communicate data directly with the underlying hardware switch. This data may be used to instruct the hardware to perform a switch-specific action.

JTAPI also defines call model objects that work together to describe telephone calls and the endpoints involved in a telephone call:

PROVIDER This object might manage a PBX connected to a server, a telephony/fax card in a desktop machine, or a computer networking technology such as IP. It hides the service-specific aspects of the telephony subsystem and enables Java applications and applets to interact with the telephony subsystem in a device-independent manner.

CALL This object represents a telephone call. In a two-party call, a telephone call has one Call object and two connections. A conference call is three or more connections associated with one Call object.

ADDRESS This object represents a telephone number.

CONNECTION This object models the communication link, which is the relationship between a Call object and an Address object.

TERMINAL This object represents a physical device such as a telephone and its associated properties. Each Terminal object may have one or more Address objects (telephone numbers) associated with it, as in the case of some office phones capable of managing multiple line appearances.

TERMINAL CONNECTION This object models the relationship between a Connection and the physical endpoint of a call, which is represented by the Terminal object.
Applications built with JTAPI use the Java "sandbox" model for controlling access to sensitive operations. Callers of JTAPI methods are categorized as "trusted" or "untrusted," using criteria determined by the runtime system. Trusted callers are allowed full access to JTAPI functionality. Untrusted callers are limited to operations that cannot compromise the system's integrity. In addition, JTAPI may be used to access telephony servers or implementations that provide their own security mechanisms, such as user name and password.

SUMMARY

The latest application of CTI is the integration of traditional call centers with the Internet, enabling Web site visitors to use the Internet access connection for initiating voice calls to a company representative to request help or place an order. As corporate intranets become more popular, large companies are leveraging their global IP networks with call centers that offer help desk and other types of support services aimed at trading partners as well as internal staff. These call centers take voice-data integration to a higher level by integrating multimedia messaging capabilities such as

e-mail, Web messaging, facsimile, real-time text chat, dynamic help, escorted Web browsing, shared page markup, and collaborative form completion. Some even offer the option for video communication, if the customer has a video camera and uses an H.323-compliant Web browser.
***See Also* Middleware, Multimedia Documents, Unified Messaging**

CPE-Based WAN Restoral

Many fundamental business operations have become entirely dependent on the consistent, reliable operation of the network infrastructure. Although carriers offer restoral solutions for their services, there are many that can be implemented by customer premises equipment (CPE) and used efficiently and economically in conjunction with the switched digital services offered by major carriers, such as AT&T, MCI World Com, and Sprint. Typically, several restoral capabilities are required, since most organizations typically rely on multiple types of lines, services or networks for their voice and data requirements.

A new element in the world of WANs is the Internet. The basic forms of CPE-based restoral/recovery processes and facilities for the Internet are typically the same as for conventional WANs, differing only in the type of CPE used on each type of network and the protocols involved. With a TCP/IP-based intranet, for example, routers are capable of diverting traffic around failed nodes or points of congestion.

Importance of WANs

Despite the increasing reliance on wide-area networks of all types in nearly every aspect of business, many corporations remain ill-prepared to recover from the loss of these assets. Most businesses can survive a 1 or 2 day network outage, but after that serious consequences start to set in: customer service deteriorates and cash flow slows.

If primary business functions are interrupted for three to seven days, other adverse impacts are felt: possible negative publicity about the company, lost customers, and legal liability. Beyond one week, the impact is enormous: increased litigation, staff burnout and high turnover often follow extended outages.

The consequences of an outage are more severe for businesses that rely on their networks to deliver financial services to customers. Brokerage houses, banks, and insurance companies—to name a few—cannot afford even a few minutes of network downtime without experiencing adverse impacts, including financial liabilities. Likewise, the impact of an outage of any length can be severe for a telemarketing firm because the network is tightly integrated with the core business operation.

Restoring T1 Lines

Despite the growing popularity of public network services, T1 lines are still in high demand among companies with private networks. A variety of means are available to protect T1 networks from congestion and link failures. Traffic can be offloaded to standby links or the switched digital services of various carriers, including ISDN. The CPE that handles this is usually the T1 multiplexer. The advanced transport management systems of some multiplexers can provide instantaneous restoral of T1 facilities, either by rerouting traffic to spare capacity on the private network or by calling up the required amount of bandwidth from the public switched network. There are several restoral options for T1 lines:

T1 BACKUP VIA ISDN An ISDN PRI-equipped multiplexer can reroute traffic from a failing or congested T1 line to an ISDN facility in 10 seconds or less. With carrier arrangements already in place (i.e., access lines at each end of the circuit), the restoral process can be programmed for automatic implementation.

FRACTIONAL T1 BACKUP VIA ISDN ISDN can also be used to back up FT1 links. If a fractional link fails, dial backup to ISDN can be implemented in a manner analogous to placing a standard telephone call.

AUTOMATIC REROUTING A T1 multiplexer can reroute traffic to available bandwidth elsewhere on the private network on a node-to-node basis. The circuits to be routed are identified by the network manager on a per-circuit or user group basis and stored in a program that can be implemented automatically when a failure occurs or service degrades to a specified bit error rate.

PRIORITY REROUTING To ensure that the applications of the highest priority are always rerouted on the best paths and to avoid circuit collisions during reroutes, priority rerouting is performed on a network-wide basis, rather than on a node-to-node basis.

DOWN-SPEEDING Many automatic reconfiguration schemes result in service denial to some users; however, with a multiplexer's down-speeding capability, all applications can get enough bandwidth without bumping users off the network during a reconfiguration. While this might result in a slowdown of some high-bandwidth applications, the objective is to keep everyone online until the primary network reconfiguration is restored.

DIAL BACKUP The dial backup function of the T1 multiplexer allows management information to be conveyed between the central system controller to a remote node that has become isolated from the rest of the network due to a line cut or service failure.

NETWORK MODELING With the modeling capabilities of some T1 multiplexers' management systems, network planners can simulate various disaster scenarios on an aggregate or node level anywhere on the network. Off-line simulation allows planners to test and monitor changing conditions and determine their impact on network operations.

ONLINE DISASTER RECOVERY Many T1 multiplexers have fiber-optic interfaces. The increasing abundance of fiber lines has created parallel networks along high-density routes, which means users can protect themselves against potential disaster by allocating their traffic among several competing carriers. When the fiber of one carrier is cut or fails, traffic can be instantly diverted to the spare capacity of another carrier's fiber.

INVERSE MULTIPLEXING Through an inverse multiplexer, a company can set up the appropriate amount of switched digital bandwidth needed to temporarily replace failed leased lines. Depending on the inverse multiplexer used, bandwidth can be added and subtracted dynamically as requirements change, which minimizes the per-minute usage charges for switched services.

DDS Dial Backup

Several dial backup solutions are available for low-speed digital data services:

DIGITAL DATA SETS Some digital data sets have the ability to "heal" interruptions in transmission. Should the primary digital network facility fail, communication is reestablished over the public switched network via the data set's built-in modem and integral single-call dial backup unit.

DATA SERVICE UNITS For DDS at 56 Kbps, organizations can use data service units (DSUs) on leased lines to initiate restoral over switched 56-Kbps services such as AT&T's Switched 56 Service for dial-up, point-to-point 56-Kbps operation. Per-minute charges accrue only when the switched service is actually used.

INTEGRATED NETWORK MANAGEMENT SYSTEMS (INMSS) To address varying network recovery requirements, an INMS controller can be used which interfaces with a number of different Network Product Managers (NPMs). Each NPM functions as a specialized network management system. For example, one NPM could be dedicated primarily to analog and digital data set networks, while another could be used for high-capacity digital backbone networks. Still another NPM might specialize in the management of statistical multiplexed and packet-switched networks. The INMS simplifies network recovery by coordinating and implementing backup procedures among all the NPMs.

INTELLIGENT CALLING SYSTEMS Dial backup systems can be cumbersome to implement if many lines are knocked out. In such cases, the ability to reroute traffic rapidly over dial-up facilities can save organizations from financial disaster. For large private networks, an Intelligent Calling System (ICS) may provide the solution. Via an NPM, multiple dial backup calls can be initiated to remote sites. The ICS provides an economical means of initiating four-wire dial backup for analog and digital point-to-point or multipoint circuits. The ICS is especially suited for use on multidrop DDS lines because it reduces the number of modems and stand-alone dial backup units that would normally be required.

FAULT-TOLERANT SYSTEMS Conventional dial backup configurations require two dial-up lines for each leased-line circuit being restored—one to transmit (Tx) and the other to receive (Rx). Fault-tolerant dial backup systems are capable of single-call, full-duplex operation.

DIAL BACKUP UNITS Over the years, dial backup units (DBUs) have become a popular means for temporarily rerouting modem and digital data set transmissions from failed facilities to the public switched network. A DBU may be a stand-alone device that plugs into a modem or an optional add-in module that sits on top of the modem's main circuit board. On failure of the primary line, operation over the public switched network can be manually or automatically initiated.

In both manual and auto-originate modes, the remote location is treated as unattended; the auto-answer capability provides the means to automatically answer a dial backup call. An auto-terminate capability disconnects the backup call when the originating site ceases to transmit due to the restoral of the primary link. An auto-abort capability disallows call transfer and disconnects the call if the handshake protocol (including the "security check") is not properly completed. This prevents "wrong number" calls and hackers from disturbing the network.

On restoral of the failed line, dial backup can be terminated manually or automatically upon detection of acceptable signal quality.

Hub/Router-Based Restoral

As noted, dial backup provides a more economical solution than a second leased line when users require a backup link for disaster recovery. Accordingly, router-equipped hubs with ISDN interfaces provide connectivity on demand if the primary WAN link fails. In this case, a second dial-up ISDN line is automatically brought online. This feature also supports the activation of a second line for load-share traffic if the primary link nears congestion.

Nodal routers are used for building highly meshed wide-area internets. In addition to allowing several protocols to share the same logical network, these devices pick the shortest path to the end node, balance the load across multiple physical links, reroute traffic around points of failure or congestion, and implement flow control in conjunction with the end nodes.

SATELLITE-BASED RESTORAL Satellite carriers offer shared network restoral arrangements, whereby traffic is routed off failed terrestrial facilities and carried over previously contracted backup satellite links. There are, however, limitations associated with such arrangements, including the long lead time (a minimum of 48 hours) required to place backup satellite circuits into operation and the associated high cost of the standby circuits.

SONET-BASED RESTORAL One of the key selling points of SONET (synchronous optical network) is its integral support for disaster recovery. SONET fiber facilities are typically configured in a dual counter-rotating ring topology. This topology makes use of self-healing mechanisms in SONET-compliant equipment (i.e., the Add-Drop Multiplexers, or ADMs) to ensure the highest degree of network availability and reliability. In the event of a break in the line, traffic is automatically switched from one ring to the other, thus maintaining the integrity of the network. In the unlikely event that both the primary and secondary lines fail, the SONET-compliant equipment adjacent to the failures automatically loops the data between rings, thus forming a new C-shaped ring from the operational portions of the original two rings. When the break is fixed, the network automatically returns to its original state.

SONET's embedded management channels allow users (and carriers) to continuously monitor the links and take preemptive corrective action in response to impending trouble conditions. In private SONET networks, managers can reconfigure channels and facilities without the involvement of telephone companies. Through software programming, it is even possible to "map" SONET circuits so that they can be automatically rerouted to alternate carrier facilities should a failure occur on the primary circuit(s).

INTEGRATED RESTORAL CAPABILITIES The latest CPE-based network restoral systems feature centralized real-time alarm reporting and diagnostic support, as well as intelligent network restoral. These integrated systems enable network managers to remotely detect problems and instantly restore service from a PC. Link and equipment restoral can be set for automatic, controlled, or manual switching.

With automatic switching, the integrated system instantly restores the failed link or equipment to reestablish service. Automatic switching can also be scheduled for weekends or off-hours to satisfy routine outage period requirements. With controlled switching, the network manager can supervise the switchover of failed links or equipment. With manual switching, the network manager can override automatic and controlled switching.

CUSTOMER CONTROLLED RECONFIGURATION   Customer controlled reconfiguration (CCR) is both a carrier-provided service and a CPE-based WAN restoral solution. CCR gives businesses the means to organize and manage their own circuits from an on-premises terminal that issues instructions to the carrier's digital cross-connect system (DCS). If a circuit drops to an unacceptable level of performance or fails entirely, the network manager can issue rerouting instructions to the DCS. If several circuits have failed, the network manager can upload a pretested rerouting program to the DCS to restore the affected portion of the network.

IMPORTANCE OF RESTORAL PLANNING   With CPE-based wide-area network restoral solutions, the responsibility for disaster recovery is more on the user than the carrier to implement corrective action, unless the services of an outsourcing firm are used.

Given the potentially severe consequences of a network outage, companies of all types and sizes should consider performing a complete assessment of their network assets, giving particular attention to ascertaining the cost of failures at key points of the network.

Consideration must be given to such things as the amount of spare bandwidth that is available to handle the traffic from failed lines, the availability of spare components and subsystems to get CPE up and running after a failure, and the availability of key technical personnel to respond in a timely manner to any emergency situation.

To minimize exposure to risk and prevent adverse impacts on the organization, disaster recovery plans must be developed and tested. Over time, they must be fine-tuned to take into account changes on the network, such as the addition of new equipment, lines, technologies, and applications. Changes in carriers and services also require that the disaster recovery plan be updated and tested.

SUMMARY ■ ■ ■ ■ ■ ■ ■ ■ ■

The risks normally associated with a company taking responsibility for disaster recovery have been minimized in recent years through better network design, capacity planning, and performance baselining tools that provide a complete picture of the corporate network at every stage—from planning and testing to implementation and ongoing management. In fact, the use of these tools can head off many problems that could adversely impact network performance later.

In addition, advances in technology and the adoption of quality control processes by equipment vendors during production have combined to greatly increase the reliability of today's communications products. Most CPE is now modular in design, permitting fast isolation and easy replacement of faulty components from inventory. Some products even come with redundant power supplies, backplanes, control logic, and storage. When subsystem A fails, subsystem B takes over with minimal disruption in performance. Key components and subsystems can even be "hot-swapped"—pulled out and replaced without powering down the network.

***See Also* Carrier-Based WAN Restoral**

Dark Fiber

Dark fiber is the unused transmission capacity in a carrier's fiber-optic trunks. This capacity is leased to corporations, local exchange carriers (LECs) and Internet service providers (ISPs) who furnish their own equipment to send and receive traffic over that link.

Dark fiber puts the user in control of the growth and development of its private, facilities-based network by providing unlimited, unmetered bandwidth—whether it is a fiber ring around a metropolitan area or a point-to-point connection between far-flung nodes. The bandwidth is dedicated to the exclusive use of the leasing company, LEC or ISP—no other traffic from another customer can access the fiber.

Dark fiber interconnections can be deployed in several ways to meet a user's requirements: single entry, point-to-point connections; dual entry, diverse ring networks for multiple locations; and private network builds. Dark fiber is available in both single-mode and multimode fiber. The capacity of the fiber can be increased through the addition of Wave Division Multiplexing (WDM) or Dense Wave Division Multiplexing (DWDM) —all without increasing the lease price.

SUMMARY

The term *dark fiber* refers to fiber-optic cable without any of the electronic or optronic equipment necessary for transmission. There are a variety of applications for dark fiber. Communications carriers and ISPs gain "last mile" connectivity to the most highly populated metropolitan areas and position themselves to meet future long distance demands of their customers. Corporate and government users of dark fiber benefit from private building-to-building networks featuring the fastest transmission speeds available and the highest levels of reliability and security.
See Also **Fiber-Optic Technology, Synchronous Optical Network**

Data Compression

Data compression has become a standard feature of most bridges and routers, as well as modems. In its simplest implementation, compression capitalizes on the redundancies found in the data. The algorithm detects repeating characters or strings of characters and represents them as a symbol or token. At the receiving end, the process works in reverse to restore the original data.

Compression Efficiency

Compression efficiency tends to differ by application. The compression ratio can be as high as 6-to-1 when the traffic consists of heavy-duty file transfers. The compression ratio is less that 4-to-1 when the traffic is mostly database queries. When there are only "keep-alive" signals or sporadic query traffic on a T1 line, the compression ratio can dip below 2-to-1. Encrypted data exhibits little or no compression because the encryption process expands the data and uses more bandwidth. However, if data expansion is detected and compression is withheld until the encrypted data is completely transmitted, the need for more bandwidth can be avoided.

The use of data compression is particularly advantageous in the following situations:

- When data traffic is increasing due to the addition or expansion of LANs and associated data-intensive, bursty traffic
- When LAN and legacy traffic must contend for the same limited bandwidth
- When reducing or limiting the number of 56/64-Kbps lines is desirable to reduce operational costs
- When lowering the Committed Information Rate (CIR) for Frame Relay services or sending fewer packets over an X.25 network can result in substantial cost savings

The greatest cost savings from data compression most often occurs at remote sites, where bandwidth is typically in short supply. Data compression can extend the life of 56/64-Kbps leased lines, thus avoiding the need for more expensive fractional T1 lines or N × 64 services. Depending on the application, a 56/64-Kbps leased line can deliver 112 Kbps to 256 Kbps or higher throughput when data compression is applied.

History

Symplex Communications Corp. pioneered data compression for bridges and routers with the introduction of its Datamizer I in 1983, which offered 2-to-1 compression to achieve 19.2-Kbps throughput on 9.6-Kbps lines. Datamizer IV was the first device to provide 4-to-1 compression over leased lines and could be configured to activate additional lines as traffic volume increased. It also could be configured to automatically reroute data if a line failed.

The company's Datamizer V is adept at handling a combination of hard- and easy-to-compress data types in that it automatically adjusts compression techniques to maximize the throughput benefit on a per packet basis. If users experience compression of less than 2-to-1, the Datamizer V's default setting for easy-to-compress data can be changed to the hard-to-compress option to achieve better performance. The newer multi-port Datamizer 6 supports higher speeds, provides additional throughput performance, and offers optional encryption features. It operates seamlessly on virtually all forms of telecom services, including frame relay, dedicated lines, satellite links, and ISDN.

Of course, there are other products that implement data compression on WAN links. Fourelle Systems, for example, offers an innovative hardware and firmware solution called Venturi that increases overall bandwidth for both wired and wireless TCP/IP networks. In a typical installation, a standard client application (e.g., Web browser) communicates through a local Venturi proxy, which in turn transmits data across a bandwidth-constrained link to a Venturi compression server proxy. Venturi server then communicates to a network-based application (e.g., Web server). Each proxy communicates to its respective application using standard TCP/IP protocols, requiring no change to the application. Using an optimized IP-based transport, Venturi combines several data-dependent compression techniques, resulting in a 50-percent to 99-percent reduction in the amount of data transmitted.

Types of Data Compression

There are several different data compression methods in use today over wide area networks—among them are TCP/IP header compression, link compression and multi-channel payload compression. Depending on the method, there can be a significant tradeoff between lower bandwidth consumption and increased packet delay.

TCP/IP HEADER COMPRESSION With TCP/IP header compression, the packet headers are compressed but the data payload remains unchanged. Since the TCP/IP header must be replaced at each node for IP routing to be possible, this compression method requires hop-by-hop compression and decompression processing. This adds delay to each compressed/decompressed packet and puts an added burden on the router's CPU.

LINK COMPRESSION With link compression, the entire frame—both protocol header and payload—are compressed. This form of compression is typically used in LAN-only or legacy-only environments. However, this method requires error correction and packet sequencing software, which adds to the processing overhead already introduced by link compression and results in increased packet delays. Also, like TCP/IP header compression, link compression requires hop-to-hop compression and decompression, so processor loading and packet delays occur at each router node the data traverses.

With link compression, a single data compression vocabulary dictionary or history buffer is maintained for all virtual circuits compressed over the WAN link. This buffer holds a running history about what data has been transmitted to help make future transmissions more efficient. To obtain optimal compression ratios, the history buffer must be large, requiring a significant amount of memory. The vocabulary dictionary resets at the end of each frame. This technique offers lower compression ratios than multi-channel, multi-history buffer (vocabularies) data compression methods. This is particularly true when transmitting mixed LAN and serial protocol traffic over the WAN link and frame sizes are 2K bytes or less. This translates into higher costs, but if more memory is added to get better ratios, this increases the up-front cost of the solution.

MIXED-CHANNEL PAYLOAD DATA COMPRESSION By using separate history buffers or vocabularies for each virtual circuit, multi-channel payload data compression can yield higher compression ratios that require much less memory than other data compression methods. This is particularly true in cases where mixed LAN and serial protocol traffic traverses the network. Higher compression ratios translate into lower WAN bandwidth requirements and greater cost savings.

But performance varies because vendors define payload data compression differently. Some consider it to be compression of everything that follows the IP header. However, the IP header can be a significant number of bytes. For overall compression to be effective, header compression must be applied. This adds to the processing burden of the CPU and increases packet delays.

External Data Compression Solutions

Although bridges and routers can perform data compression, external compression devices are often required to connect to higher speed links. The reason is that data compression is extremely processor intensive, with multichannel payload data compression being the most burdensome. The faster the packets must move through the router, the more difficult it is for the router's processor to keep up.

The advantages of internal data compression engines are that they can provide multi-channel compression, lower cost, and simplified management. However, by using a separate internal digital signal processor (DSP) for data compression, instead of the software-only approach—all of the other basic functions within the router can continue to be processed simultaneously. This parallel processing approach minimizes the packet delay that can occur when the router's CPU is forced to handle all these tasks itself.

SUMMARY

Data compression will become increasingly important to most organizations as the volume of data traffic at branch locations begins to exceed the capacity of the wide area links. Multi-channel payload solutions provide the highest compression ratios and reduce the number of packets transmitted across the network. Reducing packet latency can be effectively achieved via a dedicated processor like a DSP and by employing end-to-end compression techniques, rather than node-to-node compression/decompression. All of these factors contribute to reducing WAN circuit and equipment costs as well as improving the network response time and availability for user applications.

See Also **Voice Compression**

Data Service Units

The Digital Service Unit (DSU) is a stand-alone, rack-mounted, or router-integrated device that connects various data terminal equipment via RS-232, RS-449 or V.35 interfaces with widely available digital services that offer 56/64 Kbps access, including Digital Data Service (DDS) and cell- or frame-based services such as ATM, SMDS, and frame relay. Typical applications for the DSU include LAN interconnection, dedicated Internet access, and remote PC access to local hosts.

Digital Data Service (DDS) operates at speeds of 1.2 Kbps to 56/64 Kbps in support of point-to-point or multipoint applications. Most DSUs have a built-in asynchronous-to-synchronous converter, accommodating asynchronous input devices that operate at speeds of 1.2 Kbps to 57.6 Kbps, as well as synchronous input devices that operate at either 56 Kbps or 64 Kbps. When packaged with Channel Service Unit (CSU) functions, the DSU/CSU device interfaces with T1 services at 64 Kbps and $N \times 64$ Kbps up to 1.536 Kbps.

The DSU converts the binary data pulse it receives from the DTE to the bipolar format required by the network. The DSU also supplies the transmit and receive logic, as well as timing. Any device that connects directly to a digital line (via an external or internal CSU) must perform these functions, or it needs a DSU. Any piece of network equipment that does not have a bipolar port needs a DSU to connect to a CSU. The most common type of access device is the combination unit, which offers DSU and CSU functionality, which eliminates these concerns and reduces the number of devices that must be managed.

Like the CSU, the DSU also provides the means to perform diagnostics. The front panel of the DSU provides a set of LED indicators that show the status of the V.35 DTE interface, various test modes, and loop status. The device responds to standard loopback commands from the service provider or the user side. Included with the remote loopback capability are selectable bit error rate test patterns, the results of which are displayed on the front panel.

Like CSUs, DSUs can be managed either by the vendor's proprietary network management system or by Simple Network Management Protocol (SNMP) tools. The DSU provides a non-disruptive in-band SNMP management channel over a DDS leased circuit. For frame relay connections, SNMP management is provided with a connection from the DSU's management port to a router port or external LAN adapter at the remote site. In this case, management frames are sent across the network as data, then routed to the DSU by the remote router. Typically, in-band management is lost in a frame relay network due to the packetizing and switching of the data. This management method allows diagnostic packets to be sent to the distant router and then forwarded to the DSU through the external management connection to the management port on the DSU.

There are now simple versions of DSUs that are not designed to be managed. They will automatically setup on any standard 56/64K DDS facility and provide the correct data rate to the DTE port. Remote control and testing of the distant DSU is accomplished via an external modem connection and ASCII interface.

Modular routers, such as Cisco Systems' 1600 and 3600 series routers, can be equipped with a WAN interface card that incorporates a fully managed DSU/CSU to facilitate the deployment and management of Internet and intranet connectivity. This integrated solution eliminates the need for external DSU/CSUs and allows all of these components, including the router, to be managed both locally and remotely as a single entity via SNMP or a Telnet session using the Cisco command-line interface.

SUMMARY

DSUs, as well as CSUs, not only provide an interface between DTE and the carrier's network, but such devices help network managers fine-tune their networks for performance and cost savings. For example, when using the diagnostic capabilities of a DSU/CSU connected to a frame relay network, the network manager can monitor traffic on each permanent virtual connection (PVC) to set an appropriate committed information rate (CIR) and allowable burst rate on each circuit. In addition, the delay between network nodes can be measured as well as the performance of a line between the user and local carrier to see if the carrier is actually delivering the level of service promised.

See Also **Channel Service Units**

Data Warehouse

A data warehouse is an extension of the database management system (DBMS) which consolidates information from various sources into a high-level, integrated form used to identify trends and make business decisions. For a large company, the amount of information in a data warehouse could be up to several trillion bytes, or terabytes (TB). The technologies that are used to build data warehouses include relational databases; powerful, scalable processors; and sophisticated tools to manipulate and analyze large volumes of data and identify previously undetectable patterns and relationships.

The benefits of a data warehouse include increased revenue and decreased costs due to the more effective handling and use of massive amounts of data. Data warehouse applications are driven by such economic needs as cost reduction or containment, revenue enhancement, and response to market conditions. In being able to manage data more effectively, companies can improve customer satisfaction and cement customer loyalty. This can be done by sifting through the data to identify patterns of repeat purchases, determine the frequency and manner with which customers use various products, and assess their propensity to switch vendors when they are offered better prices or more targeted features. This kind of information is important because a change of only a few percentage points of customer retention can equate to hundreds of millions of dollars to a large company.

The benefits of data warehousing can now be extended beyond the corporate headquarters to remote branch offices, telecommuters and mobile professionals. Virtually anyone with a Web browser and an Internet connection can access corporate data stores using the same query, reporting and analysis tools previously reserved for technically elite number crunchers using expensive, feature-rich client/server tools. The Web-enabled data warehouse makes it possible for companies to leverage their investments in information by making it available to everyone who needs it for making critical business decisions.

System Components

The data warehousing framework typically encompasses several components: an information store of historical events (the data warehouse), warehouse administration tools, data manipulation tools, and a decision support system (DSS) that enables strategic analysis of the information.

A key capability of the DSS is data mining, which uses sophisticated tools to detect trends, patterns, and correlations hidden in vast amounts of data. Information discoveries are presented to the user and provide the basis for strategic decisions and action plans that can improve corporate operational and financial performance. Among the many useful features of a DSS are:

- An automatic monitoring capability to control runaway queries
- Transparent access to requested data on central, local, or desktop databases
- Data-staging capabilities for temporary data stores for simplification, performance, or conversational access
- A drill-down capability to access exception data at lower levels
- Import capabilities to translators, filters, and other desktop tools
- A scrubber to merge redundant data, resolve conflicting data, and integrate data from incompatible systems
- Usage statistics, including response times

The latest trend in data warehouses is to integrate them with the corporate intranet for access by remote users with browser-enabled client software. The primary purpose of Web-enabling a data warehouse is to give remote offices and mobile professionals the information they need to make tactical business decisions.

Development Toolkits

A variety of toolkits are available that allow developers to create custom reports that can be hosted on the corporate Web site. In some cases, users can drill down into these reports to uncover new information or trends. With more sophisticated Web-based online analytical processing (OLAP) tools, users can directly access the corporate data warehouse to do simple queries and run reports. With the resulting information, reports can be culled for information remote users need for their everyday decisions.

At a retail operation, for example, a store manager might tap into a canned sales report to figure out when a specific item will run out of stock, and a business analyst at the corporate headquarters might use client/server OLAP tools to analyze sales trends at all the stores to make strategic purchasing decisions.

For users who need more than static HTML documents but like the convenience of Web browsers, there are plug-ins that dynamically query the back-end database. With such tools, employees can do things like drill down into the reports to find specific information. At an insurance company, for example, users might have the ability to drill down to a particular estimate line within a claim. This granularity might let a user determine how many claims relate to airbags and how long and at what cost it takes to repair them. Such information can be used to determine the discount drivers qualify for if their vehicle is equipped with airbags.

Data Accessibility

Making a data warehouse accessible to Web users solves a dilemma faced by many companies. On one hand, they do not want to limit users by providing only predefined HTML reports that cannot be manipulated. On the other hand, they do not want to overwhelm users with a tool they are not trained to understand. A Web-based tool that allows some interactivity with the data warehouse offers a viable alternative for users who are capable of handling simple queries. In turn, this makes corporate information more accessible to a broader range of users, including business analysts, product planners, and salespeople.

Since not everyone has the same information requirements, some companies have implemented multiple reporting options:

CANNED REPORTS These are predefined, executive-level reports that can be viewed only through the Web browser. Users need little to no technical expertise, knowledge of the data, or training because they can only view the reports, not interact with them.

READY-TO-RUN REPORTS For those with some technical expertise and knowledge of the data, report templates are provided that users fill in with their query requirements. While dynamic, these reports are limited to IS-specified field values, fill-in boxes, and queries.

AD-HOC REPORTS For technically astute people who are familiar with the data to run freeform queries, unlimited access to the data warehouse is provided. They can fill in all field values, choose among multiple fill-in boxes, and run complex queries.

Web-enabled Warehouses

Many vendors offer Web tools that support an intranet architecture comprising Web browsers, Web servers, application servers, and databases (Figure 29). The Web servers submit user requests to an application server via a gateway such as the Common Gateway Interface (CGI) or server API. The application server translates HTML requests into calls or SQL statements it can submit to the database. The application packages the result

Figure 29
A corporate intranet architecture for a Web-enabled data warehouse.

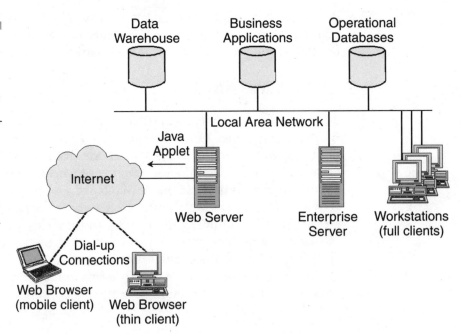

and returns it to the Web server in the proper format. The Web server forwards the result to the client.

This model can be enhanced with Java applets or other client-side programs. For example, the query form can be presented as a Java applet, rather than the usual CGI. Among the advantages of a Java-based query form is that error-checking can be performed locally rather than at the server. If certain fields are not filled in properly, for example, an appropriate error message can be displayed before the query is allowed to reach the server. This helps control the load on the server.

An all-Java approach provides even more advantages because connecting clients and servers at the network and transport layers is much more efficient than doing so at the application level using CGI scripts. This means users can design and execute queries and reports much more quickly than they can with other types of tools.

Security

Obviously, a data warehouse contains highly sensitive information, so security is a key concern for any company contemplating a Web-enabled data warehouse. Not only must the data warehouse be protected against external sources, many times it must be protected against unauthorized access from internal sources as well. For example, branch offices might be prevented from accessing each other's information, or the engineering department might be prevented from accessing the marketing department's information, while all departments might be prevented from accessing the personnel department's records. Security can be enforced through a variety of mechanisms from user names and passwords to firewalls and encrypted data transmissions.

Java offers additional levels of security. Applets that adhere to the Java Cryptography Extensions (JCE) can be digitally signed, encrypted, and transmitted via secure streams to prevent hackers from attacking applets during transmission. Specifically, JCE-based applets make use of Diffie-Hellmann authentication, a technology that enables two parties to share keys, and DES (Data Encryption Standard) for scrambling the data for transmission.

Some Web-enabled OLAP and DSS tools are designed to take advantage of the Secure Sockets Layer (SSL) protocol, which protects data transferred over the network through the use of advanced security techniques. The SSL protocol provides security that has three basic properties:

CONNECTION PRIVACY Encryption is used after an initial handshake to define a secret key.

DATA PROTECTION Cryptography algorithms such as DES or RC4 are used for data encryption. The peer's identity can be authenticated using a public key.

CONNECTION RELIABILITY Message transport includes an integrity check based on a keyed message authentication code (MAC).

Optimization Techniques

Some data warehouse products come with intelligent agents that work on behalf of users to find information stored in the data warehouse. Agents filter out low-priority information and proactively deliver personalized information back to the user. Agents can deliver results to directory listings and/or notify the user directly through alerts or e-mail with interactive report attachments. A number of agents can be launched by the user and be active at the same time within the data store. Agent requests run in the background, keeping desktops free for other tasks. Agents save the user from having to personally surf and filter through gigabytes and terabytes of warehoused and derived data. They can also be used to monitor the data store for changes to specific information and report the changes to the user when they occur.

Some Web-based development tools optimize database interactivity by giving users the ability to access back-end databases in real time. Users can submit queries against the entire database or refresh existing reports to obtain the most up-to-date data. In addition to providing for more flexible report generation, this dynamic report creation environment minimizes the need to manage and store physical reports online, which streamlines storage requirements. However, some servers can also deliver copies of pre-executed reports to users who do not need to filter existing reports and want to optimize data delivery.

Other Web tools optimize functionality by giving users the ability to manipulate returned data in real time. Users can drill up, down or across data, apply calculations, toggle between charts and tables, and reorganize the local data. With Java and ActiveX, a high degree of interactivity between the user and data is now possible over the Web.

Some vendors offer a report subscription feature that provides access to previously built report templates that can be quickly personalized with a user's parameters to save time and prevent the reinvention of popular reports. A groupware application provides seamless sharing of interactive reports and their assumptions to eliminate duplication of effort, ensure consistent analysis, and encourage collaboration.

With some products, persistence and state are maintained between client and server and server and database, allowing data to pass freely and efficiently. To improve efficiency and performance, the server caches data and only delivers small packets of information to the user as they are needed. As the number of users increases, a load-balancing capability allows additional machines to automatically start up to meet demand.

Reports can be scheduled to run when the system load is low. Designated reports can be scheduled to run daily, weekly, or monthly. This is especially useful for running large reports without tying up network and processor resources. Depending on the sophistication of the user, he or she can design, run, and view reports and charts and immediately communicate the insights gained from this analysis over the corporate intranet.

Some vendors' data warehouse products incorporate Internet push technology to automate the delivery of information to end users. Users can decide which reports they wish to subscribe to, and have those reports automatically pushed to their desktops. The automation of report delivery can be enhanced by allowing users to run reports that provide text alerts to exception conditions, such as record weekly sales or low item in stock. This feature allows users to focus on immediate business action instead of report creation.

SUMMARY

In the mainframe-centric environment of the past, employees had to submit their information requirements in writing to IS staff and then wait two or three weeks for a report. Today, with a Web-enabled data warehouse, users can have immediate access to the information they need. The Web offers unparalleled opportunity to deliver business reports to huge numbers of users without the information bottlenecks of the past. A standard Web browser provides the interface through which corporate data is accessed.

Although the rich functionality and high performance of client/server tools cannot yet be duplicated on Web-based data warehouse offerings, this is not necessarily a handicap. This is because the overwhelming majority of users only need to access pre-defined reports or execute limited queries. The relatively small number of technically elite users who require the ability to interact with large data sets or create decision-support applications for others in the corporation, will continue to use the more powerful client/server tools to which they have become accustomed.

See Also **Collaborative Computing, Database Management Systems, Intelligent Agents, Storage Servers**

Digital Power Line

Among the objectives of the Telecommunication Act of 1996 is to accelerate the entry of public utilities into the telecommunications marketplace. At the same time, deregulation in the utility industry has prompted energy companies to move into other fields to improve overall revenues. Opportunities range from automation services, telephony and cellular to Internet projects. To date, the impact of utility companies in the converging voice-data communications marketplace has been minimal. This could change soon with the introduction of digital power line technology to the U.S., which economically delivers information access services through the existing power distribution infrastructure of electric utility companies.

For many utility companies, leveraging assets is now a critical business objective—especially in the new deregulated environment. Utilities are well positioned for many of the new developments in the telecommunications marketplace, and have numerous strengths and assets that can be brought to bear in the telecom business. Some of these assets can be leased to established telecom service providers or used directly to offer new services to customers.

Utility Company Assets

Among the strengths and assets that utility companies can bring to the telecom business are:

- Extensive rights-of-way and poles, ducts, conduits, and physical assets for routing cables
- Substantial space in substations and other properties for new equipment and towers
- Space on existing towers, masts, and poles for the addition of new antennas and equipment
- Maintenance operations, including vehicles, personnel, and associated expertise
- Established customer relationships and a direct link with every potential telecom customer
- Billing and customer information systems
- Reputation for reliable service and associated brand name recognition
- Telecom technology experience for internal purposes
- Fiber-optic installation, operations, and maintenance experience

In offering new telecom services, either directly or through alliances, utility companies can open new sources of revenue, improve their relationship to customers, increase efficiencies, and discourage electric power business competition by offering value-added service.

Utility companies that lack expertise in data and voice services can acquire companies with that expertise or partner with established service providers. There are even outsourcing firms that cater specifically to the needs of utility companies, providing expertise in such areas as local number portability (LNP), signaling system 7 (SS7), rating systems, commission systems, customer care, service activation, wholesale and retail billing, convergent billing, and network management. Such firms also provide automated mapping/facilities management and geographic information systems.

Transmission Technology

Although the direct impact of utility companies in the converging voice-data communications marketplace has been minimal, the availability of digital power line (DPL) technology will enable utility companies to become major competitors in the telecommunications industry.

Developed in the UK by Nortel in the early 1990s, DPL offers 1 Mbps of bandwidth to support multiple applications. In addition to basic Internet access, it can be used for telephony (IP voice), multimedia, smart applications/remote control, home automation and security, home banking/shopping, data backup, information services, telecommuting, and entertainment.

The technology turns the low-voltage signals going between the customer premises and local electricity substation into a local-area network. Multiple substations are then linked by fiber-optic circuits to Internet switching points. By giving customers access to the Internet through their existing electricity supply system, the technology is available to virtually anybody. It offers permanent online connection with the potential for lower charges for communication services.

Although the technology's developers—Northern Telecom and British utility Norweb Contracting—have focused their attention in European and Asian markets, they are now turning their attention to building partnerships with utility companies in the U.S. To promote DPL, Nortel and Norweb Contracting have established a company called Nor.Web. The company has already signed up utility companies, mainly in Europe, to deploy DPL service.

System Configuration

The Digital Power Line solution consists of a DPL 1000 mainstation, a basestation, a coupling unit and a communications module. The total solution provides a competitively positioned access network, which integrates seamlessly into today's WAN networks.

At the customer premises, a stand-alone unit connects to the power supply by standard coaxial cable. The unit is then connected to the computer by a standard Ethernet cable. It supports laptops as well as PCs and Macs and supports a wide variety of platforms. It also supports the Universal Serial Bus (USB), enabling several machines to be connected to a single DPL box simultaneously to create a LAN in the home or office.

The DPL 1000 has the added advantage of allowing advanced network management by the utility company for the purpose of monitoring and maintenance as well as providing additional services to satisfy growing customer requirements. The system also unleashes the future potential for home automation by enabling remote operation of appliances such as lights, oven, and lawn sprinkler system.

Due to the nature of this technology, it can be rolled out in discrete, targeted phases. Utilities not wishing to operate data services themselves have the option of charging a right-to-use fee to an operating company for accessing their plant. The system is managed with Nortel's Magellan ATM management platform, which allows network operators to monitor and maintain pre-determined service levels and availability.

Regulation

To help utility companies get into the telecom business, per the Telecommunications Act of 1996, the FCC has granted them special status as "Exempt Telecommunication Companies" (ETC), which relieves them from many of the rules that govern incumbent telephone companies (i.e., RBOCs).

The rules require applicants seeking such status to file a brief description of their planned activities together with a sworn statement attesting to any facts or representations to demonstrate ETC status as defined in the 1996 Act. The procedures also provide for public comment on the application, but limits comment to the adequacy or accuracy of the information presented.

CSW Communications was the first company to file an application for ETC status. CSW is a subsidiary of Central and South West Corporation, a registered public utility holding company. Within six months following the enactment of the 1996 Act on February 8, 1996, the FCC received 15 applications for a determination of ETC status and approved them all. Many of these applications included a public utility that had acquired or maintained an interest in an ETC. Since then, many more utility companies have chosen to participate as ETCs.

As noted, many utility companies are starting to offer telecommunications services, mostly in partnership with established service providers. Progress to date has not been significant enough to worry incumbent telephone companies because market entry entails a huge investment of capital. However, this situation could change dramatically if DPL technology proves successful in the U.S. as it is in Europe. This would mean that utility companies could potentially reach every household and business with new voice and data services simply by leveraging their existing infrastructure, which entails only an incremental capital investment.

SUMMARY

Enough utility companies now offer or plan to offer telecom services that the trend is irreversible. Collectively, utility companies can make a significant impact in the converging communications marketplace and in the next few years contribute to the further erosion of telco revenues in much the same way as competitive local exchange carriers (CLECs) have done over the same time span. The use of DPL technology enables electric utility companies to fully leverage their extensive infrastructures for power distribution, enabling them to offer customers just about any kind of telecom service for which they see a market.

See Also **Next Generation Networks, Phoneline Networking, Powerline Networking, Wireless Home Networking**

Digital Subscriber Line

The regional Bell holding companies (RBOCs) are trying to move rapidly into providing long distance services, but as of mid 1999, have been stopped by the Federal Communications Commission (FCC) at every turn until they can prove that they have opened up the local loop to significant competition. Meanwhile, the RBOCs have been pursuing the competitive long-distance cellular and CATV markets with mergers and strategic partnerships. They are also looking for ways to leverage their local loops to provide value-added services, such as high-speed Internet access.

The problem is that the existing local loops were designed for voice telephony, so the amount of bandwidth for data is limited. If telephone companies want to offer advanced services, they must find an economical way to increase the bandwidth of existing twisted pairs so that data can pass reliably at multi-megabit speeds. One of the most economical ways for telephone companies to increase the bandwidth capacity of existing twisted pair wires in the local loop is by implementing new digital subscriber line (DSL) technologies.

Of the dozen or so variants of DSL technologies, the most popular among telephone companies is Asymmetric Digital Subscriber Line (ADSL), which supplies enough bandwidth to support multimedia—video, audio, graphics, and text—to the customer premises. ADSL carves up the local loop bandwidth into several independent channels suitable for any combination of services, including plain old telephone service (POTS), ISDN, interactive gaming, home shopping and video-on-demand. The electronics at both ends compensate for line impairments, increasing the reliability of high-speed data transmission. The following table summarizes the major DSL technologies:

Technology	Downstream Speeds	Upstream Speeds	Maximum Distance (ft.)
ADSL (Asymmetric DSL)	1.5 Mbps to 8 Mbps	64 Kbps to 640 Kbps	12,000 to 18,000
HDSL (High-bit-rate DSL)	512 Kbps and 2 Mbps	512 Kbps and 2 Mbps	12,000
RADSL (Rate Adaptive DSL)	128 Kbps to 7 Mbps	64 Kbps to 640 Kbps	18,000 to 25,000
SDSL (Symmetric DSL)	1.5 Mbps to 2 Mbps	1.5 Mbps to 2 Mbps	10,000
VDSL (Very-high-bit-rate DSL)	13 Mbps to 52 Mbps	1.6 Mbps to 2.3 Mbps	1,000 to 4,500

Other digital subscriber line technologies include:

CONSUMER DSL CDSL is a Rockwell proposal that can be installed without extra hardware to filter the data stream from the voice channel.

ETHERLOOP A proposal from Nortel, this DSL variant offers 10 Mbps over very short wires.

HDSL 2 This technology has the same characteristics as HDSL, except that it requires only one pair of wires instead of HDSL's two pairs of wires.

ISDN DSL IDSL is a dedicated, data-only service for which users pay a flat fee rather than the per-minute charges typical of switched ISDN, something that would be attractive to users who spend a lot of time on the Internet, for example.

MULTIPLE VIRTUAL LINES A technology from Paradyne in which one MVL line supports one modem at every phone extension. The devices can talk modem-to-modem within the same building as if on a LAN.

UNIVERSAL DSL Similar to CDSL, UDSL is a proposed international standard that allows any DSL modem to be used with any DSL service provider, as long as the provider's modem also meets the standard.

DSL Service Management

Service providers have had a difficult time provisioning DSL and tailoring the service to individual users. This has slowed the growth of DSL despite high interest among consumers and businesses. Now there are products available that make it easier for carriers to implement and manage DSL service.

Cisco Systems, for example, offers a product called the Service Selection Gateway (SSG), a dedicated system that sits in a carrier's central office or a service provider's point of presence. The system lets service providers offer users a way to instantly select additional services beyond the flat-rate DSL access service they may already have. The system even advertises services to customers so that they can instantly know what services are available and easily select a service. Once a user is authenticated—usually by username and password—access to the various network resources and services can be granted by the carrier.

Another problem with DSL has been the lack of integration between its element management system and carriers' operations support systems (OSS). An OSS is a database that is used by the carrier to support telecommunications services to its customers. Among the functions of OSS are preordering, ordering, provisioning, maintenance and repair, and billing. To address the integration problem, Cisco and other vendors offer carrier-class service policy administration systems that interface with existing operations support systems.

DSL and USB

Typically, DSL modems connect to PCs through an Ethernet cable, which requires the user to install an Ethernet adapter card into the PC to use a DSL service. Because the installation of an Ethernet adapter card is beyond the expertise of many users, service providers have had to send out a technician to get the user up and running. This raises the support costs of the carrier, while putting users through considerable inconvenience.

Now there are DSL modems that connect to a PC using the Universal Serial Bus (USB). This is an easy to use and flexible interconnect specification that enables instant peripheral connectivity external to the PC. It allows users to add peripheral devices without expensive add-in cards or configuration challenges such as DIP switches and IRQ settings. A single connector type simplifies connection of all USB-compliant devices, including telephony and broadband adapters, digital cameras, scanners, monitors, joysticks, keyboards, and other I/O peripherals.

The use of USB greatly simplifies installation of DSL modems and eliminates the need for on-site visits by a technician. The user simply plugs the modem into the existing USB port on a PC and loads some driver software. The modem connects to the DSL service provider through a standard ADSL port. Some DSL modems are upgradable through a simple software download.

Once the connections are made, a wizard-driven graphical user interface allows the user to easily get the modem ready for proper operation without technical support. Setup is as simple clicking a mouse to install a pre-configured service profile software module created by the carrier and bundled with the modem. By using service profiles, no other configuration tasks are required during installation.

SUMMARY

One problem still plagues DSL: there is no compatibility between the different versions. A high bit-rate DSL (HDSL) modem, for example, cannot talk to an ADSL modem. And a rate adaptive DSL (RADSL) modem cannot talk to a symmetric DSL (SDSL) modem. Even within a single DSL technology, vendors have not taken care to make their products interoperable.

A group of vendors, known as the Universal ADSL Working Group (UAWG), are working on a standard for DSL Lite, an inexpensive, easy-to-install version of DSL, which will be presented to the International Telecommunication Union (ITU). Many vendors will make DSL Lite modems based on the UAWG proposal without waiting for the ITU to act. In many cases, this means changing the firmware that runs their existing modems. Eventually, there will be a standard handshake between DSL modems that will let them reveal to each other what DSL technologies they support so they can negotiate how they will speak to each other.

***See Also* Cable Modems, Universal Serial Bus**

Directory Services

A directory service gives users and administrators transparent access to all computers, printers, servers and other resources throughout the network. The international standard for directory services is X.500, which specifies that a compliant directory service must provide the following network functions:

- Provide a unique naming method for all network resources to give them a clear identity, guaranteeing that all resources on the network can be easily identified

- Map network names to network addresses, giving network resources easy-to-remember names that represent actual network resources

- Provide location-independent access to network resources, giving users transparent access to resources, regardless of where the users and the requested resources are physically located

- Support extensible attribute information, allowing their objects to be described in detail so they can be easily identified in a directory service

- Provide extensive querying and searching capabilities so users can locate network resources by their attributes (for example, locating a printer by searching for the printer's type or location)

Applications

The X.500 standard offers scalability and interoperability across heterogeneous networks. There are several applications for X.500 directory services:

CORPORATE DIRECTORY A distributed corporate directory, accessible from anywhere in the company, listing networked resources and users with their related details and attributes is essential to the smooth running of an enterprise network.

GLOBAL MESSAGING In order to get the most out of a corporate mail system, it is important that users be able to find a correspondent's mail address and profile easily and quickly, from anywhere on the network.

SECURITY MANAGEMENT As companies do more business over the network, security is becoming a key concern. The double-key encryption sys-

Figure 30
The X.500 query-response architecture.

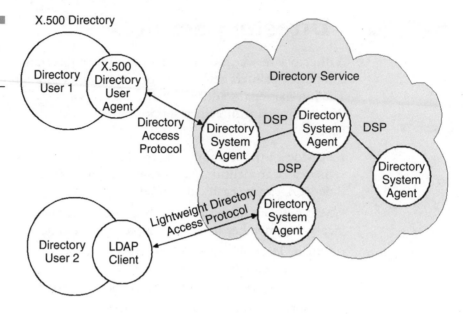

tem is the most commonly used to secure electronic transactions and requires a repository of public keys (for example, using the X.509 format). An X.500 directory provides the means to set up an encryption-key repository for integration with the X.509 security framework.

NETWORK MANAGEMENT A global directory listing resources and their attributes can simplify the jobs of network managers by providing a scalable repository of networked objects.

In the X.500 directory architecture, the client queries and receives responses from one or more servers in the server's Directory Service with the Directory Access Protocol (DAP) controlling the communication between the client and the server (Figure 30).

The Directory client, called the Directory User Agent (DUA), supports users in searching or browsing through one or more directory databases, and in retrieving the requested directory information. The DUA can be implemented in all kinds of user interfaces through dedicated DUA clients, Web-server gateways, e-mail applications, or middleware. DUAs are currently available for virtually all types of workstations, including those running DOS, Macintosh, OS/2, Unix.

Directory information is stored in a Directory System Agent (DSA), a hierarchical database designed to provide fast and efficient search and retrieval. The Directory System Protocol (DSP) controls the interaction between two or more DSAs. This is done in a way that allows users to access information in the Directory without knowing its exact location. The Directory Access Protocol (DAP) is used for controlling communication between a DUA and DSA.

Lightweight Directory Access Protocol

LDAP is based on the standards contained within the international X.500 standard, but is significantly simpler since it does not require the upper layers of the OSI stack. This makes it easier to implement, especially in clients. Because it is a simpler version of X.500, LDAP is often referred to as X.500 lite. Whereas X.500 is under the control of the International Telecommunication Union (ITU), LDAP is under the control of the Internet Engineering Task Force (IETF) so it can more easily evolve to meet Internet requirements.

LDAP runs directly over the TCP/IP stack. To do this, it had to shed many of X.500's overhead functions. However, LDAP makes up for this loss of power in the following ways:

- Whereas X.500 requires special network access software, LDAP was designed to run over TCP, making it ideal for Internet and intranet applications.
- LDAP has simpler functions, making it easier and less expensive for vendors to implement.
- LDAP encodes its protocol elements in a less complex way than X.500, thereby streamlining coding/decoding of requests.
- LDAP servers return only results—or errors—which lightens their burden.
- LDAP servers take responsibility for "referrals" by handing off the request to the appropriate network resource. X.500 returns this information to the client, which must then issue a new search request.

Although the concept of LDAP enjoys widespread industry support, there is incompatibility among LDAP-compliant applications because the standard does not specify a consistent naming scheme for accessing directories by such fields as name, address, phone number and e-mail address. So vendors have been using different ways for storing and maintaining this information. This problem is being addressed by the Lightweight Internet Person Schema (LIPS), an industry initiative headed by the Network Applications Consortium (NAC).

Lightweight Internet Person Schema

LIPS is designed to ensure easier implementation of LDAP through the definition of common terms for attribute names and content.

For example, a messaging client may want to browse an LDAP directory to retrieve a name and phone number. Without LIPS, one server could define "phone number" as a field called PHONE with a length of 10 characters, and another vendor could define the field as BUS_PHNE with a length of 20 characters to accommodate international numbers. LIPS solves this problem by defining the field name, size, and acceptable characters (syntax), for 37 common attributes.

This is not intended to be an exhaustive list of attributes; in fact, most directories have far more than 37 fields. LIPS presents a baseline schema, containing only the minimum common fields that loosely define an individual.

By adhering to these standardized attributes, client software vendors will be able to produce server-independent products using the LDAP standard. To be fully compliant, a vendor must expose all of the LIPS attributes with the given field names and minimum sizes (larger values are allowed). However, there is no requirement that the attributes contain any data.

SUMMARY ▉ ▉ ▉ ▉ ▉ ▉ ▉ ▉ ▉

The X.500 standard was first approved in 1988 and enhanced in the 1993. An ambitious standard, X.500 is still too complex for most directory implementations. The University of Michigan developed a simpler version of DAP, known as the Lightweight Directory Access Protocol, for use on TCP/IP nets, including the Internet. LDAP offers much of the same basic functionality as DAP and can be used to query data from proprietary directories, as well as from an open X.500 service. Most major suppliers of e-mail and directory-services software support LDAP, which is now the de-facto directory protocol for the Internet.

See Also **Policy-Based Management**

Document Imaging Systems

Organizations of all types and sizes are reaping efficiency improvements and productivity increases by implementing document imaging systems. LANs and WANs facilitate the exchange of imaged documents between local processing points and remote locations, thus eliminating the delay inherent in traditional paper-based systems. In fact, the automation of document workflow can be compared to an assembly line whereby repetitive, paper-intensive tasks are performed at image-enabled workstations. There, resource allocation can be monitored and productivity improved through centrally controlled, automated document distribution and routing processes.

Applications

Document imaging technology is widely used in supporting banking and financial services, government, health care, insurance, manufacturing and utilities, and retail operations. In general, document imaging technology is best applied to processes that fit any of the following situations:

- Processes that contribute to the organization's core functions and those processes whose need for improved quality and timeliness is readily apparent
- Processes where workflows involve repetitive tasks, which lend themselves to automation
- Processes in which the time spent in paper handling can be dramatically reduced
- Processes for which a significant positive return on investment is likely

System Components

A basic imaging system typically includes the following components (Figure 31):

SCANNERS (AUTO- AND MANUAL FEED) Large, high-speed production scanners are auto-feed; small, desktop systems employ a manual feeder. Many scanners integrate OCR capabilities for reading text and data.

WORKSTATIONS These are used for a variety of functions such as quality assurance, indexing, data entry, OCR character reject correction, and system administration.

SOFTWARE These manage the system's functionality and include workflow administration, compression/decompression, database storage and retrieval, communications, and other imaging-specific applications.

OUTPUT DEVICES These include printers, and are used to generate production reports and operational statistic reports and to produce hardcopy facsimiles of stored images.

STORAGE MEDIA This is determined by the nature of the data being stored and the access time required for the application. These can include online (quick access), near online (moderate access time), and offline (ac-

Figure 31
The components
of a simple
document
imaging system.

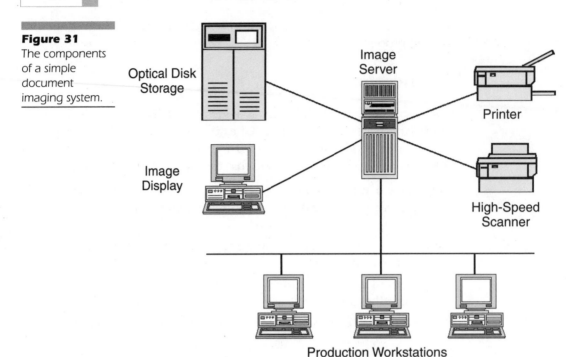

cess could range from hours to days). Common storage devices include magnetic disk, optical disc (WORM and Rewritable), and magnetic tape.

As processing requirements increase, so does the size and complexity of this simple configuration. An expanded version can include multiple scanners, multiple servers, and large numbers of imaging workstations divided across multiple LANs that connect to a high-speed backbone such as FDDI, Fast Ethernet, or ATM.

Image Processing

The process of converting paper documents to images starts with the scanning of internal business documents, including all incoming correspondence and customer forms arriving by facsimile, postal mail, and other sources. Capture software controls the speed and quality of scanning, and automates such tasks as image enhancement, document separation, document assembly, and indexing. In some cases, the capture software even recognizes barcodes, Web forms, and EDI documents. The software also stores all documents into a database in various formats, including plain text and HyperText Markup Language (HTML) for access via the Web.

Special workflow management software allows an administrator to control the flow of documents from workstation to workstation, check the work of various operators, and reroute documents for special handling. Some imaging systems even allow the documents to be updated by different users, whereupon the updated document is filed in order of last update first. Thus, a complete history can be maintained in an electronic case file.

When a wide-area network (WAN) connects geographically separate LANs, the benefits of imaging can be extended throughout the entire enterprise. The size of image files versus other types of traffic normally traversing LANs and WANs, however, can slow access and response time for all applications. In such cases, performance can be improved by creating workgroup subnets to isolate traffic, upgrading the network with more bandwidth, or by implementing a cache system that allows frequently accessed documents to reside in local storage.

Another optimization technique relies on a process called "forms removal" to improve compression ratios already applied by the capture software. When scanning forms, it is the data that is important—not the form itself. In eliminating the repetitive forms and saving only the data, compression ratios can be vastly improved. A dental claim form, for example, scanned at 300 dpi and compressed according to the international Group 4 standard, typically uses 73 Kbytes of disk storage. When the form is removed, the same image file occupies only 13 Kbytes in compressed form. When the form is called up for display on a workstation monitor, the data is overlayed on an image of an empty form. The file containing the empty form can be cached (i.e., stored locally), while the data can be retrieved from a database server.

Still another optimization technique is the pre-fetching of images through workflow software. This allows the system to batch-retrieve images overnight when the system is used the least. The images are moved from the server's optical media, where images are permanently stored, to local magnetic media, which provides a faster access time for workflow operations. Caching is implemented based on an understanding of which images are likely to be required next. This technique is effective in workflow applications where there are queues of images to be worked on. The vendors' storage management utilities are used to implement image caching.

Dual Deployment

Today's document imaging systems can be deployed in "thin client" Web browser environments, providing users such as remote offices, customers,

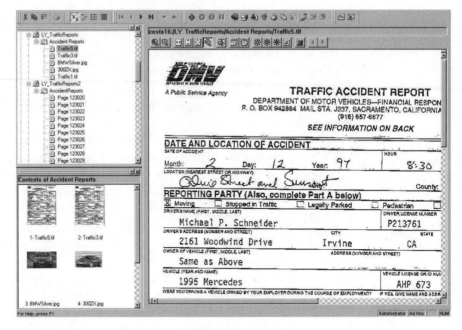

business partners and suppliers with easy access to corporate document libraries and allowing them to participate in document-centric workflow processes. Additionally, they can be implemented using the same software components in a "thick client" Microsoft Windows environment (Figure 32).

The software's dual deployment capability enables organizations to develop applications that meet specific business needs and decide on the appropriate implementation vehicle—Windows, Web browser or a mix of both client types independent of underlying technology. This approach increases user flexibility and enables organizations to react quicker to customer demands in traditional and electronic commerce business environments.

Planning Considerations

The success of document imaging often hinges on how well the system's integrators understand business processes and workflows. For them, the most obvious task is to learn how employees do their jobs and solicit employee input on how things can be improved. The participants should settle procedural problems early so that workflows can be properly scripted for automation and documents routed to the appropriate workstation operators. Mapping existing departmental procedures usually reveals processes that duplicate effort, employees working at cross-purposes, and unnecessary paperwork and filing requirements.

The key planning undertaking is an evaluation of document processing needs. At minimum, this should include the following:

- Listing all documents that are currently being processed in the course of business

- Determining how many of each type of document arrives at each business location daily, weekly, monthly, and yearly

- Finding out who provides the data in the documents, the purpose of the data, and what information systems or applications currently use the data provided in the documents

- Establishing which information in each document is most frequently used

- Determining appropriate index fields for each type of document

- Preparing a flowchart for each type of document that shows the path it follows when it is received in the office, the stops it makes, what happens at each stop, and what alternative paths exist

To properly implement document imaging, a company must first understand its business processes to ensure that the investment in technology is targeted wisely. This entails a thorough evaluation of business processes and workflows. Often third-party assistance can discover ways to streamline business processes, or validate (or invalidate) the conclusions of corporate staff. The high cost of imaging system implementation justifies these extended measures.

SUMMARY

Despite the large investments organizations have made in office technologies over the years, productivity gains have been hampered by the huge quantity of paper these technologies have tended to produce, making the benefits of the "paperless office" elusive. With document imaging systems, the problems of paper overload and workflow can finally be overcome to provide significant opportunities for reengineering work processes and streamlining management structures.

See Also **Data Compression, Groupware, Workflow Applications**

Electronic Commerce

The term *electronic commerce* (e-commerce) has been used to describe a number of technologies that have appeared over the years, the oldest being electronic data interchange (EDI), which is still used by many businesses today. Automatic teller machine (ATMs) arrived on the scene a few decades later, paving the way for electronic banking. One of the latest additions to the electronic commerce family include secure online transactions over the Internet through virtual storefronts set up on the Web.

The popularity and ubiquity of the Internet has awakened companies to the numerous commercial opportunities that entail the provision of goods and services to a vast global marketplace. Increasingly, consumers are coming to feel that their leisure time could be better spent somewhere other than a shopping mall. If they already have a computer and Internet connection and know where to go on the Web for recognizable brands, they can order products and services with only a few clicks of the mouse. What was once a full-day outing is now reduced to a few minutes of online browsing at a virtual storefront. In the process, consumers can avoid the usual problems associated with current shopping channels, including traffic; unattractive stores; indifferent, rude, or inexperienced sales help; and stores that lack clear market differentiation.

There is no standard way to build e-commerce sites on the Internet. A majority of such systems in use today have been built from scratch with PERL (Practical Extraction and Reporting Language), TCL (Tool Command Language), or some other derivative of the C programming language. Java, the latest derivative of C, may eventually overtake all the other languages for building e-commerce applications now that the security issues have been addressed. Simple e-commerce applications can even be pieced together quickly with either JavaBeans or ActiveX Controls, or some combination of the two.

Graphical Development Tools

Instead of hand-coding e-commerce applications, professional results can be achieved with shrink-wrapped graphical development tools. Easy-to-use wizards and templates even enable novice computer users to create and maintain a professional, commercial Web site without knowing any HTML or scripting languages. As the design progresses, the code is generated automatically in the background. When the site is ready to go live, the code is transported to appropriate directories on the Web server via file-transfer protocol (FTP).

Such tools can greatly lower the cost of entering the e-commerce marketplace, especially for small- to medium-size businesses whose operating budgets are already stretched to the limit. Among the vendors offering such tools are Forman Interactive Corp. The company's E-Commerce Edition of Internet Creator provides sample templates that can be used to build a professional Web site in less than an hour. In addition to a shopping cart, the product's capabilities include:

SEARCH Provides searches based on the products offered at the Web site, giving customers the ability to find what they are looking for quickly.

SECURE TRANSACTIONS Automatically configures the Web site for secure credit-card transactions, so customers do not have to worry about providing their credit-card numbers.

CUSTOMIZED ATTRIBUTES Lets the developer assign attributes to the products, thus distinguishing them by such things as model, size, or color.

ORDER FORM GENERATOR Builds custom forms to capture order information and tracks data about customers to find out exactly what they are looking for.

SHIPPING CHARGE CALCULATOR Calculates shipping charges using four different methods (shipping weight, shipping value, quantity shipped, and per-item).

To start a project, the developer opens Internet Creator (Figure 33). The left pane provides a tree view showing Web-site elements as icons in a hierarchical format. Elements such as image file, table, form, text paragraph, link, and Java applet are represented by icons in the left pane. When opened, each icon displays subitems. The right pane shows the browser view, which displays the pages of the site as they will appear in a real Web

Figure 33
The main window
of Forman
Interactive's E-
Commerce Edition
of Internet
Creator, showing
the two panes
and toolbars.

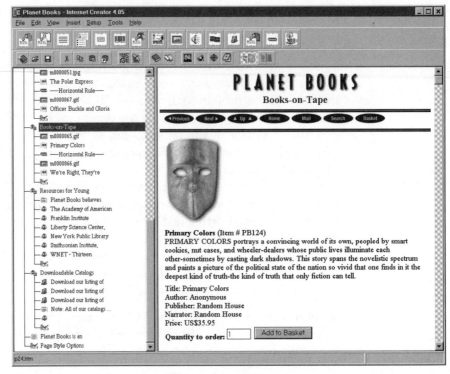

browser. An editing toolbar provides point- and-click access to the various page editing functions. A paragraph toolbar provides access to 15 controls that represent the types of available paragraphs, including text, audio or video file, table, form, applet, and external file.

The first step in setting up a site for e-commerce is to select the Commerce Wizard from the Setup menu and choose between two distinct ordering systems: E-mail order forms or Interactive Shopping Basket. For the latter, the Commerce Wizard guides users in setting up the basic options for the shopping basket, as well as the shipping options.

The wizard also asks users for such things as FTP account information, the preferred currency and weight symbols for the transaction items, the path to the shopping basket program (usually /cgi-bin/shop.cgi), and a list of item feature search fields. A bookseller, for example, might use search fields such as title, author, publisher, and category to enable visitors to search for books based on these criteria.

The wizard also asks for the credit cards and payment methods that will be accepted. A disclaimer can appear on all ordering screens. This allows the user to add text such as "All orders are shipped with 24 hours" or "All our products have a 90-day money-back guarantee."

The wizard also allows users to select the type of shipping calculation to apply to the order. Shipping charges can increase (or decrease) based on an order's total weight, total value, or total number of units shipped. A shipping charge can also be set up for each individual item, or no shipping charge can be applied whatsoever. Users can enter any number of carriers and delivery options for customers to choose from. A carrier can be selected from the drop-down list, and specific information for that carrier can be filled in. For example, it can be specified that orders under 10 pounds cost $15 to ship and orders 10 pounds and over cost $25 to ship.

Internet Creator supports a variety of server-based features, including a shopping basket, a private newsgroup for developing an online community, e-mail, and search forms that enable customers to quickly locate items of interest. To set up these options, users select the Server Wizard from the Setup menu and follow the prompts to enter server-specific information such as the host name, account name, password, and host directory. There are also options for setting up e-mail, newsgroups, and searches. For example, users can choose to implement the standard Internet Creator e-mail form, a custom e-mail form, or a simple e-mail link to the browser's e-mail facility. And they can choose to display a newsgroup button on every page or just the home page. A search button can appear on every page or home page, and be set for full-text searches or orderable item searches.

When the finished application is ready to be published, an FTP wizard allows all the files to be transfered onto a Web Server. Simply click on the tool's upload button, and Internet Creator will open an FTP connection and begin to upload the site. If the files were compressed to speed the upload process, Internet Creator will automatically decompress them when they arrive at the server.

Making Online Purchases Easier

While graphical development tools make it easier for entrepreneurs and businesses to build e-commerce Web sites, this is not necessarily the final step in the process. The online purchasing process must be made easier, which means selecting tools that integrate with other products and services.

For example, while many customers are accustomed to having their credit-card information traverse the Internet, many more would-be customers prefer not to divulge this information, no matter what protocols are used to make the transaction secure. Therefore, e-commerce developers must establisyh alternative billing mechanisms and embed them into their Web sites.

One innovative billing system allows customers to apply charges from Internet transactions to their telephone bill. This method of payment is expected to extend the convenience of electronic commerce to millions of consumers who do not feel comfortable using credit cards for online payments.

The eCharge system, offered by eCharge Corp., can be used for purchases of digitally deliverable goods and services for up to $50, including subscriptions to online media, software downloads, information services, online games, and technical support. Online shoppers inside North America will be able to use eCharge to purchase virtually anything on the Web as long as the purchase price does not exceed a credit limit set by their local phone company.

First-time users download the free eCharge software as a one-time set-up. After this step, consumers are able to immediately use eCharge for purchasing from properly equipped merchant sites. Intershop is among the first e-commerce development tool vendors to integrate the eCharge system into its product, Merchant Edition.

Other tool vendors have integrated CyberCash's Agile Wallet to expedite online purchases. This is a Java implementation that overcomes the problems associated with the company's old program—long download times and complicated installation. In addition, entering information tended to disrupt the shopping experience.

Now CyberCash uses an in-purchase registration procedure called InstaBuy that lets customers register and load a wallet. With this procedure, there is no need for customers to load credit-card information before shopping.

With Agile Wallet, consumers never need to reenter their payment and shipping information for Internet purchases (Figure 34). During their first purchase from any Agile Wallet-enabled Web site, consumers type in the necessary payment information, including their name, billing address, shipping address, credit-card number, and expiration date.

This information is saved on a secure server during a customer's first Agile Wallet purchase, and it is recalled the next time the customer visits that merchant or any other Agile Wallet-enabled merchant (Figure 35). Since consumers can use the same wallet and the same password at multiple merchant sites, with the assurance of secure and private storage of their financial information, they get the convenience of one-click shopping. The CyberCash payment system uses 1024-bit RSA and 56-bit DES encryption. Both are hard cryptographic solutions that offer industrial strength security for payment transactions.

Figure 34
For first-time purchases, the Agile Wallet securely stores payment information entered by the shopper into a Java applet.

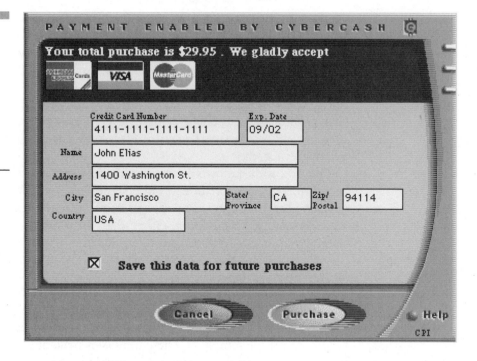

Figure 35
For repeat purchases, the shopper's payment information is automatically retrieved from a Java applet downloaded from a secure server. A single click is all that is needed to complete the purchase.

Consumers never need to enter their purchase information again—whether they are buying from the same merchant or any other merchant using Agile Wallet. Instead, the Agile Wallet applet pops up within the merchant's payment page already loaded with the shopper's purchase information. The shopper simply verifies the information and completes the purchase with a single click. When required, consumers can easily change addresses and credit cards, or even add other payment options—such as debit cards, smart cards or electronic checks—to their protected personal information.

The Java wallet motifs are consistent between merchants that use the program. This lets consumers understand that Agile Wallet is part of a common service provided by CyberCash to give them extra convenience and security when shopping, rather than an invasion of privacy perpetrated by an unscrupulous Internet merchant.

Storefront Services

For companies that do not have the resources to build their own e-commerce applications, there are services that simplify the building process by providing online templates. These templates make it easy for users to enter product descriptions, pricing information, and digitized images. Once the virtual store is built, the owner can come back any time to make changes and improvements or retrieve orders.

There is no hardware or software to buy; all users need are a computer, modem, and Internet browser. The service provider then hosts the resulting application on its own Web server for a monthly fee. Another benefit of this type of arrangement is that users can immediately take advantage of enhancements and upgrades to the software used by the service provider.

SUMMARY

Among the most important requirements for e-commerce site design are getting a site up and running quickly; creating unique virtual storefronts instead of being confined to vendor templates; and easily gathering statistics, analyzing trends, and quickly updating the storefront in response to changes in the virtual marketplace. Although today's graphical development tools allow nonprogrammers to design and implement e-commerce Web sites, they differ in their capabilities to meet these needs. While some tools allow entrepreneurs and small businesses to set up shop quickly and professionally through the use of wizards and templates, larger companies will require sophisticated tools that are more flexible and extensible.

See Also **Electronic Data Interchange, Electronic Mail**

Electronic Data Interchange

One of the oldest e-commerce technologies is Electronic Data Interchange (EDI), which has been around in one form or another for more than a decade. EDI provides a standard format for exchanging things such as inventory reports, purchase orders, shipping notices, and invoices between trading partners.

Traditionally, EDI has been used by large corporations to link suppliers to their financial systems for processing just-in-time (JIT) supply orders to keep manufacturing assembly lines moving while minimizing warehousing costs. A network of private lines provided the necessary links between suppliers and the company's order management system, which automated transaction processing.

An alternative to expensive private lines is for the participating companies to subscribe to a value-added network (VAN) service that supports EDI. With this type of service, the network provider assumes responsibility for transmitting, controlling, logging, auditing, and archiving all messages through a central clearinghouse. In addition to supporting the key standards that make EDI work—such as the Electronic Data Interchange for Administration Commerce and Transport (EDIFACT) protocol, which standardizes message formats—most VANs offer security features such as receipt notification and public key cryptography to guarantee confidentiality, authentication, integrity, and nonrepudiation of transmissions. Systems integration and user training are part of their EDI services as well.

Whether private lines or VAN services are used for EDI, many smaller companies find these methods of participation too expensive. And even though EDI has migrated from the mainframe environment to more popular UNIX servers, it is still too technically challenging for many smaller companies to implement. Rather than be bothered with getting all the components operating properly, many smaller firms do not concern themselves with EDI, preferring to put their money into traditional capital improvements.

EDI over the Web

The fundamental problem with EDI is that it tends to be exclusionary—only large firms or businesses with a high volume of transactions find EDI worthwhile; everyone else has to be dragged kicking and screaming into the fold, usually at great expense. This is precisely the problem that the Web can solve—it can make EDI easier to use, cheaper to implement, and more accessible to a broader base of potential users. In fact, the Web makes EDI worthwhile even for small, infrequent transactions. EDI can be accessed with the familiar Java-enabled browser.

Large companies in which each line of business has its own platform on which to run its order management system have found Java especially useful. Instead of trying to design complicated mapping schemes to get EDI information into multiple systems or even standardizing on one or two systems, the order-management application can be written in Java, which yields many benefits.

Because Java applications are interoperable across platforms, it is no longer necessary to implement wholesale system replacements to simplify order management. Java can even dispense with the need to develop a neutral format in order to accommodate both corporate and external customer requirements. By allowing customers to exchange business documents via the Web using Java for forms processing, any company at any location can reap the benefits of EDI with minimal disruption to internal operations.

Java for EDI

Java-based EDI applications are downloaded to the remote computer only when needed. This not only gives users access to the information they need, but simplifies the software maintenance tasks of network administrators, since emulators and other applications reside only in one place—a Web server. To update the software, only the version on the e-commerce server needs to be updated; since each business partner downloads the latest application when it has a transaction to send, there is no need to send new software to each partner.

In most cases, the EDI application is stored in hard disk cache at the client location; in others, they are stored in cache memory. Either way, the EDI application does not take up permanent residence on the client machine. Since EDI applications are delivered to the client only as needed and all maintenance tasks are performed at the server, users are assured of access to the latest application release level. This not only saves on the cost of software, it permits companies to get away with cheaper computers, since every computer need not be equipped with the resources necessary to handle every conceivable EDI application.

Hybrid Approach

Since large companies have invested considerable time and money in EDI mapping and translation software—software that converts back-end business data into a common EDI format for sharing across value-added networks—they are not likely to abandon systems that work perfectly well just to venture into the unknown territory of the Internet. In this case, a hybrid product that makes use of either VANs or the Internet offers a viable solution.

The hybrid approach can be implemented with a UNIX-based intelligent messaging hub that pores through EDI messages sent over the VAN or the Internet and automatically routes orders electronically to the appropriate department or division. The messaging hub accepts information in any format the customer uses, including Simple Mail Transfer Protocol (SMTP) or Multipurpose Internet Mail Extensions (MIME). This approach allows large businesses to expand their electronic commerce efforts to include smaller companies, but without disrupting existing trading relationships with their peers.

For companies with their own intranets, extending electronic commerce to the Internet is a natural step. They are in a good position to develop, test, and experiment with e-commerce applications and decide whether it makes sense to convert data coming off the Web into EDI format or other middleware, or send it directly into back-end systems. Since the Internet represents a potentially serious security threat, however, a firewall is necessary to prevent unauthorized access to corporate resources. In addition, X.509 certificates and encryption technology can be used with trading partners to safeguard the integrity of the transactions.

The hybrid approach will become more important as companies learn to compete in the global economy. In Eastern Europe and many countries in Asia and Latin America, where the communications infrastructure is not very advanced, the Internet is viewed as a crucial, low-cost vehicle for the transport of business information. In many cases, the easiest way to reach emerging markets in these areas is through the ubiquitous Internet. Documents can be easily exchanged and orders placed using virtually any Java-enabled Web browser.

The hybrid approach can even be used to send nonsensitive, low-priority information over the Web and sensitive, high-priority information over the VAN. For example, a company can receive shipping status information from its freight carriers by means of the Web and still send more sensitive purchase information and bank payments over a VAN.

Since VANs charge according to usage, large companies with high transaction volumes stand to save a lot of money. VAN vendors, including IBM's Advantis business unit and General Electric Information Services, can charge $150 per hour or more. For large companies, such as Mobile Corp., VAN charges can easily exceed $100,000 a year. By contrast, the cost of an Internet connection can be as low as $1 per hour.

If private lines are used for EDI, which are billed for at a fixed monthly rate, using the Web to exchange routine business documents can alleviate periodic bottlenecks that make it unnecessary to add more lines, thereby saving even more money.

Packaged Solutions

Companies large or small do not have to devise their own hybrid solutions. EDI software is now available that lets Windows users send and receive EDI data over the Internet or a VAN and integrate that data into business applications using Java-based forms.

One such product is the TrustedLink Commerce from Harbinger Corp. The product line, which runs on Windows systems, is available in three configurations. The base product, TrustedLink Commerce, is designed for small and midsized companies that exchange business documents via a value-added network. TrustedLink Commerce Internet allows customers to send and receive EDI messages securely via the Internet. TrustedLink Commerce Integrator combines TrustedLink Commerce with a mapping tool for importing and exporting data to and from user software applications, such as order entry and accounting.

The mapping tool, TrustedLink Mapping Workbench, allows users to map EDI data into a format that their business applications can understand, in which fields can be dragged and dropped from EDI documents and associated with database or application files. When used with TrustedLink Commerce, these maps automatically translate EDI data into application data, and application data into EDI forms. This eliminates the need for users to rekey data from their business applications into their EDI software.

The product is aimed at smaller companies that are typically pushed by their larger trading partners into conducting business via EDI. The appeal of Harbinger's EDI software is that users can run it on any Windows platform. Companies that send only a few hundred EDI documents per day will likely use the Windows 95 version of TrustedLink Commerce. If the EDI volume is 1,000 or 2,000 documents per day, they will probably want to go with the Windows NT Server version of the product.

Internet-Based EDI Services

GE Information Services, a traditional VAN service provider, offers a global Internet-based EDI service called GE TradeWeb. Subscribers can register for the service using online terms and conditions that conform to their country's laws and in their native language. Subscribers can also pay for their subscriptions in the currency of their country.

Subscribers to the GE TradeWeb service receive an EDI mailbox, access to the online EDI forms library, and access to the trading partner directory. These capabilities, coupled with the simple interface of GE TradeWeb, brings EDI within the reach of companies of all sizes, enabling them to exchange basic business documents online, thus lowering their costs, removing complexity, boosting their efficiency, and enabling them to compete globally.

Certain Internet-based VANs offer EDI services to particular industries. For example, MessageXpress, a subsidiary of DAC Services, provides electronic commerce to fleet operators in transportation and allied industries. One of these e-commerce services is Web Basic EDI, which enables any company with Internet access to be EDI-compliant. EDI forms are designed to hub specifications and placed on a secure MessageExpress Web site. Senders access the site by entering a user name and password. Sending and receiving EDI is intuitive and straightforward—simply a matter of filling in an online form and clicking on the send button at the bottom (Figure 36). The contents are sent to trading partners in an EDI-compliant format.

SUMMARY

The objectives of EDI implementation include replacing labor-intensive paper-based processes, reducing administrative costs, and speeding the flow of orders and payments between businesses. Traditionally, these and other benefits have been achieved by linking trading partners over private lines or through value-added networks. But these methods are expensive and complicated, which effectively excludes many smaller companies from participation. With products and services that allow EDI transactions over the Internet, however, smaller firms can now participate in EDI, which opens up new business opportunities with larger firms. Even traditional VAN providers are starting to offer Internet-based EDI services.

See Also **Electronic Commerce**

Electronic Mail

Electronic mail (e-mail) is a method for transmitting and receiving text, data, and images over a network. The advantages of e-mail include fast delivery and cost savings compared to regular postal mail, overnight carrier services, and traditional facsimile (fax).

Several kinds of e-mail systems are in use today. There are the internal client/server e-mail systems used by large corporations, which operate over

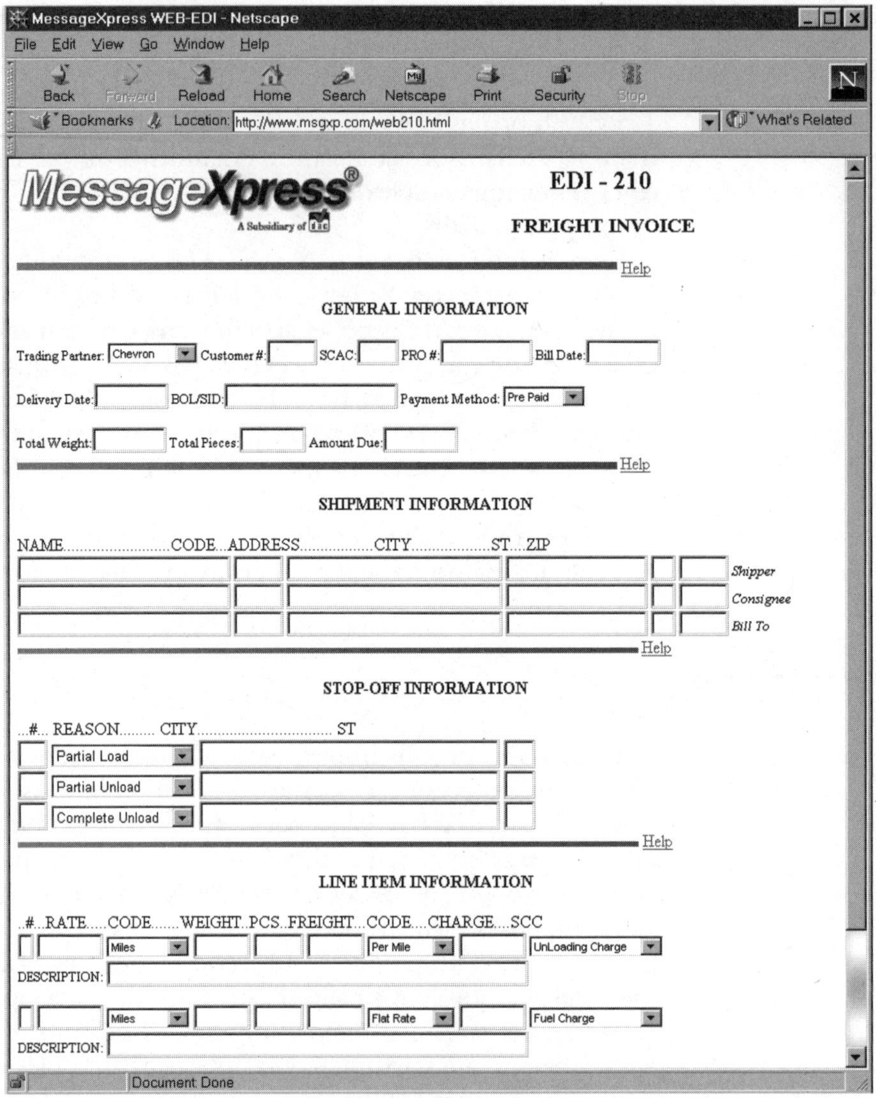

Figure 36

A MessageXpress
Web Basic EDI
form for a freight
invoice.

local area networks (LANs). There are also mainframe- and minicomputer-based e-mail systems in the corporate environment, but these have largely given way to client/server e-mail systems with the transition to LANs. On-line services and Internet service providers offer e-mail accounts as part of the subscription price for bulletin board and/or Web access. Many e-mail programs can be downloaded from the Internet as shareware (Figure 37). Public e-mail services are also offered by telephone companies and long-distance carriers, as well as wireless data service providers.

Electronic mail enhances business transactions by replacing paper forms and manual delivery systems with faster, more reliable computer

Figure 37
A popular full-
featured
shareware e-mail
program is
Pegasus Mail,
written by David
Harris.

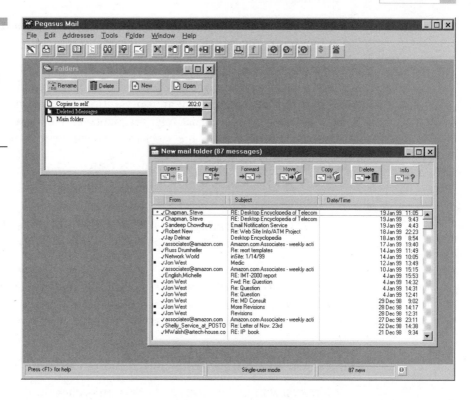

communications. With information stored in digital form—and conveyed electronically between computers and other communications devices such as personal digital assistants (PDAs), pagers, and wireless phones—messages can be easily edited and integrated into other applications. Just as important, e-mail can be accessed virtually anywhere, anytime. In addition to various wireless services that make this possible, mobile professionals can access and manage their e-mail through the Internet.

Electronic mail is also the key enabler of e-commerce on the Internet. A user might select items from an online catalog, for example, and then enter delivery and credit-card information into a form, which is e-mailed to the company's Web server for processing. Various e-commerce protocols make online purchasing secure and reliable.

Features

There are numerous e-mail products to choose from, and they are available from dozens of vendors. The most sophisticated products have reached the point where they now offer a fairly standard set of features, including:

AUTOFORWARDING Automatically redirects mail to another account. This is especially useful when recipients are away on vacation and someone else must handle their job responsibilities. Recipients can also receive a copy, even if the e-mail is autoforwarded to someone else.

AUTOMATED RESPONSES Lets recipients automate replies to messages, either selectively or globally. For every incoming message, recipients can have a reply sent indicating that they are on vacation and will not be checking e-mail, for example. Alternatively, recipients can choose to receive only high-priority messages or messages from specific e-mail addresses.

CLICKABLE E-MAIL ADDRESSES Automatically opens a message compose window when users click on an e-mail address appearing in the body of the message.

CLICKABLE URLS Automatically opens a Web browser and goes to the selected URL (Uniform Resource Locator) appearing in the body of the message.

COLOR-CODED MESSAGE LABELING Lets users apply different colors to messages listed in the inbox. Some products allow parts of a message—body text, URL or e-mail address, and reply text—to be color-coded as well.

COMPRESSION After messages are deleted from the various file folders, the e-mail program performs compression to reclaim disk space.

CONTEXT-SENSITIVE ONLINE HELP Lets users click on items in the user interface to read brief explanations of what functions they perform.

DEFAULT ROUTING Lets users specify the default address for recipients in the address book who have multiple e-mail addresses.

DISTRIBUTION LISTS Lets users create a list of addresses for mass mailings. This list has its own name and, when entered in the To: field, the message and attachment are sent to all addresses on the list without displaying the e-mail address of every person.

DRAFT Incomplete messages can be saved in a file folder, allowing users to continue working on them at a later time.

E-MAIL ALERTS The e-mail program can be configured so, when it is running in the background, an audio tone or visual alert signals users when new mail arrives.

EXCEPTIONS Allows users to exempt specified addresses from a mailing list, without having to create a new list.

EXPANDED TEXT Allows users to store abbreviations for commonly used text strings in a message, which can be expanded at any time with a single keystroke.

FILTERS Allows incoming messages to be identified by a variety of criteria so the appropriate action can be taken on them automatically. For example, nuisance messages from a specific e-mail address can be deleted automatically from the inbox. Filters can check for key words in the subject line, allowing certain messages to be saved automatically in a specific file folder or the e-mail address automatically added to a distribution list.

INCREMENTAL MAILINGS Allows users to break up very large lists to expedite mass mailings.

LIST SUPPRESSION In distribution lists, instead of having the e-mail address of every recipient listed, users can suppress them by substituting a list name.

MAILBOX MANAGEMENT Allows users to create any number of file folders to arrange incoming messages by such criteria as date, time, author, and subject.

MAPI SUPPORT A MAPI (Mail Applications Programming Interface) utility built into the e-mail program allows users to send files directly from MAPI-compliant applications, including word processors and spreadsheets, as e-mail attachments.

MULTIPLE ACCOUNTS MANAGEMENT Allows users to manage multiple e-mail accounts and switch between them via hot keys or the point-and-click method.

NESTED MAILING LISTS Allows users to embed other mailing lists (child lists) within another mailing list (parent list).

NICKNAMES Also called aliases, allows users to map the real names of individuals to their e-mail address.

POLL TIME Allows users to determine how frequently the e-mail program scans the server for new mail while the program is open and connected to the service provider.

REMINDER Also called a tickler, allows users to send messages to themselves with a specified delivery date to remind them of important occasions, deadlines, or events.

REMOVE EXPIRED MESSAGES Automatically removes messages from the server if the message is a specified number of days old and has already been read.

SCHEDULED DELIVERY Allows users to compose messages offline, save them in an outbox, and send them all at once automatically on a designated date or time.

SIGNATURE Allows users to create a standard signature or "sign-off" that is automatically placed at the end of every outgoing message. Multiple signatures can be created and used as appropriate for personal or business mail.

SPELL CHECKER Checks the spelling of text in e-mail messages before they are sent.

STYLIZED TEXT CAPABILITIES Lets users format messages using various fonts, colors, and sizes.

UNIX-TO-UNIX ENCODE/DECODE Some e-mail products include the capability to automatically UU encode/decode files without resorting to external utilities.

VOICE RECORDING Allows users to record an audio message and send it as an e-mail attachment.

Protocols

In the Internet environment, the two key e-mail protocols are the Post Office Protocol, version 3 (POP3) and the Simple Mail Transfer Protocol (SMTP). Although these are the most prevalent protocols for retrieving and sending e-mail, respectively, newer protocols overcome the limitations of POP3 and SMTP. The Internet Message Access Protocol, version 4 (IMAP4) will eventually take the place of POP3, while the Extended Simple Mail Transfer Protocol (ESMTP) will eventually take the place of SMTP.

POP3 POP3 is an Internet standard that defines a mechanism for accessing a mailbox located on a remote host machine (i.e., a server). This protocol is necessary because a user's computer, particularly a mobile computer, will not usually be online all the time and consequently will not always be available to receive mail directly from the Internet. A larger machine, such as a UNIX server, will typically be online all the time and so is better suited to receiving e-mail on behalf of its clients. When a client establishes a connection to the server, POP3 retrieves the new mail.

SMTP Whereas POP3 is used for retrieving new mail from a UNIX server, SMTP sends mail to the server and moves it among other servers on the Internet. SMTP plays two roles, sender and receiver. The sender acts as a client and establishes a TCP connection with a receiver, which acts as a server. During an SMTP session, the sender and receiver exchange a sequence of commands and responses. The receiver announces its host name and then the sender announces its host name, followed by the message originator and message recipient(s). The mail data is then transmitted until a line containing carriage-return and line-feed characters is reached, which indicates that the message is complete. SMTP adds the time the message is sent and keeps track of all the hosts that relayed the message, as well as the time that each received the message.

IMAP4 POP3 works best when only one computer is used for e-mail, since it was designed to support offline message access, wherein messages are downloaded and then deleted from the mail server. POP3 does not work well for mobile professionals who typically use a notebook or palmtop computer on the road and desktop computers at the office and home. This mode of access is not compatible with access from multiple computers since it tends to sprinkle messages across all the computers used for e-mail access, making management difficult. Unless all the computers share a common file system, the offline mode of access that POP3 was designed to support effectively ties the user to one computer for message storage and manipulation—which, for many mobile professionals, is impractical.

With IMAP4, users can request messages (both new and saved) that meet specific criteria, and manage them from more than one computer (Figure 38). Although IMAP was originally developed in 1986 at Stanford University, it has only recently become appreciated by mainstream e-mail vendors. The reason is that the problems of POP3 did not become apparent until fairly recently, with the dramatic increase in the number of mobile professionals, field personnel, telecommuters, and staff at small branch offices. Since these people typically use multiple computers, IMAP was rediscovered to address the problem of e-mail retrieval and management by a highly distributed workforce.

IMAP gives users more choices about what categories of messages to fetch from the server, where to store them, and what the server should do with specific categories of read messages. Among the key functions supported by IMAP4 that are not provided by POP3 are:

ACCESS AND MANAGEMENT OF MULTIPLE MAILBOXES This includes the ability to name and access different incoming and archive message folders, and the ability to list, create, delete, and rename them. These mailboxes can be on the same server or on different servers. An IMAP client may allow users to see the mailboxes at the same time, and move messages from one to the other.

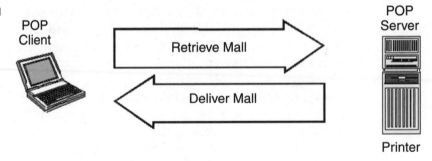

Store-and-Forward
- -
Client-Server

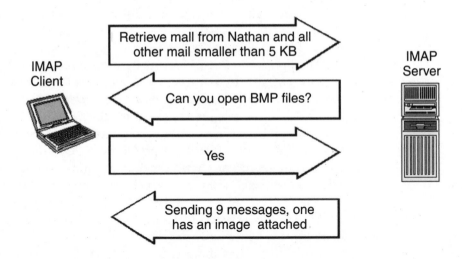

SUPPORT FOR CONCURRENT ACCESS TO SHARED MAILBOXES This capability is useful when multiple individuals are processing messages coming into a common inbox. Changes in mailbox state can be presented to all concurrently active clients.

MANIPULATION OF PERSISTENT MESSAGE STATUS FLAGS Messages in the mail repository can be marked with various status flags (e.g., "deleted" and "answered") and stay in the repository until explicitly removed by the user.

STORAGE FLEXIBILITY Users can save messages directly on either the client or server machine, or be given the choice of doing both as circumstances dictate.

ONLINE PERFORMANCE OPTIMIZATION This capability is especially useful over low-speed links and entails fetching the structure of a message without actually downloading it, selectively fetching individual message parts and using the server to search in order to minimize data transfer between client and server. With multimedia or multipart MIME (Multipurpose Internet Mail Extension) messages, for example, transferring selected parts of a message can be a huge advantage, for example when a user is in a hotel room and receives a short text message with a 5MB video clip attached.

DISCONNECT MODE In addition to allowing users to work in online and offline modes, IMAP supports a disconnect mode. Using this mode, the mail client can disconnect from the network and then synchronize with the mail server when the connection is reestablished.

ACCESS TO DATA OTHER THAN E-MAIL In addition to e-mail, users can uniformly access other classes of data, including documents in their native format.

ESMTP ESMTP is a collection of enhancements to SMTP, including those that improve its reliability and performance. For example, a feature called checkpointing reconstructs interrupted message transmissions when the connection is reestablished. Pipelining lets multiple message commands be sent at once. ESMTP also supports authentication, answering corporate concerns about the security of SMTP. These and other enhancements make SMTP more competitive with X.400 for linking business client/server e-mail systems on the LAN with those of public carriers, commercial online services and Internet service providers.

SUMMARY

While various proprietary electronic mail systems have been used for some time for internal communication within many organizations, Internet e-mail offers open communication on a truly global scale. Many businesses that have experienced the convenience and effectiveness of internal e-mail are now looking to the Internet as a way of extending its reach to local, national, and international locations. The X.400 gateway protocol solves many of the problems of linking and integrating dissimilar, geographically distributed messaging systems in support of e-mail and message-enabled applications.

See Also **Electronic Commerce, Fax Servers, Short Messaging Service, Unified Messaging**

Enterprise Networks

An enterprise network comprises all the interconnected resources that belong to a corporation, regardless of the network operating system (NOS), traffic types, protocols, transmission media, and applications at each location. Specifically, an enterprise network includes:

- All the devices connected over LANs or attached to legacy systems at every office location
- All the devices connected over LANs or attached to legacy systems
- The high-speed backbone within a building, campus, or metropolitan area that interconnects multiple LANs and legacy systems
- The facilities and services that extend these resources to remote users over a wide-area network (WAN), which can span international locations
- The management systems, help desks, directory services, security systems, and other tools that administrators use to keep the enterprise network operating at peak performance

Most enterprise networks have been cobbled together over the years to meet specific operational demands. Often data, voice, and video requirements were satisfied with different types of lines and services. This made network management difficult and inflated the cost of doing business. Now, corporate intranets that combine voice, data, and video traffic are becoming enterprise networks.

SUMMARY

In recent years, the trend has been to converge traditionally separate voice, data, and video networks over IP-based intranets, which companies had previously been using for routine tasks, such as e-mail, file transfers, and Web site access. The Internet Protocol (IP) packetizes any type of traffic, allowing voice, data, and video to share the same network and even be integrated into the same application. This convergence permits the development and implementation of new applications such as collaborative computing, unified messaging, and multimedia call centers. IP/PSTN gateways even allow traffic to pass between the intranet and public switched telephone network. *See Also* **Campus-Area Networks, Metropolitan-Area Networks, Wide-Area Networks**

Entertainment Networking

An entertainment network integrates PCs, TVs, audio/video components, and set-top devices in the home. This type of network offers all the benefits of traditional PC-to-PC home networking products, including file sharing, print sharing, and access to the Internet. In addition, it delivers entertainment to users wherever they are in the home. From any TV, users can:

- View and control VCRs, DVD decks, and video game stations
- Watch a DVD movie playing from a PC
- Move entertainment from satellite receivers or cable boxes
- Transform existing video camcorders into baby monitors or security cameras
- View and control the PC from any TV in the house, which allows them to run PC applications, check e-mail, surf the Web, and monitor the Web-surfing activities of children
- Use infrared transmission to move files between disconnected laptop computers and PCs on the network

All networked devices are interconnected through the CATV cable that provides programming to the television sets in various rooms of the home. The practical value of doing this is to be able to centrally control all of these systems through one device—either a computer or the television.

The HomeConnex system offered by Peracom Networks, for example, supports up to 16 simultaneous video sources, including PCs, on the network. Each video device becomes part of the network and each TV is capable of viewing every video device on the network. Four basic components comprise the HomeConnex system:

PC CASTER A hub device that insulates the home network from the outside world, providing high-quality pictures to every TV and ensuring that sensitive data on the network never leaves the home.

MEDIACASTER Connects video devices to the network via standard RCA audio/video connectors.

CABLECASTER Connects PCs to the network via the USB port.

WIRELESS REMOTE Combined with an easy-to-use graphical interface that pops up on every TV, the wireless remote controls each device on the network.

With central control at the television, for example, a different channel can be assigned to each video source, whether that source is a PC, security camera, VCR, game station, or another television in the house. Turning to that channel enables the assigned video source to be viewed and controlled. Not only can a PC's screen be viewed, for example, but any application or resource running on the PC can be viewed and controlled from the television, including Web browsing, e-mail, a multicast session, and chat. In addition, a DVD movie playing on a PC can be picked up by any television in the house, eliminating the need for separate DVD units at every television.

SUMMARY

In the future, audio sources will also be interconnected throughout the home using the same CATV cable. This will permit the creation of home intercom systems, piped music, and audible alerts for wake-up calls, schedule reminders, and system malfunctions. As more CATV networks are upgraded for bidirectional communication, it will even become possible to place telephone calls over the cable network and receive notifications of incoming calls at the television or PC through the display of caller ID information. And with the addition of a video camera, videoconferences can be set up with family, friends, or colleagues at the office.

See Also **Digital Power Line, Phoneline Networking, Powerline Networking, Small Office/Home Office LANs, Wireless Home Networking**

Ethernet

Ethernet is a type of local-area network (LAN) that operates at 10 Mbps. It originated out of experimental work done by Xerox Corporation in the 1970s. The co-inventor of Ethernet was Dr. Robert Metcalf, who founded 3Com Corp. in 1979. During 3Com's 20-year history, Ethernet has become the industry's most widely deployed computer connectivity solution. Much of the original Ethernet design was incorporated into the 802.3 standard developed in 1980 by the Institute of Electrical and Electronic Engineers (IEEE).

Ethernet is based on a bus topology that is contention-based, meaning that stations compete with each other for access to the network, a process that is controlled by a statistical arbitration scheme. Each station "listens" to the network to determine if it is idle. Upon sensing that no traffic is on the line, the terminal is free to transmit. If the channel is already in use, the station backs off and tries again.

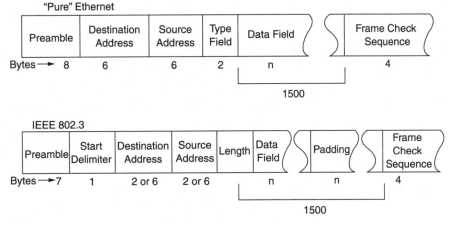

Figure 39
A comparison of
Ethernet frame
formats: "pure"
Ethernet (top) and
IEEE 802.3
Ethernet (bottom).

Frame Format

The IEEE 802.3 standard defines a multifield frame format, which differs only slightly from that of pure Ethernet (Figure 39):

PREAMBLE The frame begins with an eight-byte field called a preamble, which consists of 56 bits having alternating 1 and 0 values. These are used for synchronization and to mark the start of the frame. The same bit pattern is used for the pure Ethernet preamble as in the IEEE 802.3 preamble, which includes the one-byte start-frame delimiter field.

START FRAME DELIMITER The IEEE 802.3 standard specifies a start-frame delimiter field, which is really a part of the preamble. This indicates the start of a frame.

ADDRESS FIELDS The destination address field identifies the station(s) that are to receive the frame. The source address field identifies the station that sent the frame. If addresses are locally assigned, the address field can be either two bytes (16 bits) or six bytes (48 bits) in length. A destination address can refer to one station, a group of stations, or all stations. Pure Ethernet uses 48-bit addresses, while IEEE 802.3 permits either 16- or 48-bit addresses.

LENGTH COUNT The length of the data field that follows is indicated by the two-byte length count field, which determines the length of the information field when a pad field is included in the frame.

PAD FIELD To detect collisions properly, the frame that is transmitted must contain a certain number of bytes. The IEEE 802.3 standard specifies that if a frame does not meet this minimum length, a pad field must be added to make up the difference.

TYPE FIELD Pure Ethernet does not support length and pad fields, as does IEEE 802.3. Instead, two bytes are used for a type field. The value specified in the type field is meaningful only to higher network layers and is not defined in the original Ethernet specification.

DATA FIELD The data field is passed by the client layer to the data link layer in the form of eight-bit bytes. The minimum frame size is 72 bytes, while the maximum frame size is 1,526 bytes, including the preamble. If the data to be sent uses a frame that is smaller than 72 bytes, the pad field stuffs the frame with extra bytes. In defining a minimum frame size, there are less problems to contend with in collision handling. If the data to be sent uses a frame that is larger than 1,526 bytes, it is the responsibility of the higher layers to break it into individual packets in a procedure called *fragmentation*. The maximum frame size reflects practical considerations related to adapter-card buffer sizes and the need to limit how long the medium is tied up in transmitting a single frame.

FRAME CHECK SEQUENCE A properly formatted frame ends with a frame check sequence, which provides the means to check for errors. When the sending station assembles a frame, it performs a cyclical redundancy check (CRC) calculation on the bits in the frame. The sending station stores the result of this calculation in the four-byte frame check sequence field before sending the frame. At the receiving station, an identical CRC calculation is performed and compared with the original value in the frame check sequence field. If the two values do not match, the receiving station assumes that a transmission error has occurred and requests that the frame be retransmitted. In pure Ethernet, there is error correction; if the two values do not match, a notification of the error is simply passed to the client layer.

Media Access Control

Several key processes are involved in transmitting data across the network, among them data encapsulation/decapsulation and media access management.

DATA ENCAPSULATION/DECAPSULATION Data encapsulation is performed at the sending station. This process entails adding information to the beginning and end of the data unit to be transmitted. The added information is used to perform the following tasks:

- Synchronize the receiving station with the signal

- Indicate the start and end of the frame
- Identify the addresses of sending and receiving stations
- Detect transmission errors

The data encapsulation function is responsible for constructing a transmission frame in the proper format. A CRC value for the frame check sequence field is calculated and the frame is constructed. When a frame is received, the data decapsulation function is performed. The receiving station recognizes the destination address of the frame, performs error checking, and then removes the control information that was added by the data encapsulation function at the sending station. If no errors are detected, the frame is passed up the protocol stack to the client application.

MEDIA ACCESS MANAGEMENT A process called *media access management* or *link management* is responsible for several functions, starting with collision avoidance and collision handling, which are required for the proper operation of contention networks.

COLLISION AVOIDANCE Collision avoidance entails monitoring the line for the presence or absence of a signal (carrier). This is the "carrier sense" portion of CSMA/CD. The absence of a signal indicates that the channel is not being used and that it is safe to transmit. Detection of a signal indicates that the channel is busy and that transmission must be withheld. If no collision is detected during the period of time known as the *collision window*, the station acquires the channel and can safely transmit data.

COLLISION HANDLING When two or more frames are offered for transmission at the same time, a collision occurs, which triggers a sequence of bits called a "jam." All stations recognize the jam as a collision. At that point, all transmissions in progress are terminated. Retransmissions are attempted at staggered intervals. If there are repeated collisions, link management "backs off," which involves increasing the retransmission wait time following each successive collision.

On the receiving side, link management is responsible for recognizing and filtering out fragments of frames resulting from a transmission that was interrupted by a collision. Any frame less than the minimum size is assumed to be a collision fragment and is not reported to the client layer as an error.

New methods have been developed to improve the performance of Ethernet by reducing or totally eliminating the chance for collisions without having to segment the LAN into smaller subnetworks. Special algorithms sense when frames are on a collision course and temporarily block one frame, while allowing the other to pass.

Ethernet is the most popular type of local-area network, and its success has spawned some interesting innovations. 10BaseT Ethernet, for example, allows LANs to operate over unshielded twisted pair (UTP) wiring instead of thick (5BaseT) or thin (10Base2) coaxial cable. There are also high-speed versions of Ethernet, including Fast Ethernet at 100 Mbps and Gigabit Ethernet at 1 Gbps.

See Also **Fast Ethernet, Gigabit Ethernet**

Event Monitoring and Reporting

The ability to monitor network activity and receive alarms or reports of the occurrence of selected events helps managers keep the network operating at peak performance. Typically, an event manager agent tracks network activity, logs selected events, and automatically alerts the person responsible for responding to certain network occurrences. The following occurrences may be designated as events for reporting purposes:

- Changes to a system's hardware inventory
- Changes to a system's software inventory
- Application access failures
- Failed log-on attempts
- Job failures
- Connection failures
- No device response to polls
- Disabled protocols

Notification Methods

Event monitoring and reporting capabilities are integral features of most help desk, LAN administration, and enterprise management products. On a LAN, for example, the administrator can specify the network activity to be tracked. A notification feature can be set to alert the LAN administrator how each type of event should be reported. The following methods of event notification typically are available:

CONSOLE MESSAGES Display the name of the event and color-code the alarm conditions to indicate their importance. Some products use detailed graphical displays that help administrators manage alarms at the enterprise level (Figure 40).

Figure 40
Cabletron Systems
offers a highly
descriptive and
functional alarm
management
capability as part
of its SPECTRUM
Enterprise
Manager.

Figure 40
Cabletron Systems
offers a highly
descriptive and
functional alarm
management
capability as part
of its SPECTRUM
Enterprise
Manager.

E-MAIL The event level and name is sent in an e-mail message.

PAGING The event name, nature of the problem, and time of the event can be sent to a pager (Figure 41).

TWO-WAY PAGING Sends event notifications to a pager and accepts a response that can acknowledge the receipt of an event, escalate or demote the status of a problem, run diagnostics, or take corrective action.

TEXT-TO-TELEPHONE (OR CELL PHONE) Event messages are translated from text to voice for notification over the phone.

Figure 41
Alarms can be
reported, along
with a brief
description of the
event, to a pager.

INTERACTIVE VOICE RESPONSE   Allows events to be reported via telephone and acknowledged via a menu of responses, which may include launching a diagnostic routine or declining responsibility.

Different notification methods can be set as appropriate for each type of network event and the person responsible for that type of event. Events related to disk storage, for example, can be directed to a database administrator. Events related to unsuccessful log-on attempts can be sent to the security administrator for investigation of possible unauthorized entry. Events that show jabber on the network—indicating a faulty transceiver—can be sent to the help desk for resolution.

Some event management products, such as Telamon's TelAlert system (Figure 42), can do all this, plus be interfaced with motion detectors, water detectors, temperature and humidity monitors, and smoke detectors. Depending on the event, notifications can be sent to a security office, maintenance facility, or building manager.

Notification Priorities

Event notifications can also be processed based on priority level. For example, if three network events occur simultaneously, notification of the event with highest priority is sent first. A priority level may be a number from 1 to 9—with 1 indicating the lowest and 9 indicating the highest priority. When the LAN administrator specifies the network activity to monitor, a priority level for each event is assigned based on how crucial the activity is and whether or not someone has to be notified when the event occurs.

The LAN administrator can choose one or more contacts to receive notification of each event level. For example, a technician can be specified to receive a pager message when high-priority events occur, and an e-mail message can be sent to a help desk operator when routine application-related events occurs. Acknowledgment of the receipt of event data can be sent to the console to help ensure the proper response to events.

Some network monitoring tools use distributed, intelligent agents to gather protocol and network activity data on Ethernet and token-ring LANs. The data gathered by the agents is stored in a relational database where it is correlated for traffic analysis, billing, and report generation. With the ability to identify traffic loads, including which nodes are generating the most traffic, the resulting information can be used to charge departments for their share of the resources, including dial-out connections.

Even if such information is not used for charge-back, the network monitoring tool can still reduce costs and help administrators determine

Figure 42
The paging administration interface of Telamon's TelAlert system allows the network administrator to specify individuals or groups to receive event notifications and create appropriate text or numeric messages that will be automatically sent to their pagers.

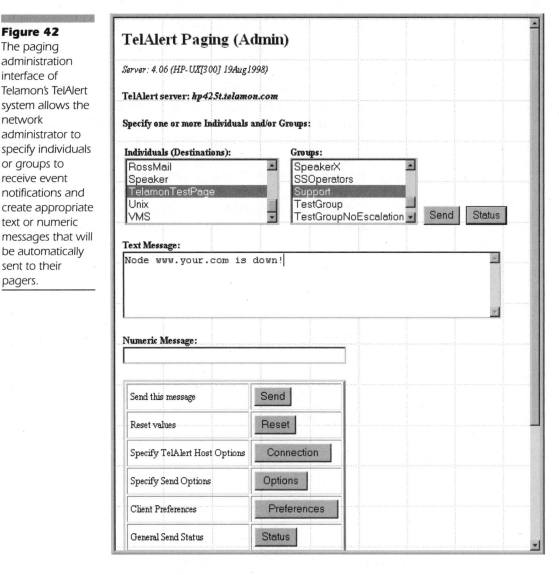

policies for more efficient network use. In addition, monitoring the network for predefined traffic thresholds on a particular LAN segment gives administrators the means to identify traffic patterns that could cause the network to crash. Traffic reports can even identify the need to change the network. If too many users on one or more network segments are logging on to different servers or using resources in another building, for example, a lot of backbone traffic can be created. With the aid of traffic reports, the network can be redesigned to alleviate backbone traffic and make sure bottlenecks do not occur.

SUMMARY

Event monitoring and reporting tools simplify problem notification, allowing responses to be more timely and corrective action to be more effective. They are most useful when integrated into help desk, LAN administration, and enterprise management systems. This allows a high level of automation in problem resolution. For example, trouble tickets can be automatically opened, populated with event information, updated, and even closed when an automated response resolves a problem. By providing interactive response capabilities, these tools increase the geographic range of technical support personnel. The result is greater network and systems availability and increased productivity throughout the organization.
See Also **LAN Management, Network Management Systems.**

Fast Ethernet

As computing becomes more distributed and accommodates multimedia applications, bandwidth becomes more in demand. Traditional Ethernet, which runs at 10 Mbps, has difficulty in meeting the demand for more bandwidth while providing acceptable response times. Although desktop computers have increased in performance and capabilities, the LAN to which they are attached has remained the same. A category of technologies called Fast Ethernet, which operates at 100 Mbps, has become available to break the LAN performance bottleneck. Of these, 100BaseT provides the smoothest migration path from 10BaseT Ethernet—the dominant network in use today.

Compatibility

100BaseT Ethernet uses the same contention-based media access control (MAC) method—carrier sense multiple access with collision detection (CSMA/CD)—that is at the heart of 10BaseT Ethernet. The 100BaseT MAC specification simply reduces the "bit time"—the time it takes for each bit to be transmitted—by a factor of ten, providing a 10× boost in speed. The packet format, packet length, error control, and management information in 100BaseT are identical to those in 10BaseT.

In the simplest configuration, 100BaseT is a star-wire topology with all stations connected directly to a hub's multiport repeater. The repeater detects collisions. An input signal is repeated on all output links. If two inputs occur at the same time, a jamming signal is sent out on all links. Stations connected to the same multiport repeater are within the same collision domain. Stations separated by a bridge are in different collision domains. The bridge would use two CSMA/CD algorithms, one for each domain.

Since no protocol translation is required, data can pass between 10BaseT and 100BaseT stations via a hub equipped with an integral 10/100 bridge. Both types of LANs are also capable of full duplex, can be managed from the same SNMP management application, and can use the same installed cabling. However, 100BaseT goes beyond 10BaseT in terms of media, offering extensions for optical fiber.

100BaseT also includes the media-independent interface (MII) specification, which is similar to the 10-Mbps attachment unit interface (AUI). The MII provides a single interface, which can support external transceivers for any of the 100BaseT media specifications. The following table summarizes the characteristics of 10BaseT Ethernet and 100BaseT Fast Ethernet:

	10BaseT	100BaseT
Speed	10M b/s	100M b/s
IEEE Standard	802.3	802.3u
Media Access Control	CSMA/CD	CSMA/CD
Topology	Star	Star
Cable Support	Coaxial UTP optical fiber	UTP optical fiber
UTP Cable Support	Categories 3, 4, 5	Categories 3, 4, 5
UTP Link Distance (max)	100 meters	100 meters
Network Diameter	500 meters	210+ meters
Media Independent	Yes (via AUI)	Yes (via MMI)
Full Duplex Capability	Yes	Yes

AUI: Attachment unit interface
CSMA/CD: Carrier-sense multiple access with collision detection
MMI: Media-independent interface
UTP: Unshielded twisted pair

Media Choices

To ease the migration from 10BaseT to 100BaseT, Fast Ethernet can run over Category 3, 4, or 5 UTP cable, while preserving the crucial 100-meter segment length between hubs and end stations.

Using fiber allows more flexibility in distance. For example, the maximum distance from a 100BaseT repeater to a bridge, router, or switch using fiber-optic cable is 225 meters (742 feet), but the maximum fiber distance between bridges, routers, or switches is 450 meters (1485 feet). And the maximum fiber distance between a fiber bridge, router, or switch when the network is configured for half-duplex is two kilometers (1.2 miles). By connecting together repeaters and other internetworking devices, large well-structured networks can be easily created with 100BaseT.

The types of media used to implement 100BaseT networks can be summarized as follows:

100BASETX A two-pair system for data grade (Category 5) unshielded twisted pair (UTP) and STP (shielded twisted pair) cabling.

100BASET4 A four-pair system for both voice and data grade (Category 3, 4, or 5) UTP cabling.

100BASEFX A two-strand multimode fiber system.
Together, the 100BaseTX and 100BaseT4 media specifications cover all cable types currently in use in 10BaseT networks. In addition, all of these cabling systems can be interconnected through a hub. This helps organizations retain their existing cabling infrastructure while migrating to Fast Ethernet, as shown in Figure 43.

A new development is wireless Fast Ethernet, which offers full-duplex Fast Ethernet links over point-to-point radio connections. While point-to-point wireless has been around for years, what is new is the speed and the transparent connection to Ethernet. Devices, including routers and switches, plug into the wireless system with standard RJ-45 connectors. The wireless system connects to a small roof-mounted dish to create 100-Mbps links between buildings located up to six miles apart. Security is maintained through proprietary authentication, whereby one system can only talk to another specified system.

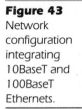

Figure 43
Network configuration integrating 10BaseT and 100BaseT Ethernets.

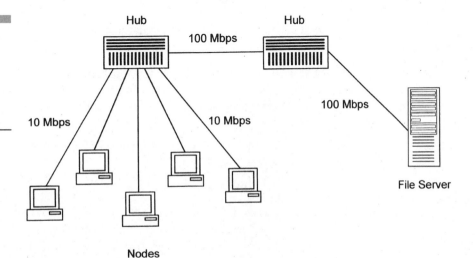

Network Interface Cards

Fast Ethernet NICs are available for both the client and server sides of a network. Both have a variety of features that affect price, but the server NICs usually cost more due to dedicated processors and increased onboard RAM.

Server-side NICs with onboard processors generally have better performance and lower host CPU usage. Some feature RISC-based microprocessors that run portions of the NIC driver, leaving more CPU cycles for the host. Server NICs should have at least 16KB of RAM. Adding more RAM reduces the number of dropped packets, which increases the network's performance. Some server NICs can be equipped with as much as 1GB of RAM. The choice depends on how busy the server is with other processes.

Fault tolerance and load balancing are two important considerations when selecting a server-side NIC. A fault-tolerant NIC comes with a failover driver that diverts traffic to another NIC when it becomes disabled. Some vendors have taken fault tolerance one step further, balancing the load across multiple NICs.

Server-side NICs and hub modules can be equipped to support a vendor's proprietary features. For example, 3Com Corp.'s server-specific Fast Ethernet cards can encapsulate token-ring traffic, allowing a Fast Ethernet backbone to carry token-ring traffic between stations. Bay Networks (now owned by Nortel) offers a Fast Ethernet module for its Multimedia Switch that maps Ethernet-based QoS levels to ATM QoS so Ethernet end stations can access voice and video already traveling over an ATM backbone.

Competing Solutions

While 100BaseT is the most popular implementation of Fast Ethernet, there are two competing alternatives: 100BaseVG and 100VG-AnyLAN.

100BASEVG Another standard for 100-Mbps transmission over copper wire is 100BaseVG (voice grade), originally developed by Hewlett-Packard and AT&T. It does away with the MAC Carrier Sense Multiple Access/Collision Detection layer and replaces it with another technique called *demand priority* and a signaling layer that ostensibly makes the network more secure and efficient.

Advocates of 100BaseVG note that CSMA/CD was originally designed for a bus topology and so it had to have the collision-detection mechanism that today's Ethernet networks provide. However, because most users have moved to the hub-based star topology such collision detection is outdated. Demand priority accommodates this reality by making the hub a switch instead of a repeater, which makes for a more efficient and secure network.

With CSMA/CD, nodes contend for access to the network. Each node "listens" to the network to determine whether it is idle. Upon sensing that no traffic is currently on the line, the node is free to transmit. When several nodes try to transmit at the same time, a collision results, forcing the nodes to back off and try again at staggered intervals. The more nodes that are connected to the network, the higher the probability that such collisions will occur. 100BaseVG centralizes the management and allocation of network access within the network hub.

Unlike the contention-based CSMA/CD method, demand priority is deterministic; the 100BaseVG hub serves as a traffic cop, providing equal access to each of the attached nodes. The hub scans each port to test for a transmission request and then grants the request based on priority. Once the hub gives the node access to the network, it is guaranteed a full 100 Mbps of bandwidth. 100BaseVG includes two levels of access priority, and high-priority requests jump ahead of low-priority requests. Applications and drivers issue requests for high-priority access, based on the type of data queued up to cross the LAN. Advocates of 100BaseVG claim this method reduces overhead at each node and permits greater bandwidth use than CSMA/CD.

In addition to replacing CSMA/CD with demand priority, 100BaseVG replaces 10BaseT's Manchester coding with 5B/6B coding. Manchester coding requires 20 MHz of bandwidth to send information at a rate of 10 Mbps. With 100BaseVG, data and signaling are sent in parallel over all four pairs at rates of 25 MHz each using 5B/6B, which is a more efficient coding scheme. This quadrature (or quartet) signaling entails the use of all four pairs of the cable used for 10BaseT to send and receive data and access signals, whereas 10BaseT uses one pair to transmit and one pair to receive. 100BaseVG runs over any grade of UTP wiring, including Category 3 and Category 5.

100VG-AnyLAN IBM and HP expanded the 100BaseVG specification to include support for token-ring networks as well as Ethernet. The result is 100VG-AnyLAN, which allows Ethernet and token-ring packet frames to share the same media. It supports Category 3 unshielded twisted pair, Type 4, Type 5, and other cable types. However, AnyLAN does not make Ethernet and token-ring networks interoperable—it can only unite them physically through the use of common wiring and a common hub. For true interoperability, a router is needed with the capability to logically merge the two protocols into a common packet and addressing scheme.

SUMMARY

Throughout the 1980s, network bandwidth was delivered by shared low-speed LANs—typically Ethernet and token ring. Devices such as bridges and routers helped solve bandwidth problems by decreasing the number of users per LAN, a solution known as *segmentation*. In the 1990s, a richer set of options for solving bandwidth problems has become available. These alternatives include several varieties of Fast Ethernet, each offering 100 Mbps. At an average of 10 times the speed of today's shared-bandwidth LANs, these solutions can eliminate server bottlenecks, handle the needs of bandwidth-intensive applications, and meet the growing need for sustained network usage. On the heels of 100BaseT's success, Ethernet has been scaled further, offering 1 Gbps over twisted pair wiring.

See Also **Asynchronous Transfer Mode, Fiber Distributed Data Interface, Gigabit Ethernet, High-Speed Token Ring, Wireless LANs**

Fax Servers

A fax server is simply a computer that has special software installed on it to enable all clients on a network to send and receive imaged documents. Instead of equipping all clients with a modem and fax software, users take advantage of the server's capabilities to schedule and route faxes. In not having to equip every client machine with its own modem and fax software, the organization simplifies administration and saves money on software licenses.

Network Capabilities

Faxes can be delivered either directly to the recipient on the LAN or to another fax machine using the Public Switched Telephone Network (PSTN) or private WAN. For the cost of a local call, faxes also can be delivered to destinations worldwide using the Internet, an intranet, or an extranet.

Some fax servers offer redundancy and load-balancing features. Redundancy means that when one fax server fails, another fax server takes over its jobs. Load balancing means that jobs are directed to servers with available phones lines. This improves network performance by dynamically allocating bandwidth and fax channels among servers to prevent congestion at any single point on the network.

An optional fax-on-demand feature lets users dial into the fax server to retrieve documents. Users can call the server directly and request fax documents via a touch-tone telephone or request a fax document from a Web site by clicking on a hyperlink set up for that purpose.

Electronic mail gateway modules that can be added to fax servers to allow users to send, receive, and manage faxes through their e-mail boxes. They can gain access to their e-mail boxes remotely and use them as a universal collection point for faxes, as well as voice and e-mail messages. Some gateways support the most commonly used mail systems such as Lotus cc:Mail, Microsoft Mail, HP OpenMail, and Qualcomm's Eudora, as well as the standard Internet mail server protocols SMTP, POP3, and IMAP.

A Web client module can be added to the fax server to permit corporate users to manage their fax mail boxes through all popular Web browsers, such as Netscape Navigator and Microsoft Internet Explorer. Whether they are at work, at home, or on the road, they can simply enter the Web address of the fax server for immediate access to all of their sent, received, and stored faxes.

A corporate fax server can be connected to an IP/PSTN gateway to allow documents to be sent over low-cost IP connections (Figure 44). IP faxes are like traditional faxes, except the Internet is used instead of the telephone network for at least some part of the transmission. The ITU T.38 standard defines a secure way to set up IP fax transmissions in real time.

A fax server can also be equipped for optical character recognition (OCR). With this module, fax documents are scanned for recognizable characters and converted into text files that can be imported into any application. Once there, users can edit the documents.

Administrative Features

Among the administrative features commonly found on fax servers are the following:

APPLICATION PROGRAMMING INTERFACE (API) Allows the administrator to integrate the fax capability with any of the desktop applications.

Figure 44
With a fax server connected to an IP/PSTN gateway, documents that start on the IP network are sent to the remote gateway where a local number is dialed to make the PSTN connection to a traditional fax terminal.

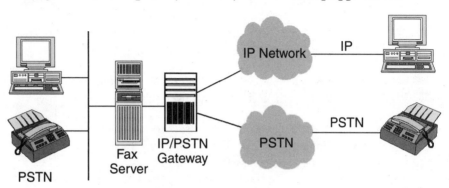

ACTIVE DIRECTORY INTEGRATION   Supports ADSI-compatible (Active Directory Services Interface) networks, such as Windows NT 4.0, eliminating the need to maintain separate user databases.

WINDOWS NT AND NETWARE INTEGRATION   Uses Windows NT and NetWare file and print services without burdening the server's CPU.

SCALABILITY   Allows the administrator to cascade (and load-balance) multiple fax servers to increase the number of available phone lines.

PARTIAL RETRANSMISSION   When a fax transmission fails due to poor line quality or communication errors, the fax server automatically resends only the remaining pages.

FLEXIBLE INBOUND ROUTING   Incoming faxes can be delivered directly to users' mailboxes using direct inward dial (DID), ISDN, T.30 subaddressing, DTMF, line group routing, or manual routing.

CONFIDENTIALITY   The administrator can grant users various privilege levels to ensure fax confidentiality.

TRANSACTION LOGS   Provides real-time event information, including user name, recipient, fax number, destination, time sent, number of attempts, and duration of call.

BILLING   Allows the administrator to configure and verify bill-back codes for accurate cost-analysis and tracking within the organization.

User Features

Among the user features commonly available from fax servers are the following:

FAX FROM ANY APPLICATION   Enables users to fax from any application as easily as printing using the fax printer driver.

NATIVE ATTACHMENT SUPPORT   Lets users fax any file in its native format by attaching it to a fax cover page.

OLE SUPPORT   Allows users to fax documents in their native format. Object Linking and Embedding (OLE) prevents loss of data and ensures correct formatting of documents.

COVER-PAGE MANAGEMENT   Users can select from multiple cover pages on the fly while typing the message. A cover-page editor lets users create and manage their own personal cover page.

MULTI-INDEXED PHONE BOOKS   In addition to the corporate phone book kept on the server, users can maintain multiple personal phone books on their own computer, with all entries searchable by name, company, and fax number.

BROADCAST AND SCHEDULING   Allows users to choose the recipients of a fax broadcast and set a time to start the broadcast.

INCOMING FAX NOTIFICATION   Users receive notification of a new incoming fax through a pop-up message window that enables them to quickly view and print the fax.

FOLLOW-ME   When users are out of the office, incoming faxes are automatically forwarded to another fax number or a designated colleague.

FAX STATUS   Users are kept informed about the status of incoming and outgoing fax transmissions via color codes.

MIXED DISTRIBUTION   Once a message is composed, it can be sent to a distribution list comprised of both fax and e-mail recipients.

SUMMARY

Fax servers make sending and receiving imaged documents as easy as sending e-mail. The fax capability can be added to an existing application server with appropriate software, or it can be a dedicated device that simply attaches to the LAN in UNIX, Windows, or Novell environments. Fax servers come in workgroup or enterprise configurations and functionality can be added with optional software modules available from the vendor. In handling faxes through a server, the organization can offer sophisticated capabilities more economically than equipping every client with its own hardware and software.

See Also **Electronic Mail, Unified Messaging**

Fiber Distributed Data Interface

The Fiber Distributed Data Interface (FDDI) is a high-speed network that employs a counter-rotating token-ring technology for fault tolerance. It provides interconnectivity between computers and peripherals, including the interconnection of LANs, within a building or campus environment. Originally conceived to operate over multimode fiber-optic cable, FDDI can run over single-mode fiber-optic cable, shielded twisted pair copper, and even unshielded twisted pair copper wiring.

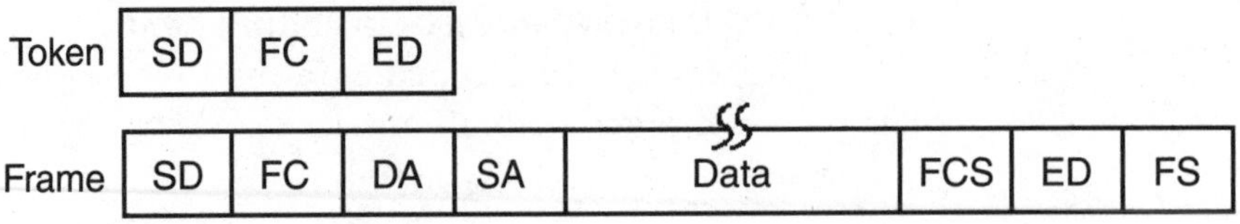

Figure 45
Frame format of
FDDI.

Token | SD | FC | ED

Frame | SD | FC | DA | SA | Data | FCS | ED | FS

AC	Access Control
DA	Destination Address
ED	Ending Delimiter
FC	Frame Control
FCS	Frame Check Sequence
FS	Frame Status
SA	Source Address
SD	Starting Delimiter

FDDI uses two counter-rotating rings: a primary ring and a secondary ring. Data traffic usually travels on the primary ring. The secondary ring operates in the opposite direction and is available for fault tolerance. If appropriately configured, stations may transmit simultaneously on both rings, thereby doubling the bandwidth of the network to 200 Mbps.

Frame Format

The ANSI X3T12 standard defines a multifield frame format for FDDI (Figure 45), and is composed of the following:

PREAMBLE Stations need to synchronize with the signal's frequency, which is done by means of 64 preamble bits.

STARTING DELIMITER This one-byte field indicates the start of a frame.

FRAME CONTROL This one-byte field is used for access and frame control.

DESTINATION ADDRESS This six-byte field contains the destination MAC address.

SOURCE ADDRESS This six-byte field contains the source MAC address.

INFORMATION FIELD This variable-length field (up to 4478 bytes) contains the user data.

FRAME CHECK SEQUENCE This four-byte field contains the checksum for error control.

ENDING DELIMITER This four-bit field indicates the end of a frame.

FRAME STATUS This 12-bit field indicates the status of a frame, such as error detected, addressed recognized, and frame copied.

Token Passing

FDDI uses a timed token protocol, while 802.5 token ring uses a priority/reservation token access method. This means there are some differences in frame formats and in how a station's traffic is handled. Management of the rings is also different.

With FDDI, a timed token-passing access protocol passes frames of up to 4,500 bytes in size, supporting up to 1,000 connections over a maximum multimode fiber path of 200 km (124 miles) in length. When FDDI is run over copper wire, such as Category 5 cabling, the distance between connections must be less than 100 meters.

Each station along the path serves as the means for attaching and identifying devices on the network, regenerating and repeating frames sent to it. Unlike other types of LANs, FDDI allows both asynchronous and synchronous devices to share the network. FDDI stresses reliability and its architecture includes integral management capabilities, including automatic failure detection and network reconfiguration.

Any change in the network status—such as power-up or the addition of a new station—leads to a "claim" process during which all stations on the network bid for the right to initialize the network. Every station indicates how often it must see the token to support its synchronous service. The lowest bid represents the station that must see the token most frequently. That request is stored as the Target Token Rotation Time (TTRT). Every station is guaranteed to see the token within two times the number of TTRT seconds of its last appearance.

This process is completed when a station receives its own claim token. The winning station issues the first unrestricted token, initializing the network on the first rotation. On the second rotation, synchronous devices may start transmitting. On the third and subsequent rotations, asynchronous devices may transmit, if there is available bandwidth. Errors are corrected automatically via a beacon-and-recovery process during which the individual stations seek to correct the situation.

Standards

These processes are defined in a set of standards (X3T12) sanctioned by the American National Standards Institute (ANSI). The standards address four functional areas of the FDDI architecture:

PHYSICAL MEDIA DEPENDENT Data is transmitted between stations after converting the data bits into a series of optical pulses. The pulses are then transmitted over the cable linking the various stations.

The PMD sublayer describes the optical transceivers, specifically the minimum optical power and sensitivity levels over the optical data link. This layer also defines the connectors and media characteristics for point-to-point communications between stations on the FDDI network. The PMD sublayer is a subset of the physical layer of the OSI reference model, defining all of the services needed to transport a bit stream from station to station. It also specifies the cabling requirements for FDDI-compliant cable plant, including worst-case jitter and variations in cable plant attenuation.

PHYSICAL LAYER The Physical Layer (PHY) protocol defines those portions of the physical layer that are media independent, describing data encoding/decoding, establishing clock synchronization, and defining the handshaking sequence used between adjacent stations to test link integrity. It also provides the synchronization of incoming and outgoing code-bit clocks and delineates octet boundaries as required for the transmission of information to or from higher layers. These processes allow the receiving station to synchronize its clock to the transmitting station.

MEDIA ACCESS CONTROL FDDI's data link layer divides is divided into two sublayers. The Media Access Control (MAC) sublayer governs access to the medium. It describes the frame format, interprets the frame content, generates and repeats frames, issues and captures tokens, controls timers, monitors the ring, and interfaces with station management.

The Logical Link Control (LLC) sublayer, while not part of the FDDI standard, is required for proper ring operation and is part of the IEEE 802.2 standard. In keeping with the IEEE model, the FDDI MAC is fully compatible with the IEEE 802.2 Logical Link Control (LLC) standard. Applications that can currently interface to the LLC and operate over existing LANs, such as IEEE 802.3 CSMA/CD or 802.5 token ring, should be able to operate over an FDDI network.

The FDDI MAC, like the 802.5 token-ring MAC, has two types of protocol data units, a frame and a token. Frames carry data (such as LLC frames), while tokens control a station's access to the network. At the MAC layer, data is transmitted in four-bit blocks called 4B/5B symbols. The symbol coding is such that four bits of data are converted to a five-bit pattern, therefore the 100-Mbps FDDI rate is provided at 125 million signals per second on the medium. This signaling type is employed to maintain signal synchronization on the fiber.

STATION MANAGEMENT The Station Management (SMT) facility provides the system management services, detailing control requirements for the proper operation and interoperability of stations on an FDDI ring.

It acts in concert with the PMD, PHY, and MAC layers. The SMT facility manages connections, configurations, and interfaces. It defines services such as ring and station initialization, fault isolation and recovery, and error control. SMT is also used for statistics gathering, address administration, and ring partitioning.

Equipment Classes

Three classes of equipment are used in the FDDI environment: single attached stations (SASs), dual attached stations (DASs), and concentrators (CONs).

A DAS physically connects to both rings, while a SAS connects only to the primary ring via a wiring concentrator. In the case of a link failure, the internal circuitry of a DAS can heal the network using a combination of the primary and secondary rings. If a link failure occurs between a concentrator and a SAS, the SAS becomes isolated from the network.

These equipment types may be arranged in any of three topologies: dual ring, tree, and dual ring of trees. In the dual ring topology, DASs form a physical loop, in which case all the stations are dual-attached. In a tree topology, remote SASs are linked to a concentrator, which is connected to another concentrator on the main ring. Any DAS connected to a concentrator performs as a SAS. Concentrators may be used to create the network hierarchy known as a dual ring of trees. This topology offers a flexible hierarchical system design that is efficient and economical. Devices requiring highly reliable communications attach to the main ring, while those that are less crucial attach to branches off the main ring. Thus, SAS devices can communicate with the main ring, but without the added cost of equipping them with a dual-ring interface or a loop-around capability that would otherwise be required to ensure the reliability of the ring in the event of a station failure.

Failure Protection

FDDI provides an optional bypass switch at each node to overcome a failure anywhere on the node. In the event of a node failure, it is bypassed optically, removing it from the network. Up to three nodes in sequence may be bypassed; enough optical power will remain to support the operable portions of the network.

In the event of a cable break, the dual counter-rotating ring topology of FDDI allows the redundant cable to handle normal 100-Mbps traffic. If both the primary and secondary cables fail, the stations adjacent to the failures automatically loop the data around and between rings, thus forming a new C-shaped ring from the operational portions of the original two rings. When the fault is healed, the network reconfigures itself.

FDDI concentrators normally offer two buses corresponding to the two FDDI backbone rings. Fault tolerance is also provided for stations that are connected to the ring via a concentrator because the concentrator provides the loop-around function for attached stations.

SUMMARY

An extension of FDDI, called FDDI-2, offers Hybrid Mode, which uses a 125 μsec cycle structure to transport isochronous (i.e., real-time) traffic, in addition to synchronous and asynchronous frames. However, FDDI is limited by distance, while ATM is a highly scalable broadband networking technology that spans both LAN and WAN environments. This means that ATM will eventually dominate corporate networks, especially for running multimedia applications.

See Also **Asynchronous Transfer Mode, Fast Ethernet**

Fiber-Optic Technology

Fiber-optic transmission systems have been in commercial use for about 25 years. With rare exception, all long-distance traffic is carried over fiber-optic cable. In addition, most metropolitan areas are served by fiber-optic rings that provide high-bandwidth, fault-tolerant communications for businesses. Fiber is widely used because it is more economical than traditional copper cable. A few ounces of optical fiber can transport the same amount of data as several tons of copper wire. Fiber-optic cable is also more efficient for data transport because the media does not have to contend with the physical limitations of copper wire, such as its high attenuation and susceptibility to various forms of interference.

Light Sources

Fiber-optic transmission circuits typically use light-emitting diodes (LEDs) or injection laser diodes (ILDs) as their light source. The transmitter circuitry translates electronic signals into optical signals by modulating the drive current to the emitter. Both LEDs and ILDs emit an intense beam of monochromatic light in the form of visible red light or invisible infrared light. Monochromatic light ensures a uniform frequency whereby emitted photons travel at a constant speed; ordinary light is unacceptable for fiber transmission because it contains a range of frequencies that creates "chromatic dispersion"—a data-corrupting condition caused when photons travel at slightly different speeds and cause the energy bursts to disperse at varying distances.

Most LEDs supply power at 0.5 to 11.5 milliwatts (mW), whereas ILDs supply power at three to 10 mW. Because ILDs have a short rise time (the time required to go from off to on) of only one to two nanoseconds, they are useful in applications that require high data rates. Similarly, because ILDs have a higher bias current range (100 to 500 milliamps), they are best suited for long-distance communications. By comparison, most LEDs have a rise time of 3 to 20 nanoseconds, and a bias current range of 50 to 150 milliamps, which is adequate for short communications links such as LANs, interoffice trunks, and local loops.

In addition to supporting higher data throughput rates than LEDs, ILDs have lower signal distortion ratios. Therefore, ILD signal repeaters can be spaced as much as 40 to 60 miles apart. In some cases, amplifiers can now be used in place of repeaters. Amplifiers boost light signals without converting them to electrical energy and back, as is required by repeaters. Technology is changing at such a pace that repeaterless links of 9,000 miles and longer are now possible over transoceanic routes.

Another diode technology is vertical-cavity surface-emitting laser (VCSEL), of which there are two types. One emits 850-nm (nanometer) light and can launch signals down a fiber at gigabit rates. However, a poor match with the optical properties of fiber limits the effective distance of transmission. As a result, the shorter-wavelength diodes have been limited to Ethernet and Fibre Channel applications, where distances traveled are less than a few hundred meters. The other version of VCSEL operates efficiently at 1,300-nm wavelengths, which makes it a good match for the fiber used in long-haul networks. VCSELs have become popular in optoelectronic design because they emit light through the top layer of the diode rather than laterally through the edge of the device. This makes it much easier to package parallel arrays of the devices, which reduces costs.

In hybrid fiber/coax (HFC) systems, which are currently deployed in the local loop by CATV operators and RBOCs to bring video and other broadband services to the home, three laser-based transmitters have emerged. All three optimize the optical transmission of both analog and digital signals. The most widely deployed optical transmitter technology in the HFC environment is the distributed feedback (DFB) semiconductor laser diode, which emits 1310-nm wavelength light. DFB-based optical transmitters offer good performance for links as long as 22 miles and cost less than the other two technologies.

Technically, lasers should transmit only monocromatic light, but they do emit light at other frequencies. The goal is to minimize the emission of such light. YAG (yttrium-aluminum-garnet) laser-based optical transmitters emit 1310-nm wavelength light like the DFB-based transmitters, but deliver a higher performance level with less light at other frequencies. In addition to offering enhanced optical power output, YAG lasers have better distortion characteristics; this makes them especially suited for fiber architectures that require cascaded links.

The 1550-nm transmitter technology is a blend of DFB and YAG. It produces only 0.25 dB/km optical fiber loss, which is 30 percent less than does 1310-nm light, and also exhibits YAG's excellent low-distortion properties. These characteristics make 1550-nm optical transmitters well suited to long-distance transmission—as required by microwave replacement and head-end consolidation applications—and dense broadcast architectures, which require high optical power and multiple optical splits to serve many nodes. The 1550-nm transmitter technology also facilitates the design of longer fiber runs without the need for regeneration. Several key component technologies, such as erbium-doped fiber amplifiers (EDFAs), which operate within the 1550- nm window, eliminate regeneration.

Light-Detecting Devices

At the receiving end of a fiber link, equipment with a light-detecting device is used to create an electrical current from the transmitted light. Commonly used light-detecting devices include:

- Positive-Intrinsic-Negative-Channel (PIN) Photodiodes, which produce an electrical current in proportion to the amount of light energy projected onto them
- Avalanche Photo Diodes (APDs), which operate in much the same way as PIN photodiodes but require more complex receiver circuitry and provide faster response times
- Phototransistors, which function as amplifying detectors

Amplification is useful for fiber-optic systems with long spans between amplifying stages. Each of the three light-detecting devices has its drawbacks. For example, phototransistors require receiver circuitry more complex than PIN photodiodes but less complex than the circuitry required by APDs. In addition, the phototransistor's response time is slower than that of the photodiodes.

Transmission Modes

There are two types of optical fiber: single-mode and multimode. Single-mode fibers transmit only one light wave along the core, while multimode fibers transmit many light waves. Single-mode fibers entail lower signal loss and support higher transmission rates than multimode fibers, so they are the type of optical fiber most often selected by carriers for use on the public network. Over 90 percent of fiber-optic cable installed by carriers is single-mode.

Multimode fibers have relatively large cores. Light pulses that simultaneously enter a multimode fiber can take many paths and may exit at slightly different times. This phenomenon, called *intermodal pulse dispersion*, creates minor signal distortion and thereby limits both the data rate of the optical signal and the distance that the optical signal can be sent without repeaters. For this reason, multimode fiber is most often used for short distances and for applications in which slower data rates are acceptable.

Multimode fiber can be further categorized as step-index or graded-index. Step-index fiber has a silica core encased with plastic cladding. The silica is denser than the plastic cladding; the result is a sharp, step-like difference in the refractive index between the two substances. This difference prevents light pulses from escaping as they pass through the optical fiber. Graded-index fiber contains multiple layers of silica at its core, with lower refractive indices toward the outer layers. The graded core increases the speed of the light pulses in the outer layers to match the rate of the pulses that traverse the shorter path directly down the center of the fiber.

The fibers in most of today's fiber-optic cable have an outside (or cladding) diameter of 125 microns. (A micron is one-millionth of a meter.) The core diameter depends on the type of cable. The cores of multimode fibers comprise many concentric cylinders of glass, and each cylinder has a different index of refraction. The layers are arranged so light introduced to the fiber at an angle will be bent back toward the center. The bending results in light, which travels in a sine-wave pattern down the fiber core, and allows an inexpensive noncoherent light source, such as an LED, to be used. Almost all multimode fibers have a core diameter of 62.5 microns. Bandwidth restrictions of 200 to 300 MHz/km limit the maximum length of multimode segments to a few kilometers. Wavelengths of 850 and 1300 nm are used with multimode fiber-optic cable. Single-mode fiber consists of a single 8- to 10-micron core. This means that a carefully focused coherent light source, such as a laser, must be used to ensure that light is sent directly down the small aperture. Single-mode fiber is normally operated with light at a wavelength of 1300 nm. Because possible light paths through the fiber are restricted, there is essentially no bandwidth limit for single-mode fiber. Frequencies of many gigahertz can be carried tens of kilometers without fiber-optic repeaters.

Performance Advantages of Fiber

Optical fiber provides many performance advantages over conventional metallic cabling. These advantages include the following features:

- High bandwidth capacity
- Low signal attenuation
- High data integrity
- Immunity to electrical and radio interference
- High levels of security
- Greater durability

These advantages make optical fiber the most advanced transmission medium available today, and a low-cost alternative to satellite communications.

HIGH BANDWIDTH   The laser components at each end of the optical-fiber link allow for high encoding and decoding frequencies. For this reason, optical fiber offers much more bandwidth capacity than copper-pair wires. Data can be transmitted over optical fiber at multigigabit speeds, as opposed to a maximum transmission speed of only 44.736 Mbps over T3 copper-based digital facilities. The speed of transmission over fiber facilities is further increased by the Synchronous Optical Network (SONET), which permits speeds of up to 13 Gbps. As noted, experimental techniques have demonstrated the feasibility of transmitting data at one trillion bits per second.

When a photonic technology called Dense Wave Division Multiplexing (DWDM) is applied to SONET links or rings, the transport capacity can by increased by a factor of 10 without the carrier having to lay any additional fiber-optic cable.

However, DWDM works only with the newer grade of optical fiber called nondispersion-shifted fiber, which can support up to 80 channels on a single hair-thin strand. While next-generation carriers use the newer grade of fiber, incumbent service providers—AT&T, MCI, and Sprint—use mostly dispersion-shifted fiber, which suffers from signal bleeding. This problem limits the older fiber to eight channels per strand, which is derived from the older wavelength division multiplexing (WDM) technology.

LOWER SIGNAL ATTENUATION Signal attenuation, measured in decibels (dB), refers to signal "loss" during transmission (i.e., when the received signal is not as strong as the signal transmitted). Signal attenuation is attributed to the inherent resistance of the transmission medium.

For transmissions over metallic cable, loss increases with frequency and limits the amount of available bandwidth. The characteristics of optical fiber, however, are such that little or no inherent resistance exists. This low resistance allows the use of higher frequencies to derive enough bandwidth to accommodate thousands of voice channels. Whereas an analog line on the local loop has a frequency range of up to 4 KHz for voice transmission, a single optical fiber has a range of up to 3 GHz.

Although fiber exhibits much lower attenuation than metallic cable, glass fibers are not perfectly clear. The amount of light coming out of the distant end of a fiber is always somewhat less than the amount transmitted into the fiber. Fiber loss may be due to several causes, among them:

- Absorption by impurities
- Scattering by impurities or by the defects at the core-cladding interface and Rayleigh scattering by the molecules of the medium (i.e., silica)
- Fiber bends and micro-bends
- Scattering and reflection at splices

All of these factors contribute to the degradation of the fiber transmission over long distances. Excessive attenuation results in a received signal that is too weak to be useful. It is usually a linear quantity, increasing in direct proportion to the length of a fiber-optic cable run. The minimum attenuation of a standard telecom fiber occurs around the 1550 nm wavelength; it is of the order of 0.2 dB/km.

HIGH DATA INTEGRITY Data integrity refers to a performance rating based on the number of undetected errors in a transmission. Once again, fiber optics surpass metallic cabling. A typical fiber-optic transmission system produces a bit-error rate of less than 10^{-9}, while metallic cabling typically produces a bit-error rate of 10^{-6}.

Because of their high data integrity, fiber-optic systems do not require extensive use of the error-checking protocols common in metallic cable systems. Because the error-checking overhead is eliminated, the data transmission rates are enhanced. In addition, because the required number of retransmissions is reduced with fiber, overall system performance is greatly improved.

IMMUNITY TO ELECTRICAL AND RADIO INTERFERENCE Electromagnetic and radio frequency interference (EMI and RFI) are the principal sources of data errors in transmissions over metallic cable systems. The immunity of fiber facilitates installation because fiber-optic cables need not be rerouted around elevators, machinery, auxiliary power generators, fluorescent lighting, or other potential sources of interference.

Fiber's immunity to interference makes it more economical to install, not only because less time is required to route the physical cables but because there is no need to build special conduits to shield fiber from the external environment. In addition, because optical fibers do not generate the electromagnetic radiation that often causes crosstalk on metallic cables, multiple fibers can be bundled into a single cable to further simplify installation.

HIGH LEVELS OF SECURITY Fiber is a more secure transmission medium than unshielded metallic cable because optical fibers do not use signal-radiating electromagnetic or radio frequency energy. In order to tap a fiber-optic transmission, the wire core must be physically broken and a connection fused to it. This procedure is routinely used to add nodes to the fiber cable, during which it prohibits the transmission of light beyond the point of the break, and therefore makes unauthorized access easily detectable.

GREATER DURABILITY Optical fiber is not a delicate material; in fact, fiber's pull strength (the maximum pressure that can be exerted on the cable before damage occurs) is 200 pounds—eight times that of Category 5 unshielded twisted pair (UTP) copper wire. Fiber-optic cables are reinforced with a strengthening member inside the cable and a protective jacket around the outside of the cable. These reinforcements produce the same tensile strength as steel wire of an equal diameter. Figure 46 illustrates the basic construction of fiber-optic cable.

The inherent strength of fiber, combined with the added reinforcement of being bundled into cable form, gives fiber-optic cables the durability necessary to withstand being pulled through walls, floors, and underground conduits without being damaged. In addition, fiber cables are designed to withstand higher temperatures than copper, which makes fiber networks better able to survive potentially disastrous fires. In typical operating environments, fiber is also more resistant to corrosion than copper wire and, consequently, has a longer useful life.

SUMMARY

Improvements in both the optical fiber itself and the electronic components at each end of the cable constantly increase the data-carrying capac-

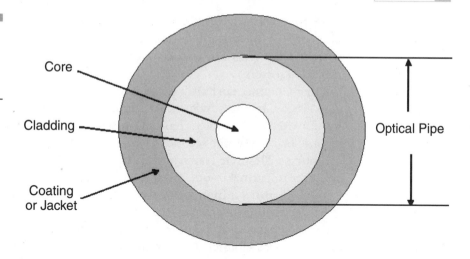

Figure 46
Cross-sectional view of basic optical fiber cable construction.

ity of fiber optics. In fact, the capacity of optical fiber doubles approximately every 18 months. Experimental work done in Japan by Fujitsu, NTT, AT&T, and Lucent Technologies shows that it is now possible to transmit a trillion bits of data per second. This is the equivalent of transmitting the contents of 300 years' worth of daily newspapers in a single second, or conveying 12 million telephone conversations simultaneously. *See Also* **Fiber Distributed Data Interface, Fibre Channel, Next Generation Networks, Synchronous Optical Network**

Fibre Channel

Performance improvements in storage, processors, and workstations—along with the move to distributed architectures such as client/server—have spawned increasingly data-intensive and high-speed networking applications. The interconnection between these systems and their input/output devices demands a new level of performance in reliability, speed, and distance. These performance needs can be met with Fibre Channel (FC), a highly reliable, gigabit-per-second interconnect technology that allows concurrent communications among workstations, mainframes, servers, data storage systems, and other peripherals—even those that use other interfaces such as the High-Performance Parallel Interface (HPPI) or Small Computer System Interface (SCSI).

Key Advantages

An ANSI standard (X3T11) since 1994, FC supports multiple topologies using optical fiber, coaxial cable, a copper wire—each with its own distance limitation. For single-mode fiber-optic cable, the distance is limited to 10 kilometers (km) with a repeater. For multimode fiber-optic cable, the distance is 500 meters. For copper wire and coaxial cable, the distance is limited to 30 meters.

Fibre Channel currently operates over 133 Mbps, 266 Mbps, 532 Mbps, and 1.0625 Gbps bandwidths. At the gigabit-per-second speed—accounting for overhead—Fibre Channel's peak data transfer rate is 100MB per second, or 200MB per second in duplex mode. A Fibre Channel adapter can burst a 2048-byte frame at the link rate. Future enhancements will enable Fibre Channel to operate at 2 and 4 Gbps.

Fibre Channel is highly efficient for real-time data transport because it is hardware-intensive rather than software-intensive. Fibre Channel simply takes what is in the sending device's buffer and transports it to the receiving device's buffer. It does not care what the data is or how it is formatted. Once the cells, packets, or frames reach the receiving buffer, what happens to them after that is inconsequential.

One reason LANs are so software-intensive is that each station, or node, must be capable of recognizing error conditions on the network and take appropriate action to recover from them. These are tasks that individual Fibre channels do not have to perform. If a data transfer fails because congestion causes a busy condition, the channel retries immediately without checking with software as to whether it should or should not. Also, there are no complex routing algorithms to compute to send data to another port far away. The switch fabric relieves each Fibre Channel port of the responsibilities for station management associated with error recovery and routing. Freed from these burdens, the individual channels are able to move huge volumes of data quickly from one point to another.

Fibre Channel can be deployed as dedicated point-to-point links, shared loops, and highly-scalable switched topologies to meet virtually any data application requirement. It even maps several interfaces such as HPPI and SCSI and common transport protocols, including the Internet Protocol (IP) and Asynchronous Transfer Mode (ATM). These and other interfaces and transport protocols all map onto the Fibre Channel physical layer protocol, allowing it to merge high-speed I/O and networking functionality in a single connectivity solution.

These and other characteristics make Fibre Channel a compelling solution for storage, video, graphic, and mass data transfer applications. It is particularly useful for mission-critical data applications that demand high availability.

Protocol Stack

The five layers of the Fibre Channel standard define the physical media and transmission rates, encoding scheme, framing protocol and flow control, common services, and the upper-level application interfaces:

FC-0   This layer specifies the physical characteristics of the media, transmitters, receivers, and connectors that can be used with Fibre Channel, including electrical and optical characteristics, transmission rates, and other physical components of the standard.

FC-1   This layer defines the 8B/10B encoding/decoding scheme used to integrate the data with the clock information required by serial transmission techniques. Fibre Channel uses 10 bits to represent each 8 bits of "real" data, requiring it to operate at a speed sufficient to accommodate this 25-percent overhead. This scheme is the same one used in IBM's ESCON (Enterprise Systems Connection).

FC-2   This layer defines the rules for framing the data to be transferred between ports, the different mechanisms for using Fibre Channel's circuit and packet switched service classes, and the means of managing the sequence of a data transfer.

FC-3   This layer defines the common services required for advanced features such as striping to multiply bandwidth and hunt groups to allow more than one port to respond to the same alias address.

FC-4   This layer provides seamless integration of existing protocol standards, including the Fiber Distributed Data Interface (FDDI), HPPI, Intelligent Peripheral Interface (IPI), SCSI, IP, Ethernet, Token Ring, and ATM. It also integrates IBM's Single Byte Command Code Set (SBCCS) of a mainframe's Block Multiplexer Channel (BMC).
Fibre Channel is recognized by ANSI as meeting OSI (open system interconnection) standards, even though its five layers do not relate directly to the seven OSI layers.

Classes of Service

To accommodate a wide range of communications needs, Fibre Channel defines three different classes of service (CoS):

CLASS 1   This CoS is a circuit-switched connection that functions in the same manner as a dedicated physical channel. When a host and device are linked, that path is not available to other hosts. When the time needed to make a connection is short or data transmissions are long, Class 1 is the preferred CoS. Two interconnected supercomputers would use this CoS to ensure rapid and smooth communication.

CLASS 2   This CoS is a connectionless, frame-switched link that provides guaranteed delivery with an acknowledgment of receipt. There is none of the delay required to establish a connection as in Class 1, and none of the uncertainty over whether delivery was achieved. If delivery cannot be made due to congestion, a busy signal is returned and the sender tries again. The sender knows it has to retransmit immediately without waiting for a time-out to expire. As with traditional packet-switched systems, the path between two devices is not dedicated, allowing better use of the link's bandwidth. Class 2 is ideal for data transfers to and from a shared mass-storage system physically located at some distance from several individual workstations.

CLASS 3   This CoS is a connectionless service that allows data to be sent rapidly to multiple devices attached to the fabric, but no confirmation of receipt is given. It is most practical when it takes a long time to make a connection. By not providing confirmation, the one-to-many Class 3 service speeds the time of transmission. However, if a single user's link is busy, the system will not know to retransmit the data. Class 3 service is very useful for real-time broadcasts, such as weather visualizations, where timeliness is key and information not received has little value after the fact.
Fibre Channel also provides an optional mode called Intermix, which reserves the full Fibre Channel bandwidth for a dedicated (Class 1) connection, but also allows connectionless traffic to share the link if there is available bandwidth.

Topologies

As noted, Fibre Channel supports multiple interoperable topologies—including point-to-point, arbitrated-loop, and switched. The simplest topology is point-to-point, where two nodes are connected together directly.

In the arbitrated loop, any number of nodes may be connected together, but only 127 nodes may be active at any one time due to addressing limitations. Multiple loops can be connected together in a switched topology to greatly extend the addressing range, or to reduce the bandwidth limitations of a single shared loop.

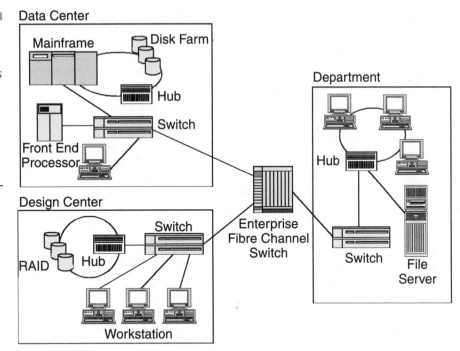

Figure 47
Fibre Channel meets the connectivity needs of diverse distributed computing environments through the use of hubs and switches.

Switches furnish the backbone for all connected devices, with one or more switches providing the switching fabric. The switch fabric allows the attachment of thousands of nodes. The devices can be attached directly to Fibre Channel switches or to hubs that in turn connect to the switches (Figure 47), depending on the application. Accessing the services available from the switching fabric requires a network interface card (NIC) for each node, which enables it to connect to the fabric, as well as to the operating system and the applications.

Fibre Channel's discovery mechanism—Simple Name Service (SNS)—learns the address, type, and symbolic name of each device in the switching fabric. This configuration knowledge is important to the performance levels of channels. SNS information resides in Fibre Channel switches. NICs and storage controllers request SNS data from the switches.

For error recovery and fault isolation, Fibre Channel offers an optional feature called Registered State Change Notification (RSCN). This mechanism issues updates on configuration changes, including device and link failures, and is particularly useful when disk arrays, tape devices, and hosts are directly attached to a switching fabric rather than to shared-access arbitrated loops. RSCN allows switched networks to recover from faults much faster than shared-media networks because problem devices or links can be isolated until fixed and returned to service. This isolation allows other devices and links to operate normally. Many FC products do not rely on RSCN alone—they have load-balancing and automatic failover capabilities that keep critical applications running without interruption.

Implementation Issues

Moving to Fibre Channel requires considering an entire infrastructure that includes chips, disks, mechanical components, device drives storage systems, adapters, and management software—all of which have to work well with one another. Fortunately, Fibre Channel can be phased in, starting with a simple point-to-point link and going with a loop, and eventually expanding to a switched fabric when demand warrants.

While Fibre Channel allows data to be delivered as fast as the destination buffer is able to receive it, several factors can limit performance. For example, in TCP/IP environments, if an unmodified protocol stack is used on the host system, the protocol stack overhead may become the limit to sustainable performance. Host bus issues can also limit performance. When selecting adapter cards, buyers must ensure that they have been optimized specifically for Fibre Channel applications.

High-availability applications work well with redundant systems and links. For redundancy, Fibre Channel dual loops can be created. This involves cabling two fully independent loops to provide two independent paths for data with fully redundant hardware. Most disk drives and disk arrays targeted for high-availability environments have dual ports specifically for this purpose.

If full redundancy is excessive, consideration can be given to another option for preserving resiliency in Fibre Channel: cabling an arbitrated loop through a hub or concentrator to isolate/protect the rest of the nodes on the loop from the failure of an individual node(s). Unfortunately, the hub or concentrator represents a single point of failure, unless redundancy is added at extra cost.

SUMMARY ▬▬ ▬ ▬ ▬ ▬ ▬ ▬ ▬ ▬

Fibre Channel combines the best attributes of a reliable channel with the seamless transparency of a flexible network. Regardless of what the data is, how much there is of it, and how it is formatted (cells, packets, or frames), Fibre Channel is capable of moving it from the sending device's buffer to the receiving device's buffer at full wire speed. And with anticipated enhancement to 4 Gbps, the technology is positioned to meet the high-availability needs of organizations far into the future.

Although Fibre Channel solutions have been around for several years, they mostly filled a niche market because all of the pieces from major system vendors were not in place and readily available. Today, with second sources in place, potential users are more comfortable migrating to the technology because it means they can choose different components from different vendors to meet their individual needs, rather than be shackled to a dominant vendor. FC's open interoperability standards make this possible. *See Also* **Asynchronous Transfer Mode, Fiber Distributed Data Interface, Fiber-Optic Technology, High-Performance Parallel Interface, Small Computer System Interface**

Financial Management of Technology

In recent years, financial management has become a crucial function of technology procurement for both telecommunications and information systems (IS). The more a company understands how much it is spending for equipment, software, and services, the better it will be positioned to use existing technology more efficiently and take advantage of opportunities for cost savings—both of which provide competitive advantages.

However, financial management is not so simple because companies tend to vacillate between two approaches to managing their technology investments: asset management and expense management. Under asset management, technology is viewed in strategic terms and is seen to have a direct correlation with improving customer service, growing the customer base and entering new markets. Under expense management, technology is viewed merely as a necessary cost of doing business—something that can be cut back on, or at least not improved or expanded, to save money. Such companies try to cope with competitive pressures by reducing their technology budgets, extending the use of current systems, buying used equipment, forgoing hardware and software upgrades, and skimping on maintenance.

Despite the current booming U.S. economy, today's telecom and IS managers may not have the luxury of straight-line budgeting, whereby a simple increase of 5 or 10 percent is tacked on to the previous year's budget. This is because companies are under unrelenting pressure to improve financial performance, regardless of how the economy is doing. As a result, most telecom and IS budgets are either cut or held in check, even while top management and end users continue to expect more from existing technology and incremental investments. Thus, there is no room for mistakes in choosing new equipment and vendors or services and providers.

Financial Management Structure

To get a firm handle on technology costs, a financial management structure for communications must be put in place. Although this is primarily the responsibility of the Chief Information Officer (CIO) or other executive with equivalent responsibilities, the telecom and IS managers often provide the key inputs that makes such a structure work. The financial management of technology consists of several functional areas:

COST REPORTING Many times companies do not have an accurate picture of what they are spending for communications because the reporting structure is flawed. For example, management reporting includes capital expense reporting and operating expense reporting. However, all too often the capital and operating expenses for communications are distributed among several financial accounts and are not organized for effective use. Unless these expenses are discretely identified and properly summarized, decision-making could be hampered. For example, recording capital asset depreciation, communications equipment leasing, personnel occupancy, and salaries under non-communications related expense accounts usually results in underestimating the true size of the communications department budget.

More mundane factors also throw off communications budgets. Consider the simple telephone, for example. The mentality of looking at the telephone as a $39 instrument not worthy of any expertise—internal or external—is still common. What is often overlooked is the fact a $39 telephone really may cost several hundred dollars in ongoing maintenance charges in the form of moves, adds, and changes.

BUDGETING AND CONTROL Budgeting and control involves capital and operating budget preparation, performance monitoring and forecasting, and project analysis. Financial management expertise in these areas is more essential in today's operating environment than in the past. The failure to properly calculate capital asset depreciation and taxes, for example, could reveal that a project is not as cost-effective as it initially appeared. Economic conditions also must be factored into the cost equation. For example, a project that increases fixed costs may put the company at a disadvantage when current market conditions actually favor financial flexibility and variable costs.

FINANCIAL OPERATIONS The financial operations area includes service order processing, inventory management, and capital asset management. Since service orders involve operating and/or capital fund expenditures, they demand appropriate controls. If not controlled, service order processing can consume an inordinate portion of the communications budget.

Inventory management is necessary not only for tracking equipment and cabling, but to prevent excess equipment purchases. It is not uncommon to find businesses spending more than 10 percent of their annual communications budget on unnecessary services, equipment and cabling just because they did not have accurate inventory records.

ACCOUNTING Accounting includes maintaining a proper chart of accounts, performing bill reconciliation, and implementing a user charge-back system. A chart of accounts accumulates and reports financial information on various subclassifications of assets and expenses. To be effective, however, it must provide an adequate number of accounts and an adequate definition of accounts to manage the various components of the communications expense—and in a way that allows corporate management to react accordingly. Without a properly organized chart of accounts, it is often difficult to decide where to report different product and service charges. It will also be difficult to generate appropriate financial management reports. The CIO, for example, must have access to usable expense information with the right level of detail.

Bill reconciliation ensures that vendors are paid only for the products and services they deliver. The fact that there continues to be a lucrative business in helping companies recover money lost to carrier and vendor billing errors on a contingency fee basis, demonstrates that bill reconciliation is either not being done, or that it is being done poorly. Thorough bill reconciliation can save corporations millions of dollars a year.

User charge-back allocates communications expenses to the appropriate business units using the services. However, there are potential roadblocks. Some companies get bogged down in the complexity of their charge-back systems in the quest to become accurate. Sometimes the charge-back system gets sidetracked by political issues. Other times the problems are technical, as in the way voice and data are integrated. There must be a balance between the level of effort that goes into this process and what is going to be done with the information. Success hinges on the accumulation of only as much detail as can be reasonably managed.

AGREEMENT MANAGEMENT Relationships with vendors and carriers are defined by contracts and agreements. Customized network services contracts, such as those offered under AT&T's Tariff 12, require extensive financial analysis. These and other contracts must be reviewed periodically to ensure that defined performance standards, prices, payment schedules and other provisions are still compatible with corporate objectives.

The potential pitfall of these and other types of agreements is that companies often abdicate their responsibilities for financial management to the carriers and vendors, who are probably not in a position and should not be trusted to pick up that responsibility. Furthermore, when communications expenses lose their visibility they tend to get out of control: either the company pays too much or it gets too little. The challenge is to maintain the right level of control so that the company can make decisions that are in its best interests.

Depending on the dollar amount involved and whether corporate objectives are being achieved, corporate managers may want to consider alternative scenarios, renegotiate terms and conditions, or limit the length of service agreements to minimize exposure to financial risk.

PRODUCT MANAGEMENT   Product management for communications focuses on such activities as product definition, unit cost analysis, and comparative product pricing analysis. It also involves the company knowing how its communications costs compare with those of its competitors.

When a company defines products for delivery to users, it also develops a more accurate framework for comparative pricing. For example, in calculating loaded costs, such variables as vendor costs, depreciation, occupancy expenses, personnel salary costs, and other overhead expenses are taken into account. By comparing current loaded costs to equivalent alternatives, the company can more actively manage its communications requirements and associated expenses.

NEEDS ASSESSMENT   One way telecom and IS managers can improve the financial management process is to perform a needs assessment and consider the infrastructures they manage before making any decisions. Needs assessment means reevaluating facilities and services, bridges and routers, LANs and servers, and other key components of the network to determine if they still meet user needs. One important cost-saving measure might be to look for opportunities to replace only the hardware and software that has reached or is near the end of its useful life, while upgrading other components.

Needs assessment entails quantifying the potential losses of nonimplementation, perhaps in competitive terms, as well as the benefits of implementation, relating them to the organization's overall business objectives. This information can be used to develop an appropriate network topology, expansion plan, or upgrade policy.

Many companies building or adding on to large networks can improve the cost/performance ratio of their operations by devoting more time to needs assessment. This can be made somewhat easier with such tools as project management, security analysis, and network analysis software.

These tools can generate summary and detail reports, often in graphic form, that can be used to validate network expansion, upgrades, and configuration changes.

There are now some very advanced tools that can help managers ask "what if" questions as they plan or make changes to their networks. These tools automate many of the tedious tasks involved with planning and design, including estimating network costs based on traffic patterns, usage, tariff information, and equipment depreciation. There are even tools that help network managers sort through the advantages of buying or leasing equipment. Some tools price out the network or upgrade according to various criteria and provide a bill of materials.

A needs assessment from a network topology perspective can go a long way toward identifying the true costs of a network or upgrade. The local-area network topology should identify workgroup and departmental networks and the connections between them; communities of interest and their local subnetworks; the number of attached hubs and workstations; and the location of mission-critical databases and whether they are located on minicomputers, mainframes, or file servers.

The wide-area network topology should identify all switching and feeder nodes; the speeds and locations of the lines and/or carrier services; the LAN interconnection equipment; any special transmission requirements that will improve performance and safeguard important data, such as compression and encryption; and network management systems, both primary and subordinate, located at domestic and international locations.

Other cost items for LANs and WANs may include provisions for disaster recovery, spare bandwidth to handle congestion, and system modules supporting specific protocols for various interoperability requirements.

The final part of this analysis includes backward tracing to determine how well the proposed network or upgrade meets specific requirements. From each workstation, local subnetwork, workgroup, department, and the backbone network itself, the data traffic must be traced to the appropriate network elements to see if they meet the performance objectives.

Implementing client/server systems requires a careful cost-benefit analysis because of the potentially large investments in both new equipment and maintenance, and other less tangible trade-offs from the decision to adapt an entirely new network architecture. In fact, the move from mainframes to the client/server architecture is most frequently cited as the reason for projected increases in training. Additionally, support costs may also increase as more components are brought into the office environment and such concerns as security and software maintenance and distribution become increasingly important.

Given the "do more with less" budgetary climate that afflicts many organizations, a staggered implementation based on lower-cost systems and servers is more practical than adding mainframe capacity. Simultaneously, building client/server networks may require substantial organizational adjustments. Many such adjustments will involve additional expenses, counterbalanced by the fact that client/server applications are often faster and more economical to develop than applications for traditional host systems, especially when object-oriented programming tools are used for creating reusable software modules.

A cost-benefit analysis of new technology should be based on the intention to increase revenue due to the technology, not simply to save costs. Companies should learn from their client bases what technological investments will solve client needs and return profits at the same time.

For this and other reasons, companies of all types and sizes are turning to large systems integration firms to manage their projects' risks. However, expecting the systems integrator to act as a deep-pocket partner who absorbs the financial loss if the project is late or fails is not realistic. Managing the risk together is a realistic expectation. A firm is more likely to build quality systems by having clear goals, managing with those goals in mind, hiring the right people, creating a sense of teamwork, and devising a good plan based on mutually agreed upon cost estimates.

RETURN ON INVESTMENT   Return on investment (ROI) refers to the anticipated cost savings, productivity gains, or other benefits that will accrue to the organization as a result of implementing a new technology or service. The ROI is typically used to help cost-justify a capital investment. To help top management make confident and informed decisions, the network manager, together with other department heads, should prepare an executive report explaining each option and its associated risks and benefits.

This report should address an organization's strategic business objectives, identifying potential targets for improvement and providing a high-level cost-benefit analysis. Its objective is to outline a preliminary plan for implementing the network, upgrade, or expansion plan within the existing work environment. The report should include which departments would gain the most benefit from the proposed plan based on such parameters as traffic volume; geographic diversity; application requirements in terms of bandwidth, reliability, speed, connectivity, protocols, and delay; and customer-supplier linkages via such means as electronic mail, electronic data interchange, document imaging, and CAD/CAM.

In addition, the report should address enterprisewide requirements in an expandable, modular fashion that protects existing investments while building for future needs, including the requirement for interconnectivity among operating groups, subsidiary companies, and trading partners.

To ensure the plan's effective implementation, it is advisable to stress an environment that is both structured and flexible. Industry standards provide the structure, while flexibility results from building the solution on standard platforms that can be tailored to specific user needs. In addition to maximizing integration potential, adhering to standards facilitates the incorporation of technological advances as they occur, regardless of their origin. Thus, as new technologies emerge to better support business strategies, standards will permit the organization to take an early advantage without obsoleting current investments. These new technologies may be refinements to existing network elements, such as advances in bridging and routing technology incorporating support for frame relay or switched multi-megabit data services (SMDS), or they may be complementary technologies, such as voice annotation and multimedia support for electronic mail systems operating on LANs.

SUMMARY

Financial decisions are not made in a vacuum. The needs of the various departments and workgroups must be considered, as well as top management's orientation to the corporate network—whether it is viewed as an asset or an expense. Moreover, new purchases must be justified in terms of their return on investment, which should not only include the point at which the purchase price can be recovered through cost savings over previous equipment and services, but also less tangible benefits such as improving customer service and goodwill.

See Also **Technology Migration Planning, Technology Procurement Alternatives**

Firewalls

As more computers become networked, often providing access to the public Internet, the possibility of break-in attempts increases. Corporate networks today contain valuable resources that provide access to services, software, and databases. By occupying a strategic position on the network, firewalls work to block certain types of traffic and allow the rest of the traffic to pass. They can also reconfigure themselves to stop successive break-in attempts and provide tools that can be used to track down the source of the attacks.

Types of Firewalls

There are several types of firewalls: packet filters, stateful packet inspectors, circuit-level gateways, and application gateways. Some firewall products combine all four into one firewall server, offering organizations more flexibility in meeting their security needs.

PACKET FILTERS With packet filters, all IP packets traveling between the internal network and the external network must pass through the firewall. User-definable rules allow or disallow packets to pass. The firewall's graphical user interface allows system administrators to implement packet filter rules (Figure 48).

STATEFUL PACKET INSPECTORS This type of device examines the packets it sees, just like packet filters, but goes a step further: it remembers which port numbers are used by which connections and shuts down access to those ports after the connection closes.

CIRCUIT-LEVEL GATEWAYS All the firewall's incoming and outgoing connections are circuit-level connections that are made automatically and transparently. The firewall can be configured to enable a variety of outgoing connections such as Telnet, FTP, WWW, Gopher, AOL, and user-

Figure 48
Like other products, CyberGuard Corp's Firewall 3 allows the network administrator to control packet filtering, which permits or denies connections using criteria based upon the source and destination host or network and the type of network service.

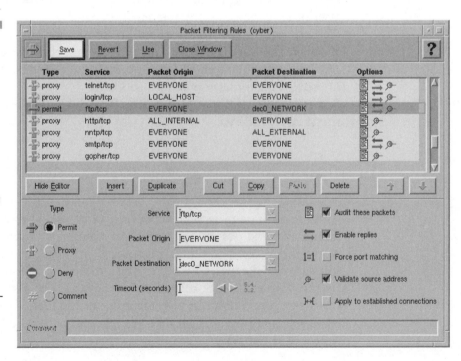

defined applications such as Mail and News. Incoming circuit-level connections include Telnet and FTP. Incoming connections are permitted only when authenticated by one-time password tokens.

APPLICATIONS SERVERS   Some firewalls include support for several standard application servers, including Mail, News, WWW, FTP, and DNS (Domain Name Service). Security is enhanced by partitioning these applications from other firewall software, so if an individual server is under attack, other servers/functions are not affected.

To aid security, firewalls offer logging capabilities as well as alarms that are activated when probing is detected. Log files are kept for all connection requests and server activity. The files can be viewed from the management console, displaying the most recent entries first. The log scrolls in real time as new entries come in. The log files include a variety of useful data, including connection requests, mail and news traffic, server activity, FTP session activity, and error conditions.

An alarm system watches for network probes, and can be configured to watch for TCP or UDP probes from either the external or internal networks. Alarms can be configured to trigger e-mail or pop-up windows, they can send messages to a local printer, or they can halt the system if a security breach is detected.

Another important function of firewalls is remapping all internal IP addresses for the purpose of hiding them from public view. The source IP addresses are written so outgoing packets originate from the firewall. The result is that all the organization's internal IP addresses are hidden from users on the greater Internet. This also provides organizations with the important option of being able to use nonregistered IP addresses on their internal network. In not having to assign every computer a unique IP address and not having to register them for use over the greater Internet (which would result in conflicts), administrators can save hundreds of hours of work.

Other Security Features

Firewalls often provide other security features, such as encryption, that enhance the primary capabilities already discussed. With the increasing use of Java applets and ActiveX controls on Web sites, more firewalls are also able to deny access to Web pages that contain these elements. Java applets and ActiveX controls are self-executing programs capable of not only hiding harmful viruses or agents, but also potentially giving remote users access to resources on private networks once they get inside.

Another security feature is the log file, which records connection requests and server activity. The information compiled in the log file can be used to identify possible security breaches. The file can be viewed from the console displaying the most recent entries and scrolled in real time as new entries come in.

One of the latest features to be added to firewalls is automatic reconfiguration upon the detection of a break-in attempt. While terminating the attack, the firewall is automatically reconfigured to repel further traffic from the attacking system, thereby preventing further attempts at unauthorized access. This combination of intrusion detection and automatic reconfiguration enhances the security of corporate networks by making it more difficult for hackers to penetrate the firewall.

Many firewalls allow the system administrator to install, configure, and monitor network security from remote and local sites. Some even issue pager alerts (alarms) to notify the administrator of any remote or local security policy violation so prompt action can be taken to track down the source.

An optional feature of some firewalls is virtual private network (VPN) implementation. The VPN option implements IPSec-compliant encryption software to establish VPNs over the Internet. By deploying VPNs, companies can create their own secure networks on the Internet without the cost of expensive leased-line networks. VPNs permit companies to migrate their legacy systems to the Internet, enabling them to communicate securely and efficiently via Web-based applications to departments, business partners, and customers in remote locations.

LDAP Support

Some firewalls support the Lightweight Directory Access Protocol (LDAP), which enables the storage and management of user-level security information in any LDAP-compliant directory server. This allows crucial user information to be shared among multiple network applications, including firewalls and VPNs, throughout the enterprise. This eliminates the need for network managers to maintain redundant user information across multiple proprietary user data stores.

Check Point Software, for example, offers a module that enables its Fire-Wall-1 to become a full LDAP client, capable of communicating with one or more LDAP directory servers to obtain identification and security information for network users. This information includes:

IDENTIFICATION   User name, login name, e-mail address, directory branch, and associated template

AUTHENTICATION Authentication scheme, authentication server, and password

ACCESS CONTROL Authorized sources and authorized destinations

TIME RESTRICTIONS Time and day access privileges

ENCRYPTION Key negotiation scheme, encryption algorithm, and data integrity method

GROUPS Group membership

The firewall uses this information to enforce the enterprise security policy. When a user requests a connection to a protected host, the firewall can query the LDAP server for the appropriate user-level security information. All communications between the LDAP server and the firewall are protected by the Secure Sockets Layer (SSL) protocol. The LDAP module includes a Java-based GUI client to define and manage security information in the LDAP directory (Figure 49).

By storing user-level security information in a centralized LDAP directory, organizations can separate user-management responsibilities and security management. This significantly reduces the burden on network security managers to maintain current and accurate user information. In addition, the centralization of user information improves efficiency and overall enterprise security by eliminating redundant data.

Figure 49
Managing
FireWall-1 users is
done within
Check Point
Software's Java-
based LDAP
Account
Management
Client.

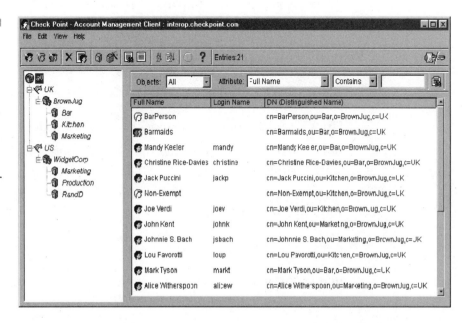

SUMMARY

While many companies are focused on thwarting attacks from external sources, a more dangerous threat comes from disgruntled or temporary employees. To address this problem, firewalls can be used between departments and other network security devices, software, and policy-based management can be used to prevent unauthorized access to data. Some vendors offer users a central source for all of their security needs—including firewalls, intrusion detection, content filtering, virus scanning, logging and reporting, and policy management.

***See Also* LAN Security, Security Risk Assessment**

FireWire

FireWire is a cross-platform high-speed serial bus developed by Apple Computer with some assistance from Texas Instruments. Defined in the IEEE 1394 standard, FireWire moves large amounts of data between computers and peripheral devices at speeds of 100, 200, and 400 Mbps. A follow-on standard, IEEE 1394B, defines FireWire at 800 Mbps, 1.6 Gbps, and 3.2 Gbps.

FireWire is specifically designed to connect digital consumer products—including digital camcorders, digital video tapes, digital video disks, set-top boxes, and music systems—directly to a PC or Macintosh computer. It features simplified cabling, hot swapping, and multiple speeds on the same bus, and allows up to 63 devices to be connected on a single bus. Connecting to a device is as easy as plugging in a telephone jack. In addition, FireWire allows users to instantly connect devices without first turning off their machines.

Features

FireWire has a number of compelling features, including:

DIGITAL INTERFACE Eliminates the need to convert digital data into analog for better signal integrity.

THIN SERIAL CABLE A shielded twisted pair cable replaces today's bulky and expensive interfaces.

EASE OF USE Eliminates the need for terminators, device IDs, screws, or complicated set-ups.

HOT PLUGGABLE Devices can be added and removed while the bus is active, eliminating the need to shut down the computer.

SCALABLE   The standard defines multiple speeds on a single bus.

FLEXIBLE   The standard supports free-form daisy-chaining and branching for peer-to-peer implementations.

FAST, GUARANTEED BANDWIDTH   The standard supports guaranteed delivery of time-critical data, which permits the use of smaller buffers to lower the cost compared to other interfaces.

FireWire supports two types of data transfer: asynchronous and isochronous. Asynchronous is for traditional load-and-store applications where data transfer can be initiated and an application interrupted as a given length of data arrives in a buffer. Isochronous data transfer ensures that data flows at a preset rate so an application can handle it in a timed way. For multimedia applications, this method of data transfer reduces the need for buffering and helps ensure a continuous presentation to viewers.

FireWire offers a standard, simple connection to all types of consumer electronics, including digital audio devices, digital VCRs, and digital video cameras, as well as traditional computer peripherals such as optical drives and hard disk drives. The protocols also include device-specific commands to start and stop camcorders and VCRs.

Comparison with USB and SCSI

Another approach to connecting devices, the Universal Serial Bus (USB), provides the same "hot plug" capability as FireWire. Although USB is a less expensive technology, its data transfer rate is limited to 12 Mbps. As opposed to handling high-speed multimedia, USB is better used for connecting medium- and low-speed peripherals to the computer. This broad category includes digital cameras, modems, keyboards, mice, digital joysticks, some CD-ROM drives, tape and floppy drives, digital scanners, and specialty printers.

The Small Computer System Interface (SCSI) offers a high data transfer rate (up to 40 Mbps), but requires address preassignment and a device terminator on the last device in a chain. FireWire can work with the latest internal computer bus standard, PCI, but higher data transfer rates may require special design considerations to minimize undesired buffering for transfer rate mismatches.

A SCSI bus requires that devices be serially daisy-chained together, with each device having a nonconflicting, preassigned address, and that the final SCSI device be terminated. But FireWire devices can be connected in multiple configurations. These can include a star or tree pattern with its own daisy-chained branches. Device terminators are not required. And FireWire addressing, unlike SCSI, is done dynamically; there is no need for address preassignment.

SUMMARY

FireWire has the bandwidth capacity to replace and consolidate most other peripheral connection communication methods in use today. Hot plugging and dynamic reconfiguration make FireWire a user-friendly alternative to today's interconnects. By providing an inexpensive, high-speed method of interconnecting digital devices, FireWire is a versatile I/O connection. Its scalable architecture and flexible peer-to-peer topology make FireWire ideal for connecting audio, video, and computer devices. Its isochronous support allows low-cost implementations of multimedia interfaces.

See Also **Small Computer System Interface, Universal Serial Bus**

Flow Management

To build Web sites for e-commerce and other mission-critical applications, administrators are mirroring site content at additional points of presence (POP). This provides redundancy in case one POP goes down, and enables traffic to be routed between the sites to increase overall response time. Flow management software determines where to send a request so the fastest service can be provided to clients.

Resonate Inc.'s Global Dispatch, for example, integrates multiple POPs into a single Web site resource. The company's flow management software uses the three most important factors in determining where to send a request: POP availability, POP load, and the Internet latency between client and POP.

As requests are received, the Global Dispatch scheduler instructs the agents at each Web server to measure the latency between the POP and the client's local Domain Name System (DNS). The results are sent back to the Global Dispatch scheduler and combined with current load and availability information in order to return the IP address (or virtual IP address) of the POP best suited to respond to the client. Global Dispatch stores this information in cache to enable faster response to future requests.

This arrangement can be especially beneficial to large organizations with many servers. Eventually, servers must be taken offline for repairs or upgrades. Flow control software allows a machine to be removed from the server mix, and the traffic is routed to the other Web servers so users can continue to access various services.

SUMMARY

Both the level of traffic and the need for reliable Internet services is increasing. Flow management is the next logical step in increasing the performance and management of Web sites and corporate intranets. Optimizing performance at the WAN level results in enterprise-wide increases in availability and speed.

See Also **Load Balancing, Network Caching**

Frame Relay

Frame relay is a fast-packet technology for transmitting data in high-speed bursts across a digital network. Frame relay can be used for integrated data, voice, and fax applications, as well as legacy Systems Network Architecture (SNA) and mission-critical intranet applications. Available since 1992, frame relay also supports multiple protocols, offers compatibility with other networking technologies, and costs less than other wide-area networking solutions.

The need for frame relay arose partly out of the emergence of digital networks, which are faster and less prone to transmission errors than older analog lines. Although the X.25 protocol overcomes the limitations of analog lines, it does so with a significant performance penalty, due mainly to its error-checking capability, which relies on the store-and-forward method of transmission. In being able to do without this and other functions, and be able to support different types of virtual circuits, frame relay offers more efficient usage of available bandwidth and, as a result, more configuration flexibility.

X.25, the only popular, nonproprietary method for wide-area, multi-protocol data communications, included error-correction and flow-control capabilities because at that time the public network was not dependable in terms of noise-free transmission between switches. X.25 became a popular protocol, but extensive switch-processing times and packet queuing at the switch locations caused packet transmission delays.

As noise-free fiber-optic lines came into widespread use, an improved protocol started being used for switch communication. With networks being faster and more reliable, the devices associated with those networks also became faster and more reliable, allowing them to handle end-to-end error correction and flow control. A new fast-packet technology called frame relay produced a higher level of reliability by eliminating switch-to-switch error-correcting procedures that were no longer necessary.

Frame relay is currently in use in various industries for different applications. Airline reservation systemsuse frame relay technology in their networks to improve performance, especially during busy traffic periods, and to lower online transmission costs. Manufacturing companies use frame relay to alleviate the periods of hectic activity when large amounts of data transmission occur between remote LANs. Banks are one of the biggest users of the frame relay technology, using it to connect LAN data traffic, teller machine transactions, and branch office terminals to SNA-based computer mainframes at bank data centers.

Technology

Frame relay is a fast-packet technology. A *fast packet* is transmitted without any error-checking points along the route. The receiver is responsible for ensuring that the packet arrives without error and for initiating a request for retransmission if needed. Fiber-optic media enables fast-packet transmission to work with a high degree of reliability, making internodal error correction of the kind used in X.25 packet-data networks unnecessary. Without this error-correction process, the queuing and processing of packets within the network is greatly reduced, allowing for faster transmission of data.

Bandwidth is used very efficiently with frame relay technology through the use of statistical multiplexing. In time-division multiplexing (TDM), which is typically used on private leased lines, each device is assigned a fixed-bandwidth channel. If a device has no data to send and none is received, the available bandwidth goes unused. Technologies using packets are able to perform statistical multiplexing, allowing the available bandwidth to be shared by many devices. This is more efficient because any unused bandwidth can be immediately used by another device to transmit data, resulting in faster response times and significant cost savings.

Lower operating costs result from several factors. The cost per transmitted byte is less because frame relay transmits data faster over the same bandwidth than other WAN protocols. Frame relay's ability to combine data packets from various sources via one access point minimizes the number of required access points in a network. The faster frame relay network and more responsive applications allow users to be more productive.

Advantages

The most compelling advantages of a carrier-provided frame relay service include:

NODE-TO-NODE CONNECTIVITY Any node connected to the frame relay service can communicate with any other node via a preprogrammed PVC or dynamically via an SVC. The need for multiple private lines is eliminated for substantial cost savings.

HIGHER SPEEDS Frame relay service supports transmission speeds up to 44.736 Mbps.

IMPROVED THROUGHPUT/LOW DELAY Frame relay service uses high-quality digital circuits end to end, making it possible to eliminate the multiple levels of error checking and error control in an X.25 network. The result is higher throughput and less delay compared to conventional packet-switching.

COST SAVINGS Multiple permanent virtual circuits can share one access link, eliminating the cost of multiple private-line circuits and their associated customer premises equipment (CPE), for substantial cost savings.

FLAT-RATE CHARGES Once a customer location is connected to the frame relay service "cloud," charges are insensitive to distance.

SIMPLIFIED NETWORK MANAGEMENT Customers have fewer circuits and less equipment to monitor. In addition, the carrier provides proactive monitoring and network maintenance 24 hours a day.

PROTOCOL TRANSPARENCY Frame relay service supports any higher-layer protocol that uses LAPF as the underlying data-link layer protocol.

INTERCARRIER COMPATIBILITY Carrier-provided frame relay service is compliant with the Frame Relay Forum standards associated with frame relay network implementation, namely FRF.2.

CUSTOMER-CONTROLLED NETWORK MANAGEMENT This allows customers to obtain network management information via in-band SNMP queries launched from their own network management stations. Information available includes performance monitoring, fault direction, and configuration information about frame relay service.

PERFORMANCE REPORTS Customers can manage their frame relay service to maximum advantage. Available online network reports include those for usage, errors, health, trending, and exceptions.

Types of Circuits

The two primary types of virtual circuits supported by frame relay are Permanent Virtual Circuits (PVCs) and Switched Virtual Circuits (SVCs).

PVCs are like dedicated private lines: once set, the predefined logical connections between sites stay in place. This allows logical channels to be dedicated to specific terminals. SVCs are analogous to dial-up connections, which require path setup and teardown. A key advantage of SVCs is that they permit "any-to-any connectivity." The SVC requires fewer logical channels at the host because the terminals (endpoints) contend for a lesser number of logical channels. Of course, it is assumed that not everyone will require access to the host at the same time.

SVCs can lower the expense of frame relay backbone networks by more efficiently supporting endpoints that infrequently communicate with each other. By requiring less backhaul and increased bandwidth usage through dynamic connectivity, SVCs become a very cost-effective solution.

Another type of virtual circuit, not widely implemented, is the Multicast Virtual Circuit (MVC), which can be used to broadcast the same data to a group of up to 64 users over a reserved data link connection. This type of virtual circuit might be useful for expediting communications among members of a single workgroup dispersed over multiple locations, or to facilitate interdepartmental collaboration on a major project.

The same frame relay interface can be used to set up SVCs, PVCs, and MVCs. All three may share the same digital facility. It is even possible to bundle SVCs within PVCs to help avoid delays that can time-out SNA sessions and disrupt voice traffic. In supporting multiple types of virtual circuits, frame relay networks provide a high degree of configuration flexibility, as well as more efficient usage of available bandwidth.

Standards

The Frame Relay Forum (FRF) is a nonprofit organization composed of corporate members dedicated to the implementation of frame relay in accordance with national and international standards. The forum was started in 1991 and is made up of carriers, consultants, users, and vendors. The FRF currently has 300 members worldwide. All of the major network equipment vendors are members of the Forum, including 3Com, Alcatel, Ascend Communications, Bay Networks, Cabletron, Cisco Systems, Ericcson, Hewlett-Packard, IBM, Lucent Technologies, Motorola, Nortel, and Siemens.

The Forum develops the growth of frame relay by promoting an understanding of the applications, benefits, and user experiences of frame relay technology. The Forum also concentrates on the development and approval of implementation agreements (IAs) that designate how standards will be applied for support of the frame relay protocol.

The Forum is continually evaluating the development of new applications related to the frame relay market, always seeking interoperability among the various equipment manufacturers and service providers. The FRF's evaluation of these developments will lead to future IAs for designating how the standards will be applied for the support of this protocol. Frame relay IAs to date consist of the following:

FRF.1.1 User-to-Network Implementation Agreement

FRF.2.1 Network-to-Network Implementation Agreement

FRF.3.1 Multiprotocol Encapsulation Implementation Agreement

FRF.4 Switched Virtual Circuit Implementation Agreement

FRF.5 Frame Relay/ATM Network Internetworking Implementation Agreement

FRF.6 Frame Relay Customer Network Management Implementation Agreement

FRF.7 Frame Relay PVC Multicast Service and Protocol Description Implementation Agreement

FRF.8 Frame Relay ATM/PVC Service Interworking Implementation Agreement

FRF.9 Data Compression over Frame Relay Implementation Agreement

FRF.10 Frame Relay Network-to-Network Interface Switched Virtual Connections Implementation Agreement

FRF.11 Voice over Frame Relay Implementation Agreement

FRF.12 Frame Relay Fragmentation Implementation Agreement

FRF.13 Service Level Definitions Implementation Agreement
The American National Standards Institute (ANSI) T1S1 committee issues standards describing frame relay (T1.606, T1.618, and T1.617) and the ITU publishes similar standards (I.233, Q933, and Q922 Annex A). The Frame Relay Forum expands the capabilities of the technology and works with these standards organizations.

Implementation Issues

For any network service, cost is the primary issue that usually merits the most consideration, and frame relay is no exception. The cost of implementing frame relay technology varies with each network and organization. Frame relay service providers can invoice users for a wide range of items, including:

- Access rate, which is the cost of the total bandwidth needed at the access point
- Burst characteristics, usually determined by a committed burst size and burst excess size
- Committed Information Rate (CIR), the anticipated, average rate of data flow through the access point
- Customer Premises Equipment, which includes the frame relay or internetworking access equipment leased by the service providers
- Port fee, the cost for connection to a port on the service provider's switch
- Required Private Virtual Circuits (PVCs) and Switched Virtual Circuits (SVCs)
- Usage amount, normally invoiced at a cost per megabyte of data moved over the network

For companies moving from IBM's host-centric (mainframe) environment to a distributed computing environment (local-area networks), the major carriers offer wide-area network services that address the specific needs of SNA users. Carriers have become quite adept at prioritizing traffic so SNA can be handled reliably end to end.

The attraction of frame relay is that it reduces transport costs by eliminating more costly leased lines. A carrier-managed service relieves customers of having to continually monitor and fine-tune network performance to handle delay-sensitive SNA traffic. If implemented from a central office, there are savings in the cost of equipment as well, since the carrier provides it. Carriers also provide service-level agreements that guarantee latency, frame discard rates, network availability, and response to trouble calls..

For a legacy SNA shop that does not have the expertise or resources, a managed SNA service can be an economical interconnectivity option. Although frame relay networks generally are much more difficult to configure, administer and troubleshoot than private lines, the carrier assumes these responsibilities. Subscribers can even obtain periodic reviews of their network's performance and efficiency from the carrier. This helps ensure that mission-critical applications are providing the highest level of performance, reliability and availability.

National carriers like AT&T, MCI WorldCom, Sprint, and Cable & Wireless Communications offer managed frame relay services for SNA traffic. Some regional carriers such as U S WEST and Ameritech also offer managed SNA service. SNA has traditionally been carried over dedicated point-to-point wide-area links. It is estimated that 60 percent of the data traffic over private lines consists of SNA traffic, which explains why carriers want to address this market with managed SNA services over frame relay.

SUMMARY

Networking professionals today must deal with a variety of difficult networking issues, including improving performance and manageability, reducing costs, providing high levels of availability, and ensuring their network's future by providing a path for new technologies. Frame relay is a proven technology with many demonstrable benefits. It integrates multiple protocols and applications, and has proven to be fast, reliable, and cost-effective.

***See Also* Asynchronous Transfer Mode, Frame Relay Access Devices, Integrated Services Digital Network, Video over Frame Relay, Voice over Frame Relay**

Frame Relay Access Devices

Frame relay offers a more economical alternative to wide-area networking than private leased lines, delivering high availability and predictable response times. Connecting a site to a frame relay network requires the use of a Frame Relay Access Device (FRAD) or router. Both types of devices assemble data into frames suitable for transmission over a frame relay service and return them to their original format at the destination end. Some routers can do this as well, but FRADs are generally more adept at congestion control and prioritization than routers. Some FRADs have a built-in CSU/DSU.

Basic Features

FRADs perform basic routing functions. Like routers, FRADs are primarily used for data transport and provide multiprotocol support to address legacy and LAN networks. Through a modular approach, voice and dial backup can be added as well. Both types of devices can be managed with the Simple Network Management Protocol (SNMP) and both provide additional management features. Generally, FRADs are a more economical way to connect branch offices to corporate centers than routers, and they are easier to configure.

For small remote sites, there are fixed-configuration FRADs that come in one- and two-port versions for connecting to Ethernet or token-ring LANs. Traffic from LAN-attached devices and locally attached serial devices can all be combined and transported over a single high-speed Frame Relay line. Typically, they come with embedded SNMP, which operates in-band through either the DTE port or the network link.

Among the protocols commonly supported by FRADs are SNA/SDLC, BiSync, SLIP, PPP, IP Routing, Telnet, Transparent Async, and Transparent Bit-Oriented Protocol (TBOP). The devices are also available with an integrated 56/64-Kbps DSU/CSU, and both EIA- 232 and V.35 are supported on the DTE and DCE interfaces.

FRADs increase application availability and ensure guaranteed response time in several ways. Priority output queuing enables network administrators to prioritize traffic by protocol, message size, physical port, and SNA device. Custom queuing offers the same level of prioritization granularity and ensures that mission-critical traffic receives a guaranteed minimum amount of bandwidth. Low-volume traffic streams receive preferential service, transmitting their entire offered loads in a timely fashion. High-volume traffic streams share the remaining capacity, obtaining equal or proportional bandwidth.

Some FRADs combine compression, routing and bridging technology with an access rate of up to 2 Mbps to the wide-area frame relay service. Service availability is further enhanced by automatic alternative Data Link Connection Identifier (DLCI) routing.

More functions are continually being packed into FRADs. Depending on vendor, a unit may also include a bit error rate tester, dialed management connections, alarm dial-out, ISDN backup, and bandwidth-on-demand options. With more vendors supporting voice-over frame relay, FRADs are offering features such as software-configurable interfaces for connecting to PBXs and key telephone systems, including support for:

FXO (FOREIGN EXCHANGE OFFICE)   A trunk loop start connection that emulates a single-line telephone to central office lines or PBX stations

FXS (FOREIGN EXCHANGE STATION)   A station loop start connection to a standard, single-line telephone instrument, or a loop start trunk circuit of a PBX

2/4-WIRE E&M (EAR AND MOUTH)   A tie-line trunk circuit that connects multiple PBXs or a PBX to another voice-switching system such as an automated call distributor (ACD)

SNA Application Support

When used to support SNA applications, FRAD performs the functions of a LAN-to-host gateway, a multiprotocol router, a bisync-to-frame relay access device, and a polled async-to-frame relay device. Traffic is consolidated and transported over a frame relay network to various hosts at the central site (Figure 50).

Most FRADs perform local SDLC polling acknowledgment of keep-alive frames (sometimes called "spoofing") to minimize the risk of timing-

Figure 50
The role of a FRAD in supporting legacy SDLC/SNA data on a frame relay network.

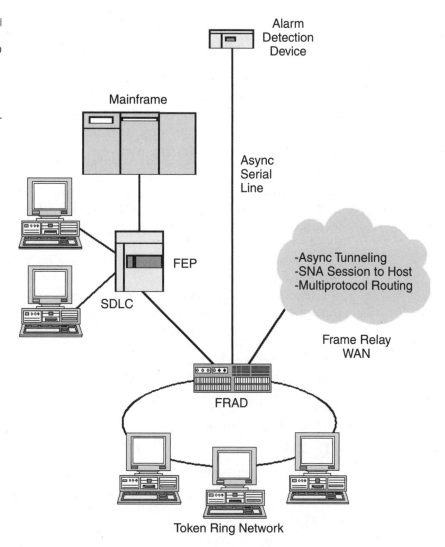

out SNA sessions. Some FRADs combine traffic from multiple SNA/SDLC devices onto a single virtual circuit, instead of requiring a separate virtual circuit for each attached device, which results in even greater cost savings.

Some vendors have optimized their FRADs so they can provide advanced features and still support SNA traffic. They equip their FRADs with a higher-speed CPU to handle processing-intensive features such as protocol translation, address resolution, and compression. Some vendors use proprietary techniques, such as cell-based processing, to divide LAN frames into cells. This enables cell-based prioritization of transmitted traffic, which enhances SNA response times and minimizes the impact of LAN traffic in mixed environments.

Some routers and FRADs come with an integral CSU/DSU. These integrated devices greatly simplify installation and management, but may not include all the functionality of separate products. However, for a delay-sensitive application like SNA, an integrated solution may be the best approach. This is because the more devices the traffic must pass through and the more processing they must do to perform advanced functions, the more variable delay is increased on the network.

Management

FRADs can interoperate with a central site router. The need for on-site trained personnel is eliminated by the FRAD's self-learning functionality, which complements the integrated SNMP in-band management. The FRAD "learns" its frame relay environment and downloads all configuration parameters from the central site, requiring no interaction at remote sites to establish connectivity.

Carriers that offer frame relay services provide their customers with the option of purchasing or leasing FRADs. Carrier-provided FRADs offer management reports that provide such information as:

- Traffic correlation
- Frame relay usage
- Peak period usage
- Throughput analysis
- Detailed port statistics
- Detailed VC (Virtual Circuit) statistics
- Virtual circuit usage to CIR (Committed Information Rate)
- Port usage
- Virtual circuit usage

- Port usage

- Discard eligible counts (transmit)

- Discarded frame—Cause (transmit)

However, CPE-based FRADs usually provide more management capabilities and reporting granularity. For example, Sync Research offers a management package called SNA Outlook for its FrameNode multiservice FRAD. It gathers extensive data on the availability and performance of SNA flows and attached devices using industry standards and Sync's own enterprise management information bases (MIBs). SNA Outlook leverages this data to provide an extensive view of SNA over frame relay performance on a controller or port basis. Specific features of SNA Outlook include:

- Proactive identification of specific controllers, connections, or events that are outside of performance norms using "Situations to Watch" reporting

- Enhanced SNA and WAN capacity planning using hourly and daily volumes versus historical baseline comparisons

- Up-to-date and historical views of SNA QoS (Quality of Service) factors down to a specific controller

- End-to-end network-response time monitoring

- Online, hardcopy, or Web-based reporting

SUMMARY

Frame relay access devices perform many of the same functions as routers, but usually support a smaller number of protocols, making them very economical for tying in branch offices to the corporate network. The two types of devices are often used together on the same network, with FRADs connecting to a central router, where a network manager can monitor and control the remote devices. The functionality of routers and FRADs is rapidly converging. Both support voice, video, and legacy data. Eventually, little will distinguish the two types of devices on the frame relay network.

See Also **Data Service Unit/Channel Service Unit, Frame Relay, Simple Network Management Protocol**

G

Gateways

Gateways make the translations that permit interoperability between dissimilar networks or systems. They consist of protocol conversion software that usually resides in a server, minicomputer, mainframe, or front-end device. Gateways process the various protocols so information from the sender is intelligible to the receiver, despite differences in their networks or computing platforms.

Gateways are often used for LAN workstation connections to legacy host environments, such as IBM SNA 3270 systems and IBM midrange systems. In some cases, the gateways can even be used to consolidate hardware and software. For example, an SNA 3270 gateway shared among multiple networked PCs can be used in place of IBM's 3270 Information Display System. Although the IBM system is a standard means of achieving the micro-host connection, it is expensive when used to attach a large number of stand-alone microcomputers. The relatively high connection cost per PC discourages host access for occasional users and limits the central control of information.

If the PCs are on a LAN, however, one gateway can emulate a cluster controller and thereby provide all users with host access at a very low cost. Cluster controller emulators use an RS-232C or compatible serial interface to a host adapter or communications controller, such as an IBM 3720 or 3745. They can support up to 254 simultaneous sessions.

Gateways are also used on the Internet. Possible scenarios in which gateways are used include:

E-MAIL GATEWAYS Vendors such as Lotus, Microsoft, and Novell each rely on a proprietary interface to send messages throughout a network of server-based post offices and clients. To connect to the rest of the world via the Internet, they provide gateways to make the translations to other message routing protocols like the Simple Mail Transfer Protocol (SMTP).

IP/IPX GATEWAYS　IPX is the proprietary networking protocol of Novell's NetWare. Starting with version 4.11, NetWare includes IntranetWare and a built-in IP/IPX Internet gateway that gives users access to applications in both environments at the desktop.

WEB-LEGACY GATEWAYS　The integration of Web applications and legacy data on mainframes is implemented with gateways that translate HTML (HyperText Markup Language) and HTTP (HyperText Transfer Protocol) code into 3270 data streams. IBM, for example, bundles its CICS[1] Gateway for Lotus Notes and its CICS Internet Gateway with its Transaction Server products. The CICS Internet Gateway interfaces Web servers and host (mainframe or AS/400) legacy CICS applications, translating between HTML and 3270 data streams to let Web browsers display 3270 screens as if they were Web pages. The Gateway makes Transaction Server applications available to anyone running a Web browser.

Gateway Services

Carriers and value-added network (VAN) service providers offer gateway services that let companies connect disparate messaging platforms such as e-mail, fax, and legacy systems via the Internet infrastructure or dedicated IP backbones.

AT&T, for example, offers IP gateway services that encompass multiple messaging infrastructures, letting companies retain their existing systems and still take advantage of AT&T's high-performance, managed IP backbone for cost savings. The services help IS managers overcome the difficulties of integrating messaging protocols such as SMTP, UUCP, and X.400. The gateway also provides a single point of access to send messages, whether the recipients use faxes, telex, or proprietary mail. The service lets users connect through a browser plug-in or through their existing e-mail clients.

SUMMARY ■ ■ ■ ■ ■ ■ ■ ■

Gateways operate at the highest layer of the OSI reference model—the Applications layer. They go beyond the capabilities of bridges and routers in that they not only connect disparate networks, but ensure that the data

---

1. CICS is IBM's Customer Information Control System, a suite of client-server transaction processing products that allows organizations to exploit applications and data on different platforms.

transported from one network is compatible with that of other networks at the application level. In addition to its translation capabilities, a gateway can check on the various protocols being used, ensuring that there is enough protocol processing power available for any given application. It also can ensure that the network links maintain a level of reliability for handling applications in conformance to user-defined error rate thresholds.

See Also **Bridges, IP-to-PSTN Gateways, Routers**

Gigabit Ethernet

Gigabit Ethernet is an extension of the 10-Mbps version of Ethernet. Offering a raw data rate of 1000 Mbps or 1 Gbps, Gigabit Ethernet maintains full compatibility with the huge installed base of Ethernet nodes through the use of LAN switches or routers. Because the frame format and size are the same for all Ethernet technologies, no other network changes are necessary.

Gigabit Ethernet supports full-duplex operating modes for switch-to-switch and switch-to-end-station connections and half-duplex operating modes for shared connections using repeaters and the CSMA/CD access method. Figure 51 shows the functional elements of Gigabit Ethernet.

Gigabit Ethernet was ratified as a standard by the Institute of Electrical and Electronic Engineers (IEEE) in mid-1998. The 802.3z standard draws heavily on the use of Fibre Channel and other high-speed networking components. Fibre Channel encoding/decoding integrated circuits and optical components are readily available and are specified and optimized for high performance at relatively low costs. Gigabit Ethernet employs Fibre Channel's high-speed, 780-nm (short wavelength) optical components for signaling over optical fiber and 8B/10B encoding/decoding scheme for serialization and deserialization. Fibre Channel technology operating at 1.063 Gbps has been enhanced to run at 1.250 Gbps, thus providing the full 1000-Mbps data rate for Gigabit Ethernet. For longer link distances—up to at least 2 km using single-mode fiber and up to at least 550 meters on 62.5-micron multimode fiber—1300-nm (long wavelength) optics are specified.

Initially operating over optical fiber, there is now a 1000BaseT specification (IEEE draft 802.3ab) for Gigabit Ethernet over copper wire. This gives businesses that rely on the easy-to-use, high-speed Gigabit Ethernet technology the flexibility of deploying it over both their installed copper and fiber infrastructures. The 1000BaseT specification enables Gigabit Ethernet to operate over distances of up to 100 meters using Category 5

Figure 51
Functional
elements of
Gigabit Ethernet.

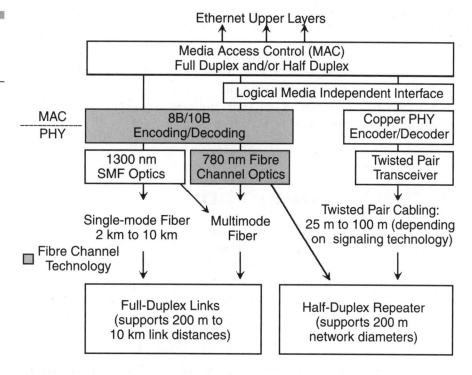

balanced copper wiring, which constitutes the majority of the cabling inside buildings.

The following table summarizes the standardization efforts of Gigabit Ethernet over various media:

Specification	Transmission Facility	Purpose
1000BaseLX	Long-wavelength laser transceivers	Support links of up to 550 meters of multimode fiber or 3,000 meters of single-mode fiber.
1000BaseSX	Short-wavelength laser transceivers operating on multimode fiber	Support links of up to 300 meters using 62.5-micron multimode fiber or links of up to 550 meters using 50-micron multimode fiber.
1000BaseCX	Shielded twisted pair (STP) (STP) cable spanning no more than 25 meters	Support links among devices located within a single room or equipment rack.
1000BaseT	Unshielded twisted pair (UTP) cable	Support links of up to 100 meters using four-pair Category 5 UTP.

Source: IEEE 802..3z Gigabit Task Force

SUMMARY

The seamless connectivity to the installed base of 10-Mbps and 100-Mbps equipment, combined with Ethernet's scalability and flexibility to handle new applications and data types over a variety of media, makes Gigabit Ethernet a practical choice for high-speed, high-bandwidth networking. *See Also* **Ethernet, Fast Ethernet**

Groupware

Groupware is a category of software that allows people to work together over a network to share ideas and information, keep everyone informed, coordinate and schedule activities, improve the quality of work, use their time more effectively, and increase productivity. Some examples of products that fall into the groupware category are:

- Videoconferencing applications that allow participants to see each other and share documents at the same time

- Scheduling programs that examine the electronic calendars of each person for the purpose of arranging meetings at a time and place that is convenient for everyone

- Meeting applications that allow participants to view the same documents on their computer screen for the purpose of editing them together

- Discussion group programs that allow users to carry on multithreaded discussions in text form

- Mail lists that enable people to post messages that are viewed by all list subscribers

- Chat programs that enable people to carry on text conversations in real-time

- Workflow management systems that assign documents to various workstations for multistep processing.

Several groupware products are aimed at business users on large-scale enterprise networks. Examples include Lotus Notes and Domino, Microsoft Exchange Server, Netscape SuiteSpot, and Novell GroupWise. All four can be used over corporate intranets and provide messaging and/or publishing capabilities over the Web.

Shared Web Space

There are also dedicated Web-based groupware products. One of these is involvFree Web Teaming, which is offered by Changepoint Corp. As a free service offered in partnership with U S WEST, involvFree Web Teaming provides an easy way for people to collaborate on projects, regardless of their location, using only a Web browser. The service provides step-by-step instructions that help users instantly create and use an interactive team space on the Web. It also includes a personal workspace for each team member to create memos, post calendar items, comment in discussions, add Web links, assign tasks, and take polls (Figure 52).

Users can instantly create a password-protected dedicated interactive site on the Web for their team or group. Based on a "lite" version of Changepoint's corporate teaming product involved Intranet, the free site allows invited participants to:

- Customize a personal page
- Use a discussion board
- Post files of any type for others to download (pictures, word processor files, spreadsheets, etc.)
- Assign and track to-do lists or tasks
- Post Web links
- Vote on issues

Figure 52

Changepoint's involvFree Web Teaming is a Web-based groupware product that facilitates interactive collaboration.

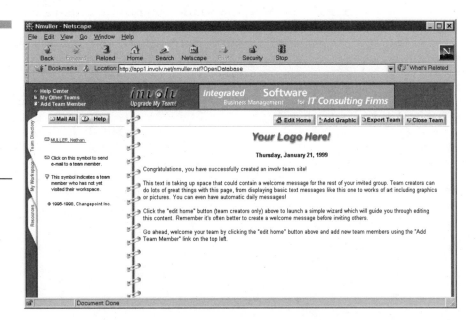

Users need only a Web browser; no plug-ins or additional software are required. There is a customizable team or group home page that greets users when they enter. Single teams of up to 15 members are supported and up to 3MB of storage is provided. If more features, users, and storage space are required, upgrade options are available.

The advantages of this free Web space are compelling. For example, messages with attachments are posted once for access by the entire team; with regular e-mail, a separate attachment must be sent to every team member, which wastes server and bandwidth resources. With information posted once in a central location, team members know they have the latest version. And when new team members are added, they can catch up on things by viewing the entire team history. Finally, team information can be accessed wherever members happen to find a browser, without having to carry around a laptop and phone line to log onto the Internet.

SUMMARY

Groupware technology is designed to facilitate the work of project teams. This technology may be used to communicate, coordinate, solve problems, compete, and negotiate. While traditional technologies such as the telephone may also qualify as groupware—especially if used for teleconferencing—the term is applied primarily to a specific class of technologies that rely on computer networks.

See Also **Electronic Mail, Videoconferencing Systems, Workflow Management**

High-level Data Link Control

High-level Data Link Control (HDLC) is a bit-level protocol for data transmission over point-to-point links. It was written in the 1960s and standardized by the International Organization for Standardization (ISO). Despite its age, HDLC has influenced all subsequent frame formats, including IBM's Synchronous Data Link Control (SDLC), which was introduced in 1973, and the X.25 Link Access Procedure Balanced (LAPB) link layer.

The job of HDLC is to ensure that data passed up to the next layer has been received as transmitted—error free, without loss, and in the correct order. It also performs flow control to ensure that data is transmitted only as fast as the receiver can accept it.

HDLC sends serial data as a clocked stream of bits, which are organized into frames. Each frame is bounded by the special bit pattern 01111110. To prevent this flag pattern from appearing within the user's data, the sending device inserts a 0 after any five consecutive 1s in the data, once the opening flag is sent. This procedure is called zero-bit insertion or "bit stuffing." At the receiving end, after detecting the start of a frame, the device removes any zero that appears after five consecutive 1s within the frame. Three categories of frames are used in HDLC:

- Information frames transport data across the link and may encapsulate the higher layers of the OSI architecture.
- Supervisory frames perform the flow control and error recovery functions.
- Unnumbered frames provide link initialization and termination.

The HDLC information frame consists of a header, user data, and Frame Check Sequence (FCS), as shown in Figure 53. The FCS is the result of a mathematical computation performed on the frame at its source. The

Figure 53
An HDLC frame
with flag
delimiters at each
end.

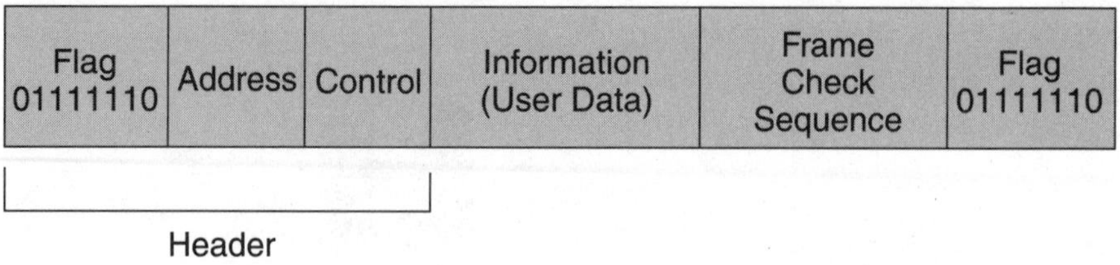

Flag 01111110	Address	Control	Information (User Data)	Frame Check Sequence	Flag 01111110

Header

same computation is performed at the receiving side of the link. If the answer does not agree with the value on the FCS field, this means some bits in the frame have been altered in transmission, in which case the frame is discarded.

The HDLC frame header contains a destination address field, which is used for multipoint versions of HDLC, such as SDLC. This field enables many systems to share a single transmission line. Each system is assigned its own address, and traffic is sent to a particular system when its address is inserted in the header.

The HDLC frame header also has a control field, where some protocols such as SDLC and LAPB put frame numbers and acknowledgement numbers for supervisory functions. These link protocols retransmit numbered frames that are not acknowledged within a timeout period. Other protocols such as IP, IPX, and DECnet do not require numbering and acknowledgment, in which case the control field is set to identify it as an unnumbered information frame.

SUMMARY

Many options are defined in the HDLC standard, and there are many vendor implementations of HDLC. Consequently, for many years there was no standard for point-to-point communication, which made it difficult to internetwork equipment from different vendors. This problem was solved by the Internet Engineering Task Force (IETF), which came up with the Point-to-Point Protocol (PPP). This protocol could be used to carry the protocol data units of IP, IPX, DECnet, ISO, and others over any full-duplex circuit—either synchronous bit-oriented or asynchronous start/stop byte-oriented—from slow dial-up to ISDN and fast leased lines to SONET fiber-optic lines.

See Also **Synchronous Data Link Control**

High-Performance Parallel Interface

The High-Performance Parallel Interface (HPPI) is a gigabit-per-second technology that was originally developed in the 1980s to allow mainframes and supercomputers to communicate with one another, and with directly attached storage devices.

HPPI is spelled out in a series of ANSI standards: X3.183-1991 (physical layer), X3.222-1993 (switch control), X3.218-1993 (link encapsulation), X3.210-1992 (framing), ANSI/ISO 9318-3 (disk connections), and ANSI/ISO 9318-4 (tape connections).

HPPI can be configured for either of two speeds—800 Mbps or 1.6 Gbps, both either simplex or full duplex. These high-speed connections can be set up with just three messages:

REQUEST   This message, issued from the source, asks for a connection.

CONNECT   This message, issued from the destination, indicates that the connection has been established.

READY   This message, issued by the destination, indicates that it is ready to accept a stream of packets.

HPPI technology is available in switches, routers, mainframe, and supercomputer channels, interfaces for mass storage devices, frame buffers, and adapters for workstations that use the Peripheral Component Interface (PCI) bus.

Characteristics

Through Physical-layer flow control, a HPPI switch offers reliable and efficient communications between devices operating at different speeds. In essence, the source keeps track of the Ready signals and sends data only when the destination can handle it. Connection-oriented circuit switches allow multiple conversations to take place currently, enabling the aggregate bandwidth to be very high—equal to 800 Mbps or 1.6 Gbps times the number of ports.

HPPI is protocol-independent, which means that its channels can handle raw data formatted with its own framing protocol (without any upper-layer protocols), as well as IPI-3 (Intelligent Peripheral Interface) framed data used to connect high-speed peripherals like RAID (redundant array of inexpensive disks) devices to computers.

Most vendors of computers with HPPI channels also have software drivers for TCP/IP. Although HPPI is a circuit-switched technology, it can handle datagram traffic efficiently, given the small size of most TCP/IP packets. Furthermore, HPPI's streamlined signaling sequences allow connections to be set up and torn down in less than one microsecond. A single port on a HPPI switch can deliver hundreds of thousands of IP datagrams every second, outperforming IP routers or hosts, which would typically be the sources of this traffic. The latency of a HPPI switch averages 160 nanoseconds—so fast that any bottleneck would be associated with protocol processing and data handling at the end devices, not the switch itself.

For short distances, HPPI links computers via two 50-pair copper cables. This configuration furnishes a full-duplex (or dual-simplex) 800-Mbps channel that can extend up to 25 meters. HPPI also works with single-mode and multimode optical fiber across the campus or metropolitan area. SONET links are used for longer-distance connectivity. HPPI-Serial interfaces allow runs of 300 meters over fiber; with fiber extenders, devices can be as far apart as 10 kilometers.

Applications

HPPI is used in a variety of networking environments. It can play a key role in linking workstations and other hosts, connecting workstations with storage systems at higher speeds than any other currently available technology, attach display peripherals for real-time visualization, and provide connectivity to other networks.

HPPI not only provides connectivity to other networks, it can be complementary to those networks. With regard to ATM, for example, a gateway from a HPPI workstation cluster to ATM would give end users the best of both worlds: the high throughput of HPPI in the local area and the wide-area connectivity of ATM. With such a gateway, even multimedia data could fan out from HPPI-attached servers to multiple ATM desktops. The source gateway would encapsulate HPPI data and ship it over the ATM network in 53-byte cells. At the destination gateway, the data would be put back into its original format. This process, called *tunneling*, is implemented at ATM Adaptation Layer 5 (AAL5)—the same layer that supports LAN emulation.

HPPI is often compared to Fibre Channel. Of the two, Fibre Channel is more complex because it specifies four data rates, three kinds of media, four transmitter types, three distance categories, three classes of service, and three possible fabrics. While the intent was to make Fibre Channel rich in features, it also delayed the process of bringing Fibre Channel products to market.

Despite their differences, HPPI and Fibre Channel can be complementary technologies. There are ANSI standards that specify how to send upper-layer Fibre Channel protocols over the lower-layer HPPI media, as well as how to ship the HPPI upper-layer protocol over Fibre Channel lower-layer media. IP-level routing offers another way to map HPPI to Fibre Channel, and vice versa.

As noted, HPPI takes advantage of SONET to extend connectivity over long distances. In this case, the HPPI network is terminated at a HPPI-SONET gateway, which frames HPPI data for transport over SONET. This can be done fairly easily, since HPPI maps well both to a single SONET OC-12c (622 Mbps) circuit or multiple SONET OC-3 (155 Mbps) circuits.

HPPI-6400

The High-Performance Networking Forum (HNF) has developed a standard for Gigabyte System Network (GSN), also known as HPPI-6400. The draft standard proposes a one-gigabyte-per-second parallel copper interface capable of supporting link distances of up to 40 meters. An optical standard for GSN is also in development.

Users of processor-intense data or low-latency applications such as HDTV and film postproduction, storage area networks, satellite imagery, and seismic modeling are expected to be among the first to benefit from the eventual 6.4-Gbps speed of GSN technology. With more than six times the capacity of Gigabit Ethernet or Fibre Channel, HPPI-6400 promises to radically change the conception of a LAN. The technology enables a cluster of commodity computers, servers, and storage devices—located throughout a building or across a campus—to operate as the equivalent of a massively parallel supercomputer. This is made possible because HPPI-6400 is as fast or faster than most system buses. In the process, HPPI-6400 blurs the distinctions between LANs, I/O, and storage interconnects, as well as the buses and backplanes found inside computers.

HPPI-6400 employs a fixed-length cell of 32 bytes, giving it many of the efficiency advantages of ATM. Software overhead is greatly reduced by providing a reliable link level in the hardware, along with multiplexing, retransmission, and flow control.

SUMMARY

HPPI was created in the 1980s, when Los Alamos National Laboratory in New Mexico needed a circuit-switched connection that would run at 800 Mbps between its Cray supercomputers and visualization devices. Unlike packet-switched interconnections, circuit switching creates a dedicated pathway between nodes for the duration of a session. The throughput rate was close to Gigabit Ethernet. HPPI-6400 throughput towers above to-day's networking technologies, with at least six times the bandwidth. Fibre Channel supports 1.06 Gbps today, but will not be able to approach the speed of HPPI-6400 for many years. Gigabit Ethernet is similarly bandwidth-limited. ATM can surpass its commercial limit of 622 Mbps, experimentally reaching 2.488 Gbps, but it lacks flow control and retransmission capabilities.

See Also **Asynchronous Transfer Mode, Fibre Channel, Gigabit Ethernet, Synchronous Optical Network**

High-Speed Serial Interface

The High-Speed Serial Interface (HSSI) is a full-duplex synchronous serial interface capable of transmitting and receiving data at up to 52 Mbps. HSSI began life as a de facto specification developed by Cisco Systems and T3plus Networking. Cisco, a router vendor, and T3plus, a manufacturer of DS3/E3 DSUs and bandwidth managers, were concerned about the bandwidth bottleneck created by lower-speed serial interfaces such as V.35 and RS-232 when interfacing data communications equipment (DCE) to data terminal equipment (DTE) for wide-area network communications. In 1993, HSSI was ratified by the ANSI EIA/TIA TR 30.2 committee as EIA SP-2796 for the physical interface specifications, and as EIA SP-2795 for the electrical specifications.

Applications

Among the applications for HSSI are T1 consolidation, LAN interconnection, IBM channel extension, and multimedia networking. HSSI comprises the physical and electrical interface between DTE (such as computers, routers, channel extenders, and peripherals) and DCE (such as multiplexers and DSU/CSUs) that, in turn, establish links over the wide-area network. This arrangement supports the transmission of digital data at rates up to 52 Mbps in full-duplex mode over shielded cable at distances of up to 50 feet. HSSI can connect through a DSU, for example, to such WAN services as:

- Carrier-provided T3/E3 services (44.736 Mbps)
- Fractional T3 (FT3) and switched T3
- Switched Multimegabit Data Service (SMDS)
- Frame relay at T3 rates
- Point-to-point T3 interconnection between remote Fiber Distributed Data Interface (FDDI) LANs using a router/DSU equipment configuration
- Interconnection of multiple 16 Mbps token-ring and 10 Mbps Ethernet LANs over the wide-area network
- ATM (Asynchronous Transfer Mode) and SONET (Synchronous Optical Network) through the use of special equipment designed to support these connections

Working with a bandwidth manager (also known as an inverse multiplexer), HSSI can run at any speed to support bandwidth allocation among input devices up to 52 Mbps. This capability makes HSSI useful on managed transmission facilities. Bandwidth managers can dynamically allocate bandwidth to many devices and applications at varying speeds across broadband connections on the wide area network.

Other Interfaces

Before HSSI, standard inputs to communications devices, such as routers, multiplexers, and DSU/CSUs, were generally limited to V.35, EIA-422/449, and T1 connections. While T1 specifies a maximum data rate of 1.544 Mbps, the upper limits of V.35 and EIA-422/449 are more difficult to define because the data rate varies with distance: the shorter the connections between DTE and DCE, the higher the data rate.

V.35 was originally intended to support no more than 6 Mbps, while EIA-422/449 tops out at 10 Mbps. Although there are ways to increase the data rates of these interfaces to as high as 45 Mbps, the drawback is that they become, in essence, proprietary interfaces. Furthermore, increasing the speed of V.35 and EIA-422/449 can also increase electromagnetic emissions, which can cause interference problems with other nearby equipment and connections.

Although no single device is likely to need the full bandwidth of a T3 or OC-1 line, HSSI makes it possible to allocate the 52 Mbps among different DTE. Other interfaces such as V.35 and EIA-422/449 are not designed to do that. With HSSI, for example, a single port can provide connections to six routers. In contrast, six separate V.35 interfaces would be needed to provide connections to the same six routers.

SUMMARY

HSSI is now widely adopted as a high-speed interface in numerous types of broadband data communications systems. The driving forces behind HSSI growth is the increasing use of T3, ATM, and SONET network services and high-bandwidth applications. HSSI overcomes the limitations of V.35 and EIA-422/449 interfaces, thereby providing an important means of reducing bottlenecks between LAN interconnect and network access devices.

See Also **High Performance Parallel Interface**

High-Speed Token Ring

With 16 Mbps token ring connections between switches easily become congested at busy times and high performance servers becoming less able to deliver their full bandwidth potential over a 16 Mbps token ring connection, the need for a high-speed solution for token ring has become readily apparent in recent years. Other high-speed technologies were already available for inter-switch links and server connections—FDDI, Fast Ethernet and ATM—but they are inadequate for the token ring environment. Consequently, several token ring vendors teamed up to address this situation by forming the High Speed Token Ring Alliance (HSTRA) in 1997. A year later, the alliance issued a specification for High Speed Token Ring (HSTR), which offers 100 Mbps and preserves the native token ring architecture.

However, to keep costs to a minimum and to shorten the time to completion of a new standard, HSTR is based on the IEEE 802.5r standard for Dedicated Token Ring, adapted to run over the same 100 Mbps physical transmission scheme used by dedicated Fast Ethernet. HSTR links can be run in either half-duplex or full duplex mode, just like Dedicated Token Ring.

Alternative Solutions

Although FDDI, Fast Ethernet and ATM had been available well before the HSTR specification was issued, these high-speed technologies were not adequate for supporting token ring traffic.

FDDI provides 100 Mbps capacity and a high degree of fault tolerance, but is viewed by many as obsolete because of its cost, complexity, and lack of an upgrade path to speeds beyond 100 Mbps. In addition, FDDI and token ring use different packet formats and different address format conventions, requiring the translation of every packet that passes between them. This translation requires substantial amounts of processing power, which has a negative impact on the price/performance equation.

Fast Ethernet offers low cost and simplicity, but does not make an ideal interconnection medium for token ring environments. In the packet formats of the two technologies, the bit order of every packet address is the reverse of the other, which requires that address translation be carried out on every packet. In addition, Ethernet does not support source route bridging (SRB), making it impossible to support many of the fault tolerance and load-sharing mechanisms on which enterprise LAN users depend. For instance, SRB lets users set up multiple or parallel paths to off-load traffic from congested or failed routes. Finally, Ethernet supports a smaller maximum packet size than token ring (1526 bytes with preamble vs. up to 18.2 Kbytes[2], making it necessary to reconfigure the networking software in every token ring end station to limit the packet size. This would result in a corresponding drop in the actual throughput performance achieved at each end station.

ATM offers a better high-speed solution for token ring environments than FDDI or Fast Ethernet without compromising the special features of token ring. The ATM Forum specification for LAN Emulation (LANE) supports the handling of token ring packets over ATM networks, providing full compatibility with token ring's address format, maximum packet size, and source routing capabilities. ATM is also very scalable—offering link speeds of 155 Mbps and 622 Mbps as well as 2.4 Gbps and 10 Gbps—and can be used to build fault tolerant networks. However, ATM is complex and too expensive to introduce into the token ring environment.

By preserving token ring's native packet formats, handling the full range of token ring packet sizes, and supporting token ring's source routing and packet priority capabilities, HSTR provides the means to expand the capacity of token ring networks without the compromises imposed by other solutions, and without users having to learn a new technology.

---

2. Although token ring permits a maximum frame size of 18 Kbytes, most practical applications of token ring use a maximum frame size of 4096 bytes, which is still 37 percent larger than the maximum Ethernet frame size.

Implementing HSTR

Unlike alternative technologies, HSTR uses existing switches, hubs, bridges, routers, network interface cards (NIC) and cabling. This introduces greater throughput where the enterprise needs it most—at the server and backbone. Upgrading these connections with HSTR only requires that an HSTR uplink be plugged into a token ring switch and that the existing 16 Mbps server network NIC be replaced with a 100 Mbps HSTR NIC. To complete the upgrade, the two devices are connected with appropriate cabling. The 100 Mbps HSTR operates over both Category 5 UTP and IBM Type 1 STP cable, as well as multimode fiber-optic cabling.

It is also possible to connect desktop systems to token ring switches on dedicated 100 Mbps HSTR connections. Token ring vendors offer 4/16/100 Mbps adapter cards that enable companies to standardize on a single network adapter and prepare their infrastructure for the eventual move to HSTR. While the HSTR standard does not define an auto-negotiation algorithm, individual vendors have a number of ways to implement the feature while adhering to the standard. With this feature, HSTR products operate at the maximum connection speed, automatically determining whether to transmit at 4 Mbps, 16 Mbps or 100 Mbps. Many corporations install auto-negotiating 4/16/100 Mbps NICs in today's desktops, even though the need for 100 Mbps throughput to the desktop might be years away. When the hub or switch at the other end of the connection is later upgraded to 100 Mbps HSTR, the token ring desktop will automatically adjust transmission to 100 Mbps.

Since Ethernet packets can be carried over token ring links, HSTR makes an ideal choice of backbone medium for the mixed-technology LAN. With support for the maximum token ring frame size, an HSTR backbone segment will be able to handle Ethernet and token ring frames on the same VLAN connection, which Fast Ethernet would not be able to do without a lot of processing to break down the larger token ring frames.

SUMMARY

HSTR is the next step in the evolution of token ring network technology, which began in 1985 as a 4 Mbps transmission method, in 1989 increased its speed to 16 Mbps, and now has made the jump to 100 Mbps. Just as Fast Ethernet has been adapted to run 10 times faster, work is underway on an even faster version of token ring to run at 1 Gbps. Despite these advancements, ATM is likely to remain the preferred choice for those requiring a WAN backbone to accommodate mixed Ethernet and token ring traffic. ATM meets the demand for guaranteed Quality of Service (QoS) in the LAN, and provides a way to extend the LAN backbone environment seamlessly across high-speed WAN links.

See Also **Fast Ethernet, Gigabit Ethernet, Token Ring**

Hubs

Hubs consolidate LAN segments at a central point. A collapsed backbone within the hub interconnects the various segments. This simplifies wiring and facilitates network management. Through the hub, the network is segmented so the failure of one segment—shared among several devices or dedicated to just one device—does not affect the performance of the other segments.

Hubs come in many sizes, from four-port workgroup hubs to stackable department hubs to enterprise-level hubs with hundreds of ports interconnected through a gigabit-per-second backplane. The largest hubs interface to the wide-area network, including router modules that provide access to corporate intranets or the public Internet. Some hubs even accommodate server modules, which can host a Web site or a data warehouse that can be accessed over a corporate intranet.

The larger hubs are modular in design to support networks that combine different LAN topologies and media types in a single chassis. Ethernet, token ring, ATM, and FDDI modules enable multiple networks to coexist within a single hub. LAN segments using twisted pair wiring, coaxial cable, and optical fiber also can be interconnected through the hub. The various segments can be isolated from each other, allowing NetWare to be run on one segment, NetBEUI on a second, and TCP/IP on a third.

Hub Components

Enterprise-level hubs contain four basic components: chassis, backplane, plug-in modules, and a network management system.

CHASSIS The chassis contains an integral power supply and/or primary controller unit, and varies in the number of available module slots. The modules insert into the chassis and are connected by a series of buses, each of which may constitute a separate network or be integrated into one or more backbone networks. The various modules fit into a chassis backplane socket, providing connectivity to other modules across the common backplane.

Stackable hubs provide a dedicated uplink port, which allows the hub to be connected to other hubs as a means of accommodating growth. The uplink port looks like any other RJ45 port, but may require a special crossover cable that reverses the transmit and receive pairs. Some hubs use special cascade cables to join multiple hub units into a single integrated stack and carry all information between units (Figure 54).

BACKPLANE   The main artery of the hub is its backplane, a board that contains one or more buses carrying all communications between LAN segments. The hub's backplane is analogous to a PC bus through which various interface cards are interconnected. The data path that carries traffic from card to card is often called a *channel*; unlike the PC, however, a hub's backplane typically consists of multiple physical or logical channels. Minimally, the hub accommodates one LAN segment for each channel on the backplane.

Segmenting the backplane this way allows multiple independent LANs or LAN segments to coexist within the same chassis. The segmented backplane typically has dedicated channels for Ethernet, token ring, and

Figure 54

3Com's SuperStack II PS Hubs, for example, require a special cascade cable to join multiple hubs into a single integrated stack. Source: 3 Com Comp.

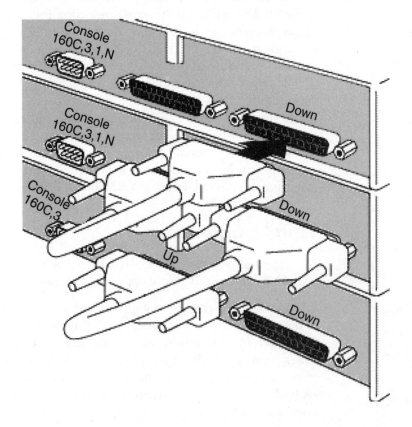

FDDI. A separate backplane channel usually carries management information. Some hubs employ a multiplexing technique across the backplane to divide the available bandwidth into multiple logical channels. Some hubs support load sharing that allows network modules to select the backplane channel that will transport the traffic. Other hubs allow backplanes to be added or upgraded to accommodate network expansion and new technologies.

The potential bandwidth capacity of newer backplane designs supporting ATM switching is quite impressive, reaching several gigabits per second—more than enough to accommodate several Ethernet, token ring, and FDDI networks simultaneously.

MODULES The functionality of hubs is determined by individual modules. Typically, the vendor will provide multiport Ethernet and token-ring cards, LAN management, and LAN bridge and router cards. The use of bridge and router modules in hubs overcomes the distance limitations imposed by the LAN cabling and facilitates communication between local- and wide-area networks. Several router modules support data link switching (DLSw) for tunneling SNA traffic over TCP/IP for multiprotocol connectivity in IBM environments. There are even plug-in terminal servers, communications servers, and file/application server modules.

Hub vendors also offer a variety of WAN interfaces, including those for X.25; frame relay; ISDN; fractional T1, T1, and T3; SMDS; and ATM. As many as 60 different types of modules may be available from a single hub vendor, many of them provided under third-party OEM, technology-swap, and other vendor-partnering arrangements.

These modules plug into vacant chassis slots. Depending on the vendor, the modules can plug into any vacant slot or slots specifically devoted to their function. Hubs supporting any-slot insertion automatically detect the type of module that is inserted into the chassis and establish the connections to other compatible modules. In addition, many vendors offer a "hot-swap" capability that permits modules to be removed or inserted without powering down the hub.

The modules provide ports to which LAN segments are connected. A key capability of today's hubs is port switching, which permits users to be allocated and reallocated easily from one segment to another via software to suit operational needs.

This capability increases bandwidth by permitting completely flexible allocation of ports to any network segment—even across multiple inter-linked hubs. It also makes it easy to alter workgroups—adds, moves, or changes to user groups are accomplished with a few keystrokes or mouse clicks from a management console, rather than physically rewiring the connections at a patch panel.

MANAGEMENT SYSTEM   Hubs occupy a strategic position on the network, providing the central point of connection for workstations, servers, hosts, bridges, and routers on the LAN and over the WAN. The hub's management system is used to view and control all devices connected to it, providing information that can greatly aid troubleshooting, fault isolation, and administration.

Hub vendors typically provide proprietary management systems that offer value-added features that can make it easier to track down problem-causing workstations or servers. Most of these management systems support the Simple Network Management Protocol (SNMP), enabling them to be controlled and managed through an existing management platform such as IBM's SystemView for AIX, Hewlett-Packard's OpenView, and Sun's Solstice SunNet Manager. Some hubs have Remote Monitoring (RMON) embedded in the hub, making possible more advanced network monitoring and analysis up to the application layer. The latest trend is Web-based management, which allows administrators with password access to manage the hub-based network from any workstation with Web browser functionality.

SUMMARY ▪ ▪ ▪ ▪ ▪ ▪ ▪ ▪

Hubs provide a central point of interconnection and management for the components that comprise workgroup, departmental, and enterprise networks. Hubs, which were developed in response to the need for structured wiring as networks became bigger and more complex, allow the wiring infrastructure to expand in a cost-effective manner as the organization's computer systems grow and move, and as local- and wide-area interconnectivity and integration requirements become more sophisticated. *See Also* **Bridges, Gateways, Routers**

I

Integrated Services Digital Network

The Integrated Services Digital Network (ISDN) made its commercial debut in 1980. As a digital service, it offers many advantages over the previous generation analog telephone network, including:

- Faster call setup and network response times
- Support for such multimedia applications as videoconferencing
- Integration of voice and data, thus reducing the complexity and cost of cabling and equipment
- Increased network management and control
- Improved configuration flexibility and additional restoral options

Although the growth of ISDN had been slow for many years, the increasing popularity of the Internet has sparked new interest in ISDN as a high-speed access alternative for navigating graphic-intensive Web sites and using multimedia applications such as streaming video and collaborative computing. However, ISDN is still not available everywhere—customers must be located within 18,000 feet of the nearest central office or remote terminal that supports ISDN.

ISDN Channels

ISDN is a circuit-switched digital service that comes in two varieties. The basic rate interface (BRI) provides two bearer channels of 64 Kbps each, plus a 16-Kbps signaling channel. The primary rate interface (PRI) provides 23 bearer channels of 64 Kbps each, plus a 64-Kbps signaling channel. Any combination of voice and data can be carried over the B channels. Additional channels can be created within the B channels through the use of compression.

The D channel can access the control functions of various digital switches on the network. It provides message exchange between the user's equipment and the network to set up, modify, and clear the B channels. The D channel also gathers information about other devices on the network, such as whether they are idle, busy, or off. In being able to check ahead to see if calls can be completed, network bandwidth can be conserved. If the called party is busy, for example, the network can be notified before network resources are committed.

The D channel goes unused most of the time. With a new feature called Always On/Dynamic ISDN (AO/DI), the D channel can maintain continuous access to the Internet, regardless of how the B channels are used. The D channel provides ample capacity for e-mail notification and automated information delivery from channel subscriptions on the Web. It also keeps both B channels open for voice calls and faxing, which is especially beneficial to telecommuters.

With regard to ISDN PRI, there are two higher-speed transport channels called H channels. The H0 channel operates at 384 Kbps, while the H11 operates at 1.536 Mbps. These channels carry multiplexed data, data and voice, or video at higher rates than that provided by the 64-Kbps B channel. The H channels are also ideally suited for backing-up corporate Fractional T1 (FT1) and T1 leased lines. Eventually, other high-speed transport channels will be implemented in support of Broadband ISDN. Multirate ISDN lets users select appropriate increments of switched digital bandwidth on a per-call basis. Speeds, in increments of 64 Kbps, are available up to 1.536 Mbps. Multirate ISDN is used mostly for multimedia applications and video conferencing.

Call-Handling Features

A number of call-handling features can be bundled into carrier-provided basic and primary-rate ISDN services. Depending on the carrier, there may be as many as 30 features bundled into the ISDN lines. For example, Calling Party Identification (CPID) delivers the originating phone number with each SETUP message carried over the D channel. This information has a number of uses such as routing calls, identifying calling numbers, and trapping calling numbers. Numerous other custom-calling features are available with ISDN, including:

- Call pickup features provide capabilities for stations to pick up calls directed to other stations. This includes barge-in to allow a station to pick up a call that has already been answered at another station and night line support.

- Abbreviated dialing features include manual line definition, generation of an attendant request by any off-hook condition, direct-connect placement of a call to a particular number, and speed dialing via short codes that represent local or long-distance numbers.

- Attendant service features provide equivalent PBX attendant console features to the ISDN/centrex user. The attendant operator can originate and receive calls, receive and transfer calls, create and drop multiparty calls, clear calls, and hold calls.

- Restriction features enable users to deny a station the capability of originating or receiving calls. This feature group also includes the capability of restricting calls to a local centrex number and diverting a toll call-access attempt to an attendant.

- Multiway calling features include call transfer, call forwarding (if not included in the call-waiting feature group), three-way calling, call hold, and conference calling.

- Call-forwarding features can forward incoming calls only; forward only outside or in-group incoming calls; and forward calls over private facilities. These capabilities can be used in conjunction with those in other feature groups.

- Call-waiting features include alerting a busy station that an incoming call is waiting (with optional tones that distinguish an intragroup call from an outside call); alerting a calling station that the destination is busy but has waiting capability; call waiting-call type identification; cancel and institute call waiting; and automatic call-back.

- Call-hunting features assign a single address to multiple lines so calls made to a main number are transferred on busy to a list of secondary numbers without a billing/routing penalty to the caller. This feature group includes the capability to busy a number in the hunt group; queue calls in that group for the next available line; and generate announcements to the queued callers.

Equipment Requirements

Although ISDN BRI and PRI services consist of different configurations of channels, both require specific equipment to provide network connectivity. One of these elements is the Network Terminator 1 (NT1), which resides at the user's premises and performs the four-wire to two-wire conversion required by the local loop. Aside from terminating the transmission line from the central office, the NT1 device is used by the telephone company for line maintenance and performance monitoring.

Network Terminator 2 (NT2) devices include all NT1 functions in addition to protocol handling, multiplexing, and switching. These devices are usually integrated with PBX and key systems.

ISDN terminal equipment (TE) provides user-to-network digital connectivity. TE1 provides protocol handling, maintenance, and interfacing functions, and supports such devices as digital telephones, data terminal equipment, and integrated workstations—all of which comply with the ISDN user-network interface. The large installed base of non-ISDN TE2 devices (e.g., telephones and microcomputers) can communicate with ISDN-compatible devices when users attach or install a Terminal Adapter (TA) to/in the non-ISDN device. A TA takes the place of a modem. Users can connect a maximum of eight TE/TA devices to a single NT in a multidrop configuration. Some vendors have combined ISDN terminal adapters and 56 Kbps modems in a single device, providing users with more communications flexibility.

SUMMARY

Today, the telephone companies claim they can reach about 95 percent of their serving areas with ISDN. Despite this impressive figure, the phone companies generally have not been able to extend ISDN to customers located on the perimeter of the local loop. Due to changing demographics over the past decade, there are now almost as many people located around the perimeter of the local loop and beyond as there are within the local loop itself. This situation can be remedied through the introduction of certain Digital Subscriber Line (DSL) technologies. One of these technologies—Asymmetrical Digital Subscriber Line (ADSL)—not only provides Internet access and supports emerging new applications such as video-on-demand programming and interactive gaming at multimegabit-per-second speeds, but supports plain old telephone service (POTS) and ISDN service as well.

See Also **Digital Subscriber Line Technologies**

Intelligent Agents

Intelligent agents are autonomous and adaptive software programs that accomplish their tasks by executing commands remotely. What makes agents so smart is the addition of programming code that tells them exactly what to do, how to do it, and when to do it.

System administrators, network managers, and software developers can create and use intelligent agents to execute crucial processes including performance monitoring, fault detection and restoral, hardware and software asset management, virus protection, and information search and retrieval.

Network Management Application

With intelligent agents, management hierarchies can be established among distributed network operations personnel. Problems can be sent from one level in the hierarchy to another, as necessary, for rapid resolution. This capability, based on a set of predetermined policies and rules, helps organizations take full advantage of their management personnel's expertise, regardless of where they may be located.

For example, an international bank might need to control computing services for hundreds of branch offices throughout Europe. To eliminate the high expense of maintaining local operators at each branch office, the bank can deploy intelligent agents at each branch, each of which is monitored by operators at one of five regional management centers. To complete the management hierarchy, the bank can establish one European escalation center.

Any problem that the intelligent agents cannot resolve locally is sent automatically to the appropriate operator at a regional management center. Where a problem is sent is based on its nature. If a problem relates to data storage, for example, it can be sent to a storage specialist. If the regional operator cannot solve the problem, it can be forwarded, along with supporting data, to the European escalation center for resolution.

In addition to coordinating problem resolution among levels of a management hierarchy, intelligent agents allow organizations to establish management centers of expertise in different geographic locations. The intelligent agents differentiate between the various types of problems and send them to the most appropriate experts, regardless of an expert's location. For example, if an organization's database specialists reside in a management center in Chicago, all database-related problems occurring nationwide in the U.S. could be directed to the management consoles at the Chicago site. Alternatively, all network-related events could be directed to a site that employs network specialists.

Management solutions based on the use of intelligent agents also allow organizations to orchestrate shifting management control and the flow of information among multiple management centers. A large company, for example, may have major management centers in New York and Los Angeles that share responsibility for keeping a nationwide network running. This type of organization can configure its management environment so, based on the time of day, control is automatically shifted from one management center to another. Applied to a global corporation, this capability provides around-the-clock worldwide support with only one operations shift working at any given time.

At 5 P.M. in New York, for example, control would be shifted from New York to Los Angeles. With this action, each distributed intelligent agent would stop reporting status information to New York and begin reporting status information to the Los Angeles management center, which then would be responsible for problem resolution.

The capability to shift control from one management center to another also provides organizations with a backup management server in the event of a server failure. In the previous example, a sitewide power failure at the New York site could be handled by shifting control to the Los Angeles site until power in New York was restored. In the interim, open communication from Los Angeles to all distributed intelligent agents throughout the network would be maintained.

Intelligent agents can also help network managers maintain consistent service delivery levels. By integrating network performance and availability information within a network management environment, a profile can be provided about how the network is behaving so decisions can be made before a problem disrupts service. For example, the information gathered by intelligent agents can identify impending bottlenecks quickly, allowing more bandwidth to be put into service or traffic to be rerouted before users experience delays.

Data Collection

Many data collection tasks can be automated with intelligent agents. Specifically, intelligent agents can provide valuable automated assistance in the following areas:

PERFORMANCE MANAGEMENT Network performance monitoring can help determine network service level objectives by providing measurements to help managers understand typical network behavior and normal/peak periods.

FAULT MANAGEMENT When faults on the network occur, it is imperative that problems be resolved quickly to decrease the negative impact on user productivity. Intelligent agents can be used to gather and sort the data needed to quickly identify the cause of faults and errors on the network.

CAPACITY PLANNING AND REPORTING Allows for the collection and evaluation of information to make informed decisions about future network configurations to accommodate growth in client/server computing environments.

SECURITY MANAGEMENT A properly functioning and secure corporate network plays a key role in maintaining an organization's competitive advantage. Intelligent agents can help discover holes in network security by continuously monitoring network access points for possible security breaches and reconfiguring the network to prevent further attacks.

APPLICATIONS MANAGEMENT Client-side agents continuously monitor the performance and availability of applications from the end user's perspective. A just-in-time applications performance management capability captures detailed diagnostic information at the precise moment when a problem or performance degradation occurs, pinpointing the source of the problem so it can be resolved immediately.

DISK SPACE CONSUMPTION Data collection agents can be used with configuration and reporting tools to enable administrators to set alerts and thresholds to monitor user disk space consumption and quotas, ensure storage availability, and reduce the amount of time spent managing disk space. Administrators can configure the agent to monitor servers and workstations and be alerted when disk space consumption across the entire network or across selected partitions exceeds a specified quota. Alerts are automatically sent to the administrator, the user, or both so that disk space consumption can be controlled without affecting productivity.

REMOTE USER SUPPORT To deal with the problems of providing remote users with support, client-side agent software gives IT administrators a presence on each machine, regardless of its location. The agent allows IT administrators to define profiles for these out-of-reach PCs and how they should be configured, what software must be on the hard drive, and how often they should run programs like diagnostics and virus checks. When the client logs onto the corporate network, the agent facilitates an exchange of information with the remote access server, which continues until the client exactly fits the defined configuration. The agent can even be instructed to log onto an FTP server on its own, in order to download the required files. If the transmission is interrupted, the agent can pick up the installation right where it left off during the next log-in.

Internet Tasks

Intelligent agents can also be used for Internet-related tasks. They can monitor information logged by servers on the Web, for example. When the log entries exceed a designated threshold, it may indicate a high demand for applications and impending congestion if the logging rate continues. The intelligent agent can act on this information to redirect traffic to another server to balance the load.

There are now Java-based agents capable of monitoring and reporting on key health parameters of Internet systems, services, and applications. Since Java is a cross-platform development tool, agents built with Java can provide a single, unified management system to support any mix of IP-based desktop, server, and network resources that also run Java including hubs, switches, and routers. In addition to relieving the burden of front-line managers, who usually must cope with a collection of unrelated tools while demands on them are accelerating, the Java agents can self-populate through the network to add new resource support and functional enhancements.

Some mobile agents can even be launched over wireless IP services, such as CDPD (Cellular Digital Packet Data), to carry out specialized information gathering and processing tasks on private intranets. Since the agent performs the monitoring and reporting functions, the mobile IT manager does not have to waste time calling into the network to check log files, view performance indicators, or retrieve management reports. When the agent gathers the appropriate information, the results are delivered to the IT manager the next time he or she calls into the network. The use of agents on wireless networks minimizes airtime charges and allows IT professionals to work more productively while away from the network management console.

Agent Toolkits

Many vendors offer toolkits that accelerate the development of the agent and manager components, which is normally a significant and time-consuming activity. Without a toolkit, each agent must be hand-coded, or built from scratch—a process that can take weeks or months. The use of toolkits can reduce development time to only minutes, allowing developers to spend more time on the value-added components of their application, such as processing data gathered by the agent or communicating with and controlling external devices.

The agent-creation process is further simplified because developers can now use an intuitive C++ interface that insulates them from the complexities of application programming interfaces (APIs). For example, without using a toolkit, a developer might have to write more than 200 lines of code to create a simple "get" request. With a toolkit, such agent development can take as few as four lines of code, with the rest of the code being generated automatically. By drastically reducing the amount of manual coding, developer errors are reduced, and quality and productivity are increased. In addition, the code-generation process provides greater code consistency, thus improving code quality and maintainability.

There are also test kits that allow developers to verify the proper operation of agents before they are let loose in the live environment. Through the use of interactive and regression tests, the agent can be exercised during various stages of development. The interactive test method provides the ability to incrementally test the customization of the agent, while the regression-testing method allows for a complete suite of tests to be executed, with the results being verified against the expected results. The agent tester toolkit also gives developers the flexibility to customize generated test programs, incorporating event-handling and response, error-handling and complex SNMP MIB definitions.

Agent Templates

Some vendors offer rules-based templates to modify the behavior of intelligent agents. The network manager or system administrator can bring up a representation of the template used for monitoring a particular application and edit the rules concerning responses to various alerts without having to actually rewrite the agent in its native development language.

For instance, when a firewall issues an error message, under the rules described in its template, it sends all alerts to a particular system administrator. The network manager can change the rule so an automated response is initiated instead, allowing agents to resolve problems and perform routine tasks (e.g., backups, batch jobs, file maintenance) locally. This prevents system administrators from being overwhelmed by warning and informational messages, freeing them to focus only on potentially service-disrupting conditions that cannot be resolved locally.

SUMMARY

Intelligent agents are proven, indispensable tools for providing network management assistance. Problems can be identified and resolved locally by agents rather than by harried operators at a central management console or by sending technicians to remote locations, which is expensive and time-consuming. In many cases, intelligent agents can implement restoral actions automatically in response to certain performance threshold indications, or reconfigure the network to prevent unauthorized access attempts. Agents will become even more indispensable as networks

continue to expand and companies continue to minimize their personnel requirements.

See Also **Performance Baselining, Performance Monitoring, Remote Monitoring**

Internet-Enhanced Mobile Phones

Internet-enhanced mobile phones potentially represent an important
communications milestone. These devices look much like the common
cell phones in use today, but incorporate design features and functions
that allow them to be used to access the Internet.

One vendor that has been particularly active in this area is Nokia, the
world's biggest maker of mobile phones. The company's model 7110 is in-
dicative of the types of new cell phones from about 70 other manufacturers that are aiming at the world's 200 million cellular subscribers. It
displays Internet-based information on the same screen used for voice
functions. It also supports Short Message Service (SMS) and e-mail, and includes a calendar and phonebook.

The phone's memory can also save up to 500 messages—SMS or
e-mail—sorted in various folders such as inbox, outbox, or a specially defined folder. The phonebook has memory for up to 1000 names, with up
to five phone and fax numbers and two addresses for each entry. The user
can mark each number and name with a different icon to signify home
or office phone, fax number or e-mail address, for example. The phone's
built-in calendar can be viewed by day, week, or month, showing details
of the user's schedule and calendar notes for the day. The week view
shows icons for the jobs the user has to do each day. Up to 660 notes in the
calendar can be stored in the phone's memory.

Nokia has developed several innovative features to make it faster and
easier to access Internet information using a mobile phone:

LARGE DISPLAY   The screen has 65 rows of 96 pixels (Figure 55), allowing
it to show large and small fonts, bold or regular text, and full graphics.

MICROBROWSER   Like a browser on the Internet, this feature enables
users to find information on the Internet by entering a few words to
launch a search. When a site of interest is found, its address can be saved
in a "favorites" folder, or input on the keypad.

NAVI ROLLER   This built-in mouse looks like a roller (Figure 56) that is
manipulated up and down with a finger to scroll and select items from
an application menu. In each situation, the Navi Roller knows what to do
when it is clicked—select, save, or send.

Figure 55
Display screen of the
Nokia 7110.
Source: Nokia.

Figure 56
Close-up of the Navi
Roller on the Nokia
7110.
Source: Nokia

PREDICTIVE TEXT INPUT As the user presses various keys to spell words, a built-in dictionary continually compares the word in progress with the words in the database. It selects the most likely word to minimize the need to continue spelling out the word. If there are several word possibilities, the user selects the right one using the Navi Roller. New names and words can be input into the phone's dictionary.

However, the phone cannot access just any Web site. It can access only Web sites meant to give users access to important information that has value when they are mobile. For the most part, these would be Web sites that comply with the Wireless Access Protocol (WAP), a standard for Internet-enhanced cell phones. CNN and Reuters are among the content providers that have developed WAP-specific news and information services for delivery to cell phones. The WAP Forum envisions cell-phone access to news, weather reports, stock prices, flight schedules, and wireless banking, plus access to corporate and ISP e-mail.

The WAP standard supersedes the Handheld Device Markup Language (HDML) developed by Unwired Planet. However, the 30 or so sites that built content with HDML can easily convert their content for the new breed of Internet-enabled phones. HDML failed to catch on largely because there were no phones with big enough displays, which are only now becoming available.

SUMMARY

The convergence of voice and data, as well as wireline and wireless networks, is manifested in today's Internet-enabled mobile phones. Larger screen sizes, simple browsers, sophisticated messaging, and easy-to-use navigation controls make this type of phone appealing to the large majority of users who want a single, convenient device for all of their voice, messaging, and Internet access needs while they are on the road.

See Also **Electronic Mail, Mobile Phones, Short Message Service**

Internet Protocol

The Internet Protocol (IP) routes packets of data over the Internet and other TCP/IP-based networks. Although 30 years old, IP stands poised to dramatically extend and enhance communications in the 21st century. The capabilities built into IP in the 1960s to allow the exchange of data between different computers interconnected over low-speed lines are also advantageous when used over advanced high-speed internets. These next-generation networks are comprised of ATM switches and routers capable of supporting data transmission at 622 Mbps and beyond over SONET-based fiber-optic backbones and rings.

Despite its age, IP remains important for several reasons. It is nonproprietary, open, and offers cost-effective ways to merge voice and data traffic on a common platform. All things considered, IP networks meet the business-critical requirements for interoperability and integration, scalability, reliability, mediation, manageability, security, and global reach.

Interoperability and Integration

IP nets are superior to PSTN for voice-data convergence because they include all the advantages of interoperability and integration afforded by internationally recognized Internet standards. Although PSTNs also adhere to internationally recognized standards for the transmission of voice, as well as for the signaling systems used for circuit management, the standards do not address applications other than voice to the extent that the Internet community does.

Scalability

A network that easily scales has become a fundamental business requirement in recent years. It entails the ability to expand (or contract) the network as business needs change. There are several related aspects of scalability that are being addressed by next-generation networks: users, gateways, routers, and bandwidth capacity.

USERS In terms of users, scalability is illustrated by the expansion potential of VPNs. The closed user-group capability provided with the VPN service lets businesses limit incoming or outgoing voice-data calls to only members of the group. At the same time, businesses can quickly and efficiently add (or remove) users to their IP services without having to install remote systems such as access servers and modem pools.

Furthermore, the built-in security features of the VPN service—usually packet filters applied to the source address—allow for dial-in and dedicated access without the need for expensive firewalls on company premises.

When a data packet reaches a VPN access router, it is checked against a table of authorized source addresses. If there is no match, the packet is discarded. Similar filters are applied to connection points between the VPN service provider's IP backbone and the public Internet to ensure that only authorized traffic originates and terminates on the service provider's IP backbone.

The same packet-filtering capabilities of VPN can enable businesses to extend their reach to allow users within other organizations to have access to certain data sources and applications. With an IP-based infrastructure, companies can seamlessly go from being an intranet to an extranet using the same development tools, systems, and services. For example, a business can offer extranet service to its strategic partners, which gives their users access to its warehouse inventory and an Electronic Data Interchange (EDI) application that allows them to order and pay for selected items online, as well as track the status of order fulfillment. Assuming that the VPN provides effective security, extending this extranet further to include other strategic partners is basically a matter of providing them with an appropriate IP address and reconfiguring the packet filters to grant them access to specific network resources.

GATEWAYS Scalability also applies to gateways, which must accommodate incremental upgrades to handle increasing traffic and processing demands. Until relatively recently, one significant barrier to offering telephony services to a mass market via next-generation IP networks had been gateway scalability. There is the requirement for IP-PSTN gateways that is comparable in capacity to that of today's central office switches, which are capable of supporting not merely tens of hundreds of concurrent sessions, but tens of thousands.

The gateway scalability problem has been addressed by major central office switch vendors as well as CPE vendors. Nortel's IPConnect product line, for example, is a family of IP gateways that provides an interface between a managed IP network and traditional voice network. IPConnect supports the delivery of business-class enhanced services over a variety of deployments, ranging from small-office 24-port installations to carrier-class installations capable of carrying 100,000 voice-over-IP connections.

ROUTERS   Not only are next-generation service providers using fiber-optic networks for high speed and reliability, they are increasing router capacity to handle more traffic and adding more routers to eliminate excessive router hops that cause delay. Current router capacity is becoming CPU-limited. In other words, the current capacity is less than 500 kilo-packets per second, and three to six OC-3 trunks running at 155 Mbps each is enough to congest a router. This leads to excessive packet loss and, of course, degraded throughput. The problem gets worse as more router hops are involved in a TCP/IP session. As networks scale in size and logically interconnect, the span of the network also increases, which magnifies performance problems.

Excessive router hops can be eliminated by employing high-speed switching or routing technologies in the core of the network and invoking the routing function only at its edges—that is, at the entry and exit points of the network. By reducing the number of hops to a fixed number—perhaps as little as two—performance gets much more predictable, regardless of the size of the network. The high-speed technologies used to accomplish this will include ATM, emerging gigabit switch-router technologies, and the use of new routing techniques such as tag and label switching, which speed up routers by compressing the information they need to determine where to send traffic. Combined, these methods can result in a 10- to 20-times increase in packet throughput, effectively eliminating the router CPU bottleneck and overcoming the effects of network scaling.

BANDWIDTH CAPACITY   With regard to capacity, a network must be able to handle baseline and peak load, and accommodate long-term increases in traffic. Next-generation IP networks meet the capacity needs of businesses by using optical fiber and deploying capacity-enhancement technologies such as Dense Wave Division Multiplexing (DWDM).

When DWDM is applied to SONET links or rings, transport capacity can be increased by a factor of 10, without the carrier having to lay any additional fiber-optic cable. The use of SONET and DWDM is important for another reason—they go a long way in overcoming a point of contention about ATM's inordinately high overhead relative to its cell size.

ATM's cell structure is fixed at 53 octets, of which 5 are reserved for the header and 48 for data. This translates into roughly 10 percent of the traffic being devoted to overhead functions, which becomes a cost issue with incumbent carriers. By contrast, IP over SONET incurs only 2-percent overhead.

With DWDM to increase fiber's already high capacity, ATM's overhead is no longer a serious issue. Instead, the focus is on ATM's unique ability to provide an appropriate quality of service for any given application—even IP-based applications—if IP is carried over ATM[3].

Interestingly, DWDM works only on the newer grade of optical fiber being installed by next-generation network service providers. These carriers all rely on nondispersion-shifted fiber, which can support up to 80 channels on a single hair-thin strand. The incumbent service providers—AT&T, MCI, and Sprint—use mostly dispersion-shifted fiber, which suffers from the problem of signal bleeding. This problem limits the older fiber to eight channels per strand, which are derived from the older wavelength division multiplexing (WDM).

Reliability

There are at least two aspects of reliability being addressed by next-generation networks: the reliability of applications and the reliability of the physical infrastructure.

APPLICATIONS RELIABILITY   Voice and other real-time applications have their own unique set of challenges, which affect reliability. While it is technologically easy to pipe real-time applications over IP networks, it does not necessarily result in a viable service. The service provider must be able to design and manage the network to control performance parameters such as as latency, throughput, and availability. Of these, controlling latency is the most problematic.

Latency is the amount of delay that affects all types of communication links. Delay on telecommunication networks is usually measured in milliseconds (ms), or thousandths of a second. A rule of thumb used by the telephone industry is that the round-trip delay for a telephone call should be less than 100 ms. If it is much more than 100 ms, participants think

---

3. At this writing, the Internet Engineering Task Force (IETF) is considering a quality of service protocol for IP called Differentiated Services (Diff-Serv), which would be implemented by routers. Briefly, Diff-Serv provides a standard way to label packets with a type of service (ToS) to ensure that real-time applications get preferential treatment over routine applications that are not sensitive to delay. Eight types of service are defined in the Diff-Serv protocol. They are encoded in the header of an IP packet to reflect their priority or class. Although supported by a few vendors, until Diff-Serv becomes commonplace, ATM's QoS can be used to assign IP traffic various priorities.

they hear a slight pause in the conversation and begin speaking. But by the time their words arrive at the other end, the other speaker has already begun the next sentence and feels that he or she is being interrupted. When telephone calls go over satellite links, the round-trip delay is typically about 250 ms, and conversations become full of awkward pauses and accidental interruptions.

Latency affects the performance of applications on data networks as well. .The problem of latency on IP networks can be addressed by resource management protocols, such as RSVP (Resource Reservation Protocol), which was developed within the Internet Engineering Task Force (IETF) community. RSVP improves the quality of service by making enough bandwidth available on a priority basis, end to end, to support telephony or multimedia applications

There are also many proprietary techniques for dealing with latency. Among them is priority output queuing, a technique used by Cisco Systems in its routers. With this technique, network managers can classify traffic into four priorities and provide the available bandwidth to the queues in the order of their priority. The highest-priority queue gets as much bandwidth as it needs before lower-priority queues get serviced.

Alternatively, IP can be run over ATM with a quality of service (QoS) assigned to the application, which identifies it as being time-sensitive and requiring priority over other less time-sensitive data types. QoS can be handled by the network (i.e., routers, hubs, and switches), the operating system, or a combination of both hardware and operating system working together.

Potentially, IP-based next-generation networks are able to offer higher-quality voice than traditional circuit-switched networks, which are engineered for 64K-bps signals with a frequency bandwidth of less than 3 KHz. By contrast, the bit rate and frequency range of next-generation networks are limited only by the capabilities of the devices and transducers at the communications endpoints.

These inherent advantages can be leveraged to deliver two-way voice communications with a frequency of 7 KHz, which is more than two octaves better than conventional circuit-switched voice. This level of audio quality also greatly enhances integrated audio-video applications, such as distance learning, telemedicine, videoconferencing and collaborative work sessions, the playback of archived television programs, online advertising, and interactive gaming. It could also be useful for certain electronic commerce applications.

For example, potential customers could run content from music CDs and video DVDs online before choosing to buy, or preview new movie releases before spending time and money at a local theater. Although such sampling is routinely available today over the Internet, the user experience can be greatly enhanced with a wideband capability that incorporates the latest audio, acoustic, and digital signal processing (DSP) technologies. This would allow potential customers to evaluate audio-video products based on technical quality as well as content.

INFRASTRUCTURAL RELIABILITY The reliability of the physical infrastructure is determined by the quality of the optical fiber and the survivability of the links in case of disaster. The use of newer nondispersion-shifted fiber not only supports many more channels per strand than older dispersion-shifted fiber, but it offers a lower error rate as well because it does not suffer from the problem of signal bleeding.

The capacity of nondispersion-shifted fiber is illustrated by the Macro Capacity Fiber Network of Qwest LCI. The network is able to transmit two trillion bits of multimedia information per second—the equivalent to transmitting the complete contents of the Library of Congress across the country in 20 seconds. In terms of errors, it is estimated that nondispersion-shifted fiber offers less than one bit of error in every quadrillion bits—the equivalent of one grain of sand out of place on a 20-mile stretch of beach.

However, it is not enough for a network to have great capacity and offer error-free transmission—it must be able to survive virtually any type of disaster. SONET's built-in management functions enable link integrity to be continually monitored in real time and on a nonintrusive basis. Thus, when a link experiences even the slightest performance degradation, a sequence of events is automatically triggered that will result in the transfer of traffic to another link without end users being aware that anything different has happened. Typically, the cut-over takes about 150 milliseconds to implement.

Mediation

Another advantage of IP is that it can translate or mediate whatever form of information it receives, regardless of what media (copper wire, optical fiber, wireless) or service (native IP, frame relay, ATM) it runs on. Some examples of the types of network situations mediation handles include:

- Incompatible client/server software
- Incompatible LAN environments (i.e., Windows and UNIX)
- Different communications systems (i.e., IP and POTS)

- Different QoS mechanisms (i.e., IPv4, IPv6, and ATM)
- Different bandwidth capabilities (i.e., caching, mirroring, and load distribution)

This mediation capability was built into the TCP/IP protocol suite from the start. As a platform-independent set of standards, TCP/IP bridges the gap between dissimilar computers, operating systems, and networks. It is supported on nearly every computing platform, from PCs, Macintoshes, and UNIX systems to thin clients and servers, legacy mainframes, and the newest supercomputers. In supporting both local- and wide-area connections, TCP/IP also provides seamless interconnectivity between the two environments.

Although ATM also provides seamless interconnection between local- and wide-area networks, it is far more expensive to implement than TCP/IP, especially at the desktop, where each machine must be equipped with a special network interface card and middleware to make existing applications "ATM aware." Without this awareness, existing applications cannot take advantage of ATM's QoS features. Of course, the applications can be run and metered at the server, in which case only that device needs to have the middleware and an ATM interface for server-to-server or WAN interconnectivity. A possible drawback to this approach is that it may entail changing the way many users are accustomed to working.

Next-generation networks will use ATM on their fiber-optic backbones, however, while some competitors like AT&T will use ATM to implement on-ramps to these new dynamic networking environments. The approach favored by AT&T is to place ATM switches on the customer premises, which consolidate voice, frame, and IP traffic on the same access line. This allows multiple traffic types to be fed into the switch in their native formats.

Manageability

In an effort to match the performance of PSTNs, next-generation service providers rely on managed IP backbones to support time-sensitive applications. The public Internet, which has no central point of management, suffers from delays and congestion that often degrades the performance of real-time applications. To ensure the peak performance of their IP networks end to end—even across different carrier's transport facilities—next-generation service providers as well as incumbent carriers like AT&T are turning to an international standard that enables multivendor network elements to be overseen from a single platform.

To ensure common operational capabilities and a common architecture, the ITU Telecommunication Management Network (TMN) standard architecture is being pursued. Among other things, TMN specifies the use of standard interfaces through which internal network management functions are made available. This allows groups of service providers to enter into business-level agreements and deploy resource sharing arrangements that can be administered automatically through interoperable interfaces. This is of particular value to next-generation network operators because it allows them to extend the reach of their networks globally through strategic partnerships rather than physically laying more fiber-optic cable. It also allows them to provide a consistent level of performance end to end so that they can attract new customers through service level guarantees.

By making all internal network management functions available through standardized interfaces, the TMN architecture allows service providers to achieve more rapid deployment of new services, both domestic and international, and make maximum use of automated functions. Network equipment vendors can offer specialized management systems known as *element managers*, which can integrate readily into a service provider's larger management hierarchy. In turn, this architecture allows service providers to align the development of their operations systems (OS) with current and future transport technologies in a way that is disassociated from the vendor-specific aspects of network element implementation.

As networks worldwide become more advanced and as service providers continue to engage in mergers and alliances in an attempt to better serve customer needs, adopting the interoperable management solution offered by TMN standards becomes a key factor in determining their market success.

Security

All types of networks are vulnerable to attacks of one kind or another. In the voice world, the telecommunications industry has been dealing with security breaches like toll fraud for decades, starting with "phone phreaking," the term given to a method of payphone fraud that originated in the 1960s that employed an electronic box held over the speaker. When a user was asked to insert money, the electronic box played a sequence of tones, indicating to the billing computer that money had been inserted. Since then, theft-of-service techniques have been applied to mobile phones as well. The estimated cost of toll fraud ranges from $3.5 to $5 billion a year in the U.S. alone.

In the data world, there is the potential for not only the theft of service, but also the theft of intellectual property. A joint survey by Computer Security Institute, Inc. (CSI) and the Federal Bureau of Investigation (FBI), for example, showed that "cybercrime" is on the rise and that security-related financial losses totaled $137 million in 1997. This is attributable only to the amounts reported; actual losses range from $7 to $10 billion a year.

Security solutions for the Internet and corporate intranets include encryption, tunneling, certificate services, route authentication, denial of service detection, and virus scanning. These security features are implemented by proxy servers, firewalls, routers, and remote access servers—alone or in combination. As the circuit-switched and packet-switched worlds converge in next-generation networks, the best security features from both will likely be applied to this new hybrid environment.

Effective security will be mandatory for companies seeking to take advantage of the economies and efficiencies of IP-based next-generation networks for mission-critical applications and processes. Accordingly, some next-generation and incumbent carriers have chosen to deal with these concerns head-on by adopting every industry standard for security as it emerges. This includes the adoption of competing standards for secure electronic commerce.

Global Reach

Because the Internet is ubiquitous, it immediately gives any business worldwide reach. Although global reach is provided by PSTNs, the current rates for international calls include very high access and settlement charges that greatly inflate the cost of telecommunications.

Among the top global communications issues for businesses are remote access and the costs associated with providing global reach. Remote workers, for example, must be able to dial into the corporate data network from international locations. And for employees trying to access sensitive corporate information, a secure service is required.

With global IP roaming, companies can build a logical network footprint around the world so remote workers can access corporate resources from virtually any location. There are two major benefits to this: a business can give its employees access to its network anywhere and anytime, and a company can run its business globally without huge investments in infrastructure.

Some Internet Service Providers (ISPs) already offer IP roaming through alliances with other ISPs. Next-generation networks will also support global roaming through alliances. IP roaming gives telecommuters, mobile employees, and travelers dial-up access to the Internet with a local call from virtually anywhere in the world. From the Internet, they can access resources on their corporate intranet. They can also access e-mail, transfer files, and browse the Web without having to worry about long-distance call charges. Roaming services also allow users to maintain just one Internet account and log-on with one user name and password.

Corporations benefit in being able to significantly reduce remote access costs by eliminating all the Internet and intranet access challenges experienced by their IS staff, such as maintaining and upgrading modem banks. Their business travelers and telecommuters no longer need multiple Internet accounts and e-mail addresses, and do not have to rely on long-distance and toll-free services, which can reduce access expenses by as much as 90 percent.

Packet Structure

The Internet is comprised of a series of autonomous systems, or subnetworks, each of which is locally administered and managed. The subnetworks may consist of Ethernet LANs, X.25 packet networks, ISDN, or frame relay networks. IP delivers data between these different networks through routers that process packets from one autonomous system (AS) to another.

Each node in the AS has a unique IP address. The Internet Protocol adds its own header and checksum to make sure the data is properly routed over the network (Figure 57). This process is aided by the presence of routing update messages that keep the address tables in each router current. Several different types of update messages are used, depending on the collection of subnets involved in a management domain. The routing tables list the various nodes on the subnets as well as the paths between the nodes. If the data packet is too large for the destination node to accept, it will be segmented into smaller packets.

The IP header consists of the following fields:

IP VERSION (4 BITS)   The current version of IP is 4; the next generation of IP is 6.

IP HEADER LENGTH (4 BITS)   This indicates header length; if options are included, the header may have to be padded with extra 0s so it can end at a 32-bit word boundary. This is necessary because header length is measured in 32-bit words.

PRECEDENCE AND TYPE OF SERVICE (8 BITS)   Precedence indicates the priority of data packet delivery, which ranges from 0 (lowest priority) for normal data to 7 (highest priority) for time-critical data (i.e., multimedia

Figure 57
IP packet
structure.

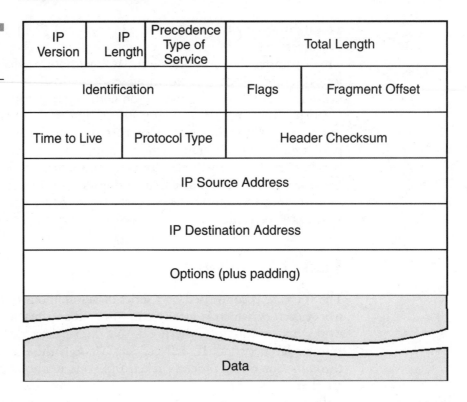

IP Version	IP Length	Precedence Type of Service	Total Length	
Identification			Flags	Fragment Offset
Time to Live		Protocol Type	Header Checksum	
IP Source Address				
IP Destination Address				
Options (plus padding)				
Data				

applications). Type of service contains quality of service (QoS) information that determines how the packet is handled over the network. Packets can be assigned values that maximize throughput, reliability, or security and minimize monetary cost or delay. This field will play a larger role in the future as internets evolve to handle more multimedia applications.

TOTAL PACKET LENGTH (16 BITS) This is the total length of the header plus the total length of the data portions of the packet.

IDENTIFICATION (16 BITS) This is a unique ID for a message, used by the destination host to recognize packet fragments that belong together.

FLAGS (3 BITS) This indicates whether or not the packets can be fragmented for delivery; if a packet cannot be delivered without being fragmented, it will be discarded and an error message will be returned to the sender.

FRAGMENTATION OFFSET (13 BITS) If fragmentation is allowed, this field indicates how IP packets are to be fragmented. Each fragment has the same ID. Flags are used to indicate that more fragments are to follow, as well as indicate the last fragment in the series.

TIME TO LIVE (8 BITS)   This field indicates how long the packet is allowed to exist on the network in its undelivered state. The hop counter in each host or gateway that receives this packet decrements the value of the time-to-deliver field by one. If a gateway receives a packet with the hop count decremented to zero, it will be discarded. This prevents the network from becoming congested by undeliverable packets.

PROTOCOL TYPE (8 BITS)   This specifies the appropriate service to which IP delivers the packets, such as TCP or UDP.

HEADER CHECKSUM (16 BITS)   This field determines whether the received packet has been corrupted in any way during transmission. The checksum is updated as the packet is forwarded because the time-to-live field changes at each router.

IP SOURCE ADDRESS (32 BITS)   This is the address of the source host (e.g., 130.132.9.55).

IP DESTINATION ADDRESS (32 BITS)   This is the address of the destination host (e.g., 128.34.6.87).

OPTIONS (UP TO 40 BYTES)   Although seldom used for routine data, this field allows one or more options to be specified. Option four, for example, time-stamps all stops that the packet made on the way to its destination. This allows measurement of overall network performance in terms of average delay and nodal processing time.

SUMMARY

Although 30 years old, Internet Protocol is capable of extending and enhancing communications in the 21st century. The capabilities that were built into IP to allow the exchange of data between different computers interconnected over low-speed lines are also advantageous when used over advanced high-speed internets. With built-in security and reliability features, IP-based next generation networks are capable of converging voice and data to handle telephone calls, videoconferencing, priority messaging, interactive gaming, telemedicine, distance learning, and on-demand information and entertainment services—in addition to streaming audio/video and multicast delivery. The use of IP makes the global distribution of these and other applications more efficient and economical than using traditional public switched telephone networks, while permitting their integration with legacy data and applications.

See Also **Asynchronous Transfer Mode, Latency, Next-Generation Networks, Synchronous Optical Network, Transmission Control Protocol, Virtual Private Networks, Voice-Data Convergence**

Internet Servers

The key to delivering Internet services is a properly equipped Web server. The Internet is a true client-server network and integrating into this environment requires servers suitable for high-traffic and mission-critical applications. The server must allow corporate constituents to access information quickly and easily. The server must have security features that enable constituents to share confidential information or conduct encrypted electronic transactions across the Internet or corporate intranet. Finally, the server must be able to support the many applications that have become the staple of the Internet, including electronic mail, file transfer, and newsgroups. These applications are used internally on corporate intranets as well.

Processor Architecture

There are basically two choices of processor architectures: RISC-based (reduced instruction set computing) or CISC-based (complex instruction-set computing). RISC processors are usually used on high-end UNIX servers, while CISC processors, such as Intel's Pentium, are used on Windows NT machines. Of note is that the performance of the Pentium now approaches that of many RISC processors and costs less.

Because of the volume of service requests—sometimes tens of thousands a day—the server should be equipped with the most powerful processor available. The more powerful the processor, the greater the number of service requests (e.g., page lookups, database searches, transactions, and forms) the server will be able to handle at any given time. In addition, if processing-intensive tasks such as encryption, packet filtering, and hostile code screening are to be performed for security purposes, a high-powered server is mandatory—these functions alone can diminish processor performance by as much as 20 percent.

Servers with symmetric multiprocessing (SMP) enable the operating system to distribute different processing jobs among two or more processors. All the CPUs have equal capabilities and can handle the same tasks. Each CPU can run the operating system as well as applications. Not only can any CPU execute any job, but jobs can be shifted from one CPU to another as the load changes. This can be very important at high-traffic sites, especially those that do a lot of local processing to fulfill service requests.

Operating System

When choosing a server, the operating system deserves particular attention. The choices are usually between a UNIX variant and Windows NT. Although some vendors offer server software for Windows 3.1 and Windows 95, they are usually intended for casual rather than business use.

The most compelling features of UNIX are its support of multiple tasks, its support of multiple users, its networking capabilities via integral support of TCP/IP, and its scalability. Although most Internet servers are based on UNIX, Microsoft's Windows NT is growing in popularity, particularly among smaller companies. A Windows NT server offers nearly the performance, reliability, and functionality of a UNIX server and is much easier to set up and administer than a UNIX server. Windows NT also provides integral TCP/IP.

Like UNIX, Windows NT is a multitasking, multithreaded operating system. And like UNIX, Windows NT supports multiple processors. While Windows NT is gaining in power and reliability, it does not scale nearly as well as UNIX on networks. UNIX can scale to hundreds of interconnected servers. Windows NT 4.0 does not scale well beyond eight processors, while Windows NT 5.0 (now known as Windows 2000) increases that number to only 16. Microsoft's NT clustering software is also limited, while multinode clustering that increases scalability through load balancing is not available at this writing. These may be serious limitations for large companies that may want to tie in all of their sites together over a private intranet.

Fault Tolerance

If the server is supporting mission-critical applications over the Internet or an intranet, there are several levels of fault tolerance that merit consideration. Fault tolerance must be viewed from both systems and subsystems perspectives.

SITE MIRRORING   From a system's perspective, fault tolerance can be implemented by linking multiple servers together. When one system fails or must be taken offline for upgrades or reconfigurations, the standby system is activated to handle the load. This is often called *site mirroring*. An additional level of protection can be obtained through features of the operating system that protect read and write processes in progress during the switch to the standby system.

LOAD BALANCING   Another means of achieving fault tolerance is to have all hardware components function simultaneously, but with a load-balancing mechanism that reallocates the processing tasks among surviving components when a failure occurs. This technique works best with a UNIX operating system equipped with vendor options that continually monitor the system for errors and dynamically reconfigures the system to adapt to performance problems.

At the subsystem level, there are several server options that can improve fault tolerance, including ports, network interfaces, memory expansion cards, disk and tape drives, and I/O channels. All can be duplicated so an alternate hardware component can assume responsibility in the event of a subsystem failure. This procedure is sometimes referred to as a *hot-standby solution*, whereby a secondary subsystem monitors the tasks of the primary subsystem in preparation for assuming such tasks when needed.

UNINTERRUPTIBLE POWER SUPPLY   To guard against an onsite power outage, an uninterruptible power supply (UPS) can provide an extra measure of protection. The UPS provides enough standby power to permit continuous operation or an orderly shutdown during power failures or to change over to other power sources such as diesel-powered generators. Some UPSs have SNMP capabilities, which lets network managers monitor battery backup from the central management console. For instance, via SNMP, every UPS can be instructed to test itself once a week and report back the results.

DATABASE ROLLBACK   Because large amounts of data may be located at the server, the server must be able to implement recovery procedures in the event of a program, operating system or hardware failure. For example, when a transaction terminates abnormally, the server must have the capability to detect an incomplete transaction so that it is not left in an inconsistent state. The server's rollback facility is invoked automatically, which backs out of the uncompleted transaction. The transaction can then be resubmitted by the program or user. A roll-forward facility recovers completed transactions and updates in the event of a disk failure by reading a transaction journal that contains a record of all updates.

Application Software

An Internet server must be equipped with software that enables it to run various applications. Some server software supports general communications for document publishing over the Web. Often called a *communications server* or *Web server,* this type of server can be enhanced with software that is specifically designed for secure electronic commerce. Server software is available for performing many different functions, including implementing newsgroups, facilitating message exchange (e.g., e-mail), improving the performance and security of communications, and controlling traffic between the Internet and the corporate network.

Communications Software

A communications server enables users to access various documents and services that reside on it and retrieve them via the HyperText Transfer Protocol (HTTP). These servers support the standard multimedia document format— HyperText Markup Language (HTML)—for presenting rich text, graphics, audio, and video. Hyperlinks connect related information across networks, creating a seamless web. Hyperlinks can also provide access to various services such as e-mail and file transfer, and even telephony and videoconferencing. Some vendors offer servers preconfigured with all the necessary Internet protocols, allowing them to be quickly put into operation.

Commerce Software

A commerce server conducts secure electronic commerce and communications on the Internet. It permits companies to publish hypermedia product catalogs formatted in HTML, and provides connections to back-end databases and other supporting applications. There are many toolkits available to help companies create and maintain electronic commerce applications, including Web-enabling traditional Electronic Data Interchange (EDI) applications.

Among the most widely used protocols available to ensure the privacy of sensitive data is the Secure Sockets Layer (SSL), which prevents eavesdropping, tampering, or message forgery. Specifically, SSL provides:

SERVER AUTHENTICATION Any SSL-compatible client can verify the identity of the sender using a certificate and a digital signature.

DATA ENCRYPTION The privacy of client-server communications is ensured by encrypting the data stream between the two entities.

DATA INTEGRITY SSL verifies that the contents of a message arrive at their destination in the same form as they were sent.
As with other types of Internet servers, vendors offer commerce servers that are preconfigured with the protocols necessary to support electronic

commerce. The idea is to encourage even small companies and entrepreneurs to set up shop on the Internet.

News Software

A news server lets users create secure, public, and private discussion groups for access over the Internet and other TCP/IP-based networks via the standard Network News Transport Protocol (NNTP). For intranet use, supporting NNTP facilitates workgroup collaboration, allowing colleagues to download documents sent to the group, mark them up, and send them back for central editing. The news server's support of NNTP enables it to accept feeds from the popular Usenet news groups and allows the creation and maintenance of private discussion groups. Most newsreaders are based on NNTP and some support SSL for secure communication between clients and news servers.

Mail Software

Client-server messaging systems are implemented by special mail software installed on a server. In essence, this software enables users to easily exchange information within a company as well as across the Internet. Mail software has many features that can be controlled by either the system administrator or each user with an e-mail account.

The mail software should conform to open standards, including POP and IMAP. The Post Office Protocol (POP) is a simple store-and-forward delivery mechanism used for retrieving new mail from a UNIX server. A newer mail protocol—the Internet Message Access Protocol (IMAP)—establishes a client-server relationship that allows the server to be used not only for storing incoming mail but for filtering it before download to the client. IMAP gives users more choices about what categories of messages to fetch from the server, where to store mail, and what the server should do with specific categories of read messages.

Proxy Software

To improve the performance and security of communications across the TCP/IP-based Internet, many organizations use a proxy server. This kind of software offers performance improvements by using an intelligent cache for storing retrieved documents. The proxy's disk-based caching feature minimizes use of the external network by eliminating recurrent retrievals of commonly accessed documents. This feature significantly improves interactive response time for locally attached clients. The result-ing performance improvements provide a cost-effective alternative to pur-chasing additional network bandwidth. Because the cache is disk-based, it can be tuned to provide optimal performance based on network usage patterns.

Some proxy servers incorporate some of the features of firewalls. For example, proxy servers can provide protection against IP spoofing, a method of attack in which an intruder spoofs or imitates the IP address of an internal computer to either send data as if they were on the inter-nal network or to receive data intended for the machine being spoofed. The proxy server guards against this type of attack by preventing any IP packets with destination addresses not found in its local address table (LAT) from entering the corporate network through the Internet.

Firewall Software

A firewall acts as a security wall and gateway between a trusted internal network and such untrustworthy networks as the Internet. Access can be controlled by individuals or groups of users, or by system names, do-mains, subnets, date, time, protocol, and service.

Some firewalls offer real-time, automated intrusion detection to stop unauthorized activities immediately, even if the network manager is not around to intervene. With such tools, network intrusions are detected and appropriate responses are taken automatically before they can cause seri-ous damage. Among the responses that can be implemented automati-cally are port reconfiguration and service denial.

Firewall security is bidirectional, simultaneously prohibiting unautho-rized users from accessing the corporate network while also managing in-ternal users' Internet access privileges. Some firewalls even periodically check their own code to prevent modification by sophisticated intruders. Vendors of firewalls (and proxy servers) include third-party virus and ap-plet scanning functionality into their products. The scanning software loads the suspect code into a protective buffer and compares it with a li-brary of known viruses and hostile Java applets and ActiveX compo-nents. If a match or near match is found, the user can eliminate the code before it is allowed to run.

Common Gateway Interface

At a minimum, an Internet server should support the Common Gateway Interface (CGI). This is a standard for interfacing external applications with information servers, such as HTTP or Web servers. Gateways can be used for a variety of purposes, the most common being the processing of form requests, such as database queries or online purchase orders. Gateway programs handle requests and return the appropriate document or generate a document on the fly. With CGI, a Web server can serve information that is not in a form readable by the client (such as an SQL database), and act as a gateway between the two to produce something which clients can interpret and display.

Gateways conforming to the CGI specification can be written in any language that produces an executable file. Among the more popular languages for developing CGI scripts is PERL (Practical Extraction and Report Language) and TCL (Tool Command Language), both derivatives of the C language.

A key advantage of using PERL and TCL is that they can be used to speed the construction of applications to which new scripts and script components can be added without requiring recompiling and restarting, as is required when the C language is used. Of course, the server on which the CGI scripts reside must have a copy of the program itself—PERL, TCL, or alternative program.

Increasingly, Java applets provide the interface between the Web server and back-end databases and applications. Java differs from the traditional CGI approach in that the applets, once downloaded, can link directly with a server on the Internet. The applets do not need a Web server as an intermediary and, consequently, do not degrade its overall performance. In most cases, the applications are stored in cache on a hard disk at the client location and in others, they are stored in cache memory. Either way, the application does not take up permanent residence on the client machine. Since applications are delivered to the client only as needed, administration is done conveniently and more economically at the server, ensuring that users have access to the latest version of the application.

SUMMARY

The client-server architecture of the Internet and its use of open protocols for information formatting and delivery allows businesses to extend communications and services globally. The type of communications and services that are available is dependent upon the application software that runs on one or more servers, as well as the access privileges of each user. A server may be dedicated to a specific Internet application or multiple applications, depending on such factors as system resources and the specific needs of the organization.

See Also **Client-Server, Firewalls, Gateways, Proxy Servers, Uninterruptible Power Supplies**

Internetwork Packet Exchange

Internetwork Packet Exchange (IPX) is a Layer 3 network protocol used by Novell's NetWare operating systems. It was adopted by Novell from the Xerox Network System (XNS). Like XNS and other datagram protocols, IPX is a low-overhead protocol used for connectionless communications, allowing individual packets to be sent to and received from user processes.

In connectionless communications, protocols do not require an acknowledgment for each packet sent. Packet acknowledgment, or connection control, must be provided by protocols above IPX, such as the Sequenced Packet Exchange (SPX) protocol. Since IPX does not support the concept of a connection or reliable delivery, it is used in situations where a guaranteed service is not required or where an occasional lost packet is not crucial.

Packet Structure

IPX packets are comprised of a variable-length data field and a 30-byte header (Figure 58). Although the maximum packet size is 65,535 bytes, the network enforces a more practical packet size of approximately 1,500 bytes. The contents of the packet's 11 fields are as follows:

CHECKSUM This two-byte field validates the contents of the packet.

PACKET LENGTH This two-byte field indicates the length of the packet in bytes.

TRANSPORT CONTROL This one-byte field provides the number of routers the packet can go through before it is discarded.

PACKET TYPE This one-byte field identifies the service that created the packet (such as SPX).

Figure 58
Structure of an IPX
packet on an
Ethernet network.

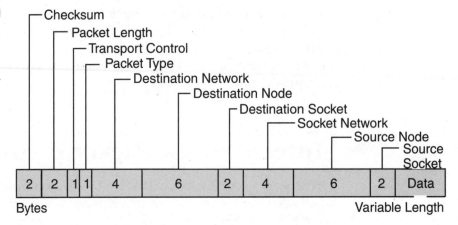

DESTINATION NETWORK This four-byte field provides the network address of the receiving network.

DESTINATION NODE This six-byte field provides the MAC address of the receiving node.

DESTINATION SOCKET This two-byte field provides the address of the process running at the receiving node.

SOURCE NETWORK This four-byte field provides the address of the sending network.

SOURCE NODE This six-byte field provides the MAC address of the sending node.

SOURCE SOCKET This two-byte field provides the address of the process running at the sending node.

DATA This variable-length field contains the user information.

Addressing

As noted, the network, node and socket addresses for both the destination and the source are held in the header of the IPX packets. The network addresses are assigned during the NetWare server installation and configuration process. The network node addresses come from the permanent numbers burned into each Network Interface Card (NIC) by the vendors. The socket addresses come from the processes running in software at each computer. The complete 12-byte address of each IPX packet is a hexadecimal number that takes the following form:

3B56A467 21972DF113AF 0811

Since several processes normally operate within a node, socket numbers provide a means by which each process can distinguish itself to IPX. When a process needs to communicate on the network, it requests that a socket number be assigned to it. Any packets IPX receives that are addressed to that socket are passed on to the corresponding process. In essence, socket numbers provide a quick method of routing packets within a node. Because socket numbers are internal to each node, several workstations can use the same socket number at one time without causing any confusion. Novell has reserved several socket numbers for specific purposes, which are listed in the following table:

Socket Number	Description
451h	NetWare Core Protocol (NCP)
452h	Service Advertising Protocol (SAP)
453h	Routing Information Protocol (RIP)
4000h to 6000h	Sockets used for interaction with file servers and other network communications.
455h	NetBIOS
456h	Diagnostics

SUMMARY

As a connectionless datagram protocol, IPX is used for simple messaging functions across the network. Reliability is provided by higher-level protocols, such as SPX at Layer 4, which is a connection-oriented protocol. With SPX, packets are not exchanged until a session is set up between the source and destination nodes. SPX uses IPX for packet delivery and adds guaranteed packet delivery and flow control. Together, IPX/SPX provides connection services similar to those provided by TCP/IP.

See Also **Internet Protocol, Transmission Control Protocol, User Datagram Protocol**

IP-to-PSTN Gateways

IP-to-PSTN gateways are used to pass telephone calls between IP nets and the public switched telephone network by performing the necessary translations between the two types of networks. When a standard voice call is received at a near-end gateway, the analog voice signal is digitized, compressed, and packetized for transmission over the IP network. At the far-end gateway the process is reversed, with the packets decompressed and returned to analog form before the call is delivered to its intended destination on the PSTN.

History

In a demonstration of the feasibility of originating calls on the Internet and receiving them at conventional phones, the Free World Dial-up (FWD) Global Server Network (FWD) project went online in March 1996. Organized by volunteers around the world, the noncommercial project was entirely coordinated in cyberspace via Internet telephony, e-mail, and chat software.

Using popular Internet telephony software, users were able to contact a remote server in the destination city of their call. This server "patched" the Internet phone call to any phone number in the local exchange area. This meant, for example, that a user in Hong Kong could use an Internet-based server in Paris to effectively dial any local phone number and talk with a friend or family member. A global server kept a list of all servers and the real-time status of each.

The specific steps required to place a call through the FWS Global Server Network and answer it with a conventional phone were as follows:

1. Connect to the Internet as usual with a PPP or SLIP connection.
2. Start the phoneware and register with the vendor's server, if necessary.
3. Start the FWD client software and connect to a FWD server in a select city.
4. Click on the Connect button and enter the domain name of the server.
5. Once connected to the server, a message indicates that the connection is made, and a ring signal is sent.
6. Upon receiving the ring signal, the user enters the telephone number in the FWD client of the person in the local calling area, leaving out the area code.
7. The phoneware dials the number entered.
8. When the called person answers, both parties can start conversing.

This procedure was certainly more complicated than just picking up the phone and dialing the long-distance number, and its first implementation was limited to processing calls in only one direction—from the Internet to conventional phones. Critics cited these limitations as the reason why Internet telephony would not pose a threat to long distance carriers for many years to come.

Today's Gateways

The technology demonstrated by the project has been continually improved to the point where it is now used in commercial products. Among the dozen or so companies that offer such products is VocalTec. The company's Telephony Gateway not only streamlines the calling process, it also provides many advanced features which make IP telephony services commercially viable.

To place a call, the user dials the nearest gateway from any phone or PC and then enters the destination number. After the gateways at each end establish the connection, the local gateway digitizes and compresses the incoming voice signals into packets, which travel over the Internet or intranet to the remote gateway where they are decompressed and reconstructed into their original form, making them suitable for transmission over the PSTN. From there, the call is then routed over the local loop connection to the destination telephone or PBX (Figure 59).

Calls also can be routed from the gateway to mobile phones on cellular networks and to computers attached to cable television networks. For the portion of the trip a voice call travels over the Internet/intranet, there are no costs incurred beyond that of the network connection. Per-call phone charges apply only for the portion of the trip the call travels over the PSTN or cellular network. On cable networks, operators charge a flat monthly fee for IP phone calls.

Figure 59
Typical IP-to-PSTN
gateway operation.

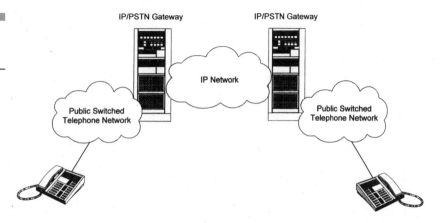

IP-to-PSTN Gateways

The IP/PSTN gateways offer many features that expedite administration, including features for call monitoring, security, and billing. They can even be equipped with an Interactive Voice Response (IVR) application, which acts as the interface between the PSTN/PBX and the IP network. The application includes an auto attendant, which guides users through the calling process.

SUMMARY

IP-to-PSTN gateways provide an interface between the conventional circuit-switched telephone network—or an existing PBX—and public or private IP nets, enabling voice calls to be terminated at multimedia-equipped PCs or ordinary telephones. The purpose of placing calls across IP nets is to minimize long distance charges. In addition, the convergence of voice and data over IP networks opens new opportunities for the provision of value-added services, such as call centers (inbound and outbound), help desks, reservation centers, distance learning, telecommuting, and telemedicine. Although these services have been available for many years over the traditional circuit-switched PSTN or private leased lines, the ability to mix voice and data over IP networks enables users to take advantage of new features and changes the cost-benefits equation in a way that can stimulate much broader usage.

See Also **Internet Protocol, Voice-Data Integration**

K

KVM Switches

When there is a need to control multiple computers from a single keyboard, video monitor, and mouse (KVM), a special device is used called a KVM switch. Such switches are commonly used in stock exchanges, banks, and financial institutions where users require access to several computers at the same time. In addition, system administrators and test lab technicians at large companies can use KVM switches to access any computer in a rack-mount configuration. KVM switches are also used for managing large-scale data centers, server farms, and help desks.

Through simple keyboard commands ("hot keys") or via the select button on the KVM switch's front panel, users can access any computer connected to the switch (Figure 60). The switch also provides a monitor connection and full keyboard and mouse emulation. With a dedicated microprocessor for each connected computer, the KVM switch enables each computer to "see" its own keyboard and mouse.

Some switches feature onscreen menus to let the user identify and select attached computers quickly and easily. The computer name window may be placed anywhere on the screen. Colors of text and background for all menus are user defined.

The KVM switches are independent of the operating system, allowing them to be used with a wide variety of operating systems, including DOS, Novell NetWare, Windows, Windows 95, Windows NT, and OS/2. Some switches also support Macintosh and Sun platforms.

Some KVM switches incorporate a "keep-alive emulation" function so that the dedicated processors that emulate keyboard and mouse are powered directly by the connected computer. As long as the connected computers remain powered and connected to the KVM switch, the dedicated processors will continue to run.

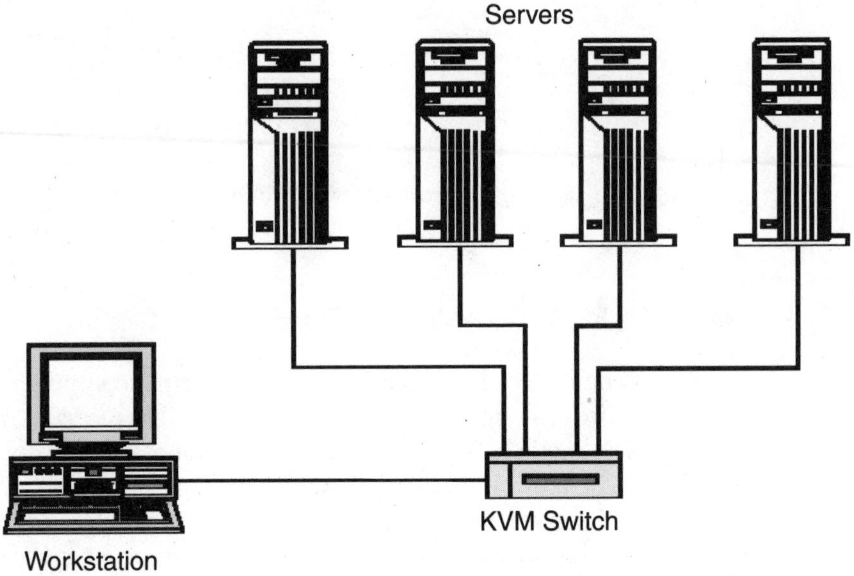

Figure 60

A KVM switch allows multiple computers to be controlled by a single keyboard, video monitor, and mouse.

Servers

KVM Switch

Workstation

Types of KVM Switches

There are two types of KVM switch—intelligent and nonintelligent. Nonintelligent switches simply provide a mechanical switch of the keyboard, video and mouse signals. Although relatively cheap, they can cause problems, since the keyboard and mouse could become out of sync with the computers.

With intelligent KVM switches, the switch emulates the existence of keyboard, monitor and mouse for each computer that is attached. As each computer sends configuration details to the keyboard and mouse, the switch remembers these configurations for later use.

The emulation of keyboard and mouse allows any computer to be booted at any time without fear of boot problems, whether it is the "selected" machine or not. The storage of configuration information allows the switch to reconfigure the keyboard and mouse each time a new computer is selected, such that they exactly match the configuration expected by that computer.

SUMMARY

The primary differences between KVM switches are in port capacity, expandability, number of access points, and user interface. KVM switches are available from 2 to 256 ports. Port capacity can be expanded beyond the initial hardware configuration by daisy-chaining or cascading multiple KVM switches together. Some switches offer multiple access points, which allows several users to independently access and operate multiple computers. Some offer advanced features like an onscreen user interface with password protection for each computer.

See Also **Hubs**

L

LAN Backup

With the increasing reliance on information for strategic advantage, corporatifons must implement procedures for backing up data, which may be difficult or impossible to replace if it is ever lost, stolen, or damaged. Yet data backup is not a simple matter for most businesses. One reason is that it is difficult to find a backup system capable of supporting different network operating systems and data, especially if midrange systems and mainframes are involved.

Backup Procedures

Protecting mission-critical data stored on LANs requires backup procedures that are well defined and rigorous. These procedures include backing up data in a proper rotation, using the correct media, and testing the data to ensure it can be easily and quickly restored in an emergency. Enterprise-wide backups are especially problematic. This is because typically multiple servers and operating systems, as well as isolated workstations, often hold mission-critical data. Moreover, the network and client/server environments have special backup needs—back up too often and throughput suffers; back up too infrequently and data can be lost.

Deciding which files to back up can be more complicated than picking the right storage media. The most thorough backup is one in which every file on every server is copied to one or more tapes or disks. However, the size of most databases makes this impractical to do more than once a month. Incremental backups copy only files that have changed since the last backup. Although this is faster, it requires careful management because each tape may contain different files. To restore a system made with incremental backups requires all the incremental backups (in the right order) made since the last full backup.

Differential backups split the difference between full and incremental techniques. Like an incremental backup, a differential backup requires a tape with the full set of files. However, each differential tape contains all the files that have changed since the last full backup, so restoration requires just the full set and the most recent differential.

Scheduling and Automation

Scheduling backups is determined by several factors, including the importance of applications, network availability, and legal requirements. Network backup software with calendar-based planning features allows system administrators to schedule the weekly archiving of all files on LAN-attached workstations. The backup can be scheduled for nonbusiness hours, to avoid both disrupting user applications and congested network traffic.

Some scheduling tools allow system administrators to set precise parameters with regard to network backups. For example, the backup can target only files that have not been accessed in the past 60 days, with the objective of freeing at least 100 megabytes of disk space on a particular server. When the backup is complete, a report is generated, listing the files that have been archived to tape, along with their file size and date of last access. The total number of bytes are also provided, allowing the system administrator to confirm that at least 100 megabytes of storage has been freed on the server.

Event-based scheduling allows system administrators to run predefined workloads when dynamic events occur in the system, such as the close of a specific file or the start or termination of a job. With regard to network backups, administrators can decide what events to monitor and what the automated response will be to those events. For example, they can decide to archive all files in a directory after the last print job or update the database when a particular spreadsheet is closed.

Although most network backup programs can grab files from individual workstations on a LAN, thousands of users might have similar or identical system configurations. Instead of backing up 1,000 copies of Windows 95, for example, the network backup program can be directed to copy only each user's system configuration files. That way if a workstation experiences a disk crash, a new copy of Windows 95 can be downloaded from the server along with the user's applications, data, and configuration files.

With the right management tools, network backup can be automated under centralized control. Such tools can go a long way toward lowering operating and resource costs, reducing the time spent on backup and recovery. These tools enhance media management by providing overwrite protection, log file analysis, media labeling, and the ability to recycle backup media. In addition, the journaling and scheduling capabilities of some tools relieve operators of the time-consuming tasks of tracking, logging, and rescheduling network and system backups.

Another useful feature of such tools is data compression, which reduces media costs by increasing media capacity. This automated feature also increases backup performance while reducing network traffic.

When these tools are integrated with high-level management platforms—such as Hewlett-Packard's OpenView or IBM's NetView, or operating systems such as Sun's Solaris—problems or errors that occur during automated network backup are reported to the central management console. The console operator is notified of any problem or error via a color change of the respective backup application symbol on the network map. By clicking on the symbol, the operator can directly access the network backup application to determine the cause of the problem or correct the error to resume the backup operation.

Image-Based Backup

Although file-by-file backup allows control over individual files, backing up entire servers is often slow and difficult to complete during network downtime. An image-based system speeds the backup process because it bypasses the file system and takes an "image" of the physical disk—including boot and data volumes, and bindery or directory information—and treats them as objects to be stored.

This method of backup is much faster in comparison to conventional file-by-file methods—in fact, the backup can occur as fast as the tape drive can run. In addition to backup, such products restore systems more easily. At any point an entire hard drive can be restored without regard for what the boot sector, operating system, or files look like—and the process is totally automated.

Hierarchical Storage Management

Hierarchical Storage Management (HSM) systems use two or more levels of storage to meet the demand for increased storage space. HSM came about because of the need to move low-volume and infrequently accessed files from disk, thus freeing up space. Although disks can be added as storage requirements increase, budget constraints often limit the long-term viability of this solution. In an HSM scheme, data can be categorized according to its frequency of usage and stored appropriately: online, near-line, or offline. Different storage media come into play for each of these categories and migration operations are under control of an HSM management system.

Frequently used files are stored online on local disk drives installed in a server or workstation. Occasionally used files are stored near-line on secondary storage devices such as rewritable optical disks installed in a server-like device called an *autochanger* or *jukebox*. Infrequently used files are usually migrated offline to tape cartridges that are stored in a tape jukebox or a library facility capable of holding hundreds or thousands of tapes. The library facility uses sophisticated robotics to retrieve individual bar-coded cartridges and inserts them into a tape drive so the data can be migrated to local storage media. The exchange time can be several minutes.

For organizations with mixed needs for online and offline storage, a near-time automated tape library offers the best compromise between price and performance. These systems bridge the gap between fast, expensive online disk storage and slow, high-capacity offline tape libraries. The exchange time is about 30 seconds.

A management system determines when a file should be transferred or retrieved, initiates the transfer, and keeps track of its new location. As files are moved from one type of media to another, they are put into the proper directory for user access. The management system automatically optimizes storage usage across different media types by removing files from one to the other until they are permanently archived in the most economical way, usually a tape library. At the same time, individual files or whole directories can be excluded from migration.

Data migration can be controlled according to criteria such as file size and last date of access. Files can also be migrated when the hard disk reaches a specified capacity threshold. For example, when magnetic disk storage reaches the established threshold of 80 percent, files are migrated to optical storage, freeing up magnetic storage until it reaches another specified threshold, say 60 percent.

Off-Site Backup

For added protection, some services move data off a LAN to a third-party storage site. This protects data from being destroyed in natural disasters, and protects data from theft or misconduct. Such services offer several features, including:

- Real-time replication of data to off-site vaults
- Replication at the file system level, rather than disk mirroring, allowing selective replication of key data

- Highly efficient file delta transmission over LAN and WAN links
- Loosely coupled asynchronous replication, preventing WAN delays from slowing production systems
- Minimal production server overhead
- Open file protection and compatibility with virtually all tape backup software

Such services implement backups over dialup modem connections or dedicated lines, or via the Internet using a secure TCP/IP connection. The service provider backs up a corporate LAN server onto hard drives at its location at night when usage, if any, is low. In the event that a company suffers data loss, the service provider can deliver a new copy via a network connection or burn a CD and physically deliver it to the company.

SUMMARY

As LANs continue to carry increasing volumes of crucial data in varying file formats, vendors continue to push the limits of backup technology. On the software side, the trend is toward increasing levels of intelligence. Backup systems must ensure not only that files are backed up, but that they can be easily located and restored. System intelligence has already progressed to the point where users need not know the tape, the location on the tape, or even the exact name of a lost file in order to restore it. This intelligence makes hierarchical storage management systems a viable option for large organizations.

See Also **Data Warehouse, Storage Area Networks**

LAN Emulation

Developed by the ATM Forum and released in 1995, the LAN Emulation (LANE) specification enables existing applications to access an Asynchronous Transfer Mode (ATM) network via protocol stacks such as TCP/IP, IPX, and APPN as if they were running over traditional LANs. LANE is generally implemented as server software or in the driver software of ATM network adapters and switches.

LANE works at the media access control (MAC) layer and enables legacy Ethernet and token-ring traffic to run over ATM with no modifications to applications, network operating systems, or desktop adapters. The legacy-end stations can use LANE to connect to other legacy systems, as well as to ATM-attached servers, hubs, routers, and other networking devices. Although LANE does not directly support the Fiber Distributed Data Interface (FDDI), a router or switch can bridge FDDI traffic onto an ATM LANE service after converting the packets to either the Ethernet or token ring.

LANE provides a translation layer between higher-level connectionless protocols and lower-level connection-oriented ATM protocols (Figure 61). At the ATM host side (left), the ATM layer manages the header for the ATM cell. It accepts the 48-byte cell payload from a higher layer, adds a 5-byte header, and passes the fixed-length 53-byte cell to the physical layer below. At the ATM-to-LAN converter, the cells are received from the physical layer, the header is stripped off, and the remaining 48 bytes are passed to the higher-layer protocols. Although the ATM layer is unaware of the types of traffic it carries, it can distinguish Quality of Service (QoS) through information learned during connection setup.

The ATM adaptation layer (AAL) sits above the ATM layer. It formats data into the ATM cell payload. The network does not know that the cells being carried within it are LAN Emulation frame cells—the ATM switches merely transport the cells in the same manner as any other connection. Once the ATM cells reach their destination, they are reconstructed into higher-level data and transmitted to the respective local devices. Since ATM can carry multiple traffic types, several adaptation protocols—each operating simultaneously—can exist at the adaptation layer. AAL Type 5 is used for LAN Emulation.

LANE sits above AAL5 and masks the connection setup and handshaking functions required by the ATM network from the higher proto-

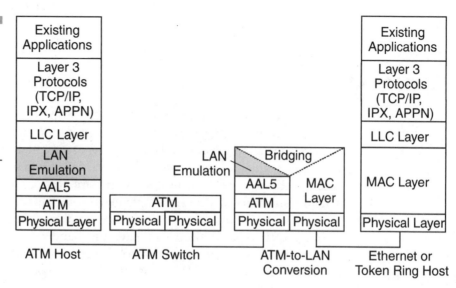

Figure 61
Through LAN emulation, both the ATM LAN and the legacy LAN are made to believe they are on the same bridged LAN.

col layers, making it independent of upper-layer protocols, services, and applications. It maps the MAC address-based data networking protocols into ATM virtual connections so the higher-layer protocols think they are operating on a connectionless LAN.

Address Resolution

The LAN emulation server is either a host or a switch running software that resolves MAC addresses to ATM addresses. ATM devices register with the LAN emulation server and become part of the same emulated LAN—essentially a MAC-layer virtual LAN that defines a common broadcast domain among end stations. In the LANE environment, there is also a device known as a broadcast/unknown server (BUS). Like a LAN emulation server, a BUS is a host or switch running software, but its job is to broadcast packets to all members of an emulated LAN. The BUS also handles packets whose destination ATM addresses are not known.

When an IP workstation communicates with another IP workstation on an emulated LAN, the originating host issues an IP address resolution protocol (ARP) packet, which goes to the BUS to be broadcast to all end stations on the emulated LAN. The target end station responds by furnishing its MAC address. To get an ATM address, the host issues another ARP packet—this time a LAN emulation packet requesting the ATM address that corresponds to the receiving host's MAC address. The LAN emulation server responds with the ATM address, which may be broadcast to every member of the emulated LAN. After all this, the originating host finally has the information it needs to set up an ATM virtual circuit call request to the target.

SUMMARY ▬ ▬ ▬ ▬ ▬ ▬ ▬ ▬

LANE is a layer-2 bridging protocol that causes a connection-oriented ATM network to appear to higher-layer protocols and applications as a connectionless Ethernet or token-ring LAN segment. It provides a means to migrate today's legacy networks toward ATM networks without requiring that the existing protocols and applications be modified. The scheme supports backbone implementations, directly attached ATM servers and hosts, and high-performance, scalable computing workgroups. Using LANE, organizations can have the bandwidth benefits of ATM without modifying existing protocols, software, or hardware. By defining multiple emulated LANs across an ATM network, they can create switched virtual LANs for improved security and greater configuration flexibility. Additional benefits include minimal latency for real-time applications and some QoS capability for emulated LANs.

***See Also* Asynchronous Transfer Mode, Quality of Service**

LAN Management

Most businesses have complex networks composed of both new and legacy PCs and servers running a variety of operating systems. LAN management tools help reduce the cost of supporting those heterogeneous environments by providing the means for IT administrators to diagnose and resolve problems with PCs and servers running on a broad range of operating systems. In addition, such tools allow the administrator to take remote control of all client PCs from a single point of administration. Some tools are even accessible from the Web, providing an additional level of management flexibility for mixed network environments. Today's LAN management tools offer a management console from which the following capabilities can be implemented:

- Software distribution
- Software metering
- Inventory
- Remote control
- Server performance monitoring
- Client virus protection
- Security administration
- Reporting via the Web

Servers

LAN management tools provide the ability to monitor crucial server health and performance parameters. Trending data, server diagnostics, and real-time monitoring assists LAN administrators in proactively managing, repairing, and enhancing the performance of servers anywhere on the network.

With many remote servers operated from central data centers, LAN administrators are often away when their servers crash or exceed defined operating parameters. When visiting a server site is inconvenient or impractical, remote management capabilities allow administrators to get servers up and running from any remote location with access via any telephone line or network, including the Web.

Desktops

In addition to servers, LAN management tools also address the needs of IT managers to control all client PCs—on both Windows NT and NetWare networks—from a single console (Figure 62). Among the capabilities such tools offer are software distribution, remote control, hardware and software inventory, software metering, virus protection, and health monitoring (Figures 63 and 64).

As with server management, client management capabilities can be accessed via the Web, allowing the administrator to remotely control clients, view crucial inventory information, and create custom queries on essential hardware or software needs.

Wake on LAN

To avoid interrupting users during prime business hours, IT organizations often prefer to run tasks such as installing software, taking inventory, backing up disk drives, and running virus scans during off hours.

Figure 62
Intel LANDesk Client Manager allows IT administrators to view and manage PCs from a single console.

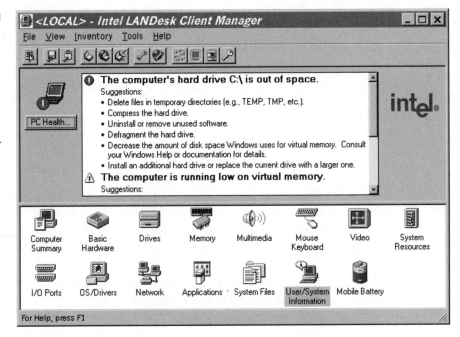

Figure 63

Intel's LANDesk Client Manager, which comes bundled with Wired for Management (WfM)-enabled PCs, provides health monitoring for parameters such as temperature, voltage, chassis intrusion, and boot processes.

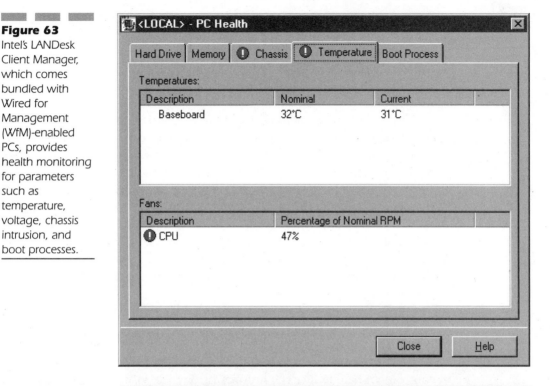

Figure 64

The system's hard drive and memory are included in the Intel LANDesk Client Manager health-monitoring capabilities.

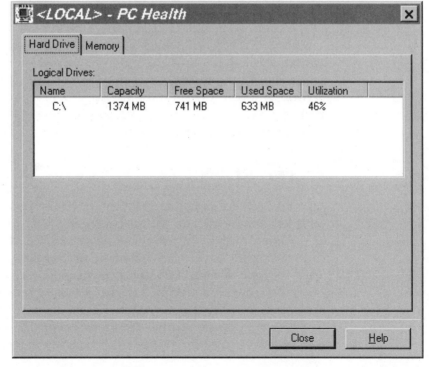

That usually requires having technicians work during off-hours to handle these chores. Even if the environment is automated enough to have remote software installation tools available, if users have left their PCs off, technicians have to physically visit the machines to turn them on. The alternative is to perform such tasks during the day, which negatively affects user productivity.

Wired for Management (WfM) systems provide a remote wake-up feature that, combined with other remote control capabilities, gives technicians the ability to turn machines on remotely and automatically during off hours in order to perform maintenance activities. The system can be automatically transitioned from a sleep state to a fully powered state over the network. Once the system is awake, it can be directed to run utilities such as virus scans, disk backups, or software upgrades, and then return to a sleep state. The combination of off-hours wake-up and remote access allows IT to save time on automated software installations and upgrades, and increases end-user productivity by avoiding disruptions during work hours.

The WfM remote wake-up is provided by Wake on LAN technology, which resides in a managed network adapter and on the system motherboard. The information provided by this instrumentation allows a much greater degree of remote sensing and alerting and, along with software control, can help reduce the number of physical visits, problem resolution times, and the probability of repeat calls—all of which directly affect the total cost of ownership (TCO) and user uptime.

Wake on LAN technology is a result of the Intel-IBM Advanced Manageability Alliance. When the system is turned off, the machine's managed adapter, using an alternate power source, continuously monitors the network and watches for a wake-up packet. When it receives that packet, it alerts the system, which then comes to a full power state and stands ready to perform any maintenance or other tasks.

Alert on LAN

Alert on LAN is a capability that protects hardware through 24-hour surveillance, monitoring, and proactive detection of system failures regardless of the power or state of the operating system. It provides administrators with the ability to remotely maintain and service networked PCs even when they are powered off. Management tools, such as Intel's LANDesk Client Manager and IBM's Universal Manageability Services, alert network managers to system malfunctions and helps them take corrective action before problems become catastrophic. For instance, if a PC fails to boot or locks up during routine off-hours maintenance, network administrators are not only alerted to the problem, but can take corrective action by remotely rebooting the PC into a diagnostic mode and fixing the problem.

Alert on LAN allows a management console to work directly with PCs. For example, when a PC with Alert on LAN generates an alert, the console sends an acknowledgement. This handshake process ensures that alerts have been received and prevents multiple alerts for the same failure. Additionally, the management console can poll a PC to determine whether it has Alert on LAN capability, and ascertain whether a mobile PC is connected and available for servicing.

Desktop Management Interface

The Desktop Management Interface (DMI) is a standard for managing desktop and server devices and reducing the overall cost of system management. Version 2.0 of the specification was released by the Desktop Management Task Force (DMTF) in 1996. It details a standard way for sending DMI management information across the network to a central site to enable remote manageability and impose a more tightly defined model for filtering events.

The DMI architecture includes a Service Layer, a Management Information Format (MIF) database, a Management Interface (MI), and a Component Interface (CI). The Service Layer acts as an information broker between manageable products and management applications. The MIF database defines the standard manageable attributes of PC and server products.

The MI allows DMI-enabled management applications to access, manage, and control desktop computers, components, and peripherals, while the CI allows components to be seen and managed by applications that call the Service Layer. The CI gets real-time dynamic instrumentation information from manageable products and passes it to the MI via the Service Layer. It shields component vendors from decisions about management applications, allowing them to focus on providing competitive management features and functions for their products.

Web Console

Today's IT professionals are constantly on the go, and they cannot always to go back to their offices to solve urgent problems. A Web console gives them the freedom to access system information, perform diagnostics, and resolve problems from any system on the LAN that has a Web browser and an Internet connection.

With a Web console, an administrator or technician can be in the field installing new systems and still respond to mission-critical problems in real time. The Web console can be launched from any system on the LAN to access the management database. From there, technicians can view the screen of a problem PC to determine the cause of the problem and resolve it without having to go to that PC or back to the central management station.

SUMMARY

LAN management tools can often be integrated with the major enterprise management platforms such as CA Unicenter, HP OpenView, and IBM NetView to deliver a remote management framework. This integration enables the enterprise management system to display icons that indicate warnings and alerts from managed PCs. IT administrators can launch either client or server management processes from the enterprise management system's toolbar icon, menu, or submap, and then view and manage the system with the familiar graphical user interface of the LAN management system. In addition, when a PC generates an alert, IT managers can receive Simple Network Management Protocol (SNMP) traps via a local SNMP agent at the predefined enterprise management console address.

See Also **Applications Metering, Applications Monitoring, LAN Troubleshooting**

LAN Security

There are many options in network protection, from simple user passwords to sophisticated biometric technologies and vulnerability scanners. Virus protection is also important—particularly to those with Internet connections—because rogue code can spread undetected throughout an organization and cause extensive damage.

Password Protection

Password protection is the most common means of securing network data. Passwords provide a minimum level of security and do not require special equipment such as keys or identification cards. Most passwords associate users with specific workstations and designated shifts or working hours.

The most effective passwords are long and obscure, but easily remembered by users. The primary drawback of password protection is that users do not always maintain password confidentiality. For example, the longer and more obscure a password, the greater chance the user will write it down—typically somewhere near the workstation where it can be accessed easily by anyone, including unauthorized users.

Several rules of thumb guarantee that passwords will remain private. First, they should be changed frequently. Second, a strict policy of unguessable passwords should be enforced—for example, discouraging the use of names of spouses and children as passwords. Finally, a multi-level password-protection scheme helps ensure confidentiality.

Network security administrators should keep a master password file on disk and survey it periodically. Any password in use for over one year should be changed, and passwords belonging to individuals who have left the organization should be rendered unusable immediately. Whenever a security breach is suspected, security administrators can review password use and potential risks. Many network management and security packages are designed to perform some of these tasks automatically.

Maintaining passwords in mixed-system environments is more difficult, if only because users have to remember several passwords. Several utilities now synchronize passwords necessary for accessing UNIX and Windows NT servers. Syntunix Technologies, for example, offers a utility that, when installed on the server, allows users to employ the same password for both UNIX and NT without having to install any client software. It offers the additional benefit of being able to change both passwords when one is changed, greatly easing the password management burden for administrators.

When users want to change their password, Syntunix uses existing Windows mechanisms and transparently synchronizes the passwords. The server intercepts the password change request from the operating system without user intervention, encrypts the change request, and propagates the change to one or more remote servers.

User Groups

User groups provide a security extension beyond password protection. Making network users part of user groups is a method administrators can use to control users' access to network applications, data, and LANs throughout an enterprise. Administrators can add and remove users from a group as needed. User groups are especially useful when dealing with company project teams.

User groups are also useful when implemented on virtual private networks (VPNs). The closed user group capability provided with the VPN service lets businesses limit incoming or outgoing voice-data calls to members of a group. At the same time, businesses can quickly and efficiently add (or remove) users to their IP services without having to install or change remote systems such as access servers and modem pools.

Smart Cards

Magnetic card systems allow users to access data on any available workstation by inserting a card into a reader attached to the workstation. The card key system allows access-level definition for each user, rather than for each workstation. Most automatic teller machines use this type of security combined with password protection (i.e., personal identification number or PIN).

Smart cards, which contain embedded microprocessors, accomplish a range of security tasks. For example, they perform online encryption, record time-on/time-off logs, and provide password and biometrics identification. Using smart cards for password entry means that users need not remember their password; the result is an extra degree of security. Smart cards are a viable security option for both local and remote access control. Some key/card setups require that two individuals perform an entry procedure before admittance is granted; this is similar to safe-deposit box access at most banks.

Sun's vision for the ultimate in thin-client computing includes a Java-based smart card that mobile employees would simply carry around to access all their files via the Web with a browser. The use of Java-enabled chips increases the information capacity of smart cards, allowing for more information and additional functions to be included—even allowing them to be upgraded and loaded with new applications after they are issued.

Biometrics

Biometric identification devices use an individual's unique physical attributes to secure data, such as fingerprints, handprints, voice recognition, and capillary patterns in the retina of a person's eye.

One drawback to biometrics is that pattern-recognition processing can take a long time—especially if the user database is large. Some systems decrease the processing time by using passwords before the biometric scanning occurs. When passwords are combined with biometrics, the security system compares the scan only with the image stored under the password entered.

The novelty of using biometric products to safeguard corporate networks has raised questions about how well the technology really works. To answer that question, the International Computer Security Association (ICSA) conducted a wide range of security product tests in early 1998 and approved several products after they passed a round of rigorous ICSA tests conducted in both laboratory and customer environments. Among the approved products was Touchstone, from Mytec Technologies, a fingerprint-matching device used for secure network access, electronic commerce, e-mail encryption, and database management. Another was Citadel Gatekeeper, from Intelitrak Technologies, a voice-print gateway server used to verify users' network access rights.

Data Encryption

Data encryption is a method of scrambling information to disguise its original content. It is the only practical means of protecting information transmitted over communications networks, but it can also be used to protect stored data. Since intruders cannot easily read encrypted data, intercepted information is more likely to remain safe if it falls into the wrong hands. Encryption methods come in two forms: hardware or software. Of the two, hardware-based encryption provides more speed and security.

There are two popular encryption methods: public key and private key. Public key encryption uses a publicly known key to encrypt the data, and a second private key to decrypt data. Private key methods uses a single-key algorithm known only to the sender and receiver. These methods work only as long as the key is kept secure. The advantage of public key encryption is that the private key need never be transmitted. It does not matter if the channel is insecure because the data can be decrypted only with the recipient's private key. Private key systems require that the secret key be transmitted to the recipient in order to decrypt the data.

The security of encrypted data is linked to the number of bits of the algorithm and key. The current standard is 128-bit security, but 256-bit key algorithms are available through software doubling. The larger the number of bits, the more key combinations are possible and the more computing power is required to break a key. The U.S. government closely regulates encryption methods for export. Encryption methods may or may not be allowed for export depending on the particular encryption technology and its intended use.

An example of a private key encryption system is Kerberos. With this system, every user's key is stored on a secure central server and kept secret. Kerberos authenticates the identity of every user and every network service via plain-text passwords. If one person compromises the server, all user keys are changed.

Developed for UNIX systems and slanted toward the open systems environment, Kerberos is an Internet standard. IBM, Apple, and Novell are currently developing a Kerberos-based security method for the OpenDoc specification. Because Kerberos has been adopted into commercial encryption products, it has become the de facto standard for remote authentication in client/server environments.

Firewalls

This is an umbrella term that encompasses any number of security techniques designed to prevent unauthorized access to a company's Internet-aware network. Most firewalls isolate an internal network in one of two ways: they deny the use of unsafe services on a front-line Internet server, such as an applications gateway or proxy server, or they use packet filtering to prevent traffic from passing to an internal network from anywhere other than predefined trusted sites. Some of today's firewalls can even filter out unsigned Java and ActiveX applets or viruses embedded in compressed e-mail attachments.

A typical packet filter checks IP address and service information to determine if traffic is coming from a trusted location. Packet filtering is transparent to users, provides a single point of entry and exit, and can disable services that have been inadvertently enabled on networks machines. Packet-filtering capabilities are widely available in commercial and freeware routing products; some router and firewall vendors bundle their products into a single Intel-based workstation.

Virus Scanners

The ever-increasing number of computer viruses makes virus protection a necessary security measure. Because workstations communicate with the network file server to obtain shared programs and data files, a virus can spread to every computer that logs onto the network. Viruses present a major threat to security because of the damage they can do to network information; viruses can corrupt data, delete files, slow system operations by spawning processes, and prevent applications from saving files. The new macro viruses that reside inside the document files of word processing and spreadsheet applications are especially difficult to detect.

Antivirus software packages recognize a multitude of known viruses, and most are updated frequently as new viruses are discovered. Network Associates, for example, offers Total Virus Defense, which includes an AutoImmune feature. It automatically extracts possible viruses and sends them to Network Associates for detection and cleaning. The product also includes Anti-Virus Informant, a customizable, Windows NT-based reporting tool that lets users monitor virus outbreaks and analyze data over time. It can track the number of machines running antivirus software or detail the number of virus instances and how fast they were eradicated.

Scanning e-mail for viruses is harder to do. Most antivirus programs on the market are limited to working with particular e-mail programs like Pegasus and Eudora, or with particular server platforms such as Windows NT. However, Panda Software's Antivirus Platinum software scans for viruses and malicious code at the application and IP layers. The software can detect and eradicate real-time viruses found in SMTP, POP3, HTTP, FTP, ActiveX, and Java code. The software also automatically updates itself by pulling down new signature files from Panda's Web site each day.

Antivirus software is available for firewalls and proxy servers, eliminating the need to load and maintain software on every client.

Vulnerability Scanners

Regularly scanning a corporate network for security vulnerabilities is just as important as scanning for viruses. Security vulnerability assessment tools are available from several vendors, including Network Associates. The company's CyberCop Scanner allows network administrators to proactively scan their networks for security weaknesses in much the same way that antivirus products scan for viruses. CyberCop Scanner can detect more than 500 known network vulnerabilities and security policy violations.

The product can even update itself when new security vulnerabilities are discovered by the company's researchers. Reporting and charting capabilities allows LAN administrators and managers to quickly audit their networks on a regular basis and prioritize detected vulnerabilities. In addition to detailed technical reports, CyberCop Scanner features graphical executive summary "snapshot" charts, displaying findings in several Web-based and 3-D interactive formats. Metrics can be tracked on a regular basis through summary reports, enabling managers to watch for trends and ensure that vulnerabilities are addressed in a timely manner.

Intelligent Agents

Intelligent agents are autonomous and adaptive software programs that accomplish their tasks by executing preassigned commands remotely without explicit activation at a management station. An agent can be assigned to monitor the network for a specific event or set of events, which trigger the agent to respond in a predefined manner. Agents are used for a variety of tasks, including security management. The following capabilities of intelligent agents can help discover holes in network security by continuously monitoring network access:

MONITOR EFFECTS OF FIREWALL CONFIGURATIONS   By monitoring post-firewall traffic, the network manager can determine if the firewall is functioning properly. For example, if the firewall was just programmed to disallow access of a specific protocol or external site but the program's syntax was wrong, the agent will report it immediately.

SHOW ACCESS TO/FROM SECURE SUBNETS   By monitoring access from internal and external sites to secure data centers or subnets, the network manager can set up security service-level objectives and firewall configurations based on the findings. For example, the information reported by an agent can be used to determine whether external sites should have access to the company's database servers.

TRIGGER PACKET CAPTURE OF NETWORK SECURITY SIGNATURES   Agents can be set up to issue alarms and automatically capture packets when external intrusions or unauthorized application access occurs. This information can be used to track down the source of security breaches. Some agents even allow for a trace procedure to discover a breach's point of origination.

SHOW ACCESS TO SECURE SERVERS AND NODES WITH DATA CORRELATION This capability reveals which external or internal nodes access potentially secure servers or nodes, and identifies which applications they run.

SHOW APPLICATIONS RUNNING ON SECURE NETS WITH APPLICATION MONITORING   This capability evaluates applications and protocol use on secure networks or traffic components to and from secure nodes.

WATCH PROTOCOL AND APPLICATION USE THROUGHOUT THE ENTERPRISE This capability allows the network manager to select applications or protocols for monitoring by the agent so the flow of information throughout the enterprise can be viewed. This information can identify who is browsing the Web, accessing database client/server applications, or using unauthorized software on the network, for example.

One example of an agent that is capable of taking action based on the nature of the security threat is Intruder Alert from Axent Technologies. The product uses a real-time, manager/agent architecture to monitor the audit trails of distributed systems for "footprints" that signal suspicious or unauthorized activity on all major operating systems, Web servers, firewalls, routers, applications, databases, and SNMP traps from other network devices. Unlike other intrusion detection tools, which typically report suspicious activity hours or even days after it occurs, Intruder Alert instantly takes action to alert IT managers, shuts systems down, terminates offending sessions, and execute other commands to stop intrusions before they damage important systems.

NetProwler, a network monitoring component for Intruder Alert, provides network-based intrusion monitoring. Together, the two provide host-based and network-based protection using a single management interface, allowing organizations to protect themselves against the widest range of existing and evolving security threats for the price of a single solution.

Security Services

By taking advantage of independent or vendor-supplied professional consulting services, companies can leverage their investments in security technology. Consulting services can help IT staff integrate various security products into the organization's security management practices. The integration process usually includes a simulated attack to ensure that the products function as expected. The results can be used to help fine-tune the products' settings to meet specific security needs.

Another consulting service that is becoming popular is security vulnerability assessment. Unlike security audits, which assume an organization already has a security policy in place, security vulnerability assessment helps companies determine where the security holes are. It then provides a blueprint for a policy to shore up those and other security weaknesses.

SUMMARY ▪ ▪ ▪ ▪ ▪ ▪ ▪ ▪ ▪

The ideal network protection strategy combines multiple security procedures for optimal data protection, but does not create resistance that may result in overt attempts to circumvent the security strategy. An often overlooked aspect of network security is that employees are responsible for as much as 80 percent of break-in attempts. While much of this activity is inadvertent or harmless in the overall scheme of things, the proverbial "disgruntled employee" is the real danger. These individuals, perceiving themselves to be wronged by the company in some way, often steal sensitive information or inflict significant damage to databases and other resources. Security works well only when employees are educated about the importance of safeguarding corporate information. New employees should be trained from the start to honor the security mechanisms that are in place and to identify security violations when they occur so appropriate and timely corrective action can be taken. Periodic refresher sessions for supervisors and managers can help leverage corporate investments in security tools and reinforce top management's security policies.

See Also **Firewalls, Intelligent Agents, Public Key Encryption, Security Risk Assessment**

LAN Telephony

LAN telephony integrates voice and data over the same medium, enabling automated call distribution, voice mail, and interactive voice response, as well as voice calls and teleconferencing between workstations. However, LAN telephony need not be confined only to LANs. Like data, voice can also go out over wide-area data networks; in some cases, it may not involve a local-area component at all. One of the most common uses of LAN telephony will be in the corporate intranet environment, where it is referred to as *IP telephony*.

Applications

With LAN telephony, users working away from their offices—at home or in a hotel—can use a single phone line to carry both data and voice traffic. Users dial up to access the corporate intranet, which is equipped and engineered to carry real-time voice traffic. Such a system provides an integrated directory view, enabling remote users to locate individuals within the corporation for voice- or e-mail connection in a unified way. Likewise, phone callers (internal or external to the corporation) can locate the mobile workers connected to any part of the intranet. Thus, LAN telephony allows users to work seamlessly from any location.

By using the LAN-based conferencing standards, transparent connectivity of different terminal equipment can be achieved; the media used by conference participants is limited only by what is supported by their terminal equipment. Connectivity to room-based conference systems or analog telephone can be achieved by means of gateways, which perform the required protocol and media translations. The building block of LAN telephony is the international H.323 standard, which specifies the visual telephone system and equipment for packet-switched networks. H.323 is an umbrella standard that covers a number of audio and video encoding standards. Among these standards is H.225 for voice packetization, which is based on the Internet Engineering Task Force's (IETF) Real Time Protocol (RTP) specification and the H.245 protocol for capability exchange between terminals.

On the sending side, the uncompressed audio/video information is passed to the encoders by the drivers, then given to the audio/video application program. For transmission, the information is passed to the terminal management application, which may be the same as the audio/video application; the media streams are carried over RTP/UDP, and the call control is performed using H.225 to H.245/TCP.

Gateways provide the interoperability between H.323 and the public switched telephone network (PSTN), as well as networks running other teleconferencing standards such as H.320 for ISDN, H.324 for voice, and H.310/H.321 for ATM. An example H.323 deployment scenario involves terminals interconnected in the same local area by a switched LAN. Access to remote sites is provided by gateways, routers, or integrated gateway/router devices. The gateways provide communication with H.320 and H.324 terminals remotely connected to the ISDN and PSTN, respectively. H.323-to-H.323 communication between two remote sites can be achieved with routers that directly carry IP traffic over the Point-to-Point Protocol (PPP) running on ISDN. For better channel efficiency, gateways can translate H.323 streams into H.320 to be carried over ISDN lines, and vice versa.

Vendor Implementation

NBX Corporation, now a unit of 3Com, is among the growing number of vendors offering LAN telephony systems. The company's NBX 100 Communication System leverages the ubiquity of Ethernet—once customers install NBX 100, they have a LAN infrastructure for connecting PCs, printers, servers, and other network devices. NBX 100 employs a packetizing technology that delivers high-quality voice reliably over the Ethernet. The system includes a network call processor for call control, voice mail, auto attendant, browser administration, and connectivity to the PSTN and WAN/Internet.

With connectivity to the WAN/Internet, NBX 100 also supports the growing demand for telecommuter offices. Using a standard multiprotocol router (ISDN, ATM, or frame relay), remote users gain access to the full suite of NBX 100 voice communication features, just as if they were in the office headquarters. Simultaneously, remote users also gain access to all the resources on their company's LAN, including e-mail, file servers, and intranet/Internet access.

SUMMARY

Unlike traditional PBXs, a LAN-based telephony system has no central point of failure. Calls are routed between remote locations over a company's wide-area data network, providing significant savings on long-distance calls, especially if the company has offices in many countries. Using a switched 100-Mbps Ethernet, network engineers can design telephone networks with essentially unlimited capacity. When the need arises for more workstations (extensions), another Ethernet switch is added. With H.323, these systems offer a high degree of interoperability. These systems can be administered locally through a Windows graphical user interface or remotely through a Web browser.

***See Also* Computer-Telephony Integration, IP-to-PSTN Gateways**

LAN Topologies

Topology is the physical and logical arrangement of a network. The physical arrangement of the network refers to how the workstations, servers, and other equipment are joined together with cables and connectors. The logical arrangement of a network refers to how the workstations, servers, and other equipment relate to each other in terms of traffic flow. There are three primary LAN topologies: linear bus, ring, and star. Another network topology is hierarchical in nature, which may incorporate elements of the bus, ring, and star. The appropriate physical and logical topology for a LAN is determined by reliability and cost objectives as well as by the connectivity requirements of users.

Bus

In a linear bus topology, stations are arranged along a single length of cable, which can be extended at either end or at both ends to accommodate more nodes (Figure 65). The network consists of coaxial cable, such as the RG-58 A/U cable used with 10Base2 Ethernet LANs. The nodes are at-

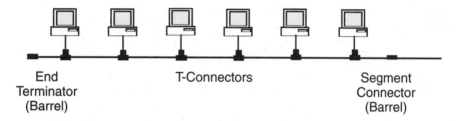

Figure 65
The linear bus
topology.

End
Terminator
(Barrel)

T-Connectors

Segment
Connector
(Barrel)

tached to the cable with a BNC (Bayonet Nut Connector) T-connector (Figure 66), the stem of which attaches to the network interface card (NIC). A BNC barrel connector attaches cable segments and a BNC terminator connector caps the cable ends. Of course, twisted pair wiring is most often used for Ethernet LANs, in which case RJ45 connectors provide the connections between devices.

A linear bus network can be further extended. For example, a tree topology is actually a complex linear bus in which the cable branches at either or both ends, but offers only one transmission path between any two stations.

Ring

In a ring topology, nodes are arranged along the transmission path so data passes through each successive station before returning to its point of origin. As its name implies, the ring topology consists of nodes that form a closed circle (Figure 67).

In token-ring LANs, a small packet called a *token* is circulated around the ring, giving each station in sequence a chance to put information on the network. The station seizes the token, replacing it with an information frame. Only the addressee can claim the message. At the completion of

Figure 66
BNC T-connectors
are used to
connect two cable
segments to a
node's network
interface card
(NIC).

Figure 67
The ring topology.

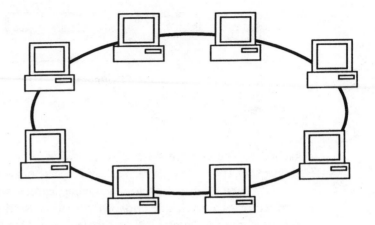

the information transfer, the station reinserts the token on the ring. A to-ken-holding timer controls the maximum amount of time a station can occupy the network before passing the token to the next station.

Star

A star network has a central node that connects to each station by a sin-gle, point-to-point link (Figure 68). All communications between nodes

Figure 68
The star topology.

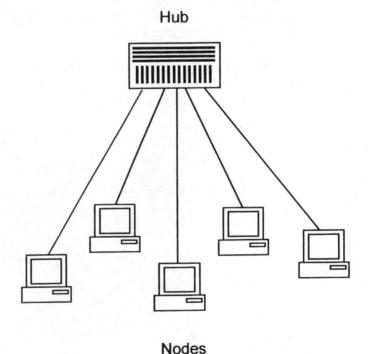

pass through the central node, which acts as a processing and coordinating point for the network. This central node is generally referred to as a *hub*. Information addressed to one or more specific nodes is sent through the central node and switched to the proper receiving station(s) over a dedicated physical path.

Hierarchical

More complex LAN topologies can be created from the basic bus, ring, and star topologies. One of these is the "dual ring of trees" on Fiber Distributed Data Interface (FDDI) networks that is created with special categories of equipment. These equipment types may be arranged in any of three topologies: dual ring, tree, and dual ring of trees (Figure 69).

Figure 69
With FDDI, a dual ring of trees can be used to create a hierarchical topology to enhance network reliability.

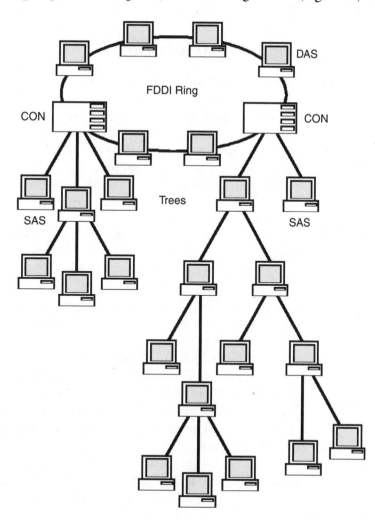

In the dual ring topology, dual attached stations (DASs) form a physical loop, where all the stations are dual attached. In a tree topology, remote single attached stations (SASs) are linked to a concentrator, which is connected to another concentrator on the main ring.

Any DAS connected to a concentrator performs as a SAS. Concentrators can be used to create a network hierarchy, which is known as a dual ring of trees. This topology offers a flexible hierarchical system design that is efficient and economical. Devices requiring highly reliable communications attach directly to the main ring, while those that are less crucial attach to branches off the main ring. Thus, SAS devices can communicate with the main ring, but without the added cost of equipping them with a dual-ring interface or a loop-around capability that would otherwise be required to ensure the reliability of the ring in the event of a station failure.

Topology Selection

Each topology has advantages and disadvantages. The bus topology characteristic of Ethernet LANs is the most economical and easiest to install. The ring is slightly more expensive and complicated. In both types of topologies, when one node malfunctions or becomes inoperable, the nodes on either side of it cannot communicate. This can be overcome by adding a hub. The nodes communicate with each other over separate cable segments via the collapsed backbone within the hub. If one node become inoperable, the other nodes are not affected since they are no longer directly connected.

In the case of Ethernet, although the physical topology has changed from a linear bus to a star, the logical operation remains unchanged in that Ethernet's Carrier Sense Multiple Access with Collision Detection (CSMA/CD) protocol still governs access. In the case of token ring, although the physical topology has changed from a ring to a star, the logical operation remains unchanged in that token ring's circulating "token" still governs access.

When it comes to link availability, the star topology is highly reliable. In this topology, all network devices connect to a central hub through dedicated or shared LAN segments. Although the loss of a link prevents communication between the hub and the affected node(s), all other nodes continue to operate as before unless the hub itself suffers a catastrophic failure.

To ensure a high degree of reliability, the hub has redundant control logic, backplane, and power supply. The hub's management system can enhance the fault tolerance of these redundant subsystems by monitoring their operation and reporting any problems. With the power supply, for example, monitoring may include hotspot detection and fan operation to detect trouble before it disrupts hub operation. Upon the failure of the main power supply, the redundant unit switches over automatically or manually under the network manager's control without disrupting the network. If a fan goes out, an alarm can be sent to the management console as well as to a technician's pager.

The flexibility of the hub architecture lends itself to varying degrees of fault tolerance, depending on the importance of the applications. For example, workstations running financial modeling applications may share a link to the same LAN module at the hub. Although this configuration might seem economical, it is problematic in that a failure in the LAN module will put all of the workstations on that link out of commission.

A slightly higher degree of fault tolerance can be achieved by distributing the workstations among two LAN modules and links. That way, the failure of one module will affect only half of the workstations. A one-to-one correspondence of workstations to modules offers an even greater level of fault tolerance in that the failure of one module impacts only the workstation connected to it. However, this configuration is also the most expensive solution.

A mission-critical application may demand the highest level of fault tolerance. This can be achieved by connecting the workstation to two LAN modules at the hub with separate links. The ultimate in fault tolerance can be achieved by connecting one of those links to a different hub. In this arrangement, a transceiver is used to split the links from the application's host computer, enabling each link to connect with a different module in the hub or to a different hub. In each case, the physical topology changes, but the logical topology remains the same.

SUMMARY

With the introduction of switching equipment into LANs, it is now possible to fine-tune the topology of smaller subsections of an organization's network. Network planners can provide the advantages of one topology over another to meet the specific needs of individuals, workgroups, or departments.

See Also **Logical Network Management, Physical Network Management**

LAN Troubleshooting

LANs often experience performance problems that can be attributed to faulty hardware. Indications of faulty hardware are that network performance is poor or that one or more devices cannot communicate. One way to identify hardware problems on a network is to track communication errors. Errors occur on a network when a device receives a transmission but is unable to make sense of it.

Types of Errors

On an Ethernet network, for example, several distinct types of errors can appear, including:

COLLISIONS These occur when two devices attempt to place a packet on the network at the same time. Collisions are detected when the signal on the cable is equal to or exceeds the signal produced by two or more transceivers that are transmitting simultaneously.

JABBERS This is a frame that is greater than 1518 bytes and has a bad CRC (checksum value). If a transceiver does not halt transmission after 1518 bytes, it is considered to be a jabbering transceiver.

FRAGMENTS These are packets that appear at a receiving station, contain fewer than the required minimum of 64 bytes, and have a bad CRC. Fragments are generally caused by collisions.

OVERSIZE These are frames that are larger than the maximum 1518 bytes and have a good CRC.

UNDERSIZE These are frames that contain less than the minimum of 64 bytes and have a good CRC.

CRC/ALIGNMENT ERRORS These are packets that do not contain the proper CRC. In addition, if the frame does not end on an 8-bit boundary, an alignment error will occur. Both of these types of errors are grouped and counted as CRC/alignment errors.

Tools

If a technician could identify what types of errors occur on the network, along with where they occur, the problems could be resolved more easily. This requires the use of special tools, such as RMON probes, that collect performance data on the network segments, as well as a monitoring application to interpret the results.

Briefly, a typical troubleshooting procedure begins by opening the monitoring application's statistics window for a RMON probe. For the selected segment, the items of interest for an Ethernet network are CRC/alignment errors, undersizes, oversizes, fragments, jabbers, and collisions. A network segment that has faulty hardware would show significant errors of one or more of these types.

Once it is determined that a network segment has excessive errors of a particular type, it might be helpful to track the error to a single device or a few devices. A technician can accomplish this by opening the Top N window of the monitoring application, which reveals the top 10 devices transmitting errors (Figure 70). The device at the top of the graph generates the most errors and thus requires the most attention.

At this point, a problem with errors in the network traffic has been discovered and the type of error or errors has been determined, as well as the possible source for the errors. The next step is to find out what is causing the errors.

If collisions are high, it might mean that the network segment is being overused and that some workstations need to be reassigned to other less-used segments. Another cause of high collision rates are cable segments that are too long.

If there is jabber, a faulty transceiver is the cause. The transceiver is a chip, which is responsible for transmitting bits to and receiving bits from

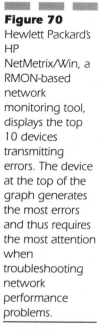

Figure 70
Hewlett Packard's HP NetMetrix/Win, a RMON-based network monitoring tool, displays the top 10 devices transmitting errors. The device at the top of the graph generates the most errors and thus requires the most attention when troubleshooting network performance problems.

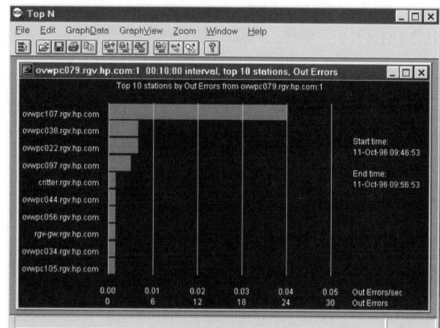

the medium. Transceivers are typically located on the network interface card (NIC). Eliminating jabber entails replacing the transceiver, which usually means replacing the entire NIC.

Packet fragments are generally the result of collisions, and the two types of errors usually go hand in hand. Taking corrective actions for collisions eliminates fragments as well.

Illegal length frames—oversizes and undersizes—can easily be traced to the transmitting station, since the frame is well formed and contains the source address in the header. Such frames are generally caused by a faulty LAN driver. The driver may be out of date or the file on disk may be corrupt.

CRC/alignment errors are usually due to cabling problems. Any one of the following cabling problems can cause this type of error:

- Segment too long
- Damaged cable
- Segment not properly grounded
- Improper termination
- Taps too close
- Noisy cable (electromagnetic interference)
- Faulty controller cards

SUMMARY

Network troubleshooting is usually more involved than what is discussed in this section, and it varies with token ring, FDDI, and other types of networks. Regardless of the type of network, the speed and efficiency of troubleshooting is dependent on proactive measures, such as maintaining baseline measurements of network parameters, e.g., usage and errors. With performance parameters for comparison, network administrators can more easily detect when something has gone wrong, ascertain the probable cause, and take corrective action.

See Also **LAN Analyzers, Network Monitoring, Network Probes, Performance Baselining, Performance Monitoring**

LAN/WAN Integration

Companies grappling with issues such as downsizing, distributed computing, and participation in a competitive global economy are faced with the necessity of integrating their local-area networks (LANs) over the wide-area network (WAN) to facilitate the timely flow of information among business units. As enterprise networks continue to grow and expand to include telecommuters, small branch offices, and far-flung international locations, the need to interconnect dissimilar LANs and diverse equipment over public and private WANs becomes even more urgent. Often, integration is complicated by the need to tie in legacy networks as well, such as IBM's Systems Network Architecture (SNA).

Integration Methods

There are several methods for integrating different types of LANs and combining legacy systems with LANs, including translation, encapsulation, and emulation. It's necessary to understand these methods in order to successfully implement any integration effort.

TRANSLATION   One way to integrate different types of LANs is through translation. Since the packet structures of Ethernet, token-ring, and Fiber Distributed Data Interface (FDDI) LANs are fairly similar—differing mainly in terms of length—translation is a fairly straightforward process. A special kind of bridge, a translating bridge, reads the data link layer destination addresses of all messages transmitted by Ethernet devices, for example. If the destination address does not match any of the source addresses in its table, the packet is allowed to pass onto an adjacent token ring.

Using a translating bridge for this purpose has a serious drawback: such devices cannot fragment packets. A token-ring or FDDI packet of 4,500 bytes cannot be placed on an Ethernet LAN, which is limited to supporting packets of no longer than 1,528 bytes, including overhead. To make this scheme work, the token-ring or FDDI devices must be configured to transmit packets at the Ethernet packet length. However, it is not very practical to ratchet down the performance of high-speed LANs just to accommodate traffic to a slower LAN.

ENCAPSULATION   Encapsulation is the process of putting one type of data frame into another type of data frame so it can be recognized by the appropriate receiving device (Figure 71). This process allows different devices using multiple protocols to share the same network. Encapsulation entails adding information to the beginning and end of the data unit to be transmitted. The added information is used to perform the following tasks:

- Synchronize the receiving station with the signal
- Indicate the start and end of the frame

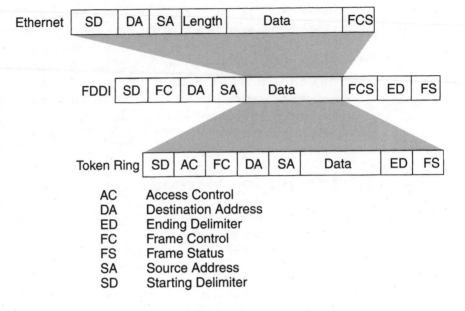

Figure 71
Encapsulation of
Ethernet and
token-ring packets
into an FDDI data
packet.

- Identify the addresses of sending and receiving stations
- Detect transmission errors

At the destination device, this envelope is stripped away and the original frame delivered to the appropriate end user in the regular manner. There is significant overhead with this solution because one complete protocol runs inside another. But any increase in transport capacity requirements and associated costs is usually offset by savings from eliminating the need for separate networks.

EMULATION One of the simplest (and oldest) ways for PCs to access SNA host applications is through terminal emulation. For a microcomputer to communicate with a mainframe, it must be made to do something it was not designed to do—emulate a terminal so it can be recognized as such by the mainframe. In the IBM environment, 3270 terminal emulation is used, which permits synchronous data transfer between microcomputers and a mainframe. With 3270 terminal emulation, data is exchanged in a format that is readily acceptable to the host.

Terminal emulation is accomplished through a micro-to-mainframe communications package consisting of software and hardware, which usually includes an interface for direct connection to a controller or local-area network. To facilitate dial-up host access, most modems sold today come with software that include 3270 and other types of emulation.

A number of 3270 terminal emulation products have become available over the years, additionally providing 3278 or 3279 terminal emulation and supporting both direct coaxial and modem connections to 3174, 3274, and 3276 controllers without requiring additional mainframe software. In addition to allowing users to save terminal screens, these emulation products allows users to hot-key between microcomputer and terminal sessions, and switch to file-transfer menus and the DOS command line. DOS and 3270 profiles give the existing keyboard dual functionality that reflects the microcomputer and 3270 terminal configurations.

While emulation has traditionally been associated with terminal emulation, there is now LAN emulation, a newer technology, which is being used to integrate Ethernet and token-ring LANs into asynchronous transfer mode (ATM) networks.

This integration is accomplished through software that breaks apart larger LAN packets and inserts them into the 53-byte ATM cells. Conversions between LAN packets and ATM cells are accomplished without generating excessive overhead. LAN emulation allows multiple LANs to coexist on the same physically interconnected ATM network. Since the emulation software resides on a network access server or bridge, no changes are required to the end-user hardware, software, or operating system.

The addresses used in ATM networks are not the Media Access Control (MAC) addresses that underlie IEEE 802.3-compatible protocols. To allow ATM-to-LAN communication, MAC addresses must be mapped to ATM addresses. This mapping is performed by the LAN emulation software. For the ATM-to-LAN link, the sending device must ask the LAN emulation server for the address of the receiving node. The ATM sender then uses that address to contact the remote LAN device.

By hiding the use of ATM from applications, LAN emulation allows legacy applications to operate unchanged over high-speed ATM networks, permitting companies to preserve their existing investments. On the other hand, because these existing applications do not understand class-of-service contracts and the other nuances of ATM, they will not gain all the benefits of ATM. LAN emulation may give way in favor of new ATM-aware applications. For now, however, LAN emulation is an important capability for integrating LANs over a high-speed WAN.

To integrate Ethernets and token rings over a 100-Mbps LAN, there is 100BaseVG AnyLAN. Developed by IBM and HP, 100VG AnyLAN allows Ethernet and token-ring packet frames to share the same media. It supports Category 3 unshielded twisted pair, Type 4, Type 5, and other cable types. However, AnyLAN does not make Ethernet and token-ring networks interoperable; it only unites them physically through the use of a common hub and wiring. For true interoperability, a router is needed, with the capability to logically merge the two protocols into a common packet and addressing scheme.

SNA-LAN Integration

To avoid the expense of duplicate networks, users are looking at ways to integrate incompatible SNA and LAN architectures over the same facilities or services. Aside from cost savings and flexibility, the consolidation of such diverse resources over a single internetwork offers several other benefits, such as:

- Eliminating the need to provide for, operate, and maintain duplicate networks (one for SNA, one for token ring, and another for non-IBM environments)
- Allowing slow leased-line SNA networks to take advantage of the higher speeds offered by LAN/WAN links
- Consolidating diverse traffic types and minimizing potential points of network failure, providing a more resilient infrastructure

While there are a number of vendor-specific solutions for integrating SNA with LANs and TCP/IP internets, there are a few standard approaches:

- APPN (Advanced Peer-to-Peer Networking)
- High-Performance Routing (HPR)
- Encapsulation or Data Link Switching (DLSw)
- Permanent virtual connections (PVCs) or RFC 1490 Frame Relay

APPN extends SNA-to-PC LANs by letting midrange processors communicate on a peer-to-peer basis, while HPR streamlines SNA traffic so routers can move the data around link failures or outages.

Generally, APPN is used when SNA traffic must be prioritized by class of service, which routes traffic directly to end nodes, or when SNA traffic must be routed peer to peer without going through a mainframe. HPR is used when traffic must be sent through the distributed network without disruptions. HPR provides link-usage features that are important when moving to packet-switched LANs, such as ATM or frame relay, and provides congestion control for optimizing bandwidth. This method offers a faster routing path than APPN because it ensures that SNA traffic is prioritized and routed to all network nodes. HPR's performance gain over APPN comes from its end-to-end flow controls, which are an improvement over APPN's hop-by-hop flow controls.

Another SNA routing technique is DLSw, which is used in environments consisting of a large installed base of mainframes and TCP/IP backbones. DLSw assumes the characteristics of APPN and HPR routing, and combines them with TCP/IP and other LAN protocols. DLSw encapsulates TCP/IP and supports SDLC and high-level data link control (HDLC) applications. It prevents session timeouts and protects SNA traffic from becoming susceptible to link failures during heavy congestion periods.

Like DLSw, RFC 1490 uses encapsulation to transport all protocols, including SNA/APPN, within frame relay frames. It provides SNA-guaranteed bandwidth through frame relay's permanent virtual circuits and, compared to DLSw, uses very little overhead.

LAN-to-Internet Integration

One of the newest trends in LAN/WAN integration is using IP-based intranets or the public Internet as a backbone between far-flung LANs for applications such as e-mail, fax, and file transfers. Hubs, switches, and servers from many interconnected vendors now provide LAN connections to IP networks. They offer a range of WAN connectivity options, including dial-up via 56-Kbps modems or ISDN BRI and direct connections via 56K DDS or fractional T1. Even full T1 and T3 connections are available. On the LAN side, the same unit can support a broad range of network topologies, including Ethernet 10/100 Mbps and token ring.

Some vendors offer preconfigured systems that make it very easy to integrate IP nets with a LAN. They come with preinstalled hardware and software, and operate totally free of the LAN operating system. With systems that offer built-in modem, ISDN, 56K DDS, and T-carrier versions, all the hardware is preset. The network administrator need only cable the system to the LAN and connect the services or lines to the WAN side of the system. Integral CSU/DSU and/or V.35 connectors facilitate the setup of router connections to the WAN. Most systems support external routers as well. Once the LAN/WAN connections are in place, the administrator can assign user access privileges to control access by time of day, day of week, IP address, domain name, and service/port number.

These systems typically have one IP address, making it unnecessary to register and maintain separate IP addresses or to install TCP/IP stacks on each individual workstation or server. Multiple systems can be connected to the network to support additional users, load balancing, and fault tolerance. Depending on vendor, these systems may even come with built-in firewall hardware and software to protect the corporate network from outside intruders, making it invisible to users on the public Internet.

SUMMARY

Throughout their development histories, LANs, WANs, and the Internet have moved along separate paths. Now with trends as diverse as corporate downsizing, distributed computing, and participation in the competitive global economy, companies are virtually forced to integrate these networks to facilitate the timely flow of information. Adding to the complexity of integrating LANs and WANs is the need to include legacy networks such as SNA into the equation. Given the vast installed base of SNA and LAN networks, the movement toward integration makes sense, especially with the advent of reliable integration techniques, such as DLSw and RFC 1490. Internet technologies continue to evolve and will play a larger role in LAN/WAN integration in the future.

See Also **Asynchronous Transfer Mode, Bridges, Routers**

Laser Communication Systems

Laser-optic transmission systems operate in the near-infrared region of the light spectrum. Using coherent laser light, these wireless line-of-sight links are well suited for campus environments and urban areas, where the installation of cable is impractical and the performance of leased lines is too slow. Unlike microwave transmission, laser transmission does not require an FCC license, and data traveling by laser beam cannot be intercepted.

Laser transmission is not a carrier-provided service, but a method for private network users to bypass the local exchange carrier for certain applications, such as point-to-point LAN interconnection. The laser system integrates into a network through a network switch, hub, bridge, router, or other interface device. Generally, the laser system is used as a direct substitute for cable, which would otherwise be needed to connect the sites. Since the laser system typically provides protocol-transparent connectivity at the physical layer, it does not provide any error-checking, bridging, routing, or repeater function on its own.

The lasers at each location are aligned with a simple bar graph and tone lock procedure. A PC can be attached to the laser units to provide operational status, such as signal strength, and to implement local and remote diagnostics.

In addition to fast, simple installation, laser systems provide wire-speed interbuilding links at 10-Mbps Ethernet, full-duplex Ethernet, and 4/16-Mbps token ring. Some products provide wire-speed connections for all network protocols from 10-Mbps and 100-Mbps Ethernet, to 100-Mbps

FDDI and 155-Mbps ATM. The latest laser systems transmit at up to 622 Mbps.

At this time, no microwave system can provide 622 Mbps of throughput in a single system. Laser units are smaller than many of the larger antenna sizes required in microwave systems, providing for easier installation. Laser is particularly advantageous where fiber-optic cabling may be difficult or impossible to complete, as when sites are separated by water or mountainous terrain. A laser system can be installed in much less time than it would take to install cable. Laser systems can eliminate the need to run fiber-optic cable for temporary connectivity needs such as conventions or sporting events.

Laser is a fairly secure method of transmission. The infrared frequencies of light used in laser systems are invisible to the naked eye and confined to a narrow path. Interception is difficult and requires that a hacker know the physical location of the beam, and also enter the beam path directly to receive the transmission using complex detection electronics. Any interruption in transmission would be detected by the user. Encryption may be added if desired.

Sometimes a laser beam will be broken by flocks of birds or some other temporary obstruction. While the signal can be disrupted when something breaks the beam path, many network protocols will handle the disruption by resending the data, and the break will not be noticeable to network users.

Although sunlight can disrupt laser communications, systems come with built-in interference filters to block out wavelengths of light other than the transmit laser wavelengths. However, the sun is still capable of interfering with data reception if its 0.5-degree disk overlaps the telescope line of sight. This is only likely to occur a few days a year for a few minutes in the worst case, but it is a potential source of errors. This can be avoided by orienting the laser system so the sun is directly behind one of the transceivers at any given time.

SUMMARY

Laser systems provide high-speed building-to-building wireless links at full network speeds. A limitation of laser systems is that transmission can be affected by atmospheric conditions that can reduce the amount of light energy picked up by the receiver and corrupt the data being sent. However, routing, switching, or bridging equipment allows traffic from the laser system to be offloaded to the wireline network until visibility improves. *See Also* **Microwave Systems, Wireless LANs**

Latency

Latency is the amount of delay that affects all types of communication systems and networks. The Internet Engineering Task Force (IETF) Benchmarking Methodology Working Group has two definitions of latency, which are discussed in RFC 1242:

FOR STORE AND FORWARD DEVICES   Latency is the time interval starting when the last bit of the input frame reaches the input port and ending when the first bit of the output frame is seen on the output port.

FOR BIT-FORWARDING DEVICES   Latency is the time interval starting when the end of the first bit of the input frame reaches the input port and ending when the start of the first bit of the output frame is seen on the output port.

Latency is usually measured in milliseconds (ms), or thousandths of a second, but it could also be expressed in microseconds, or millionths of a second (μsec). Of the two sources of latency—systems and networks—the most problematic is latency on networks. However, in both cases steps can be taken to minimize latency to improve performance.

Systems

In a computer system, latency is any characteristic that increases real or perceived response time beyond the desired response time. Specific contributors to latency include mismatches in data speed between the microprocessor and input/output devices and inadequate data buffering. Within a computer, latency can be removed or "hidden" by techniques such as prefetching—anticipating the need for data input requests—and multithreading, or using parallelism across multiple execution threads.

A similar technique, caching, is applied to servers. Under this scheme, frequently requested information is stored on a LAN server to eliminate unnecessary traffic on the wide-area network. This also has the effect of greatly reducing the delay in servicing user requests.

As applied to a network switch, systems latency is a natural by-product of store-and-forward and bit-forwarding processes. Switch latency actually accounts for a very small portion of the total end-to-end latency on a network. Today's Layer 2 switches generally use distributed architectures implemented with ASICs (application-specific integrated circuits) to reduce store and forward latencies to the 5- to 50-microsecond range.

Switch latency in this range has been shown to contribute very little to user-perceived response times. In fact, for virtually all Ethernet switching, the biggest contributor to network latency is the serialization time or packet transmission time (1.2 milliseconds per switch hop for a standard 1518-byte packet at 10 Mbps; 120 microseconds at 100 Mbps; and 12 microseconds at 1 Gbps). If a packet is stored several times between source and destination, a serialization delay is incurred for every hop along the path.

Network Latency

A rule of thumb used by the telephone industry is that the round-trip delay for a telephone call should be less than 100 ms. If the delay is much more than that, participants will think they hear a slight pause in the conversation and take the opportunity to begin speaking. But by the time their words arrive at the other end, the people there have already begun their next sentence and feel that they are being interrupted. When telephone calls go over satellite links, the round-trip delay is typically about 250 ms, and conversations become full of awkward pauses and accidental interruptions.

Latency affects the performance of applications on data networks as well. On the Internet, for example, excessive delay can cause packets to arrive at their destination out of order, especially during busy hours. The reason packets arrive out of sequence is that they can take different routes on the network. The packets are held in a buffer at the receiving device until all packets arrive, when they are put in the right order. While this does not affect e-mail and file transfers, which are not real-time applications, excess latency does affect multimedia applications because it causes voice and video components to arrive unsynchronized.

If the packets containing voice or video do not arrive within a reasonable time, they are dropped. When packets containing voice are dropped, a condition known as *clipping* occurs, which means that either the first or the final syllables in a conversation are cut off. Dropped packets of video cause the image to be jerky. Excessive latency also causes the voice and video components to arrive out of synchronization with each other, causing the video component to run slower than the voice component; for example, in a video conference, a person's lips will not match what he or she is really saying.

The problem of latency can be addressed by sending all traffic to a queue and assigning it a class of service or quality of service (QoS), so real-time applications are identified as being time-sensitive and requiring priority over other, less time-sensitive traffic. Quality of service can be implemented by the network (routers, hubs, and switches), the operating system, or a combination of both hardware and operating system working together.

With any queuing-based system, however, latency is added as packets are buffered deep within traffic-class queues. This can be avoided by using traffic management software such as Packeteer's PacketShaper, which prevents buffering by quickly assessing packet timing and directly influencing flow control rather than holding onto packets, dropping packets, and forcing retransmission. The maximum latency that PacketShaper will introduce is two milliseconds.

Enterprises who want to deploy VoIP (Voice over IP) for an existing Frame Relay network or corporate intranet, for example, would place a PacketShaper on the LAN side of a router at a branch location (Figure 72). PacketShaper can be configured with rate-control policies for Web traffic, a low-bandwidth rate guarantee for RTCP (real-time control) traffic, an 8-Kbps rate guarantee for call session (RTP) flows, and complementary latency bounds for all RTP traffic to minimize jitter over the shared frame-relay link. This would allow efficient access-link sharing without requiring new frame-relay ports or additional permanent virtual circuits.

Measuring Latency

It is often necessary to measure end-to-end latency to determine whether delays are caused by the network, an application on the user's PC, or the Internet access connection. In the past, diagnosing network slow-downs from a user's perspective was largely a matter of guesswork. Today, latency measurements can be taken with data gathered from routers across carrier and large enterprise networks using monitoring and reporting tools such as Concord's Network Health. By continuously monitoring for latency anywhere on the network, the tool can report delays before user performance is affected.

Figure 72
Setting policies within a PacketShaper IP bandwidth manager will guarantee the necessary bandwidth to Voice over IP traffic in order to ensure smooth performance, control delay-causing TCP bursts, and eliminate router-based queuing latency.

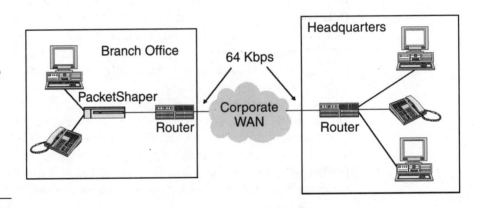

A typical method used to measure latency is to "ping" a router from a central management server. The ping command sends one datagram per second and prints (or displays) one line of output for every echo response returned. The round-trip send-response time of the packet is reported in milliseconds. A graphical management application, such as Network Health, allows network managers to ping across the same network paths the user traverses on a regular basis to more accurately measure latency. Network Health then generates reports that give network managers a clearer understanding of what the end user is experiencing and whether contracted service levels are being met by the organization's IT department or service providers.

For example, Concord's Network Health reporting tool can summarize latency information for up to 20,000 managed objects for a 30-day period. Using a simple bar chart, delay measurements are graphically displayed on a daily basis across the enterprise, showing the percentage of the day the enterprise spends in each of five latency ranges from less than 50 milliseconds to more than 150 milliseconds.

SUMMARY

Monitoring and reporting tools help network managers document and resolve network performance problems to ensure uninterrupted service. Often, this means having the capability to measure the latency of all WAN services from the network and backbone to the application server. Such tools provide the most complete view of network performance and can clearly identify where the problem is located so appropriate corrective action can be taken. Sometimes a bandwidth manager can be used to control latency.

Other possible actions network managers can take to reduce latency include rerouting traffic around points of congestion, adding bandwidth to existing routes so more traffic can get through, putting higher-speed routers or switches into the network, implementing a cache strategy, reinstalling or upgrading problem applications, and ordering higher-speed digital services or lines.

See Also **Network Availability, Network Caching, Network Congestion, Network Monitoring, Network Reliability, Ping, Service-Level Agreements**

Legacy Systems

Data communications was once a relatively simple affair, conducted in a tightly controlled mainframe environment and primarily under the auspices of IBM's Systems Network Architecture (SNA). After more than 20 years, SNA is still a stable and highly reliable architecture, and—despite the more recent trend toward distributed computing, including the advent of new architectures such as client/server—the mainframe is still valued for its ability to handle mission-critical applications.

Because of the long history of SNA, mainframes and related equipment were often referred to as "legacy systems" as a shorthand for differentiating the traditional host-centric data center from the newer distributed computing environment of desktop computers and LANs. In time, the term lost its SNA-specific connotation and was applied to any system or network that has been overtaken by technology to the point that it can't meet the needs of today's applications.

Even a previous-generation technology such as 10BaseT Ethernet, which is still widely used, is referred to as a legacy LAN when compared to Fast Ethernet at 100 Mbps and Gigabit Ethernet at 1000 Mbps. Likewise, 4-Mbps and 16-Mbps token-ring systems are referred to as legacy LANs in relation to High-Speed Token Ring (HSTR) at 100 Mbps.

Interestingly, TCP/IP is rarely if ever referred to as a legacy network, despite it being much older than SNA. One reason for this is that TCP/IP is new to most people who use it today. Until the World Wide Web (WWW) emerged in the early 1990s, few people outside of the military, science, and academic communities even knew that the Internet existed. Public access was discouraged and commercial content was illegal. All this changed when a graphical user interface called Mosaic became available, because it not only made the Internet very easy to use but it made it fun for anyone to explore. When the National Science Foundation (NFS) encouraged commercial development of the Internet so it could scale back its own funding, the Internet community kept pace by continually refining the TCP/IP suite of protocols until today it even can handle multimedia messaging, telephone calls, and videoconferences. As a result, TCP/IP is considered quite advanced.

SNA, too, has evolved to meet the needs of new application requirements. IBM's Advanced Peer-to-Peer Networking (APPN) is a decentralized version of SNA for use in distributed computing environments. Despite this, SNA systems are still referred to as legacy, mostly by competitors who want to perpetuate the idea that SNA is an outmoded technology.

SUMMARY

The huge investment in legacy systems and networks makes it impossible to cast them off in favor of new technologies—after all, IT departments are not given new budgets every time vendors come out with new products. With regard to SNA, for example, the sheer financial investment in legacy hardware and software—estimated to exceed $1 trillion worldwide—provides ample incentive to protect and leverage these assets in the new distributed computing environment of LANs and WANs. Legacy SNA traffic is now routinely carried over TCP/IP with other applications, eliminating the need for separate networks.

***See Also* LAN/WAN Integration, SNA over IP, Systems Network Architecture**

Linux

Linux is a UNIX-like operating system that was designed to run on low-end PCs. The development of Linux began in 1991 as a hobby of Linus Torvalds for his personal use while he was a student at the University of Helsinki. With assistance from a loosely knit team of programmers across the Internet—eventually numbering in the hundreds—Linux has blossomed into a feature-rich challenger to UNIX and Windows NT, with 12 million users worldwide.

Linux has all the features of UNIX, including true 32- and 64-bit multitasking, virtual memory, shared libraries, demand loading, shared copy-on-write executables, memory management, and TCP/IP networking. Among the advantages of Linux is that it is free, as are many of the applications written for it, and its graphical user interface is customizable (Figure 73). In addition, most existing X-Windows programs will run under Linux without any modification. Windows programs can also be run inside of X-Windows with the help of an emulator called WINE. Usually, Windows programs can run up to 10 times faster than on a native system, due to the buffering capabilities of Linux. MS-DOS applications can be run with the Linux DOS emulator.

Linux can now be found on a variety of hosts ranging from low-end PCs to powerful multiprocessor systems. Linux is regularly used on scientific workstations to provide network services, as a business computing platform, as a software development platform, and for personal computing at home or in the office. There is even software to connect Linux to midrange systems such as IBM's AS/400 and legacy mainframes running SNA.

Networking Capabilities

Networking support in Linux is superior to most other operating systems. Linux supports connection to the Internet and any other network

Figure 73

A graphical user interface for an SNMP (Simple Network Management Protocol) network mapping application.

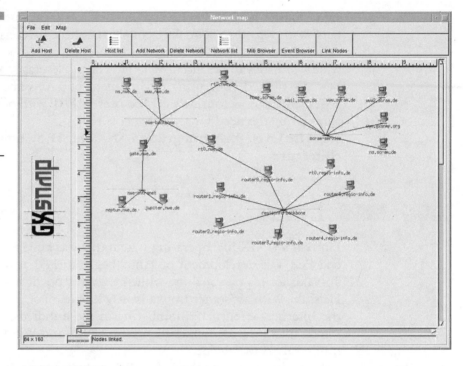

using TCP/IP, or IPX via token ring, Ethernet, Fast Ethernet, ATM, X.25, ISDN, or modem. It can also be used to directly connect two computers as peers via a modified printer cable.

As a Web server, Linux often outperforms Windows NT, Novell, and most UNIX systems running on comparable hardware. Linux supports all the most common Internet protocols, including Usenet News, Gopher, Telnet, Web, FTP, Talk, SMTP/POP, NTP, IRC, NFS, DNS, NIS, SNMP, Kerberos, and WAIS. Linux can operate as a client or server for all of these roles.

If Linux is to penetrate large enterprise networks, it must provide connectivity to SNA the way Microsoft has done with its Windows NT SNA Server package. This is now possible with Linux TN3270 communications client software, which allows Linux users to pull data from SNA applications and send it over a TCP/IP backbone for incorporation into Linux applications. An industry standard, TN3270 is a Telnet-based emulation technology that delivers SNA 3270 data over TCP/IP backbones. There is also Linux TN5250 communications software that lets Linux users access AS/400 server resources in the same way.

Commercial Distributions

Commercial distributions of Linux are available at modest cost from several companies. Red Hat Software, for example, offers Linux on CD-ROM, which includes installation documentation, installation support from Red Hat, Netscape Navigator, and application packages. The extra software allows Red Hat Linux to be set up for use as a Web server, e-mail server, DNS server, news server, and more—for multiple sites, with virtual hosting. Red Hat Linux also contains all of the software needed for client use.

Most of the networking and system administration configuration of a Red Hat Linux system can be set from an X-Windows graphical Windows-like tool called gnome or from the Web interface. Administrators can navigate through settings with gnome's Microsoft Explorer-like configuration tree, and edit selected settings in a multitabbed panel.

One hurdle that Linux must clear before it can break into mainstream enterprise use is the common belief that cheap price equals poor quality. However, those who have tried Linux in the corporate environment say it is far more stable and functional than the alternatives. With a growing number of major software vendors porting their products to Linux, it is becoming a more credible operating system.

Novell, for example, has ported its directory services technology to Linux. With Novell's Directory Services (NDS) running natively on Linux, administrators can manage users and groups, and access control rights across a network. It also enables authorized users to sign on to a corporate network just once to access resources on machines running Linux and other NDS-enabled operating systems, such as NetWare, NT, and Solaris.

Critics of Linux portray the freeware community as a semiorganized rabble of hobbyists, and question whether such a group can be trusted to act as caretakers for software used to run mission-critical applications. Despite not being developed by a single, commercial venture, the evolution of Linux has been successful and the platform has proved to be "self-governing," just like the Internet.

Like Windows NT, Linux does not support the large, multiprocessor configurations that commercial UNIX versions do. Linux's multiprocessor support is unlikely to go beyond 8- and 16-way configurations for the next five years. But future scalability will be derived by deploying such systems in clustered configurations.

SUMMARY

How much of a threat Linux will be to UNIX and Windows NT remains to be seen. Linux is somewhere in the middle of the pack in terms of capabilities compared to other flavors of UNIX. Linux needs a lot of other features before it is ready to take on UNIX in the enterprise, including better fault tolerance and system management capabilities. Against Windows NT, Linux offers a way out of the never-ending expensive upgrade cycle with Microsoft products. Most companies want to save as much money as possible and preserve their hardware and software investments over the long term, while Microsoft propagates the opposite approach, issuing regular upgrades. The way the development of Linux has been carried out, it has much more functionality than NT because the publicly available code is accessible for developers to create whatever functionality is needed.

See Also **NetWare, Windows NT, UNIX**

Load Balancing

Load balancing is a process that distributes application sessions to the best available server in a cluster or server farm. Accomplished through traffic redirection, server load balancing results in virtually unbounded server capacity and maximizes application availability. Load balancing can be implemented over LANs or over wide-area corporate intranets.

On a corporate TCP/IP-based intranet, for example, server-load balancing is implemented by a front-end redirector switch, which dynamically load-balances sessions for applications across multiple servers. This increases application performance linearly as servers are added. A software component, known as a traffic director, assures service quality by selecting the most suitable server for each session based on current knowledge of service availability, server load and processing capacity, as well as predefined maximum load threshold on individual servers.

A traffic redirector, such as Alteon Networks' ACEdirector, does not require client administration or any change to servers. Users configure a virtual IP address on the ACEdirector for each load-sharing server group. Application sessions are identified by source and destination IP and TCP port addresses. Once identified, the ACEdirector "binds" a new session to the most available server associated with the destination virtual IP address in the session request packet, and provides session address translation on all packets until the session terminates. The ACEdirector also offers special handling to meet the requirements of UDP, persistent HTTP, SSL (Secure Socket Layer), FTP, and passive FTP sessions.

The ACEdirector maintains state information on every active session from start to finish. This allows ACEdirector to provide session address translation and special session handling for applications such as FTP, and to protect servers from bogus session requests while delivering end-to-end service quality.

Most traffic redirectors, including ACEdirector, automatically bypass failed servers and applications when it distributes new sessions or redirects HTTP requests. It automatically re-enrolls the server upon detection of service restoration. In addition, a backup and overflow server can be designated for any server or virtual IP address to ensure maximum service availability.

A hot-standby traffic redirector can be configured to protect against a switch port or switch failure. Supervisory signaling and configuration updates are exchanged inband between the active and standby redirectors, eliminating the need for slow, unreliable RS-232 out-of-band connections between the switches.

SUMMARY

Some corporate intranets consist of a few hundred or a few thousand Web servers spread out among geographically dispersed locations worldwide. To maximize the performance of applications running over these intranets, new techniques such as Web caching, flow management, proxy servers, and server-load balancing are being implemented. Some vendors offer multiple capabilities in a single product, such as Web caching and server-load balancing, or proxy servers and caching. Although load balancing can be run locally as well as over wide-area networks, vendors have tended to address each environment with separate product offerings. The next generation of solutions is expected to combine wide-area and local distribution capabilities within a single product.

See Also **Flow Management, Network Caching, Proxy Servers**

Logical Network Management

Logical network management entails the automatic discovery of devices on the network, viewing their configurations, and changing the connections at a management workstation, rather than having to physically collect the information, visit the devices, and manually change the connections at a central hub or switch. Monitoring network performance and the traffic flows between network devices and on specific circuits also fall into the category of logical network management.

Auto-Discovery

A key enabler of logical network management is the automatic discovery feature provided by almost all network management systems, which automatically identifies components of the network such as routers, hubs, and switches. After the auto-discovery process determines what devices are connected to the network, it assigns them a name or IP address and a graphical symbol, and determines their relationship to the network (Figure 74a). All of this information is displayed on a network map. The auto-discovery capability can be scheduled to run at specific time intervals to ensure that network managers have the most current information about what devices are attached to the network. Information that cannot be discovered—such as trouble history, responsible technician, and lease information—is manually entered into the database.

Hewlett-Packard's OpenView, for example, offers two types of auto-discovery—basic and extended. With basic discovery, nodes having an IP address are found by requesting data from a node's ARP (Address Resolution Protocol) cache or by issuing a ping if no ARP cache is found. Since SNMP is not a connection-oriented protocol, and nodes from time to time go offline or come online, OpenView must periodically poll the gateway (see lower right of the logical network map in Figure 74) to identify new devices. With extended discovery, IPX devices are usually identified, but other management applications may perform their own discovery when extended discovery runs. DECNet nodes can be discovered this way. Basic and extended discovery are separate processes that can run concurrently.

Figure 74
A portion of the logical network map created by the auto-discovery mechanism available in Hewlett-Packard's OpenView.

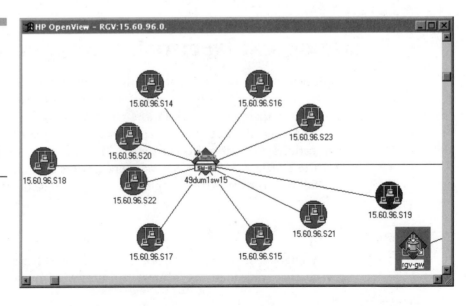

Configuration Management

Another aspect of logical network management is the configuration of both the devices on a network and the network itself. For example, a graphical display of device configurations might depict a rear view of hubs, bridges, and switches to make it easy to configure, monitor, troubleshoot, and manage the hardware through point-and-click mouse commands. For example, a network manager can pull up the back-of-the-box graphic of a particular hub to turn a port on or off. Sometimes this can even be done through a Web browser (Figure 75). And through drag-and-drop mouse commands, ports of a LAN switch can be configured into virtual network groups; that is, they can be connected together to collaborate on a project without a technician having to physically recable the switch to link the users together.

Performance Monitoring

Products such as AdvanceStack Assistant provide an integrated traffic monitor that enables network managers to view overall network performance via a graphical display of gauges that indicate usage, frames, broadcasts, multicasts, and errors. Advanced traffic monitoring allows operators to isolate network bottlenecks quickly. A "top talkers" feature reflects the most active pairs of nodes on the network. This feature helps network managers with switch placement, troubleshooting, and optimization.

Figure 75
Hewlett-Packard's AdvanceStack Assistant allows certain network devices to be viewed and controlled through a Web browser.

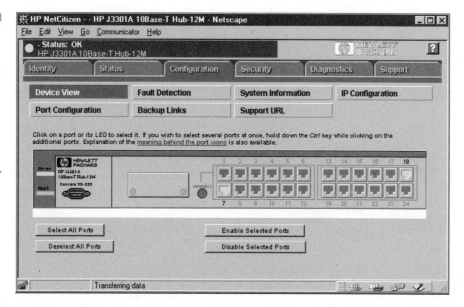

Virtual Circuits

Between the end pieces of equipment, one or more virtual circuits can be laid out, with or without the use of drawings. Sublevels of detail can also be modeled, with one circuit riding over another facility, such as a 64-Kbps service that is part of a 256-Kbps circuit, which in turn is actually made up of a series of T1 or T3 segments on the wide-area network. For each logical circuit, additional information can be tracked such as circuit identification, carrier, and contact name.

SUMMARY

Logical network management is concerned with graphical representations of the physical network and the use of centralized monitoring and control to compile the network inventory, resolve problems, and make configuration changes without having to dispatch technical personnel for on-site visits.
See Also **Physical Network Management, Virtual LANs**

Managed Applications Services

In today's fast-paced, competitive environment, companies are often overwhelmed by the complexity of technology. To assist them, carriers are forming alliances with vendors and systems integrators to develop managed applications services, so companies can focus on meeting the needs of their customers. These services come at a time when 60 percent of companies report increasing difficulty in managing telecommunications and data communications, citing systems integration and incorporating new technologies as crucial needs. Managed solutions can enable these companies to realize the significant productivity and economic gains that can result from deploying advanced technologies.

A variety of managed applications services are now offered by carriers and their partners, including managed router services, LAN monitoring and management services, and network and application help-desk services. Also offered are managed groupware services and managed electronic commerce services. These offerings contain a set of predefined features and options, usually designed for small to midsized businesses, freeing them to concentrate on their core business while reducing network operations complexity.

Managed E-Commerce Services

With regard to the Internet, companies can reach new markets and gain new revenue potential for products and services. However, taking full advantage of the opportunities offered by the Internet requires expertise that is not always available within an organization. Accordingly, carriers offer businesses the expertise needed to put their products and services in front of Internet shoppers. Solutions are available to open a new consumer sales channel, focus on business-to-business opportunities, or create a secure extranet for trading partners.

Whatever the choice, the carrier and its partners can deliver full integration of electronic catalog services, from purchase through order fulfillment, to maximize revenue potential and deliver professional customer service. The company simply supplies the products and services. The carrier and its partners provide:

- Full design of the catalog site
- Web-site hosting for providing catalog sales to customers
- Network and physical security
- Secure payment processing, including credit approval and account posting
- Order fulfillment support
- Monitoring and maintenance, 24 hours a day and 7 days a week
- Merchant reporting
- Optional consulting services

Managed Groupware Services

The two most popular groupware and messaging platforms supported by carriers in their managed offerings are Lotus Notes with Lotus Domino and Microsoft Exchange with Microsoft Outlook. These managed solutions help companies make the most of enhanced e-mail capabilities, document sharing, discussion groups, calendars and scheduling, and other applications for workflow and project management. As part of the managed groupware offering, the carrier usually provides end-to-end management, along with Internet/intranet gateways.

With a managed groupware service, organizations can avoid tying up IT staff or committing to capital expenditures to obtain the benefits of this kind of electronic workplace for their business. The carrier and its partners create, implement, and maintain a secure electronic communications solution for the organization. Expenses for services are based on the number of users each month, which means the organization pays only for the capacity and services actually needed and used. This minimizes the company's initial capital investment, enables existing assets to be used more effectively, and refocuses valuable IT resources on mission-critical priorities. The carrier and its partners provide:

- A secure environment with enhanced e-mail capabilities and protection from unsolicited e-mail (spam) and viruses
- Calendering, meeting, and project planning features
- Document sharing and discussion groups

- Round-the-clock maintenance for maximum availability and efficiency
- Scalable systems that adapt and expand to accommodate changing needs
- A network-based or premise-based solution
- Optional end-user help-desk services

Managed Router Services

Keeping a wide-area network (WAN) operating at peak performance can cause a constant drain on internal IT resources. By taking advantage of a carrier's managed router services, companies can benefit from the latest in network technology without making a major capital investment or having to search for hard-to-find qualified personnel. In addition, managed router services relieve IT professionals of routine WAN management and maintenance tasks to pursue mission-critical business areas and applications such as Y2K compliance.

Carriers offer several levels of service from basic monitoring and reporting to premium management. Available services include:

- Round-the-clock (24 × 7) fault monitoring to detect and resolve network problems
- Performance management services that track network availability, traffic levels, memory usage, protocol usage, and bandwidth usage to help identify potential problems and avert failures
- Complete router and frame-relay access device (FRAD) configuration management and maintenance from initial setup to ongoing move, add, and change management
- Inbound problem reporting 24 hours a day and 7 days a week using the carrier's help desk, as well as accurate, customizable reports delivered electronically either via e-mail or a Web browser for timely, convenient access

LAN Management Services

Maintaining optimal LAN performance can consume valuable IT resources, significantly reducing the time available to focus on projects and issues that directly affect an organization's business objectives. With LAN management services, organizations can shift responsibility for network management from internal staff to a carrier's dedicated LAN experts. Services include network monitoring, performance reporting, management, inventory tracking, security, backup and restoration, and an annual review.

Managed Help-Desk Service

A help desk provides a single resource for answering the application and technical questions of end users. Since most companies today use computer applications and technologies from many sources, when problems arise, their resolution often requires calling several companies, reviewing application help files and user's guides, and invoking trial-and-error remedies to problems. Employees can become frustrated waiting for help while productivity suffers.

A carrier's managed help-desk service can provide rapid, thorough software application and technical problem resolution 24 hours a day, 7 days a week. Billing is on a per-incident basis, which eliminates the high cost of billing by the hour or by the number of calls. Even if an incident requires multiple calls, it is still a single incident and is billed only once. A managed help-desk service provides:

- One toll-free number to call around the clock for rapid answers to all applications and technical-support questions
- A single point of contact to manage the inquiry, troubleshooting, and problem resolution process
- Support for many products, including the most popular operating systems, software applications suites, mail and groupware, databases, Web browsers, graphics, and reference programs
- Management reports that make it easy to monitor activities and allow increases in efficiency

SUMMARY

Managed applications services differ from traditional outsourcing in that the latter focuses on systems and network integration. Until relatively recently, there has been reluctance among carriers to get involved at the applications level, largely because they lacked the expertise and did not have credibility in this area among their target customers. Now with carriers facing declining revenues due to increasing competition, they must look for new business opportunities. They have been partnering with a variety of firms, especially traditional system and network integrators, who have expanded into the applications area in search of new business opportunities of their own. The resulting partnerships allow them to leverage each other's strengths to offer a comprehensive package of managed applications services.

See Also LAN Integration, Managed Applications Services, Outsourcing

Matrix Switches

A matrix switch is a multiport device that provides any-to-any connectivity between RS-232 devices such as printers, modems, and other computer peripherals at a data rate of 300 bps to 38.2 Kbps. Such switches also are used for out-of-band network management and for remote control and console port access.

Individual port parameters and distinct data rates can be set for each port. Devices connected to the matrix switch can communicate regardless of dissimilarities between data rate or parity. The administrator can connect any port to any other port for complete matrix connectivity by issuing simple ASCII commands from an attached terminal or remote terminal connected via modem.

The switch also provides a status display of all connected ports, allowing administrators to name each port, even establish hunt-groups. All parameters are stored in battery-backed memory, and the configuration of the switch can also be saved to disk for routine reconfiguration or one-time/event reconfiguration.

Matrix switches typically consist of a base unit containing the rack assembly, power supply, and control card. The ports are provided by the modules, which include a variety of host communications modules, peripheral communications modules, and peripheral data acquisition and control (DAC) modules. Multiple switches can be interconnected through dedicated link ports to create a larger switching environment.

SUMMARY

Matrix switches are an economical alternative to hubs when connectivity needs are relatively simple, and low speed but access to a lot of equipment is required, as in a test lab.

See Also **Hubs, KVM Switches**

Media Converters

As the term implies, a media converter makes the conversion from one network media type—defined by cable and connector types and band-width—to another. By performing this transition, a media converter makes it possible for organizations to extend legacy networks with the lat-est technology, instead of being tied to what the network was started with or—even worse—tearing it out and starting over. Alternatively, converters allow the use of less expensive, lower-bandwidth desktop connections from a fiber-optic backbone.

Functionally, a media converter is two transceivers or MAUs (Media Attachment Units) that can pass data to and from each other, and a power supply. Each of the MAUs has a different, industry-standard con-nector to join the different media: one medium goes in, the other comes out. The connectors themselves comply with IEEE specifications and use standard data encoding rules and link tests. Media converters sup-port connections to and from switches, hubs, routers, and even direct to servers.

Benefits and Applications

Media converters can be used virtually anywhere in the network—from the server, to the workstation (Figure 76). They can enhance the flexibility of the network by facilitating upgrades to the network to better, faster, more secure technology—as with fiber cabling—without requiring a full network retrofit. Legacy copper cabling can be left in place, while the fiber can be used for additions and extensions to the network.

Conversion devices also provide the means to extend a network. Using a media converter to integrate optical fiber allows a network to support the longer cable distances available when using fiber. In standard Ethernet and Fast Ethernet networks, for example, fiber specifications prescribe a maxi-mum distance of 2,000 meters versus the twisted pair wiring limit of only 110 meters.

The use of media converters also makes it easier to add new devices to the network—including the newer high-bandwidth switches and hubs—regardless of connector restrictions. Switches solve many of the problems common to larger networks, but the majority of Ethernet and Fast Eth-ernet switches on the market today are equipped with twisted pair con-nectors. Where the entire network is built of twisted pair wiring, the switches are easy to integrate—just plug one in. However, for many new installations, network managers are looking to optical fiber for their ca-

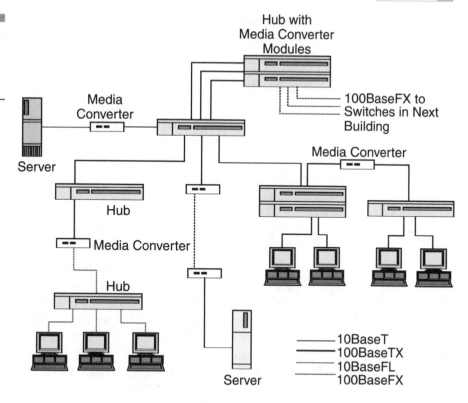

Figure 76
The use of media
converters
throughout a
network.

bling infrastructure because of its security features, bandwidth capacity, and its ability to span longer distances. For older installations that use BNC connectors, the same incompatibility arises. A simple fiber-to-twisted-pair media converter or a BNC-to-twisted-pair connector can make these devices work together on the same network. The most commonly used media converters support twisted-pair-to-fiber connections. Standard fiber connectors are typically classified as either ST (simple twist) or SC (subscriber channel). The media converters most widely used today provide quick, reliable, cost-effective connections between:

- 10-Mbps twisted pair cable segments or devices and 10-Mbps fiber-optic, single-mode or multimode (10BaseT to 10BaseFL) segments or devices

- 10-Mbps 10BaseT segments or devices and 10-Mbps Ethernet coaxial cable (10BaseT to 10Base2) segments or devices

- 100-Mbps twisted pair cable segments or devices and 100-Mbps fiber-optic, single-mode or multimode (100BaseTX to 100BaseFX) segments or devices

Newer fiber connectors are available in smaller form factors. These new fiber connectors are the MT-RJ, VF-45, and LC. These connectors are being put to use on various types of network hardware, including the latest hubs and switches. Because of their smaller size, these connectors enable more ports to be placed in a given device. A stackable 12-port Ethernet hub, for example, can now accommodate 24 ports—without increasing the size of the hub. This higher port density results in lower network costs. Media converters are available for all three of the new small form-factor connector types, providing organizations with even more flexibility in designing and expanding their networks.

SUMMARY

Media converters can be inserted almost anywhere in the network. The option of mounting media converters in a rack-mount chassis is useful where multiple converters are in use, or where they are anticipated in the future. The ability of media converters to mix media and speeds provides organizations with more flexibility in designing or extending their networks, as well as integrating legacy devices with today's advanced systems. This flexibility, in turn, enables organizations to achieve network performance goals, while containing costs.
See Also Transceivers

Metropolitan-Area Networks

As its name implies, a Metropolitan-Area Network (MAN) interconnects two or more LANs over an area larger than a campus—typically an urban area, such as a large city and its suburbs. Usually, a MAN is a high speed, fiber-optic network. An organization can build its own MAN from dark fiber leased from a carrier, lease bandwidth on a SONET ring operated by a regional carrier, or subscribe to a carrier-provided service such as Switched Multimegabit Data Service (SMDS), which runs over fiber.

Alternatively, microwave technology can be used for the MAN. However, because dark fiber may not be available to reach all locations, and permits for microwave towers are often difficult to obtain in urban environments, SONET rings and SMDS are the most viable means of implementing a MAN. Of the two, there are far more SONET rings than SMDS services, but SMDS is undergoing a resurgence after languishing for several years because of the growing popularity of the Internet and corporate intranets. SMDS can be used for high-speed Internet access in addition to LAN-to-LAN connectivity. With the availability of SMDS at DS1 and DS3 rates, plus the advent of DS0 access to SMDS, users can consolidate Internet access with the rest of their data traffic.

SMDS

Initially offered in December 1991, SMDS is a carrier-provided, connectionless, cell-switched service developed by Bellcore and standardized by the IEEE. As a connectionless service, SMDS eliminates the need for carrier switches to establish a call path between two points before data transmission can begin. Instead, SMDS access devices pass 53-byte cells to a carrier switch. SMDS cells are a fixed size of 53 bytes—the same type of cell used in ATM. The switch reads addresses and forwards cells one-by-one over any available path to the desired endpoint. SMDS addresses ensure that the cells arrive in the right order. The benefit of this connectionless any-to-any service is that it puts an end to the need for precise traffic-flow predictions and for dedicated connections between locations. With no need for a predefined path between devices, data can travel over the least congested route in an SMDS network, providing faster transmission, increased security, and greater flexibility to add or drop network sites. As a technology-independent service, SMDS offers several advantages:

SIMPLICITY Virtual connections are made as needed; there are no permanent, fixed connections between sites.

E.164 ADDRESSING SMDS addresses are like standard telephone numbers; if users know the SMDS address of another user, they can call up and begin sending and receiving data.

CALL CONTROL SMDS supports call blocking, validation, and screening for the secure interconnection of LANs and distributed client/server applications.

MULTICASTING SMDS supports group addressing.

MULTIPROTOCOL SUPPORT SMDS supports the key protocols used in local- and wide-area networking, including TCP/IP, Novell, DECNet, AppleTalk, SNA, and OSI.

MANAGEMENT SMDS is easier to manage than other services, such as frame relay and ATM, which can have a multitude of virtual circuits that makes setup, reconfiguration, and testing much more difficult.

RECONFIGURATION Sites can be connected and disconnected easily and inexpensively—usually within 30 minutes—without affecting other network equipment. This makes SMDS ideal for supporting external corporate relationships that change frequently.

SECURITY SMDS includes built-in security features that allow intracompany transmission of confidential data.

SCALABILITY SMDS is easily and cost-effectively scaled as organizations grow or application requirements change.

MIGRATION PATH Customers have a well-defined migration path to ATM, since both services use the same 53-byte frame structure.

COST Because it is a switched service, users pay only for the service when it is used, which can make it less expensive than services such as frame relay. And because SMDS has no mileage charges, the link to a provider is more economical than leased-line access.

Because SMDS can coexist with dedicated facilities, it enables customers to create hybrid public/private networks. SMDS also allows for the easy expansion of existing networks, since new sites can be quickly added to an SMDS network without total reconfiguration. Additions to an SMDS network require only a simple update to a screening database on the SMDS switch.

Architecture

SMDS defines a three-tiered architecture (Figure 77):

- A switching infrastructure comprising SMDS-compatible switches that may or may not be cell-based
- A delivery system made up of T1, T3, and lower-speed circuits called Subscriber Network Interfaces (SNIs)
- An access control system for users to connect to the switching infrastructure without having to become a part of it

LANs provide the connectivity for end users on the customer premises. The LAN is attached to the SMDS network via a bridge or router, with an SMDS-capable CSU/DSU at the front end of the connection.

T1 SNIs are used to access 1.17-Mbps SMDS offerings, while T3 SNIs are used to tap into 4-, 10-, 16-, 25-, and 34-Mbps offerings. A fractional T3 circuit can access intermediate-speed SMDS offerings. Some carriers offer low-speed SMDS access at 56 Kbps, 64 Kbps, and in increments of 56/64 Kbps. This allows smaller companies, large companies that have some small sites, and current users of frame-relay technology to also take advantage of SMDS.

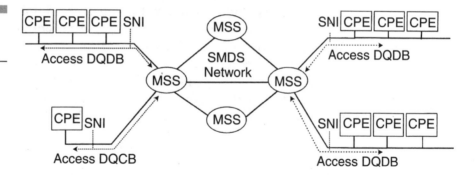

Figure 77
The SMDS
architecture.

Each subscriber has a private SNI and may connect multiple user devices (CPE) to it. At this interface point, the CPE attaches to a dedicated access facility that connects to an SMDS switch. Security is enforced, since only data originating from or destined for that subscriber will be transported across that SNI.

The SMDS Interface Protocol (SIP), operating across the SNI, is based on the IEEE 802.6 Distributed Queue Dual Bus (DQDB) media access control (MAC) scheme. The SIP consists of three protocol layers that describe the network services and how these services are accessed by the user. The SIP defines the frame structure, addressing, error control, and data transport across the SNI.

The SMDS network itself is a collection of SMDS Switching Systems (SS). The SS is a high-speed packet switch—most likely an ATM switching platform—providing the SMDS service interface. An SS will typically be located in a service provider's central office. Interconnecting several SS locations forms the foundation for a metropolitan-area or regional network.

The Inter-Switching Systems Interface (ISSI) provides communication between different switching vendors within the same network, while the Interexchange Carrier Interface (ICI) enables local telephone companies and interexchange carriers to interconnect SMDS networks.

SUMMARY

SMDS provides users with the cost-effectiveness of a public-switched network; the benefits of fully meshed, wide-area interconnection; and the privacy and control of dedicated, private networks. The key benefits subscribers can realize with SMDS include widespread current availability and increased LAN performance. It provides data management features, flexibility, bandwidth on demand, network security and privacy, multiprotocol support, and technology compatibility. SMDS also provides corporate locations with high-speed Internet access. Some national Internet service providers (ISP) even use SMDS to access major exchange points on the Internet backbone. One of these exchange points is the Commercial Internet Exchange (CIX) in the San Francisco area, where members such as AlterNet, AGIS, TCG CERFnet, MCI, NetCom, and PSInet can attach via PacBell SMDS, FDDI, or Ethernet.

See Also Campus-Area Networks, Wide-Area Networks

Microwave Systems

Microwaves are very short waves in the upper range of the radio spectrum, used mostly for point-to-point communications systems. Much of the technology was derived from radar developed during World War II. Initially, these systems carried multiplexed speech signals over common carrier and military communications networks. Today, microwave systems can handle all types of information—voice, data, facsimile, and video—in either analog or digital format.

Early technology limited the operation of these systems to the radio spectrum in the 1-GHz range. With continued improvements in solid-state technology, commercial systems are transmitting in the 40-GHz region. This spectrum can be used to provide short-range, high-capacity wireless systems that support, among other things, educational and medical applications, wireless access to libraries, and other information databases.

System Components

The basic equipment requirements for a point-to-multipoint microwave system include:

ANTENNAS   For the master, an omnidirectional antenna; for the remotes, a highly directional antenna aimed at the master station's location.

TOWER (OR OTHER STRUCTURE, SUCH AS A MAST)   To support the antenna and transmission line.

TRANSMISSION LINE   A low-loss coaxial cable connecting the antenna and the radio.

MASTER STATION RADIO   Interfaces with the central computer, transmits and receives data from the remote radio sites, can request diagnostic information from remote transceivers, and can also serve as a repeater.

REMOTE RADIO TRANSCEIVER Interfaces to the remote data unit; receives and transmits to the master radio.

PERSONAL COMPUTER Can be connected to the master station's diagnostic system either directly or remotely, for control and collection of diagnostic information from master and remote radios.

Types of Services

Microwave services include Private Operational, Common Carrier, and Broadcast Auxiliary radio services. The majority of these applications involve point-to-point systems, but may include applications for some point-to-multipoint microwave facilities such as Multiple Address Systems (MAS), which are used for point-to-multipoint communication between a master station and several remote stations located within 45 miles of the master station. Historically, MAS has been used to provide alarm, control, interrogation, and status reporting communications to companies in the power, petroleum, security, and paging industries.

PRIVATE OPERATIONAL FIXED MICROWAVE Private operational-fixed microwave systems serve many different purposes. They are meant to carry or relay voice, teletype, telemetering, facsimile, and digital communications associated with aviation, marine, public safety, industrial, and the land transportation radio services. For example, these systems operate unattended equipment; open and close switches or valves; record data such as pressure, temperature, and speed of machines; telemeter voltage and current in power lines; and perform other control or monitoring functions. Microwave systems are especially useful for controlling and monitoring various operations along installations like pipelines, railroads, and highways.

COMMON CARRIER MICROWAVE SERVICE Common Carrier microwave stations are generally used in a point-to-point configuration for long-haul backbone connections or to connect points on the telephone network that cannot be connected using standard wire line or fiber optics because of terrain. These systems are also used to connect cellular sites to the telephone network, and to relay television signals. Common Carrier microwave stations are licensed to applicants who intend to provide communications service to the public, whereas Private Operational Fixed stations are licensed to applicants for their own internal communications requirements.

BROADCAST AUXILIARY MICROWAVE Broadcast auxiliary microwave is used for relaying broadcast television signals from the studio to

the transmitter or between two points, such as a main studio and an auxiliary studio. This type of service also includes mobile TV pickups, which
relay signals from a remote location back to the studio for editing or live
broadcast.

Microwave LANs

Microwave has not been popular for LANs, since most rely on spread-
spectrum or infrared technologies. Among the few microwave LANs was
Motorola's Altair, which required an FCC license because it operated in
the crowded 18-GHz frequency band. While the license requirement may
be construed as a liability, it offered the guarantee of interference-free
communication. Unlike unlicensed spectrum, such as that often used by
spread-spectrum technologies, licensed spectrum gives the license-holder
a legal right to an interference-free data communications channel. Own-
ers of wireless LANs operating on unlicensed spectrum are continuously
at risk of unauthorized interference destroying their data communications capabilities, and do not always have legal recourse.

To make this more appealing to users, Motorola had obtained licenses
for 18-GHz operation in all U.S. metropolitan areas with populations
above 30,000. Motorola customers did not have to deal directly with the
government or wait for approval to operate Altair equipment. Upon purchasing the equipment, they filled out a simple registration form, and immediately placed their equipment into operation. Motorola's Frequency
Management Center coordinated frequencies to specific customer locations. In the event a customer moved the Altair system to another location, Motorola provided an 800 number to report that fact in compliance
with FCC requirements. Motorola handled frequency coordination for
the customer at the new location.

Although microwave has not caught on for intra-building LANs, it is
an economical alternative to leased lines for interconnecting LANs up to
30 miles apart using bridges and routers with directional antennas along
a clear line of site.

SUMMARY

Over the years, microwave systems have matured to the point that they
have become major components of the nation's public switched network
and essential mechanisms that allow private organizations to satisfy internal communications requirements and monitor their primary infrastructure. As the nation's cellular and personal communications systems grow,

point-to-point microwave facilities, serving as backhaul and backbone links, enable these wireless systems to serve less populated areas on an economical basis.

See Also **Wireless Internetworking, Wireless LANs**

Middleware

Middleware is a broad category of software that facilitates connectivity between applications, databases, or processes—usually in enterprise client/server environments—despite differences in underlying communications protocols, operating systems, platforms, database formats, and other application services. Among the most commonly used types of middleware are application programming interfaces (APIs) and remote procedure calls (RPCs).

Application Programming Interfaces

Application Programming Interfaces (APIs) define how programs interact with an operating system, other applications, communication systems, and device drivers. Vendors offer APIs to developers so they can quickly develop applications that work with their products.

Sun Microsystems, for example, offers an API for its Java Card, a smart card that offers functions implemented with the Java programming language. Smart-card applications have been written by a small cadre of professional developers using low-level, assembler-like languages. Each of these languages is proprietary to each smart-card vendor and do not interoperate.

The Java Card API offers a different approach to developing smart-card applications. It is based on an open API that is available to the entire smart-card industry. The Java Card API will run on all smart cards that are ISO 7816-compliant. Once written, the application can run everywhere a smart card is found. Instead of a small elite group of programmers, any Java programmer can now use existing development environments to write Java Card applications.

Remote Procedure Calls

Remote Procedure Calls (RPCs) increase the interoperability, portability, and flexibility of applications by allowing them to be distributed over multiple heterogeneous platforms. Using RPCs reduces the complexity of developing applications that span multiple operating systems and network protocols by insulating application developers from the details of the various operating system and network interfaces.

RPCs simplify the development of distributed applications by keeping the semantics of a remote call the same whether or not client and server share the same system. RPCs increase the flexibility of an architecture by allowing a client component of an application to employ a function call to access a server on a remote system. RPC allows the remote component to be accessed without knowledge of the network address or any other lower-level information. RPCs are usually supplied by vendors as processing modules, code libraries, and other development tools.

Object brokering is an extension of the RPC mechanism. Two popular distributed object models are Microsoft's Distributed Component Object Model (DCOM) and the Object Management Group's Common Object Request Broker Architecture (CORBA). In both DCOM and CORBA, the interactions between a client process and an object server are implemented as object-oriented RPC-style communications.

Under CORBA, for example, an Object Request Broker (ORB) is the middleware that establishes the client-server relationships between objects. Using an ORB, a client can transparently invoke a method on a server object, which can be on the same machine or across a network. The ORB intercepts the call and is responsible for finding an object that can implement the request, pass it to the parameters, invoke its method, and return the results. The client does not have to know where the object is located, its programming language, its operating system, or any other system aspects that are not part of an object's interface. In so doing, ORB provides interoperability between applications on different machines in heterogeneous distributed environments and seamlessly interconnects multiple object systems.

SUMMARY

Middleware has become a necessary ingredient for building distributed enterprise systems. This vision of integrated enterprise systems, using client/server technology as a means to make data systems more responsive to line-of-business needs, requires middleware to bridge together disparate systems.

See Also **Client/Server**

Mobile Phone Fraud

Mobile phone fraud is an extension of "phone phreaking," the term given to a method of payphone fraud that originated in the 1960s that employed an electronic box held over the speaker. When the user was asked to insert money, the electronic box played a rapid sequence of tones, which fooled the billing computer into "thinking" that money had been inserted. Since then, criminals and hackers have devoted time and money to develop and refine their techniques, applying them to mobile phones as well. Among the techniques used are cloning, rechipping, and bogus subscriptions.

Cloning

A cloned phone has been reprogrammed with details of a legitimate user's mobile identification number (MIN) and associated electronic serial number (ESN). Subsequently, any calls made using the bogus mobile phone will be billed to the legitimate user.

For analog systems, each mobile phone carries with it handshake information comprising the ESN and MIN. A service provider normally has a list of some 20,000 spare handshakes. These numbers can fall into the wrong hands in a variety of ways. For instance, lax internal security can allow disgruntled employees to obtain the handshake information from internal computer systems or customer files and sell it to cloners.

Special counterfeiting software, available for purchase through catalogs and the Internet, is used to recode the chips of other handsets with the stolen ESN. With the availability of digital systems and a variety of new technologies that work over them—including authorization, call pattern analysis, and voice verification—cloning is becoming increasingly difficult.

Rechipping

This technique is frequently used to recycle stolen mobile phones, which have been reported to network operators and barred from further services. The software gives a stolen mobile phone a new electronic identity that allows it to be reconnected to the network until it is discovered and barred again.

Sometimes new phones are reprogrammed to defraud service providers. For instance, when a corporate subscriber orders a stock of handsets, those that have not been allocated to specific individuals could be removed from stock and reprogrammed with the ESNs of other subscribers to make fraudulent calls.

A criminal does not have to steal a mobile phone to clone the ESN. With the latest scanning equipment and a laptop computer, a criminal can tune in to a call and steal the ESN from the air. Typically, the method is to sit along a busy highway, intersection, or rest area and wait for unsuspecting business people to make calls from their vehicles. Then a radio scanner is used to pick up a phone's ESN, which is broadcast at the beginning of each call. In only a few hours of scanning, a criminal can walk away with hundreds of legitimate ESNs.

Although it is illegal to sell such scanners in the U.S., ordinary police-band scanners purchased legally at retail stores can be modified to monitor mobile calls by connecting two board-level components with a strand of wire. At least one product even comes with instructions for doing this, thus allowing the manufacturer to effectively circumvent the law against selling scanners that tune into mobile calls.

In addition, the test equipment used by technicians to identify and troubleshoot RF networks can be used to capture call data, including ESNs. Dishonest technicians can then sell this information to cloners, give it out to friends or relatives, or use it themselves to obtain free service whenever they need it.

Bogus Subscriptions

Subscription fraud takes many forms: the activation of a new account to a person who uses false identification and stolen credit cards, the use of roaming privileges by a subscriber who has no intention of paying, and dealer fraud, which entails the submission of phony accounts to the carrier. Whatever the method of subscription fraud, it affects carriers in terms of billing costs, clearinghouse and settlement costs, capital costs to improve service, loss of billing integrity, and increased customer dissatisfaction. By the time the service provider catches on to a false account, the phone bill is already run up into hundreds or thousands of dollars.

This method of mobile phone fraud is by far the easiest and most risk-free. Because of competitive pressures, carriers and retail agents are signing up new customers quickly and are not able to screen all this information to weed out unqualified individuals. Often, new subscribers are allowed to begin using the service within hours of opening an account. As the carrier closes in on bad accounts when false information does not check out, criminals are already a step ahead opening new accounts. In some cases, subscription fraud actually originates from the dealer: managers or salespeople submit bogus subscription applications to receive credits from the carrier.

Fraud Management Systems

To combat cell phone fraud, a number of technologies have been implemented that can discourage cell phone cloning and stop thieves from obtaining free access to cellular networks:

PERSONAL IDENTIFICATION NUMBERS   Some service providers offer their subscribers a free fraud protection feature (FPF) to help protect against unauthorized use of their cell phones. Like an ATM bank card, FPF uses a private combination, or Personal Identification Number, that only the subscriber knows. The PIN code does not interfere with regular phone usage. Even if pirates capture the phone's signal, they cannot use it without the PIN code.

For example, the subscriber locks the phone account by pressing ˙56 + PIN + SEND. This blocks all outgoing calls, except for 911 (emergency) and the carrier's customer care number. While the account is locked, calls can still be received, and voice mail will continue to take messages. When the subscriber wants to make a call, the account is unlocked with ˙56 + 0 + PIN + SEND.

There are available database applications that help service providers make better use of PINs to detect potentially fraudulent calls and prevent their completion. One of these products comes from Sanders Telecommunications Systems, a Lockheed Martin company. The company's MicroProfile software enables the carrier to determine when a call is being placed to a number outside the subscriber's normal calling pattern. Combined with another product, an intelligent network-based application called Intelligent PIN, the MicroProfile software requires that the caller provide a PIN only for out-of-profile calls. By establishing and maintaining calling profiles of subscribers, out-of-profile call requests can be identified in real time, whereupon PINs can be requested to ensure that only legitimate calls can be placed.

Other fraud detection schemes are able to detect when more than one phone with the same personality is attempting to access the system. Upon detection, the subscriber is requested to enter a PIN prior to all new call attempts until further investigation.

PINs have become an effective fraud prevention tool, decreasing mobile phone fraud by as much as 70 percent in some areas. If a subscriber gets cloned anyway, a simple call to the service provider to obtain a new PIN is the extent of the customer's inconvenience—there is no need to change phone numbers, which also obviates the need for new business cards and letterheads. The subscriber does not even need to come into the service center. But PINs are far from bulletproof and cloners have proven particularly adept at cracking most security systems carriers have deployed.

CALL PATTERN ANALYSIS   As noted, cloned phones can be identified with a technology called call pattern analysis. When a subscriber's phone deviates from its normal activity, it trips an alarm at the service provider's fraud management system. There, it is put into queue where a fraud analyst ascertains whether the customer has been victimized and then remedies the situation by dropping the connection.

Coral Systems offers an innovative software solution called Fraud-Buster. Through the development of personalized customer profiles based on the subscriber's typical calling patterns, FraudBuster can immediately identify suspicious usage. After primary detection, a member of the carrier's fraud investigation team is immediately alerted, supplied with detailed information on the calls in question and provided with recommended actions to address the alleged fraud (Figure 78).

AUTHENTICATION   Authentication works by automatically sending a series of encoded passwords over the airwaves between the cellular telephone and the cellular network to validate a customer each time a call is placed or received. Authentication uses a complex security feature that contains a secret code and special number based on an algorithm shared only by the cellular telephone and the wireless network. Whenever a customer places or receives a call, the wireless system asks the cellular telephone to prove its identity through a question-and-answer process. This process occurs without delaying the time it takes to connect a legitimate cellular call and does not affect billing.

When a user initiates a call, the network challenges the phone to verify itself by performing a mathematical equation that only that specific phone can solve. An authentic phone will match the challenge, confirming that it and the corresponding phone number are being used by the legitimate customer. If it does not match, the network determines that the phone number is being used illegally, and service to that phone is terminated. All this takes place in a fraction of a second.

Prior to authentication, a PIN had to be entered before the call was connected. All the user needs now is an authentication-ready phone to take advantage of the service, which is usually offered free. Where authentication service is available, subscribers no longer need to use a PIN to

Figure 78
Coral System's FraudBuster software detects suspicious deviations in calling patterns, alerts a fraud investigator, and provides a detailed accounting of events that triggered the alert.

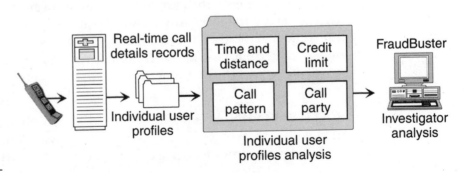

Real-time call details records

Individual user profiles

Time and distance

Credit limit

Call pattern

Call party

Individual user profiles analysis

FraudBuster

Investigator analysis

make calls, except when roaming in areas where authentication is not yet available.

RADIO FREQUENCY FINGERPRINTING   Radio frequency fingerprinting uses digital analysis technology that recognizes the unique characteristics of radio signals emitted by mobile phones. A radio frequency fingerprint can be made that can distinguish individual phones within a fraction of a second after the attempt to place a call is made. Once the fraudulent call is detected, it is immediately disconnected. The technology works so well that it has cut down on fraudulent calls by as much as 85 percent in certain high-fraud markets, including Los Angeles and New York.

VOICE VERIFICATION   Most fraud-prevention technologies—such as PINs, calling pattern analysis, authentication, and radio frequency fingerprinting—are only partially effective. Rather than verifying the caller, they merely authenticate a piece of information (PIN, ESN/MIN), a piece of equipment (RF fingerprinting, authentication), or the subscriber's calling patterns (calling pattern analysis).

Voice verification systems are based on the uniqueness of each person's voice and the reliability of the technology that can distinguish one voice from another by comparing a digitized sample of a person's voice with a stored model or "voice print."

One of the most advanced voice verification systems comes from T-NETIX Inc. The company uses a combination of decision-tree and neural network technologies to implement what it calls a "neural tree network." The neural tree is comprised of nodes, or neurons, that are "trained" through multiple repeated utterances of a subscriber-selected password or a small sample of speech. The training contrasts the acoustic features of the speaker being enrolled to features of the speakers already enrolled in the service.

During the verification process, each neuron must decide whether the acoustic features of the spoken input are more like those of the person whose identity is claimed or more like those of other speakers in the system. The neural tree network technology permits this complex decision-making, or discriminant process, to be completed in a relatively short time in contrast to other technologies. In effect, yes/no decisions are reached at each neuron of the neural tree, and a conclusion is reached after moving through five or six branches of the tree. The relative simplicity of the neural network decision path facilitates rapid analysis of spoken input with no upward limit on the number of enrollees.

The voice verification system can reside on a public or private network as an intelligent peripheral or can be placed as an adjunct serving a Private Branch Exchange (PBX) or Automatic Call Distributor (ACD). In a mobile environment, the system can be an adjunct to a Mobile Switching Center (MSC).

DATA MINING   Another method of fraud detection entails the use of data mining software that examines billing records and picks up patterns, which reveal the behavior of cloning fraud.

Two major patterns are associated with cloning fraud. One is time overlap, which means that a phone is involved in two or more independent calls simultaneously. The other is velocity, which originates from the assumption that a handset cannot initiate or receive another call from a location far away from its previous call in a short time. If this happens, there is a high probability that a clone phone is being used somewhere.

Cloning information is passed to the service provider's billing and management system where countermeasures will be executed to prevent further calls. Cloning history data are also kept in the database and can be queried by service representatives via a wide-area network.

REAL-TIME USAGE REPORTING   Several software vendors offer products that collect real-time data from cellular switches, which can be used to identify fraudulent use. Subscriber Computing's FraudWatch Pro software, for example, provides workable cases of fraud rather than just alarms. With this system, analysts can select the types of fraud, such as subscription fraud or cloning. Based on an analysis of call detail records (CDR) received from home switches and roaming CDR data feeds, the system detects fraudulent activities and provides prioritized cases to the analyst for action. The system provides investigative tools that allow analysts to ascertain whether fraudulent activity has indeed occurred. The cases are continuously prioritized according to fraud certainty factors or the probability of fraud occurring.

FraudWatch Pro receives CDR records directly from switches and roaming CDR exchanges as quickly as the data becomes available to the service provider. Profiles are created that contain details about the specific daily activity of a given subscriber. The subscriber profile contains about 40 single and multidimensional data "buckets" that are updated in real time. Analysts have a graphical interface that allows them to query the database to search for a wide range of behavior patterns as well as address queries about specific individuals.

SMART CARDS   While it is easy to intercept information from mobile phones used on analog networks, it is much harder to do so on digital networks, such as the North American Personal Communication Services (PCS) and the European Global System for Mobile (GSM) telecommunications. This is because the signals are encrypted, making them much more difficult to intercept without expensive equipment and a higher degree of technical expertise.

Although signal encryption during airtime is a standard feature of GSM networks, network operators can choose not to implement it. In this case, when a handset is turned on to access the host base station for services, the subscriber is vulnerable to the same eavesdropping attacks as with analog systems.

GSM signal encryption is done via a programmable smart card—the Subscriber Identity Module (SIM), which slips into a slot built into the handset. Each customer has a personal smart card holding personal details (short codes, frequently called numbers, etc.) as well as an international mobile subscriber identity (IMSI)—equivalent to MIN for analog systems—and an authentication key on the microprocessor. Plugging the smart card into another phone allows it to be used as if it were the original.

SUMMARY

Wireless fraud, including illegal cloning of customer phones, is being reduced by a variety of fraud containment tools rather than a single magic bullet. These tools include authentication, radio frequency fingerprinting, roamer verification and reinstatement, caller profilers, personal identification numbers, and prepaid cards. Statistics show that these tools are having the desired impact. Mobile phone revenues lost to fraud have been reduced from a high of $1.6 billion in 1995, to a low of $200 million in 1998.

See Also **Firewalls, LAN Security**

Mobile Phones

In recent years, mobile phones (also called cell phones) have emerged as a "must have" product among traveling professionals and consumers alike, growing in popularity every year since they were first introduced more than 15 years ago. Today, there are over 62 million mobile phones in the U.S., and 200 million worldwide. Their widespread use for both voice and data communications has resulted from significant progress made in their portability, network services availability, and the declining cost for equipment and network services. These factors, combined with widespread appreciation of how mobile phones can enhance productivity and security, have resulted in their sustained high popularity.

Components

Whether mobile units "permanently" mounted in a vehicle, transportable units that can be easily moved from one vehicle to another, pocket phones weighing in at less than four ounces, or wearable units, cellular telephones in their various configurations consist of the same basic components.

HANDSET/KEYPAD   The handset and keypad provide the interface between user and system. This is the only component of the system, under normal operation, with which users need to be concerned. Any basic or enhanced system features are accessible via the keypad, and once a connection is established, this component provides similar handset functionality to that of any telephone. Until a connection is established, however, the operation of the handset differs greatly from that of a conventional telephone.

Rather than initiating a call by first obtaining a dialtone from the network switching system, the user enters a number into the unit and presses the SEND function. Once the network has processed the call request, the user hears conventional call progress signals such as a busy signal or ringing. From this point forward throughout the conversation, the handset operates in a customary manner. To end a call, the user simply presses the END function key on the keypad. In addition to these functions, the handset typically contains a display that shows dialed digits as well as other features, a CLEAR key that enables the user to correct misdialed digits, functions that enable storage of numbers for future use, and other enhanced features that can vary greatly from one phone to the next.

LOGIC/CONTROL   The logic/control functions of a phone include the numeric assignment module, or NAM, for programmable assignment of the unit's telephone number by the user's carrier of choice, and the electronic serial number of the unit, which is a fixed number unique to each telephone. When signing up for service, the selected carrier makes a record of both numbers. When the unit is in service, the cellular network interrogates the phone for both of these numbers in order to validate that the calling/called cellular telephone is that of an authentic subscriber. This component of the phone also serves to interact with the cellular network protocols that determine what control channel the unit should monitor for paging signals to indicate the network's desire to connect a call coming into the phone, to determine and select the voice channels that the unit should use for a specific connection, and to monitor the received control signals of cell sites when the phone is in either standby or an in-use mode so the phone and network can coordinate transitions to adjacent cells as conditions warrant.

TRANSMITTER/RECEIVER The transmitter/receiver unit of the telephone is the heart of the radio communications component of the system, under the command of the logic/control unit. Powerful three-watt telephones are typically of the vehicle-mounted or transportable type, and their transmitters are understandably larger and heavier than those contained within lighter-weight handheld cellular units. These more powerful transmitters require significantly more input wattage than handheld units that transmit at power levels of only a fraction of a watt, and they use the main battery within a vehicle or a relatively heavy rechargeable battery to do so. A diplexer unit within the phone enables the transmitter and receiver to use a single antenna while simultaneously transmitting and receiving.

ANTENNA ASSEMBLY The antenna assembly, comprising the antenna and connecting cable, determines whether the full power produced by the transmitter is effectively coupled to free space and also whether the minute electromagnetic impulses received from the airwaves can be delivered intact to the receiver circuitry of the telephone. The antenna for a cellular telephone can consist of a flexible rubber antenna mounted on a handheld phone, an extendible antenna on a pocket phone, or the familiar curly stub seen attached to the rear window of many automobiles. Antennas and the cables used to connect them to radio transmitters must have electrical performance characteristics that are matched to the transmitting circuitry, frequency, and power levels. Use of antennas and cables that are not optimized for use by these phones can result in poor performance. Improper cable, damaged cable, or faulty connections can render the telephone completely inoperative.

POWER SOURCES Cellular phones are typically powered by a rechargeable battery. Nickel Cadmium (NiCd) batteries are the oldest and cheapest power source available for cellular phones. Newer Nickel-Metal Hydride (NiMH) batteries provide extended talk time as compared to lower-cost conventional Nickel Cadmium units. They provide the same voltage as NiCd batteries, but offer at least 30 percent more talk time than NiCd batteries and take about 20 percent longer to charge.

However, newer cellular phones may operate with optional high-energy AA alkaline batteries, which provide up to three hours of talk time or 30 hours of standby time. These batteries take advantage of the new Lithium/Iron Disulfide technology, which results in 34 percent lighter weight than standard AA 1.5-volt batteries (15 versus 23 grams/battery) and 10 years of storage life—double that of standard AA alkaline batteries.

Vehicle-mounted and handheld portable cell phones can also be powered via the vehicle's 12-volt dc battery if you use a battery eliminator that plugs into the dashboard's cigarette lighter. This saves useful battery life by drawing power from the vehicle's battery and comes in handy when the phone's battery has run down. A battery eliminator will not recharge the phone's battery, however. Recharging the battery can only be done with a special charger.

Lithium Ion batteries offer increased capacity and are lighter weight than similar sized NiCd and NiMH batteries. These batteries are optimized for particular models of cellular phones, which helps ensure maximum charging capability and long life for the battery.

Lead Acid batteries are used to power transportable cellular phones when the user wants to operate the phone away from the vehicle. The phone and battery are usually carried in a vinyl pouch.

Options and Features

In addition to the basic issues of portability, power, durability, and reliability—which are certainly key decision points when selecting a cellular telephone—the additional features and options often distinguish one unit from another.

Hands-free operation is an especially useful feature for the mobile system user, for matters of safety as well as convenience. This is equivalent to a speakerphone function on a conventional telephone. In its basic configuration, it allows users to converse without holding the handset after establishing a call.

Voice activation allows users to establish and answer calls by issuing verbal commands that the telephone in turn executes. This feature is coupled with the basic hands-free option to allow control of the unit, including establishing and answering calls without requiring users to be visually distracted by the telephone. A remote earphone/microphone combination cable is available that can be plugged into some phones. This allows for hands-free operation and provides confidentiality for at least the received side of the conversation. A portable phone can be attached to the user's belt, and the earphone/microphone cable plugs into the telephone. Memory functions allow storage of frequently called numbers in order to simplify dialing. Units may offer as few as 10 memory locations or in excess of 100.

Multiple numeric assignment module (NAM) features allow a single phone to be used with a number of different carriers. All phones allow at least one number assigned by a single carrier, but dual-NAM and multiple-NAM sets allow users to register a single phone for use on more than one network. The phone can then be used to access the best carrier for a specific location in areas where both local providers might have a number of different gaps in coverage, or to provide options at times when peak network traffic might prevent initiating a call. This feature is also useful in order to cut costs of roaming into other service areas where surcharges might apply. A frequent roamer might save considerable airtime costs by registering a single phone in the multiple service areas in which the phone is used, since roaming surcharges would not apply to a locally registered unit.

Visual status displays convey information on numbers dialed, state of battery charge, call timers, roaming indication, and signal strength. Phones differ widely in the number of characters and lines of alphanumeric information that they can display. Among the recent innovations in display technology is the use of dedicated icons. Within the quick access menu, graphic icons further enhance ease of use by visually identifying the phone's features.

Some cellular phones also feature programmable ringer tones to allow users to distinguish incoming calls by the ring emitted from the phone. Silent call alert features include visual or vibrating notification in lieu of an audible ring signal. This can be particularly useful in locations where the sound of a ringing phone would constitute an annoyance.

Voice messaging capabilities allow the phone to act as an answering machine. A limited amount of recording time is available via this option on some telephones. (Carriers also offer voice messaging services that are not dependent on the phone unit itself to function.) When the phone is left in standby mode, callers reach the answering device, which functions exactly as most voice-mail systems. Further, air time is not required for users to retrieve the messages that were left while they were in meetings or otherwise occupied.

Fax capabilities enable the user to use the telephone as an interface to a fax machine, and certain phones have the capability to receive and store fax information even when the phone is not connected to the machine or computer for which the message is destined.

Call restriction features enable users to allow other people to use the phone to call selected numbers, local numbers, or emergency numbers without permitting them to dial the world at large. Call timers provide users with information as to the length of the current call, and some telephones can also maintain a running total of airtime for all calls. These features make it easier for users to keep track of call charges.

With the increased use of cellular telephones for personal use, choice of color and styling is playing a greater role in the phone selection process. Motorola, for example, offers diverse colors, such as sunstreak (yellow), dark spruce, eggplant, teal, raspberry, regatta blue, temptation teal, and cranberry. These colors are available on Motorola's Lifestyle Series pocket phones, the Contour/Courier phone accent series, and the Star-TAC wearable series.

SUMMARY

With the integration of regional and national wireless networks into a third-generation global fabric, the industry is introducing mobile phones that can work over all major types of wireless networks. These so-called "world phones" are frequency-agile devices that accommodate both GSM in standard frequency bands (900 MHz and 1800 MHz) as well as PCS-1900 in North America, among others. Although there is unlikely to be one technical standard in the future, today's dual-mode phones are viewed as the first step in the trend toward increasing integration. Dual-mode wireless has already advanced to triple mode and more, through use of a technology called *software-defined radio*. With rapid advancements in chip technology, multimode phones and multifrequency phones represent the same design costs as a mainstream wireless phone for the consumer market.

***See Also* Cellular Telephone Networks, Internet-Enhanced Cell Phones, Mobile Phone Fraud, Software-Defined Radio, Third-Generation Wireless Networks**

Mobile Wireless Communications Initiative

The Mobile Wireless Communications Initiative is concerned with standardizing a technology that would enable users to connect their mobile computers, digital cellular phones, handheld devices, network access points, and other mobile devices via wireless short-range radio links—unimpeded by line-of-sight restrictions. The technology is known as "Bluetooth" and its objective is to eliminate the need for proprietary cables to connect various devices.

For example, Bluetooth radio technology built into both a cellular telephone and laptop would replace the cumbersome cable used today to connect the two devices when sending data over a cellular network. Printers, PDAs, desktops, fax machines, keyboards, joysticks, and virtually any other digital device can be part of a Bluetooth system. But beyond eliminating the need for cables, Bluetooth radio technology provides a universal bridge to existing data networks, a peripheral interface, and a mechanism to form small private ad-hoc groupings of connected devices away from a fixed network infrastructure.

Designed to operate in the unlicensed ISM band at 2.4 GHz, Bluetooth radio uses a fast acknowledgement and frequency hopping scheme to make a robust link, even in very noisy environments. Bluetooth radio modules avoid interference from other signals by hopping to a new frequency after transmitting or receiving a packet. Compared with other systems operating in the same frequency band, Bluetooth radio typically hops faster and uses shorter packets, making it more robust than other systems. Use of Forward Error Correction (FEC) limits the impact of random noise on long-distance links. The radio technology, a combination of circuit and packet switching, permits a data rate of up to 1 Mbps.

Functional Components

The different functional components included in a Bluetooth-compliant system are:

RADIO UNIT The maximum frequency hopping rate is 1600 hops per second. The nominal link range is 10 centimeters to 10 meters, but can be extended to more than 100 meters if the transmit power is increased.

LINK CONTROL UNIT Describes the specifications of the digital signal processing (DSP) part of the hardware—the link controller—which carries out the baseband protocols and other low-level link routines.

LINK MANAGEMENT Carries out link setup, authentication, link configuration, and other link functions.

SOFTWARE FUNCTIONS Functions defined in software include configuration and diagnosis, device discovery, cable emulation, peripheral communication, audio communication and call control, and object exchange for business cards and phone books.

Network Topology

Bluetooth supports both point-to-point and point-to-multipoint connections. Multiple users can be linked together via piconets, and several piconets can be established and linked together, forming a "scatternet."

(Figure 79). Each piconet is identified by a different frequency-hopping sequence. All users participating on the same piconet are synchronized to this hopping sequence.

A piconet starts with two connected devices, such as a portable PC and cellular phone, and may grow to eight connected devices. All Bluetooth devices are peer units and have identical implementations. However, when establishing a piconet, one unit will act as a master and the others as slaves for the duration of the piconet connection. Devices synchronized to a piconet can enter power-saving modes in which device activity is lowered. The full-duplex data rate within a multiple piconet structure with 10 fully loaded, independent piconets is more than 6 Mbps.

Before any connections in a piconet are created, all devices are in standby mode. In this mode, an unconnected unit periodically "listens" for messages. Each time a device wakes up, it listens on a set of hop frequencies defined for that unit. The connection procedure can be initiated by any of the devices, which then becomes master.

The link type defines what type of packets can be used on a particular link. The Bluetooth baseband technology supports two link types:

Figure 79
Multiple
independent and
nonsynchronized
piconets form a
scatternet.

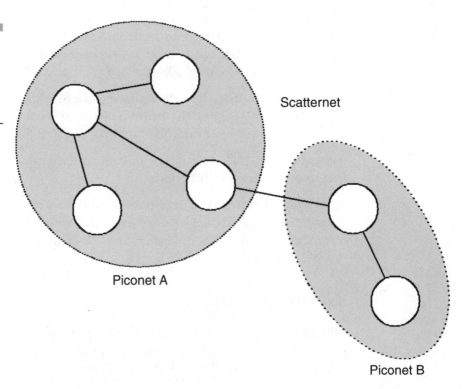

Scatternet

Piconet A

Piconet B

Synchronous Connection Oriented (SCO), which is used primarily for voice, and Asynchronous Connectionless (ACL), which is used primarily for packet data.

Different master-slave pairs of the same piconet can use different link types, and the link type may change arbitrarily during a session. Each link type supports up to 16 different packet types. Four of these are control packets and are common for both SCO and ACL links. Both link types use a Time Division Duplex (TDD) scheme for full-duplex transmissions.

Device Interoperability

Bluetooth devices must be able to recognize each other and load the appropriate software to discover the higher-level abilities each device supports. Interoperability at the application level requires identical protocol stacks. Different classes of Bluetooth devices—PCs, handhelds, headsets, cellular telephones—have different compliance requirements.

For example, a Bluetooth headset would not be expected to contain an address book. Headset compliance implies Bluetooth radio compliance, audio capability, and device discovery protocols. More functionality would be expected from cellular phones, handhelds, and notebook computers. To obtain this functionality, the Bluetooth software framework relies on existing specifications, such as vCard and vCalendar[2], TCP/IP, and others, rather than invent yet another set of new specifications.

SUMMARY

Early Bluetooth-enabled products are expected to include mobile computers, handheld PCs, digital cellular phones, and peripherals such as printers, projectors, PC Cards, and hands-free headsets. Network access points will also be available to facilitate access to LANs and WANs.
***See Also* Wireless Internetworking**

---

2. For a more in-depth discussion of vCard and vCalendar, see my book, *Desktop Encyclopedia of the Internet*, published by Artech House (ISBN: 0-89006-729-5). This 600-page reference can be conveniently ordered through the Web at Amazon.com.

Modems

Modems are used to send data between computers linked by ordinary telephone lines, allowing users to transfer files, access bulletin boards or the Internet, and connect to mainframe or midrange computers at remote sites. The communications software that comes with many modems typically includes emulation capabilities, allowing connections to a variety of legacy midrange and mainframe computers. Today's V.90-compliant modems advertise data transmission rates of 56 Kbps over analog lines and are available as stand-alone units that can be shared among several users, as cards for installation in a computer or communications server, or in the PC Card form factor for use in laptops and notebooks with Type II slots.

Modulation

A modem, or modulator-demodulator, converts the digital signals generated by computers into analog signals suitable for transmission over dial-up telephone lines or voice-grade leased lines. Another modem—located at the receiving end of the transmission—converts the analog signals back into digital form for display and manipulation by the recipient's computer.

Modems use various modulation techniques to encode the serial digital data generated by computers onto an analog carrier signal. The simplest modulation techniques rely on two signal methods to transmit information: frequency shift-keying (FSK) and phase shift-keying (PSK).

FSK is similar to the frequency modulation technique used to transmit FM radio signals. By forcing the signal to shift back and forth between two frequencies, the modem can encode one frequency as a 1 and the other as a 0. This modulation technique was widely used in the early 300-bps modems. At higher rates, FSK is too vulnerable to line noise to be effective.

PSK, another early modulation technique, uses shifts in a signal's phase to indicate 1s and 0s. The problem with this method is that *phase* refers to the position of a waveform in time; therefore, the data terminal clocks at both ends of the transmission must be synchronized precisely.

A more complex method, known as differential-shift phase keying (DPSK), uses the phase transition to indicate the logic level. With this scheme, it is not necessary to assign a specific binary state to each phase; it matters only that some phase shift has taken place. The telephone bandwidth is limited, however, and the transmission speed can go only up to 600 bps. To increase speed, it is possible to expand DPSK from a two-state to a four-state pattern represented by four two-bit symbols (known as a *dibit*) as follows:

- Maintain the same state (0,0)
- Shift counterclockwise (0,1)
- Shift clockwise (1,0)
- Shift to the opposite state (1,1)

Several other modulation techniques are much more sophisticated, with the trellis-coded technique well established as a worldwide standard since 1984. This technique entails the use of a 32-bit constellation with quintbits to pack more information into the carrier signal and offer more immunity from noise. It offers 16 extra possible state symbols. These extra transition states allow dial-line modems with trellis coding to use transitions between points rather than specific points to represent state symbols. The receiving modem uses probability rules to eliminate illegal transitions and obtain the correct symbol. This gives the transmission greater immunity to line noise. The fifth bit can be used as a redundant bit, or checksum, to increase throughput by reducing the probability of errors.

By increasing the number of points in the signal constellation, it is possible to encode greater amounts of information to increase the modem's throughput. This is because slight variations in the phase-modulated signal may be used to represent coded information, which translates into a higher throughput. In their quest to achieve greater throughput, some modem manufacturers use signal constellations of 256 or more points.

Modem Features

Most modems come equipped with the same basic features, including error correction and data compression. In addition, they have features associated with the network interface, such as flow control and diagnostics. There are also various security features that are implemented by modems.

DATA RATE Today's modems are now capable of 56-Kbps operation. However, the advertised data rate of most modems does not always coincide with the actual data rate.[3] This is because the quality of the connection has a lot to do with the speed of the modem. If the connection is noisy, for example, a modem may have to step down to a lower speed to

---

3. While line quality has a lot to do with the actual speed of the modem, due to FCC regulations on power output, receiving speeds are limited to 53 Kbps.

continue transmitting data. Some leased-line modems, however, are able to sense improvements in line quality and automatically step up to higher data rates.

With regard to dial-up 56-Kbps modems, 44 or 46 Kbps is closer to the maximum connection rates most users get. Of course, if the receiving end does not have a 56-Kbps modem, the connection rate is established at the highest mutually supported rate, which might be 33.6 Kbps, 28.8 Kbps, or even 14.4 or 9.6 Kbps.

These high-speed modems are based on technologies that exploit the fact that for most of its length, an analog modem connection is really digital. When an analog signal leaves the user's modem, it is carried to a phone company central office where it is digitized (Figure 80). If it is destined for a remote analog line, the signal is converted back to analog at the central office nearest the receiving user. The conversion is made at only one place—where the analog line meets the central office. During the conversion, noise is introduced, which diminishes throughput. But the noise is less in the other direction, from digital to analog, allowing the greater downstream throughput.

On the other hand, a fully digital ISDN Basic Rate Interface line offers up to 128 Kbps in both directions, but the service and equipment is more expensive. Although 56-Kbps modems are not meant to challenge ISDN, they will continue to fulfill a role as a cheaper alternative to ISDN.

ERROR CORRECTION   Networks often contain disturbances with which modems must deal or, in some cases, overcome. These disturbances include attenuation distortion, envelope delay distortion, phase jitter, impulse noise, background noise, and harmonic distortion—all of which negatively affect data transmission. To alleviate the disturbances encountered when transferring data over leased lines (without line conditioning)

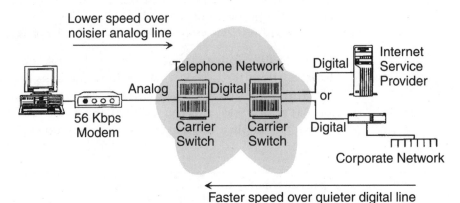

Figure 80
Most 56-Kbps modems can send data at near top speed, but only from a digital source.

and dial-up lines, most products include an error correction technique in which a processor puts a bit stream through a series of complex algorithms prior to data transmission.

The most prominent error correction technique has been the Microcom Networking Protocol (MNP), which uses the cyclic redundancy check (CRC) method for detecting packet errors, and requests retransmissions when necessary.

Link Access Procedure B (LAP-B), a similar technique, is a member of the High-Level Data Link Control (HDLC) protocol family, the error-correcting protocol in X.25 for packet-switched networks. LAP-M is an extension to that standard for modem use and is the core of the ITU error-correcting standard, V.42. This standard also supports MNP Stages 1 through 4. Full conformance with the V.42 standard requires that both LAP-M and MNP Stages 1 through 4 are supported by the modem. Virtually all modems currently made by major manufacturers conform to the V.42 standard.

The MNP is divided into nine classes. Only the first four deal with error recovery, which is why only those four are referenced in V.42. The other five classes deal with data compression. The MNP error recovery classes are as follows:

MNP CLASSES 1 TO 3   These packetize data and, Microcom claims, ensure 100-percent data integrity.

MNP CLASS 4   This achieves up to 120-percent link throughput efficiency via Microcom's Adaptive Packet Assembly and Data Phase Optimization, which automatically adjusts packet size relative to line conditions and reduces protocol overhead.

DATA COMPRESSION   With the adoption of the V.42bis recommendation by the ITU in 1988, there is a single data compression standard: Lempel-Ziv. This algorithm compresses most data types, including executable programs, graphics, numerical data, ASCII text, and binary data streams. Compression ratios of 4:1 can be achieved, although actual throughput gains from data compression depend on the types of data being compressed. Text files are the most likely to yield performance gains, followed by spreadsheet and database files. Executable files are most resistant to compression algorithms because of the random nature of the data.

Dynamic Bandwidth Allocation

Multiline modems can combine the bandwidth of two or more analog lines to achieve higher data transmission rates. This is implemented through Multilink PPP (Point-to-Point Protocol) and Dynamic Bandwidth Allocation (DBA), which work with analog lines or ISDN BRI lines.

DBA automatically provides bandwidth when needed to place an outgoing voice call. (DBA does not apply to incoming voice calls.) When Multilink PPP is enabled, both B channels are dedicated to the data call. However, while Multilink PPP is active, users might want to place a voice call. With DBA, they can place a voice call while a Multilink PPP call is active.

For example, if the user is already connected to the Internet and wants to place a voice call using an analog phone attached to the digital modem, merely lifting the handset temporarily and automatically removes one of the B channels used for Multilink PPP from the existing data link so it can be used for the voice call. The temporary removal of the channel for a voice call reduces the speed of the data call in progress to 56/64 Kbps without affecting its reliability. When the voice call ends and the handset is replaced, the channel used for the voice call is once again reinstated for the Internet connection.

Some modems address the need for more bandwidth for Internet connections. Some products integrate three 56-Kbps modems in a single unit. All three of the modems can be called into play simultaneously to access the Internet. They access Web pages as a team and share in the downloading, resulting in a throughput rate of up to 156 Kbps. The system exploits the basic nature of Web-page downloading. Pulling up an average Web page with graphics involves multiple TCP/IP sessions, and those sessions must be established sequentially. But with three modems working simultaneously and establishing unique sessions, download time can be cut by as much as two-thirds. The Internet service provider (ISP) does not have to do anything special, except support 56-Kbps modems.

Diagnostics and Other Features

Most modems perform a series of diagnostic tests to identify internal and transmission line problems. Most modems also offer standard loop-back tests, such as local analog, local digital, and remote digital loop-back. Once a modem is set in test mode, characters entered on the keyboard are looped back to the screen for verification.

Most modems also include standard calling features, such as automatic dial, answer, redial, fallback, and call-progress monitoring. Calling features simplify the chore of establishing and maintaining a communications connection by automating the dialing process. Telephone numbers can be stored in nonvolatile memory.

Other standard modem features commonly offered include fallback capability and remote operation. Fallback allows a modem to automatically drop, or fall back, to a lower speed in the event of line noise, and then revert to the original transmission speed after line conditions improve. Remote operation, as the name implies, allows users to activate and configure a modem from a remote terminal.

Security

Many businesses have become increasingly aware of the importance of implementing a sound network security strategy to safeguard valuable network data from intruders. Modems that offer security features typically provide two levels of protection: password and dial-back. The former requires the user to enter a password, which is verified against an internal security table. The dial-back feature offers an even higher level of protection. Incoming calls are prompted for a password, and the modem either calls back the originating modem using a number stored in the security table or prompts the user for a telephone number and then calls back.

Security procedures can be implemented before the modem handshaking sequence rather than after it. This effectively eliminates the access opportunity for potential intruders. In addition to saving connection establishment time, this method uses a precision high-speed analog security sequence that is not even detectable by advanced-line monitoring equipment.

SUMMARY

There was a time when it was widely believed that ISDN would eliminate the need for modems. Not only has this prediction proved to be wrong, but the technological innovations in modems have continued unabated. The modem market has witnessed a boom in recent years, which coincides with the explosive growth of the Internet. Telecommuters, mobile professionals, and branch offices all use the Internet to access corporate resources and for applications such as e-mail, file transfers, and research.
See Also Cable Modems, Digital Subscriber Line Technologies, Integrated Services Digital Network

Multicast Transmission

Multicast transmission is a method of delivering real-time applications, which are typically bandwidth intensive, to hundreds or thousands of subscribers in a way that is more efficient and economical than the traditional unicast method of transmission. Unicast delivery involves sending a separate data stream to each recipient. Multicast delivery involves sending only one data stream into the network, which is replicated only as many times as necessary to distribute the stream to the nodes (i.e., routers) that have registered subscribers attached.

Applications

Used mostly on IP networks, multicast has a number of applications. It is ideal for content providers with real-time applications such as news and entertainment events, and for the distribution of dynamic content, such as financial information and sports scores. The application itself can be audio, video, or text—or any combination of these. For content providers, IP multicast is a low-cost way to supplement current broadcast feeds. In fact, CNN is one of the biggest users of IP multicast.

Corporations can use IP multicast to deliver training to employees and keep them informed of internal news, benefit programs, and employment opportunities within an organization. They can also use IP multicast to broadcast annual meets to shareholders, or introduce new products to their sales channels. Associations can use IP multicast to broadcast conference sessions and seminars to members who would not otherwise be able to attend in person.

Political parties and issue advocacy groups can use IP multicast to keep their members informed of late-breaking developments and call them to action. Entrepreneurs can use IP multicast to offer alternative programming to the growing base of Internet access subscribers.

Performance

Many potential users of IP multicast wonder if IP multicast really works. After all, they have experienced the long delays accessing multimedia content on the Web. They see video that is slow and jerky, and hear audio that pauses periodically until something called a "buffer" has a chance to fill. When video and audio run together, the two are often out of synchronization. So they wonder how the Internet can handle a real-time multicast with acceptable quality.

The short answer is that the Internet cannot really handle IP multicast effectively—at least not yet. Currently, IP multicast works best on a managed backbone network, where a single company or carrier has control of all the equipment, protocols, and bandwidth end-to-end. This is not possible on the public Internet because there is no central management authority. While simple real-time applications may work well enough, such as text tickers, the performance of sophisticated graphically enriched real-time multimedia applications suffers.

Not only can the performance of a private network be controlled to eliminate potential points of congestion and minimize delay, but the company or carrier can place dedicated multicast routers throughout its network. This type of router replicates and distributes the content stream in a highly efficient way that does not require massive amounts of bandwidth.

For example, instead of sending out 100 information streams to 100 subscribers, only one information stream is sent. The multicast routers replicate and distribute the stream within the network to only the nodes that have subscribers. Users no longer need to purchase enormous amounts of bandwidth to accommodate large multimedia applications or buy a high-capacity server to send out all the data streams. Instead, a single data stream is sent, the size of which is based on the type of content. This can be as little as 5 Kbps for text, 10 Kbps for audio, and 35 Kbps for video.

Operation

A multicast can reach potentially anyone who specifically subscribes to the session—whether they have a dedicated connection or use a dial-up modem connection. Of course, the content originator can put distance limits on the transmission and restrict the number of subscribers that will be accepted for any given program.

A variety of methods can be used to advertise a multicast. A program guide can be sent to employees and other appropriate parties via e-mail or it can be posted on a Web site. If the company already has an information channel on the Web that delivers content to subscribers, the program guide can be one of the items "pushed" to users when they access the channel.

When people want to receive a program, they enroll through an automated registration procedure. The request is handled by a server running the multicast application, which adds the end-station's address to its subscription list. In this way, only users who want to participate will receive packets from the application.

Users can also select a multicast node from those listed in the program guide. Usually, this is the router closest to the user location. A user thus becomes a member of this particular node. Group membership information is distributed among neighboring routers so multicast traffic gets routed only along paths that have subscribers at the end nodes. From the end node, the data stream is delivered right to the user's computer.

Once the session is started, users can join and leave the multicast group at any time. The multicast routers adapt to the addition or deletion of network addresses dynamically, so the data stream gets to new destinations when users join, and stops the data stream from going to destinations that no longer want to receive the session.

Infrastructural Requirements

Several basic requirements have to be met to implement IP multicast. Most of these requirements are already met by the products on IP networks today. First, the host computer's operating system and TCP/IP stack must be multicast-enabled. Among other things, it must support the Internet group management protocol (IGMP). Fortunately, virtually all of today's operating systems can accommodate IP multicast and IGMP, including Windows NT, Windows 95, Windows 98, and most versions of UNIX.

Second, the adapter and its drivers must support IP multicast. This allows the adapter to filter data link layer addresses mapped from network-layer IP multicast addresses. Newer adapters and network drivers are already capable of implementing IP multicast.

Third, the network infrastructure—the routers, bridges, and switches—must support IP multicast. If they do not, perhaps it is only a matter of turning on this feature. If the feature is unavailable, a simple software upgrade is usually all that is needed.

Finally, applications must be written or updated to take advantage of IP multicast. Developers can use application program interfaces to add IP multicast to their applications. Although implementing IP multicast may require changes to end stations and infrastructure, many of the steps are simple—like making sure that hardware and software can handle IP multicast and making sure the content server has enough memory. Other steps are more complicated, like making the network secure.

Anyone who is serious about IP multicast should have a working knowledge of the latest protocols for real-time delivery and be familiar with the management tools for monitoring things such as quality-of-service parameters and usage patterns. This information is often useful in isolating problems and fine-tuning the network.

For companies that understand the value of multicast but prefer not to handle it themselves, multicast host services are available from sources such as GlobalCast Communications, PSINet, and UUNET. UUNET, for example, offers a multicast hosting service called UUCast, which provides six data streams of varying size to accommodate virtually any real-time data transmission:

Data Stream Size	Application Example	End-User Access Speed
5 Kbps	Ticker banner	Dial-up modem up to 56 Kbps
10 Kbps	Ticker banner or audio message	Dial-up modem up to 56 Kbps
25 Kbps	Audio/video applications	Dial-up modem up to 56 Kbps
35 Kbps	Audio/video applications	Dial-up modem up to 56 Kbps
64 Kbps	High-quality audio/video	ISDN up to 128 Kbps
128 Kbps	High-quality audio/video	ISDN up to 128 Kbps

UUCast requires a dedicated UUNET connection. The subscribing organization supplies its own content and equipment. Upon installation, the organization is given a unique multicast group address for each of the data streams. The organization's router is configured with a virtual point-to-point connection to the multicast router located in a local UUNET point-of-presence, or POP. Since these multicast routers only transmit data streams to the corresponding multicast group address, there is no interference with other traffic sources.

UUNET has equipped all domestic POPs with multicast routers, so any of its dial-up customers can also receive the data stream, if it is made available to them. A dial access router, located within each POP, recognizes the request for a particular multicast data stream and begins transmitting the appropriate content to the end user's desktop.

Costs

Telecommunications costs are always an important consideration. Companies are continually looking for ways to save money in this area and are understandably interested in the cost of multicasting.

The cost to implement multicast on an internal IP network is minimal because an existing infrastructure is simply being leveraged. There might be some upgrades to hardware and software, and possibly the need for management tools to monitor the multicasts. Organizations that are serious about IP multicast often have a dedicated full-time administrator.

An often overlooked cost comes in the form of continuously developing the multicast content. This could take a whole staff of creative people with specialized skills—writers, editors, graphic artists, audio/video production people, and a multimedia server administrator—not to mention all the expensive equipment and facilities they will need.

The cost of production hinges on the type of content to be developed. Obviously, it is much cheaper to use text only, but not very many multicast applications attract viewers if only text is involved. Production costs jump dramatically as audio and video components are added because special equipment and expertise are required. An alternative to the do-it-yourself approach is to outsource the production to a specialized firm. This can cut development costs by as much as 60 percent.

For companies that prefer to outsource IP multicast, companies such as UUNET offer a predictable price for a large-scale Internet broadcast and they take care of all the server and router management. UUNET multicast hosting is available in dedicated configurations. The monthly fee is based on a server component, starting at $3,000 per month, and a multicast stream size, starting at $2,200 per month for a 5-Kbps stream, $10,000 per month for a 25-Kbps stream, and $15,000 for a 35-Kbps stream.

The following table compares the cost of a unicast transmission to 1,000 users on a private IP network (coast-to-coast) with a multicast transmission to 1,000 users over the managed Internet backbone of UUNET:

Traditional Unicast Transmission	UUCast Transmission
1,000 dial users x 28.8 Kbps per user = 28.8 Mbps of bandwidth required	One 35-Kbps data stream required
One T-3 connection (45 Mbps). This unicast transmission will use approximately 50% of the T-3	Tremendous amounts of bandwidth are no longer needed to support a single broadcast.
Monthly cost for full T-3 = $54,000	Monthly cost $15,000

Thus, the total cost savings of multicasting over traditional unicasting is $39,000 per month or $468,000 per year, which is quite compelling. Of course, this scenario assumes that the private T-3 link is dedicated to unicasting and that the carrier does not yet offer Fractional T-3, in which case the monthly cost could be cut in half to about $27,000. Even in this case, the monthly savings for outsourcing is $12,000 or $144,000 per year.

SUMMARY

As noted, the Internet is not centrally managed, making it difficult to convey sophisticated multimedia content in real time with any consistency in performance. Congestion and delays are still obstacles that must be overcome. Other real-time applications like IP telephony experience the same problem when run over the public Internet. The next big step toward full deployment of IP multicast is convincing more Internet service providers to offer the service to customers. The argument is not hard to make, especially since multimedia applications are obviously running on the Internet anyway and multicasting actually conserves bandwidth. With only modest costs, mostly for router upgrades, wider support of multicast would improve the Internet's performance for everyone. Until then, multicast will be most effective on private intranets and carrier-managed backbones where performance can be controlled end to end.

See Also **Internet Protocol, Multimedia Documents, Voice-Data Integration**

Multimedia Documents

Multimedia documents combine two or more different media types—text, graphics, images, audio, and video—in the same file. Such documents have shown great potential for enriched communication. Studies have shown that multimedia content is absorbed faster and retained longer than text-only content.

With the increasing popularity of the public Internet and corporate intranets, many content developers are making generous use of image, video, and audio to enhance the appeal of their work. In an e-commerce application, for example, the addition of multimedia content to a product catalog can more effectively attract and retain visitors to the company's Web site. However, visitors do not have the patience to watch a blank screen for very long as the multimedia segments load. Therefore, effective distribution of multimedia documents requires applying appropriate production techniques to the content.

Development Tools

The most popular development toolset for publishing multimedia documents on the Web is Adobe Systems' Acrobat, which includes Adobe Capture, Exchange, Catalog, Distiller, and Reader. With the company's Portable Document Format (PDF)— derived from its PostScript page description language—these components are used to prepare and distribute electronic documents via the Web, e-mail, corporate networks and intranets, CD-ROM, and print-on-demand systems (Figure 81).

Among the key Acrobat components is Acrobat Capture—an optical character-recognition (OCR) package that enables users to scan in paper-based documents and convert them to the PDF file format. Capture pre-

Figure 81

Kingdom Tapes and Electronics catalog, in Adobe's PDF format, as viewed from within Netscape Navigator using the Acrobat plug-in.

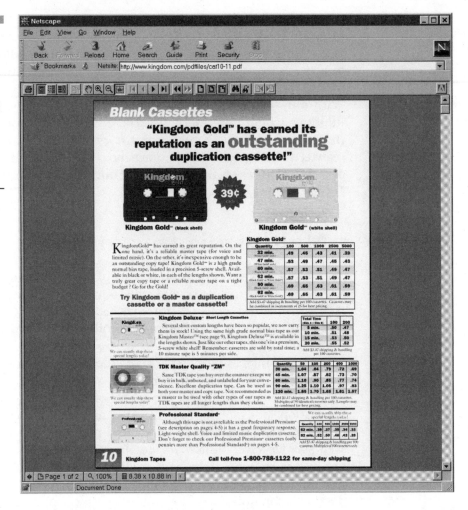

serves the layout and style of the original document, as well as any graphical elements, while translating text into its electronic equivalent. Capture includes a search-and-replace facility to help correct optical-recognition errors, a spell checker, and a document previewer.

Another Acrobat component is Exchange, which is the front end of the PDF language. It enables the user to create, view, collate, navigate, and print PDF-based documents. Multicolumn documents can even be formatted to scroll automatically to the next column when the reader reaches the end of the first column. Exchange also shows page thumbnails that can be used to move, copy, insert, delete, and replace pages in the same document or across different documents. The PDF Writer module included with Exchange is a printer driver, which allows users to transparently convert documents to the PDF format by printing from a native program, such as Microsoft Word, to the PDF "printer." For confidential documents, Exchange supports a two-level password protection scheme. One password is given to the user to open the document and another is kept by the author for changing the security options on the document.

Acrobat Search, accessible through Exchange, can perform full-text searches on PDF documents, based on an index of words and phrases extracted from the PDF documents themselves. The search engine lists all documents containing the query text and, next to each document, an icon indicates the document's relevance to the query. The search engine can also search for words and phrases contained in images, graphs, or formatted tables.

Acrobat Catalog creates full-text indexes for collections of PDF files for faster, more efficient document searching. The index can be static or dynamic. A static index is used when the content never changes. Catalog can also be configured to perform automatic indexing, so documents that have been modified, removed, or added can be properly reflected in the index. This is particularly useful in a networked environment, where PDF documents could change frequently. The frequency of indexing can be set by the network administrator.

Acrobat Distiller translates existing PostScript files to PDF format. Distiller is used when there is a need for high-quality portable reproductions of PostScript artwork, 24-bit color images, and other source documents that exploit certain features available only in the PostScript language. When Distiller is run in a networked environment, it has the ability to continuously batch process files. The network administrator can create Distiller "In" and "Out" boxes on the server. PostScript files to be "distilled" are placed in the In box, and converted PDF documents are saved by Distiller in the Out box.

Distiller employs several compression strategies to reduce the size of the resulting PDF files. Text, line art, and fonts are compressed using the Lempel-Ziv method. Color and grayscale images are compressed using the JPEG method. Monochrome images are compressed using a combination of CCITT Group 4 or Group 3, Lempel-Ziv, or Run Length Limited (RLL) encoding techniques, depending on the characteristics of the image.

The Acrobat Reader is the client application that enables users to view PDF documents. It is freely available and distributable over the Internet and versions are available for PCs, Macintoshes, and UNIX systems.

Interactive Documents

Acrobat PDF files initially lacked much of the interactivity of documents coded in the HyperText Markup Language (HTML). Adobe Systems changed that in early 1998 with the introduction of interactive PDF forms, which lets developers build dynamic, database-driven Web-based forms using Acrobat. Acrobat Forms Update is intended to help companies that want to bring paper-based forms—such as insurance claims forms or sales contracts—to the Web without having to rewrite their existing store of forms completely.

An insurance company, for example, can post its claims forms on the Web and allow policy holders to fill-in, print, save, and e-mail forms in a few minutes, instead of spending hours completing these documents by hand and waiting for the Post Office to deliver them. For the insurance company, the use of PDF reduces the time and effort to create and deploy Web-ready forms.

The Acrobat Forms Update incorporates Javascript that enables developers to design forms that retain their intelligence. This enables the forms to instantly verify data and make calculations dynamically as the form is filled in. Forms saved to a central database can be accessed and updated by many employees working on the same project, offering immediate sharing of project status and the ability to proactively track works in progress. Adobe PDF forms also support dynamic page templates within a single Adobe PDF file. This enables developers to easily create customized forms applications, catalogs, brochures, and other publications that can be personalized and delivered on the fly to users' desktops.

SUMMARY

The bandwidth limitations of the Internet, as well as its high latency and slow response time, must be overcome by both content developers and network managers. Content developers contribute by scaling down the file size of multimedia documents by limiting the use of color to only the essentials and applying appropriate data compression techniques to minimize bandwidth requirements.

See Also **Document Imaging Systems, Multicast Transmission**

Multiservice Switching

The telecommunications industry is making the transition from proprietary, circuit-switched networks to open, packet/cell-based architectures that support voice, data, and video services. This transition is referred to as multiservice switching, which entails the use of ATM-capable switching systems to support voice, video, private line, and data over a variety of services, including Asynchronous Transfer Mode (ATM), frame relay, and Internet Protocol (IP) services.

In today's networking environment, where each application has different performance requirements, a variety of services come into play. Sometimes ATM is used, and other times frame relay or IP is used. Each network provides its own service benefits, as well as costs and features. Instead of trying to use one type of network to meet all application requirements, current thinking favors an integrated network that draws upon different services as needed to support data, voice, and video. A successful multiservice network is one that uses a combination of technologies that have been around for years: ATM at the core, frame relay and IP for low-speed access, and IP performing application integration functions (Figure 82).

In essence, each of these maturing technologies are relegated to role players in the multiservice network. Together, they combine their strengths and minimize their weaknesses to create a scalable, interoperable, high-performance network that suits most enterprise network needs.

Standards

The evolution from traditional telephony-based networks to a multiservice and software-intensive architecture requires a strong consensus among all the players in the telecommunications industry. In November 1998, the major international switch vendors and service providers formed

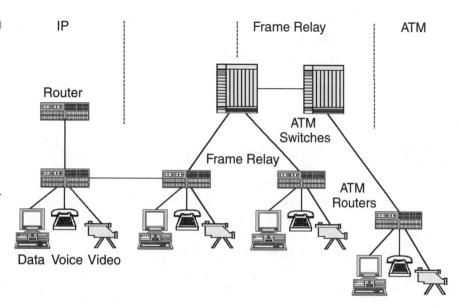

Figure 82
The multiservice network can be seen as overlapping layers, with users connecting to the layer that offers them the most appropriate service.

the Multiservice Switching Forum (MSF) to define and implement an open systems model of ATM-capable switching systems that will expedite the delivery of new integrated broadband communications services to the marketplace.

The MSF will complement the scope and activities of existing associations by enabling the integration of leading components from multiple vendors into a single multiservice switching system. The MSF will focus on protocols and interfaces used within switching systems and will use specifications of the ATM Forum, Internet Engineering Task Force (IETF), International Telecommunication Union (ITU), Frame Relay Forum, Bellcore Generic Requirements process, and other industry associations.

The goal is to speed the arrival of an open switching platform that will enable vendors and carriers to deliver new and better services at a lower cost. An important step toward this goal is to develop next-generation networks for access aggregation and enhanced quality of service features, and to separate network intelligence from switching and routing.

SUMMARY

The transition of traditional TDM networks to a packet/cell network model—which supports voice, data, and video services—is already well underway with the growth of the IP-based Internet. The MSF will enable carriers to deploy open switching systems with components from multiple vendors. This will accelerate deployment of next-generation networks supporting advanced, integrated broadband communications services. It is expected that multiservice switching systems will benefit from the same innovation and cost reductions that open systems in the computing world have achieved.

See Also **Asynchronous Transfer Mode, Frame Relay, Internet Protocol**

Multistation Access Units

A Multistation Access Unit (MAU) is a hub that connects computers and other devices to a token-ring network. The MAU physically connects computers in a star topology while retaining token ring's logical ring structure. However, every message passes through every computer, each passing it on to the next in a continuing circle until it arrives at its proper destination. This leaves the token-ring topology vulnerable in that a single nonoperating node can break the ring. The MAU solves this problem because it has the ability to bypass nonoperating nodes and maintain the ring structure.

The unit is likely to be stored in a wiring closet—mounted on a wall or installed in a standard 19-inch rack—or it can be put on a desktop. Cables from the computers are simply plugged into the MAU's RJ45 ports, which will detect the presence of a signal for each of the connections and configure the ring. Regardless of location, MAUs can be interconnected to other access units to form larger networks via their Ring-In and Ring-Out ports.

Configurations

The typical MAU is quite flexible in that it can be used in any one of three physical configurations to create or extend the token-ring network in accordance with business needs. For example, a MAU can be used as a standalone ring by attaching up to eight workstations or devices to the unit's RJ45 ports (Figure 83).

Because the ring is created by merely plugging the computers and other devices into the MAU, no technical expertise is necessary. This makes the stand-alone ring especially appropriate for small office environments such as professional offices, branch banks, stores, and classrooms.

Several MAUs can also be used together by cabling up to ten units via their Ring-In/Ring-Out (RI/RO) ports and then attaching up to eight workstations to each of the units' RJ45 ports (Figure 84). This allows a token-ring network to be expanded to keep pace with organizational growth.

Figure 83
The MAU used as a stand-alone ring.

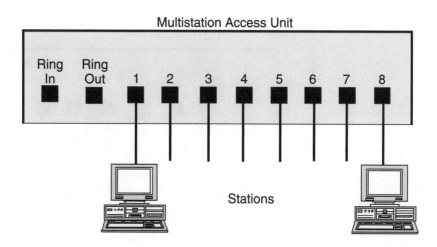

Multistation Access Unit

Ring In | Ring Out | 1 | 2 | 3 | 4 | 5 | 6 | 7 | 8

Stations

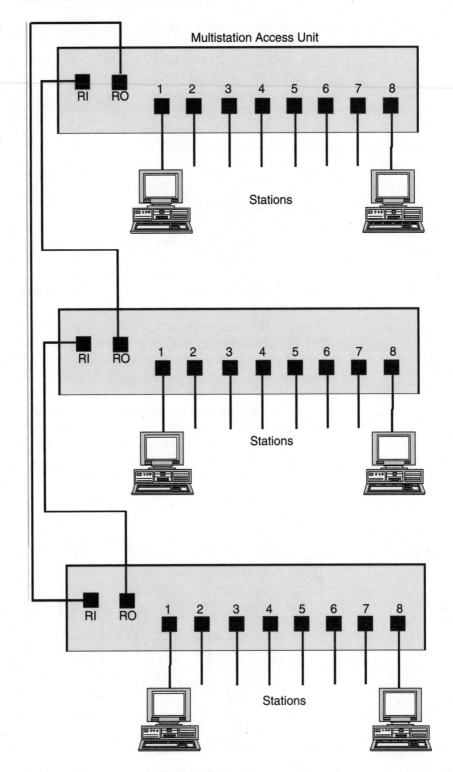

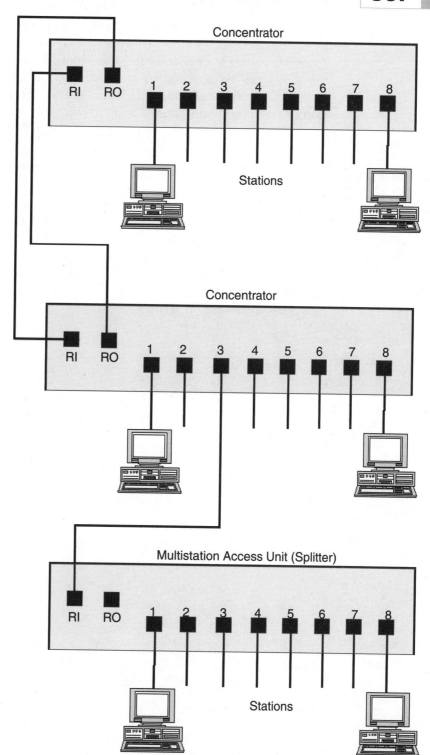

To support continued growth, a MAU can be used with a concentrator to serve additional workstations without changing the building's cabling. Instead of using a separate concentrator port for each connected device, the MAU can be attached to an existing concentrator port, with the switch on the front of the MAU set to "splitter." This allows the connection of up to eight additional devices to the MAU (Figure 85).

SUMMARY

The MAU is a low-cost, passive, token-ring access unit that provides RJ45 connections for attaching workstations and other token-ring devices to a 4- or 16-Mbps token-ring network using Category 4 or 5 unshielded twisted pair (UTP) cabling. The MAU also uses visual status indicators to make installation and use simple and quick—no manual setup or initialization is required.

See Also **Hubs, Token Ring**

NetWare

NetWare is a popular network operating system offered by Novell, the latest version of which (release 5.0) was issued in September 1998. Worldwide, there are over 80 million users of NetWare.

With NetWare 5, Novell for the first time includes the Java Runtime Environment (JRE) for Windows 95 and Windows NT in its operating system for both the client and server. The JRE includes the Java Virtual Machine (JVM), just-in-time (JIT) compiler, class libraries, and other Java extensions to decrease the complexity of running Java in Windows-based environments. NetWare 5 also includes significant enhancements from previous versions in the area of network and desktop management, Internet/intranet services, application development and deployment, and core services such as file, print and security.

NetWare 5 is the first major network to offer pure IP (Internet Protocol) in the sense that it does not require an IPX-based encapsulation—or, in the case of Windows NT Server, a NetBIOS encapsulation. As in previous versions, NetWare 5 continues to support IPX, but leaves the choice of which protocol to run to each individual user. Both IP and IPX can even be used simultaneously, allowing users to retain their existing application and routing investments, and use them for as long as it makes sense to do so.

In a pure IP environment, NetWare 5 uses the Service Location Protocol (SLP), a naming and discovery service for IP that has similar functionality to the Service Advertising Protocol (SAP) for IPX. SLP is based on IP (the SLP updates are sent via IP packets) and does not require multicast, but will take advantage of multicast if it is available. Users can still implement DNS/DHCP as a method of discovery.

Features

NetWare 5 offers many other features and major enhancements in the areas of management, file system, printing services, Web services, security and the operating system kernel itself. Among these features are:

WAN TRAFFIC MANAGER Acts as an intelligent supervisor of data going out onto wide-area network (WAN) links, giving users greater control over the movement of NDS (Novell Directory Services) data over WANs

CATALOG SERVICES Provide easier and faster access to NDS data and enable users to perform logins with no context

DOMAIN NAME SERVICE (DNS)/DYNAMIC HOST CONFIGURATION PROTOCOL (DHCP) UTILITY Is integrated with NDS for easy IP address management

JAVA DEVELOPMENT PLATFORM For easy and seamless development and deployment of Java-based applications

MEMORY PROTECTION, VIRTUAL MEMORY, AND A NEW KERNEL For improved application execution and reliability

NEXT-GENERATION STORAGE SYSTEM (NOVELL STORAGE SERVICES) Eliminates file system limitations and improves reliability, scalability, and performance

NOVELL DISTRIBUTED PRINT SERVICES (NDPS) For bidirectional, intelligent printing

CLUSTERING Supports multiple servers and offers manageability through Novell Directory Services

HOT-PLUG PCI SUPPORT Allows PCI network cards to be inserted or removed without first powering off the network device and turning it back on

Additionally, NetWare 5 ships with and supports Oracle8, the Common Object Request Broker Architecture (CORBA), VBScript-compatible Net-Basic interpreter, JavaBeans for NetWare, JavaScript, and Perl 5. Also included with NetWare 5 is the Z.E.N.works starter kit, a suite of desktop administration tools that provides application management, software distribution, software installation, desktop management and maintenance, and remote diagnostics and repair.

SNMP Support

The NetWare Management Agent (NMA) in NetWare 5 consists of several functions provided through SNMP (Simple Network Management Protocol), including:

- Real-time information about the server hardware and network operating system, including the CPU, memory, volumes, disks, connections, disk space usage per user, print Queues, open files, network interfaces, and protocols
- More than 650 types of alarms triggered by an event or a threshold
- Trends with historical data about such things as CPU usage, number of connections, and volume space

Real-time statistics (Level 1) are provided through SNMP, so they can be read by Novell's ManageWise or any other SNMP console. Levels 2 (alarms) and 3 (trends) are provided only by ManageWise or the stand-alone NMA.

SUMMARY

NetWare is optimized to run networks for small companies all the way up to enterprises with global networking needs. Windows NT, on the other hand, is a desktop operating system that has been stretched into an application server. Like UNIX, Windows NT is a good application server and performs well in that network space. However, NT was not developed to do networking, so it cannot deliver on global and centralized management nor on reliable, scalable, and high-performing Web, file, print, and security services that today's businesses need. At year-end 1998, industry analysts pegged NetWare's market share at 38 percent, more than double that of Windows NT Server at 16 percent.
See Also **Linux, UNIX, Windows NT**

Network Addresses

Regardless of the type of data network, information is transmitted in packets. Each of these packets includes a header, which indicates the point from which the data originates and the point to which it is being sent, as well as other information. With TCP/IP networks, for example, resources on the Internet or corporate intranet are defined through the use of 32-bit IP addresses, which include four address blocks consisting of numbers between 0 and 256, separated by periods (e.g., 160.130.0.252).

There are four classes of Internet addresses (Figure 86), each class designed for a different size network. Class A addresses support the largest networks, such as MILNET, allowing up to 16,777,216 hosts per network. There are 127 available Class A networks. There are 16,384 Class B networks, each supporting up to 65,536 hosts. Currently, about half of the Class B networks have already been assigned. Class C addresses provide support for up to 2,000,000 networks, each with up to 256 hosts. Class C addresses are typically assigned to the LANs of college campuses and small government contractors. Class D addresses are used for multicast traffic and are assigned dynamically. There is also a fifth designation, Class E, that is currently not assigned. Unlike Classes A, B, and C IP addresses, the last 28 bits of a Class D address have no further structure. The multicast group address is the combination of the high-order four bits of

Figure 86

IP address types.

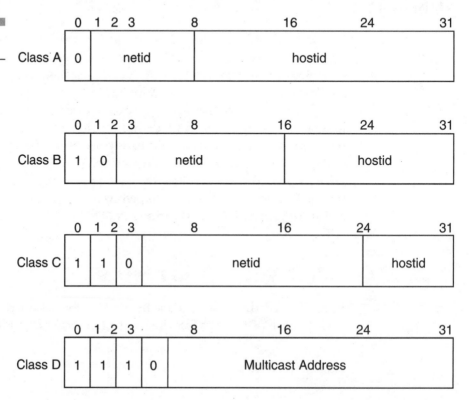

1110 and the multicast group ID, typically written as decimal numbers in the range 224.0.0.0 through 239.255.255.255.

Internet users generally do not need to specify the IP address of the destination site because they can be represented by alphanumeric domain names such as fcc.gov or ibm.com. Domain Name Systems (DNSs) throughout the network contain tables that map these domain names to their corresponding IP addresses. Current and proposed domain names are summarized in the following table:

Top Level Domain Name	Area of Emphasis
Current	
.com	Commercial service providers
.edu	Educational institutions
.gov	Government entities (nonmilitary)
.int	International treaty organizations
.mil	Military entities
.net	Network services
.org	Nonprofit and other organizations
Proposed	
.firm	Businesses/firms
.store	Goods for purchase
.web	WWW-related activities
.arts	Culture/entertainment
.rec	Recreation/entertainment
.info	Information services
.nom	Individual/personal nomenclature

Operation

DNS servers are arranged as a hierarchical database. At the top of the DNS database tree are root-name servers, which contain pointers to master-name servers for each of the top-level domains. To find out the numeric address of www.advicom.net, for example, a DNS server would ask the root-name server for the address of the master-name server that has responsibility for the .net domain.

In turn, the master-name servers for each of the top-level domains contain a record and the name-server address of each domain name. In trying to find out the numeric IP address of www.advicom.net, for example, the DNS server asks the .net server for the name of the server that handles the advicom.net domain.

The individual name servers for each domain name, such as advicom.net, contain detailed address information for the hosts in that domain. So in this example, the DNS server would then ask the advicom.net server for the name of the server that handles the advicom.net domain. Finally, this most specific name server supplies the DNS server with the IP address of the machine called www.advicom.net.

The entire process (Figure 87) takes only a few seconds. When an Internet name is submitted to a DNS server (stage 1), the server checks its information and attempts to respond with the appropriate numerical IP address (stage 2). If the server cannot respond, it directs the request to a top-level DNS server (Stage 3), which then sends the request down the DNS hierarchy (stage 4). Once an authoritative DNS server for the domain and machine is found, the response is sent to the request originator (stage 5). With this information, the client can then access the resource having that name.

DNS Configuration

There are two ways to implement DNS. One way is to use the DNS server that belongs to an Internet service provider (ISP). Many ISPs provide this as part of their service to subscribers; if not, they link to a specific DNS server on the Internet that does. The ISP will provide its subscribers with the numeric IP addresses of the primary and secondary DNS servers. Two DNS servers are required to avoid cutting off users from the Internet if the one DNS server goes down. This address information is used by subscribers to configure the TCP/IP stack on their computer.

Figure 87
Through a lookup process, the Domain Name System (DNS) maps top level domain names to IP addresses.

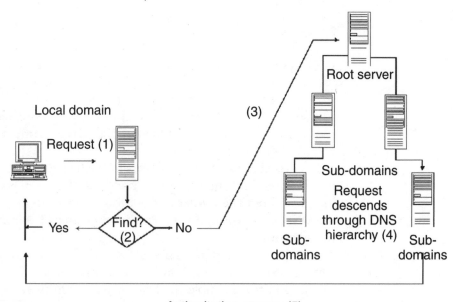

An ISP provides its subscribers with the IP addresses of its primary and secondary DNS servers. As part of the dial-up networking configuration procedure, a Windows 95 user, for example, would enter these IP addresses into the TCP/IP Properties dialog box (Figure 88), along with the host name and domain name. When configured with this and other information, the TCP/IP dialer that comes with Windows 95 can be used to access the Internet. Alternatively, third-party TCP/IP dialer can be used.

Another way to implement the DNS is to set up the primary and secondary servers on a private intranet. This gives an organization more control over the administration of IP addresses. Having a DNS server inside a private network lets staff make changes, additions, and deletions on their own schedule instead of having to wait for ISP staff to do it. And, if the server sits behind a firewall, security can be enhanced by hiding internal IP addresses from public view. This helps prevent hackers from gaining access to network resources. In addition, name resolution will be faster for internal users because the organization's DNS server will usually not be as heavily loaded as an ISP's server.

To aid in the administration of a large number of IP addresses, the Dynamic Host Configuration Protocol (DHCP) is usually implemented. This is a server program that automatically assigns IP addresses to users as

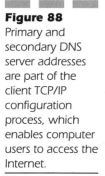

Figure 88
Primary and secondary DNS server addresses are part of the client TCP/IP configuration process, which enables computer users to access the Internet.

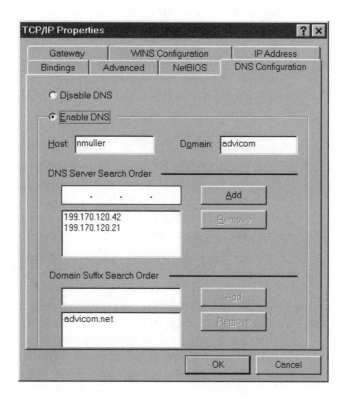

they log on the network. Each client is configured to automatically re-trieve an IP address from a pool of unused IP addresses that are assigned to the organization. This capability is referred to as dynamic IP. Many ISPs also use DHCP to manage the IP addresses of large numbers of sub-scribers.

SUMMARY

The DNS is the foundation for navigation on the Internet as well as pri-vate intranets. It enables plain text names to be assigned to various net-work resources, which are easier to remember than numerical IP addresses. Whether operated by an ISP or a company, two DNS servers are required—primary and secondary—to prevent users from being cut off from the network if the primary DNS server becomes disabled.

See Also **Firewalls**

Network Asset Management

The decentralization of computing resources from the data center to the desktop has resulted in efficiency gains for companies and productivity gains for employees—both of which contribute to enhanced competitive position. This decentralization—plus the addition of more cabling and network devices to support it—has also resulted in some loss of control of technology assets. The situation is even worse for highly distributed com-panies with many national and international operations, especially if they must also support many branch offices, telecommuters and mobile professionals.

If left unaddressed, this loss of control can greatly inflate the cost of doing business. It has been estimated that U.S. corporations alone can save $20 billion a year by implementing asset management programs that ac-count for every device on the LAN to the WAN.

Asset management tools provide a central repository for storing such information as equipment serial numbers, port configurations, and mem-ory usage—much of it collected automatically by periodic scanning of the network. Such tools also update the database whenever there are moves, adds, and changes. This and related information that is manually entered can be used to more efficiently manage the network, resolve prob-lems faster, standardize on particular hardware and software platforms, plan capacity, analyze costs and benefits, assist in budget planning, and smooth technology migrations.

The kinds of network devices that must be monitored, controlled, and accounted for in inventory include repeaters, bridges, routers, gateways, hubs, switches, remote access servers, and modem pools. These types of equipment usually are centrally managed and controlled with the element management systems vendors typically provide with their products or third-party applications—both of which can be integrated with higher level management systems, such as Sun's Solstice Enterprise Manager or Domain Manager, HP's OpenView, Tivoli's TME 10 NetView, and Cabletron's Spectrum.

A comprehensive network asset management system automates the tracking of inventory—equipment, circuits, and software—facilitates planning, reduces capital costs, improves availability, and enhances productivity. To make all this possible, a lot of information must be gathered and stored, including:

- Hardware and software configurations
- Asset locations
- Warranty information
- Lease contracts
- Maintenance agreements
- Cost-accounting information
- Chargeback information for billing purposes

With all this information stored in a database, it can be imported into billing, finance, budget, and other business systems.

Configuration Management

Among the many functions that can be incorporated into a comprehensive asset management system is configuration management, which is the process of collecting, maintaining and distributing up-to-date configuration information for evolving data, voice and video networks. The network configuration application can be used to model and store the entire network infrastructure in a relational database management system such as Oracle, including active elements such as workstations, hubs, routers, and switches, and passive elements such as cables, connectors, distribution frames, and patch panels.

The database stores the spatial location, topological information, asset attributes, and connectivity of each network object. New components can be added, or existing components changed, directly in the database using graphical tools. Graphical views of the network can be generated directly from the database, from which facility drawings and maps can be created and modified. The network map can even be overlaid on a facility drawing or map.

Making changes to a network can be time consuming and labor intensive, which can increase the chance of error. Once the network is documented, the tracking of moves, adds and changes (MACs) can be automated. The specific tasks that can be automated include:

- Moving equipment to a new location
- Adding or removing equipment
- Connecting and disconnecting equipment to or from the outlet or wall plate
- Connecting and disconnecting patch cords within one or more wiring closets
- Changing system parameters such as TCP/IP host names and addresses

These steps are managed through service requests, MAC planning, resource checking, work orders, and automatic database updates when devices are added or moved. All devices can have multiple contacts—such as user, owner, and responsible technician—and can be assigned to departments, workgroups, and projects. This improves the MAC process by making it easy to select devices for multi-device moves and to initiate, query, and report on the work required.

Any device on the network, once documented, can be selected and called up from the database, along with all of the associated asset information. For example, the HP OpenView user can have information regarding lease, maintenance, and warranty contracts, pertinent contact information, device configuration, cost of ownership, and other important asset information. Many asset management systems can be accessed by Web browsers, allowing department managers and others to use this information as well.

Related Functionality

Proactive asset management is knowing what assets are currently owned, what condition they are in, their configuration details, and how they are deployed so this information can be aligned with current business plans for expansion or downsizing. This enables strategic buying and negotiation of volume discounts, and gives the company alternatives when purchasing new equipment, so that costs can be controlled.

Keeping track of who owns what, where it is located, who uses it and making this information available to everyone who needs it in the organization, can be very useful for planning budgets, future acquisitions, retirement of assets, and anticipating needs as technology changes. Knowing where assets are deployed, their configuration, who maintains them, and when their leases expire also ensures effective and economical redeployment of existing assets.

Finance people can properly depreciate equipment and pay the appropriate taxes only if there is an accurate accounting of computer and network assets. Asset management systems provide the details of asset configuration and actual usage. As technology becomes outdated or needs change, the network managers and administrators need to know where specific assets are deployed, what components can be re-used, and who to contact for termination of maintenance agreements.

As the cost of network computing escalates it is important to track these costs and distribute them to the appropriate users for financial management and control as well as cost recovery and resale. Cost allocation also facilitates resource sharing and balancing, enabling users to pay for only the resources that they consume.

SUMMARY

Asset management is a methodology, combined with one or more software products, that helps companies gain control over what they own and operate. Asset management administers every piece of the technology puzzle—from users and IT staff, to the procurement process, to specific pieces of hardware and versions of software. An effective asset management program optimizes the use, deployment, and disposal of all assets. It can ease the recordkeeping burden of routine moves, adds, and changes. In addition to containing the cost of technology acquisitions and reining in hidden costs, such programs can improve help desk operations, enhance network management, assist with technology migrations, provide essential information for planning a reengineering strategy, and support Total Cost of Ownership (TCO) initiatives.

***See Also* Application Metering, Cable Management**

Network Booting

Network booting is the process whereby workstations start up using files located on a server instead of using files from their hard or floppy drives. Each PC or networked device has a "boot image" on the server. The boot image contains all of the usual boot files such as AUTOEXEC.BAT and

CONFIG.SYS, as well as network operating system and device drivers. When a machine equipped with special boot software is powered on, the ROM (read only memory) or code directs it to the appropriate boot image on the server.

This means network administrators can set up boot images for PCs or groups of PCs on the server, where the files can be accessed easily when they need changes. Adding, moving or deleting users is much easier when done from the administrator's desk, as is troubleshooting when a particular machine refuses to boot.

Network booting is transparent to the end user and has no effect on the user's desktop applications. If the network is down, users can boot locally, provided the LAN administrator has allowed boot files to reside on the hard drive.

Benefits

By directing PCs to boot from the server rather than from a local drive, network booting offers a number of benefits. For example, when a new driver or operating system becomes available, network clients can be upgraded from any workstation, without disrupting users or powering off their workstations. Changes made while the workstations are in use are picked up at the next boot. Typically, major upgrades mean disruptions to productivity while the LAN administrator visits every workstation and installs the new files. With boot software, a new client OS can be installed in a matter of minutes from the LAN administrator's desk. This reduces the time staff spend on upgrades during and after business hours.

Nearly half of all help desk calls are boot file-related. With network booting, the files are protected from end user modifications, reducing the likelihood of a boot problem. Where a boot file problem does occur, the LAN administrator can diagnose and correct it from their desk, instead of visiting the workstation. The LAN administrator can permit optional local booting, and decide who can and cannot boot from their local drives. Or the LAN administrator may choose to disallow local boots entirely.

By storing workstation operating systems on the server, users can try new operating systems without switching completely. For example, the LAN administrator can use this function to allow users to test Windows 95 or Windows NT before migrating to it permanently. Boot software can also be used to let users have access to two or more operating systems as their applications dictate.

Finally, boot files stored on the server are protected from end-user modifications, corruption, and piracy. Many versions of boot software also include software to detect and destroy boot sector viruses. 3Com's BootWare, for example, has this capability.

Implementation

While desktop PCs empower users, they require considerable management resources—with nearly 50 percent of PC ownership costs devoted to administration. In an effort to reduce these costs, some companies are considering NetPCs and Network Computers (NCs), or "thin clients." But while thin clients hold out the promise for lowering Total Cost of Ownership (TCO) by simplifying general PC maintenance and upgrades, most companies have been reluctant to discard their substantial existing investment in traditional PCs.

An alternative is provided with so-called "boot agent" software installed in the network interface card (NIC), which allows PCs to receive all or any part of their boot-up instructions from the network, allowing centralized control to any extent the LAN administrator decides. The use of boot agents protects existing investments in PCs, while giving administrators the same degree of management control promised by thin client architectures. 3Com's Managed PC Boot Agent, for example, is available in a number of formats:

- In a chip that can be plugged into the Boot ROM socket on 3Com NICs
- A flash upgrade for existing 3Com TriROM chip users
- Embedded in the system BIOS of select PC motherboards

Boot ROMs are programmed chips that reside on a PC's network interface card. They allow network administrators to direct PCs to boot from files on the local hard drive or from files on a network file server.

Most companies that use network booting do not even load DOS on individual workstations. By having users boot from a controlled, standard file server-resident boot image, companies are able to significantly reduce the number of problems that users cause for themselves when they either knowingly or unknowingly make changes to configuration files.

Another benefit that boot ROMs provide is faster troubleshooting. Since all users have identical boot images, hardware problems can be classified very quickly. For example, if a user calls the help desk with a problem—and nobody else is experiencing the same problem—then it is most likely a hardware problem, since everything else about their machines is the same. Without boot ROMs, help desk staff can waste a lot of time checking user configurations in an effort to identify the problem.

Although network booting software has become commercialized in recent years, the concept is not new. It originated in the Internet environment. The Bootstrap Protocol (BootP) and the Dynamic Host Configuration Protocol (DHCP) provide a framework for passing configuration information to hosts on a TCP/IP network. This means that hosts (e.g. personal computers, print servers, X-terminals, etc.) do not have to be configured before they can communicate using the TCP/IP protocol suite.

Bootstrap Protocol

For organizations that rely on the Internet Protocol (IP) for intranets, extranets and virtual private networks (VPNs), the bootstrap protocol (BootP) allows diskless clients to download configuration information from a server at the time of power up. This allows them to discover their own IP address, the address of a server host, and the name of a file to be loaded into memory and executed.

When configured as a BootP client, a computer that needs to use the Internet issues a broadcast across its local network. If there is a BootP server running on the same network, that server will hear the broadcast request, consult its reference tables, and return a packet to the client containing the following information:

- IP Host Address
- IP Gateway Address
- Domain Name Server Address(es)
- Subnet Mask

When this information is received, the client can access the Internet. The BootP protocol allows other information to be sent in the BootP reply packet, but these four elements are the only required information for a client to be able to use the Internet. Non-UNIX BootP clients usually ignore any other information sent in the reply packet. This broadcast request is only issued once per session. The configuration information is remembered until reboot.

Dynamic Host Configuration Protocol

The successor to BootP is the Dynamic Host Configuration Protocol (DHCP). Both work in the same way, except that the information on a BootP server must be manually configured. This includes matching an IP address to the MAC (Medium Access Control) address of every client.

DHCP minimizes this manual configuration process by automatically assigning IP addresses to newly attached clients.

At startup, the client sends out a BootP message on the network, which is picked up by a BootP server. The message includes the hardware address of the client's network interface card (the MAC address), since it does not yet have an IP address. The server retrieves the client's assigned configuration (i.e., the files needed to start the operating system) and sends it to the workstation via its hardware address. The information returned to the client includes its IP address, the IP address of the server, the host name of the server, and the IP address of a default router.

Reverse Address Resolution Protocol

The Reverse Address Resolution Protocol (RARP) is an old method used by diskless clients to get an IP address from the server. At startup, the client broadcasts a RARP packet on the local network, which is received by all connected devices. A broadcast is necessary because the client does not know the IP address of the server it needs to communicate with. When the server receives the RARP packet, which includes the workstation's MAC address, it does a table lookup to match the MAC address with an IP address. It then sends the IP address to the workstation via its MAC address. The limitation of RARP is that the server running it can only serve clients on a single LAN. DHCP and BootP, on the other hand, are designed so they can be routed.

SUMMARY

Booting from the network is not a new concept, but now administrators can perform "pre-OS boots" to boot PCs—even without a functional operating system installed. The PC communicates with a boot configuration server, loads an initial operating system off the network with desktop management agents and configuration details or updates, then completely exits the operating system, returning to its own hard drive to boot up normally.

See Also **Client-Server, Internet Protocol, Thin Client Architectures**

Network Caching

A cache is temporary storage of frequently accessed information. Caching has long been used in computer systems to increase performance. A cache can be found in nearly every computer today, from mainframe to PC. More recently, caching is used to improve network performance. Instead of users accessing the same information over the network, it is stored locally on a server or, in some cases, on the desktop. This arrangement gives users the information they need quickly, while freeing the network of unnecessary traffic, which improves its performance for all users.

Caching is frequently applied to the Web. When users visit the same Web site, the browser first looks to see if a copy of the requested page is already in the computer's hard disk cache. If it is, the load time is virtually instantaneous; if not, the request goes out over the Internet.

Network caching offers an effective and economical way to offload some of the massive bandwidth demand. This allows Internet service providers (ISPs) and corporations with their own intranets to maintain an active cache of the most-often visited Web sites so that when these pages are requested again, the download occurs from the locally maintained cache server instead of the request being routed to the actual server. The result is a faster download speed.

Caches can reside at various points in the network. For ISPs and backbone providers, caches can be deployed in practically every Point of Presence (POP). For enterprises, caches can be deployed on servers throughout campus networks and in remote and branch offices. Within enterprise networks, caches are on the way to becoming as ubiquitous as IP routers. Just about every large company now depends on Web caches to keep their intranets running smoothly.

Types of Caching

There are two types of caching: passive and active. With the former, the cache waits until a user requests the object again, then sends a refresh request to the server. If the object has not changed, the cached object is served to the requesting user. If the object has changed, the cache retrieves the new object and serves it to the requesting user. However, this approach forces the end user to wait for the refresh request, which can take as long as the object retrieval itself. It also consumes bandwidth for unnecessary refresh requests.

With active caching, the cache performs the refresh request before the next user request—if the object is likely to be requested again and the object is likely to have changed on the server. This automatic and selective approach keeps the cache up to date so the next end user request can be served immediately. Network traffic does not increase because an object in cache is refreshed only if it is likely to be requested again, and only if there is a statistically high probability that it has changed on the source server.

For example, the Web page of a major broadcast network might contain a logo object that never changes, while the "Breaking News" object changes often. If this page is popular among corporate users, the "Breaking News" object will be refreshed prior to the next user's request. In refreshing only content that is likely to change, users are served with the most updated information without putting unnecessary traffic on the network.

By contrast, previous generations of cache technology do not accommodate the individual nature of cached objects. They rely on global settings which treat all objects equally, thereby severely limiting the hit ratio. Since the passive cache requires frequent, redundant refresh traffic, it induces significant response time delays.

Active caches can achieve hit ratios of up to 75 percent, meaning a greater percentage of user requests can be served by the cache. If the requested data is in the cache and up to date, the cache can serve it to the user immediately upon request. If not, the user must wait while the cache retrieves the requested data from the network. Passive caches, on the other hand, typically achieve hit rates of only 30 percent. This means users are forced to go to the network 2.5 times more often to get the information they need.

Some kinds of objects in a Web page cannot be cached, and are individually marked by their Web server as such. One object of this type is a database-driven object, such as a real-time stock quote. While this particular object is not cacheable, the rest of the objects in the page usually are. For example, a Web page that delivers stock quotes may contain 30 other objects; only one of those objects—the stock ticker—may not be cacheable. If all of the remaining objects can be cached, a significant performance benefit will result.

Cache Warehouse

Canadian Telecommunications provider Teleglobe has teamed up with satellite operator Intelsat to push frequently accessed Internet content to multiple sites around the world over a satellite network. The idea is to make it easier for users to pull data from sites on the Internet in countries where bandwidth is in short supply.

Teleglobe plans to operate a "warehouse" of this cached content at one of its existing Intelsat base stations. In turn, Teleglobe intends to sell the service to ISPs around the world, which will operate "kiosks" on their own servers that store incoming data pushed from the warehouse. In the future, the warehouse itself will be able to receive pushed content—including music and sports events and real-time news.

In addition to selling its services to ISPs in developing countries, the satellite-based cache warehouse replaces a leased-line situation where an ISP would need to have a constant connection to the Internet. With Teleglobe's plan, ISPs can avoid leasing a line from the local telecom operator by installing a satellite receiver at their local hub. Because it is very expensive to lease lines in developing countries, a satellite set-up allows ISPs to save money and pass along these savings to their customers. With content cached locally in the country, users will not have to go out onto the Web and pull information from servers located around the world, which saves time and reduces network traffic.

SUMMARY ▬ ▬ ▬ ▬ ▬ ▬ ▬ ▬

Delay within the network (also known as *network latency*) is the primary reason why the Web is so slow for users. One way to make the Web provide faster response times is to bring the majority of accessed objects closer to users through active caching. Caching reduces or eliminates key elements of delay including:

- Transmission of a request from the end user to the original storage device
- Network congestion between the end user and the original storage device
- Congestion within the storage device itself
- Transmission of the data from the storage device back to the end user

In the future, there will be a global hierarchical cache structure. When a user in the U.S. requests a Web page from England, for example, the browser would examine the client cache. If the page is not there, it would then examine the network cache, then the regional U.S. cache, then the national U.S. cache, then the national U.K. cache, then the regional U.K. cache, then the network ISP cache. If the requested page is not found at any of these locations, the browser will access the server actually hosting the requested page. This hierarchical cache structure will help free up network bottlenecks and speed up the Internet for everyone. The lookups along the way would not create unnecessary network traffic because there would be a high statistical probability that the requested Web page would be found sooner rather than later.

See Also **Network Congestion, Latency**

Network Design

The scale and complexity of today's wide-area networks (WANs) requires a comprehensive tool suite that can address the numerous issues in network design and planning, covering technologies from time division multiplexing (TDM) to router, frame relay, and asynchronous transfer mode (ATM) networks.

With ATM, for example, the technology is a cost-effective option for carrying voice, data, and video across a WAN. However, network managers still need to address a variety of issues including:

- Finding the most effective way to aggregate disparate traffic types
- Balancing IP traffic, legacy data traffic, and voice requirements effectively
- Assessing the comparative value of network services pricing structures
- Retaining control of information in a virtual private network (VPN)
- Creating traffic scenarios to anticipate the needs for increased traffic and network growth

Network managers have many choices, ranging from technologies to service providers. The difficulty is making the right choice for reliability, efficiency, and growth potential. An effective modeling tool can take the guesswork out of the wide range of alternatives in the market today and greatly improve the network planning process. For example, the design tool can help network managers explore various configuration scenarios; analyze the impact of any number of failures; and check if the network has enough capacity to carry new traffic, optimally adding new trunks or deleting unneeded trunks. This modeling can be done prior to product and service acquisition so there are no surprises upon implementation.

As new carriers emerge and service offerings change, networks should be adaptable enough to take advantage of new opportunities. In addition, services are not currently available in a uniform fashion throughout the world. An effective modeling tool helps network managers stay on top of these changes. A hardware platform equipped with multiple trunk-side interfaces prevents organizations from being locked into a single technology platform if tariffs for a particular service change. For example, if frame relay makes more economic sense in a particular area than ATM, the network design tool should be able to help establish configuration recommendations that can take advantage of the best available option.

When it comes to voice traffic, the network design tool should have a module for optimal voice traffic routing design based on available capacities in the backbone network, optimal off-net routing design, and backbone design evaluation for voice traffic.

A comprehensive set of network design tools—preferably available as modules that can be purchased separately as needed—should have the following overall capabilities:

- U.S. domestic and international network links—an entire network or any pair of locations—pricing based on existing tariffs for the particular transmission line type used

- Access-network designs for frame relay—including pricing with customizable tariffs—ranging from simple homing of location to backbone multiplexer sites to secondary concentration homing designs

- Optimal topological backbone design—incorporating any mix of TDM multiplexers, routers, frame relay switches, and ATM switches—with incremental design capabilities

- Diverse designs to survive any single-node or trunk failure

- Failure simulations with customizable disaster scenarios to test network resilience

- Packet-level discrete event simulator for estimating such things as network queuing delay, throughput, and cell drop percentage

- Multidrop line design for determining the optimal locations of terminal cluster controllers

- Primary and secondary synchronization timing references assigned for every node in the network, including synchronization of multiple timing references ranked according to diversity, number of hops, and the actual distance traveled between the node and the external timing sources

- Hardware device libraries that accurately model vendor-specific TDM, router, frame relay, and ATM switches

- Network analysis capabilities to study bandwidth usage and detect network bottlenecks before any changes are made to the real network

- Generation of detailed reports concerning link cost, link usage, traffic routing, failure scenarios, design configurations, and bill of materials

- Import/export of files and reports to word processors and spreadsheets

Windows-based tools, some supporting Java, make network design relatively simple. The design process usually starts by opening a blank drawing window from within the design tool into which various network devices—multiplexers, switches, routers—can be dragged from a product library and dropped into place (Figure 89). The devices are further defined by type of components, software and protocols as appropriate. By drawing lines, the devices are linked to form a network, with each link assigned physical and logical attributes.

Rapid prototyping is aided by the ability to copy objects—devices, LAN segments, network nodes, and subnets—from one drawing to the

Figure 89

Along with network maps, device library, and device configuration windows, NetCracker Technology, a division of Advanced Visual Data, offers traffic animation features that portray a network's operation for a more realistic view of how a network design performs under a wide range of scenarios.

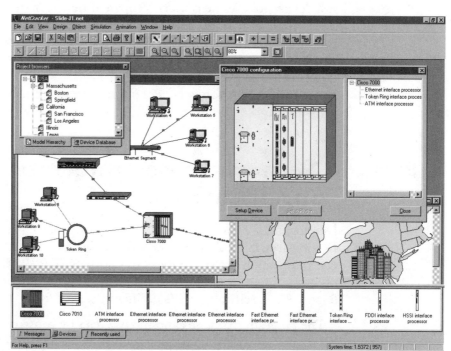

next, editing as necessary, until the entire network is built. Along the way, various simulations can be run to test the integrity of virtually any aspect of the design.

Some design products even include multimedia features such as dynamic views, which enable users to see animations of network traffic flow through the design. The user can see the impact to the network when a link is broken, when the network traffic is redirected, or when the link is restored and the traffic is reinstated to its original route. As this process takes place, the designer can view statistics, like bar graphs, charts and numbers within a new configuration in real time. Other multimedia features include voice-synthesized devices and notes, which can be embedded into the design as reference aids.

SUMMARY

Many of the issues in the design, capacity planning, and management of wide-area networks are so complex that they are intractable without effective tools that automate various processes and perform tests on network elements to ensure proper configuration and operation. Such tools can also integrate LANs and telephony services within a single design. Extensive device libraries help users compare different networking options, including specific hubs, routers, and switches from different suppliers. Increasingly, design tools are including capabilities for planning intranets and LANs that interface with the Internet, even automating IP address numbering and assignment.

See Also **LAN Analyzers, Performance Baselining, Performance Monitoring**

Network Driver Interface Specification

Developed by Microsoft and 3Com, the Network Driver Interface Specification (NDIS) is a Windows software interface that determines how communication protocol programs—such as TCP/IP, IPX, AppleTalk and others—and network device drivers should communicate with each other (Figure 90). Specifically, NDIS defines interfaces for:

■ The program that sends and receives data by constructing or extracting it from the frames, packets, or datagrams. This program,

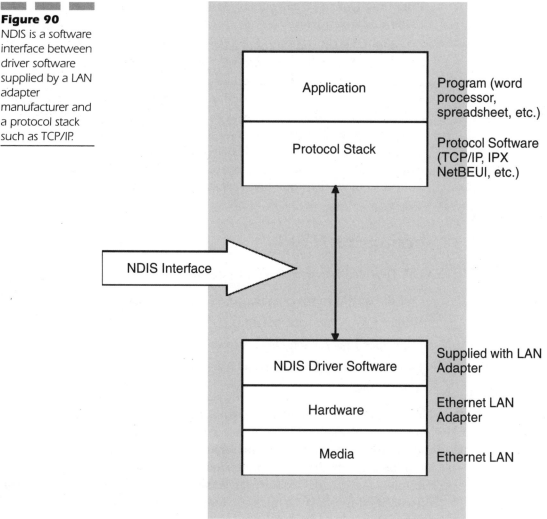

usually called a protocol stack, is layered and generally corresponds to Layers 3 and 4—the Network and Transport layers, respectively— of the Open Systems Interconnection (OSI) reference model. Examples of protocol stacks are TCP/IP and IPX.

- The program, usually called a device driver, that interacts directly with the network interface card (NIC) or other adapter hardware, which sends or receives the data on the communications line in the form of electronic signals. The driver program and the NIC interact at the Media Access Control (MAC) sublayer of OSI's Layer 2, called

Data-Link Control. Examples of MAC drivers are those for Ethernet, FDDI, and token ring.

- A program called the Protocol Manager that assists the protocol stack and MAC driver programs by informing them of the computer's location when its operating system is started or, in some cases, when a new device is added to the computer. A system file called PROTOCOL.INI identifies which protocol stacks use which MAC drivers and where each is located. A protocol stack can be bound to more than one MAC driver where a computer is connected to multiple networks, and a single MAC driver can be bound to more than one protocol stack in a computer.

Versions of NDIS

NDIS 4.0 added new features and extensions to NDIS 3.1, including:

- Out-of-band data support (required for Broadcast PC)
- Wireless WAN media extension
- High-speed packet send and receive
- Fast infrared media extension
- Media sense

The latest version of NDIS is 5.0, which is intended to increase ease of use; improve performance; enable new media types, services, and applications; and improve flexibility in the driver architecture. In the process, NDIS 5.0 introduces a new device model called the miniport driver, which facilitates plug-and-play device features. Version 5.0 consists of all the functionality defined in NDIS 4.0, plus the following extensions:

- NDIS power management, required for network power management and network wake up
- Plug and Play is now applicable to Windows 2000 network drivers
- Windows hardware instrumentation support for structured, cross-platform management of NDIS miniports and their associated adapters
- Simplified network information file (INF) format across Windows operating systems, based on the Windows 95 INF format
- Deserialized miniport for improved performance on Windows 2000 multi-processor systems

- New mechanisms for off-loading tasks such as TCP/IP checksum, IP Security, TCP message segmentation, and Fast Packet Forwarding to intelligent hardware
- Broadcast media extension, required for broadcast components
- Connection-oriented NDIS, required for native access to connection-oriented media such as ATM (including ATM/ADSL and ATM/cable modem) and ISDN support for Quality of Service (QoS) when supported by the media
- Intermediate driver support, which is required for broadcast components, virtual LANs, LAN emulation over new media (including ATM, satellite or broadcast television), packet scheduling for QoS, and NDIS support over high-speed buses such as FireWire and Universal Serial Bus (USB)

The new NDIS architecture is included in Windows 98 and Windows 2000 operating systems.

SUMMARY

Using NDIS, Windows software developers can develop protocol stacks that work with the MAC driver for any hardware manufacturer's communications adapter. At the same time, any adapter maker can write MAC driver software that can communicate with any protocol stack program. A similar interface, called Open Data-Link Interface (ODI), is provided by Novell for its NetWare local-area network operating system.
***See Also* Open Data-Link Interface**

Network File System

Originally developed by Sun Microsystems, the Network File System (NFS) is designed to give users high performance and transparent access to server file systems on global networks. NFS is the de-facto distributed file system in the UNIX community and is defined in several Internet standards. Over the years, NFS has been enhanced to support file sharing on heterogeneous systems, from PCs to mainframes. The source code is available for licensing from SunSoft and a written specification can be obtained freely via file transfer from several locations on the Internet.

Features

Since its first introduction in 1985, NFS has continued to evolve to meet the distributed file sharing requirements for the global enterprise in the 1990s and beyond, and is supported by hundreds of vendors. Some of the most important features provided by NFS include:

TRANSPARENT ACCESS Users and applications can access remote files as if they were local. They are not required to know whether the files reside on the local disk or on remote servers.

PORTABILITY Since NFS is independent of both machine and operating system, it can be easily ported to multiple OS and hardware platforms, from PCs to mainframes.

FAULT TOLERANCE NFS is designed to recover quickly from system failures and network problems, causing minimal disruption of service to users.

PROTOCOL INDEPENDENCE NFS has the flexibility to run on multiple transport protocols instead of being restricted to just one. This allows it to use existing protocols today as well as new protocols in the future.

PERFORMANCE NFS is designed for high performance so users can access remote files as quickly as they can access local files.

SECURITY The NFS architecture enables the use of multiple security mechanisms, allowing system administrators to choose the security mechanism that is appropriate for their distributed file sharing environment instead of being restricted to one solution.

These features are implemented within the NFS client/server framework, which reduces costs by enabling heterogeneous resource sharing across the global enterprise. Servers make their file systems sharable through a process called exporting. Clients gain access to these file systems by adding them to their local file "tree" via a process called mounting (Figure 91). The NFS protocol provides the medium for communication between client and server processes over the network.

Global File Access

NFS supports global workgroups by giving users transparent, fast access to server file systems connected to client systems on LANs and WANs. The following capabilities enable efficient global file access:

- The automounter keeps files continuously accessible by automatically mounting file systems on servers located worldwide.

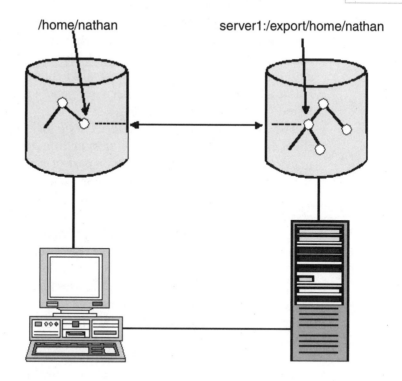

Figure 91
Through a process called mounting, a remote file system is incorporated into the client's file system tree.

- Client side caching gives clients fast access to file data transferred from the remote server and stored locally.
- File locking solves the problems posed by concurrent file access. NFS includes a file locking feature called the lock manager, which works with another process called the status monitor to guarantee that multiple readers and writers do not collide with each other.
- To exchange information between the client and the server, NFS uses efficient transport protocols such as the Transmission Control Protocol (TCP).

The automounter gives clients transparent access to server file systems, whether those servers are connected to clients over LANs or WANs. The automounter does not distinguish between different physical network technologies—it simply adds the server file system to the client file tree via the mount process, making the global network appear as if it is a single file system resource. The transparency offered by the automounter in combination with high performance file access gives global workgroups fast, easy access to file information.

Client-side caching overcomes the performance limitations of WANs. Typically, WANs are comprised of low-bandwidth connections and are prone to errors requiring retransmissions of data. This can result in significant data latency, which reduces throughput. NFS clients can offset data latency by utilizing local disk caching. By caching significant amounts of file data on the local disk, clients can greatly reduce the amount of time they would otherwise spend waiting for data to be transferred from the remote server.

The lock manager provides UNIX style locking between cooperating client and server processes, enabling synchronized access to files. The lock manager lets a process lock a file or part of a file for shared or exclusive access. When a process locks a file for exclusive access, no other process will be allowed to share the lock. When a process locks a file for shared access, other processes can share the lock but they will not be able to obtain an exclusive lock. The status monitor provides the lock manager with information on host status. It monitors the system to ensure that locks will be handled properly if a system crashes while a file is locked. If the server crashes with monitored locks in place, the locks will be reinstated when the server recovers; if the client crashes with monitored locks in place, the locks will be automatically freed by the server and must be reinstated by the client when it recovers.

Data exchange is done efficiently using TCP. Because NFS is a Layer 7 service, it relies upon lower layer transport protocols to transmit data over the network. NFS was originally designed to use the User Datagram Protocol (UDP) as the transport protocol for communication over both local and wide-area networks. UDP was chosen because in the past, it provided better performance than TCP. Although UDP works very well for LANs, it has some limitations when used for communication over WANs. One drawback is that it provides data delivery on a best-efforts basis and does not perform essential tasks such as retransmissions—it relies on the LAN protocols for these functions.

Now that there are high-performance TCP implementations available, NFS has been enhanced to use TCP. Unlike UDP, TCP is a reliable protocol that is designed to provide guaranteed data delivery. It independently performs tasks such as dynamic retransmission, packet sequencing, congestion control and error recovery, freeing NFS from having to cope with these problems. The result is an increase in the performance of NFS clients.

Because there is more overhead required to set up an initial connection using TCP than there is using UDP, all NFS client traffic can be multiplexed over one TCP connection to the server. In other words, NFS clients use one connection to each server, regardless of the number of client mounts per server. This keeps overhead to a minimum, uses fewer resources (sockets, descriptors, etc.), and makes recovery from failure faster. It also allows servers to scale to support larger numbers of NFS clients.

NFS Server

The role of the NFS server is to allow its disk file systems to be accessed by other clients on the network. The process whereby a server makes its file systems available to be shared is called exporting, and the directories that are made available are referred to as exported file systems. In order to export a file system, the system administrator must edit a configuration file and specify:

- The full pathname of the directory to be exported
- The client machines that will have access to the exported directory
- Any access restrictions

At system start up time, a system dependent program is invoked that opens and reads the information in this configuration file. This program then informs the server's operating system kernel about the permissions applicable to each exported file hierarchy. Special background processes (mountd and nfsd) are invoked and wait to receive client NFS requests to access the server's file systems. After this process is complete, the server is ready to accept client requests to access its file systems.

NFS Client

NFS clients gain access to server files by mounting the servers' exported file systems. The mount process results in integrating the remote file system into the client's file system tree. An enhanced client side service called the automounter automatically and transparently mounts and unmounts file systems on an as needed basis. From the user perspective, this creates an environment where server file systems are continuously accessible.

In order for the automounter to mount a file system it needs maps that tell it what to do. A map associates a client name (pathname) with a file system that resides on a remote server. These client names are also called mounts or mount points because they represent the points at which server file systems will be mounted. Collectively, these mount points represent the file system namespace, or the names clients use to access server file systems. After the mount process is complete, users on the client system will be able to access files on the designated server(s). To clients running MS-DOS or Windows, NFS mounts look like additional drives.

SUMMARY

NFS satisfies enterprise requirements for global file sharing by keeping distributed file systems continuously and transparently accessible to users. NFS provides fast access to file information as well as the scalability to support small to large network environments. The ability to administer NFS centrally reduces the time and effort it takes to perform a variety of routine setup and maintenance tasks and its flexible and extensible security architecture enables administrators to choose the security solution that fits their current and future needs.

See Also **Client-Server, Network Caching, UNIX**

Network Interface Cards

A network interface card (NIC) is an adapter that plugs into a computer, enabling it to communicate with other computers and devices over a local-area network (LAN). The NICs are network-specific—there are adapters for Ethernet, token ring, FDDI, ATM and other types of networks. NICs are also media-specific—there are adapters for shielded and unshielded twisted pair wiring, thick and thin coaxial cabling, and single mode and multimode optical fiber. NICs are also bus-specific—there are adapters for the Industry Standard Architecture (ISA), Extension to Industry Standard Architecture (EISA), Micro-Channel Architecture (MCA), and Peripheral Component Interconnect (PCI) architecture, among others. NICs are also available in the PC Card form factor for connecting mobile notebook users to the LAN. Major vendors offer software that enhances the capabilities of their NICs.

Client NICs

Client NICs provide the means to connect desktop computers, printers and other devices to the LAN. Today's Ethernet and token ring NICs have an auto-sensing capability that allows the NIC, when connected to a switch or hub port, to automatically sense and connect at the highest network speed. By simultaneously performing multiple processing tasks, some NICs provide the fastest data transfer speeds available for the PCI bus. Link activity and network status LEDs provide a convenient at-a-glance indication of network health and status.

NICs that feature 32-bit multimaster concurrency technology permit the card to communicate directly with the computer's CPU, bypassing sluggish interrupts and I/O channels. NICs that feature an onboard boot ROM socket allow for remote workstation bootup from the server. LEDs on the card report link status, packet activity, transmission speed, and transmission mode (half- or full-duplex).

Many NICs are optimized to work in Windows environments—specifically PCs running Windows 95, 98, or NT—and are compliant with Plug and Play. The installation software allows connection of the PC to Novell NetWare networks as well. Windows-based diagnostics and configuration utilities facilitate installation and troubleshooting.

Server NICs

Server NICs include the functions of client NICs, but have additional functionality and provide higher bandwidth. For example, NICs may be configured in a way to increase the fault tolerance of the server's LAN link. If one NIC fails, the fail-over software deactivates the faulty NIC and switches LAN traffic to an alternate card. The rerouting takes place almost instantaneously, without human intervention. The software also gives Simple Network Management Protocol (SNMP) alerts on the failed NICs. When the failed NIC begins working again, the software brings it back into the array automatically and starts balancing traffic across it again.

Many server NICs provide asymmetric port aggregation—also referred to as asymmetric load balancing or asymmetric trunking. This technology distributes outbound server traffic between two or more cards, providing a wider data pipe. The NICs operate together and appear as a single device with one network address. Asymmetric port aggregation is especially useful for Web servers, e-mail servers, and other applications where most of the traffic flows in one direction from the server to the client PCs. This method also provides fault tolerance in that if one of the NICs fails, the others take its load.

A companion technology is symmetric port aggregation (or symmetric load balancing or symmetric trunking). This method combines two or more connections into a wider pipe that can transmit data in both directions. Since combining several connections does not require replacing hubs or switches, it can put off having to invest in upgrades to gigabit LAN technologies. And it, too, provides fault tolerance.

A dual homing capability allows the server NIC to connect to different switches for additional redundancy, ensuring the server remains available even if one of the attached switches fails.

Some NICs provide the means to implement Virtual LANs (VLANs). This IP-based technology lets a single server link—one NIC or a team of NICs—carry traffic for up to 64 logical subgroups created via software. This capability provides additional bandwidth management and security features and helps reduce the administrative overhead required to manage workstation moves and changes.

There are now server NICs that allow token-ring traffic to run over a 100-Mbps Fast Ethernet backbone connection, eliminating congestion across the token-ring network backbone without the costs associated with ATM and FDDI. The NIC tunnels token-ring traffic in Fast Ethernet frames, delivering high-speed performance to token-ring clients. When installed in a Fast Ethernet server, the NIC allows token ring clients to communicate with the server at Fast Ethernet speeds via a special module installed in the token-ring switch.

SUMMARY

NICs are used to connect computers and other devices to the LAN. Although many users can get by with inexpensive "dumb" NICs, enterprise networks require "intelligent" NICs that ensure high availability to support mission-critical applications and management software to facilitate monitoring and control from a central location.

See Also **Media Converters, Transceivers**

Network Monitoring

Network monitoring tools identify usage trends, track overall network throughput, and monitor the performance of various systems, including disk and cache performance. When a problem occurs—such as a disconnected or broken cable, a hung server application, or failing disk drive—the system monitoring tool uses alarms to alert the LAN administrator.

Some network monitoring products feature an event manager agent to track network events, log network activity, and automatically alert the appropriate person responsible for responding to certain network occurrences. The following network occurrences are considered events of interest to the LAN administrator:

- Running jobs, such as network backup or a virus scan
- Recording the status of completed jobs
- Recording changes to the hardware inventory
- Logging in and out of the network
- Accessing applications
- Starting programs (successfully or unsuccessfully)

The LAN administrator can specify the network activity to be tracked, such as the times when users log in and out of the network or when certain programs are run. Network activities that may require immediate attention also can be specified. A notification feature can be set to alert the LAN administrator of the times when these events occur. The following methods of event notification typically are available:

CONSOLE MESSAGES Text messages display the name of the event and color-coded views indicate the priority level of the event. Some products use event icons to display events.

E-MAIL MESSAGES The event level and name is sent in an electronic mail message.

PAGER MESSAGES A phone number, the event name, or both, is sent to a pager.
Different notification methods can be set as appropriate for each network event. Event notifications can be processed based on priority level. For example, if three network events occur simultaneously, notification of the event with highest priority is sent first. A priority level may be a number from 1 to 9—with 1 indicating the lowest and 9 indicating the highest priority. When the administrator specifies the network activity to monitor, a priority level for each event is assigned based on how critical the activity is and whether or not someone has to be notified when the event occurs.

The administrator can choose one or more contacts to receive notification of each event level. For example, a technician can be designated to receive a pager message when high-priority events occur, and that an e-mail message be sent to a help desk operator when routine application-related events occur. An acknowledgment of receipt for event data can be sent to the console to help ensure a proper response to events by the administrator.

Some network monitoring tools use distributed, intelligent agents to gather protocol and network activity data on Ethernet and token-ring LANs. The data gathered by the agents is stored in a relational database where it is correlated for traffic analysis, billing, and report generation. With the ability to identify traffic loads, including which nodes are generating the most traffic, the resulting information can be used to charge departments for their share of the resources, including dial-out connections.

Even if such information is not used for charge-back, the monitoring tool can still be used to reduce costs and help determine policies for more efficient network use. In addition, monitoring the network for predefined traffic thresholds on a particular LAN segment gives administrators the means to identify traffic patterns that could cause the network to crash. Traffic reports can even identify the need to change the network. If too many users on one or more network segments are logging in to different servers or using resources in another building, for example, a lot of backbone traffic can be created. With the aid of traffic reports, the network can be redesigned to alleviate backbone traffic and make sure bottlenecks do not occur.

Diagnostic Tools

Many of the capabilities of stand-alone cable testers and protocol analyzers are now available as software for use by LAN administrators from a management station. Some can even be accessed over the Web from a Java-enabled browser. Included in these diagnostic tools are the capabilities to:

- Predict bottlenecks before they cause failures
- Measure LAN and WAN response time
- Determine LAN and WAN uptime
- Measure LAN segment bandwidth usage
- Trace traffic paths throughout the network
- Simulate/model impact of new applications, as well as moves, adds and changes

Some vendors offer a suite tools that allow administrators to monitor, measure, test, and diagnose performance anywhere on the entire network using a simple Web browser interface. LANQuest's NETClarity suite, for example, offers the following diagnostic tools:

NETWORK CHECKER Provides transit time, path check, remote trace route, and link integrity monitoring.

REMOTE ANALYZER PROBE Empowers up to 50 agents to capture traffic on their LAN segments. These software agents then return captured traffic traces to the NETClarity server where it can be forwarded to any Network Associates' Sniffer protocol analyzer anywhere on the network or kept for later analysis.

For performance measuring and reporting solutions, LANQuest offers:

LOAD BALANCER Measures used and available capacity on up to 50 network segments. Its graphical reports show where servers or users can be moved from one LAN segment to another without having to buy more equipment. If equipment is needed, it will help determine the minimum requirements for cost savings.

SERVICE LEVEL MANAGER Collects key RMON data and measures network response time and uptime/availability for the entire network, providing an end-user perspective on network performance. Extensive analysis and summary reports document daily/weekly/monthly uptime and performance. Response time, usage, and uptime/availability reports provide baselines, trends, and snapshots for the entire network's performance.

For modeling and simulation, LANQuest offers a Capacity Planner. This application uses flexible traffic emulation, application simulation and network monitoring tools to measure the impact of new network applications and changes to avoid crisis management. It even predicts the need to add LAN or WAN capacity to maintain desired response time and bandwidth usage, and point out where bottlenecks are most likely to occur in the future.

SUMMARY ▬ ▬ ▬ ▬ ▬ ▬ ▬

As networks evolve, network monitoring tools with more sophisticated capabilities are required. The current trend is toward seamless integration of multiple tools that provide transparent support to heterogeneous enterprise networks. Despite the growing popularity of network monitoring suites, some still neglect features LAN administrators need. These gaps are filled in by third-party application developers. For organizations committed to one of the big three management platforms—Hewlett-Packard's OpenView, IBM/Tivoli's TME, and Sun's Solstice Enterprise Manager—there is less risk in choosing third-party solutions. The products of these third-party vendors must pass a rigorous interoperability test program before they can be authorized for sale under the cooperative marketing programs of the platform vendors.

See Also **Cable Testing, Cache Management, Intelligent Agents, LAN Analyzers, LAN Troubleshooting, Performance Monitoring, Ping, Remote Monitoring**

Network Reliability

There are several basic principles involved in ensuring network reliability and achieving prompt restoral of traffic for both switched services and private lines in the event of an outage. These principles include prevention, redundancy, and restoral. A fourth principle for ensuring network reliability is repair of damaged facilities and/or equipment. Repair often plays a role in achieving network restoral.

Prevention

Prevention is the best way to avoid a network problem. Carriers generally build reliability into their networks and monitor performance on a continuing basis. With regard to data traffic, when certain performance thresholds such as bite error rate, throughput and latency are exceeded, automated processes perform such functions as rerouting traffic, activating redundant systems, performing diagnostics, isolating the cause of problems, generating trouble tickets and work orders, and dispatching repair technicians if necessary. Similar processes are triggered for voice traffic when performance thresholds such as call completion rate and dropped calls rate are exceeded.

The objective of all this activity is to maintain existing traffic flows while the problem is being fixed and to return primary routes and systems to service as soon as possible. More often than not, the carriers are able to identify and fix problems without customers ever knowing that anything out of the ordinary has occurred.

Redundancy

Many parts of carrier networks are designed with built-in redundancy, meaning that there is a duplicate or backup system immediately available that will take over the operations of a failing system. As in prevention, the objective of network redundancy is to maintain existing traffic flows while the problem is being fixed and to return primary routes and systems to service as soon as possible. AT&T, for example, has built redundancy into its network in many ways, including:

DUAL-PROCESSOR SWITCHES Central office switches are equipped with dual processors, so that if one processor fails, the second one can take over automatically. In essence, the switch—which can be viewed as a large computer—is really two computers running simultaneously, with the backup ready to take over instantly if a problem inhibits the proper operation of the primary processor.

SIGNAL TRANSFER POINTS STPs are mated pairs of dual-processor computers that route network inquiries over the signaling network—the separate data packet-switched network used to set up calls and create intelligent services. The STP pairs are not co-located, but are usually hundreds of miles away from each other, and operate at just under 50 percent capacity. With this architecture, if something happens to one STP, its mate can pick up the full load and operate until repair or replacement of the damaged STP can be made.

NETWORK CONTROL POINTS NCPs are the data bases that store the configuration details for advanced services such as toll-free or software-defined networks. They not only have dual processors, but also a backup NCP for the protection of customers' intelligent services information if the second processor should fail.

DIGITAL INTERFACE FRAMES DIFs provide access to and from the central office switches for processing calls. The digital interface units that actually handle this work have spare units available to take over immediately should there be a problem. Guiding the overall work of the DIF are two controllers running simultaneously, so that if one experiences a problem, the backup controller can take over without the customer noticing that a problem ever occurred.

POWER SYSTEMS Power systems provide direct current from redundant rectifiers fed by commercial electric power. If commercial power fails, batteries, which are kept charged by the rectifiers, provide backup power. An additional level of redundancy is provided by diesel oil powered generators, which can replace commercial power for longer periods.

NETWORK DIVERSITY Diversity is the concept of providing as many alternatives as possible to ensure survival of communications when some kind of natural or man-made disaster strikes. It includes cable and building diversity, and triversity of signaling routes, in which each pair of STPs is connected to every other pair of STPs by multiple data communications links. To ensure that connectivity will always be available, these links are deliberately sent through three geographically separated routes. Should something happen to one route, the others remain available to keep the signaling system operational.

ALTERNATE SIGNALING TRANSPORT NETWORK Within each switch there is a device that interfaces with the common channel signaling system to convey information used to set up and deliver calls. Should this interface device malfunction, the switch can use special signaling links (data circuits) that are directly connected to one or more "helper" switches to gain access to the signaling network via their interface. In this manner, the switch can continue to process long distance calls while a repair is made.

Restoral

When a problem occurs within the network there are three distinct means to restore service.

- Use currently available network capacity, usually through a third switch, to route traffic around a problem on normally available network capacity

- Bring into service specially reserved "protection" or backup facilities that are kept out of normal network usage for such emergency restoral purposes

- Repair the damaged equipment or facilities by splicing a severed fiber optic cable, for example, or replace a malfunctioning or damaged system component

On a day-to-day basis, making use of capacity that exists within the network—almost always by routing a call through a third switch for completion—is a highly effective means of balancing temporary high calling volume with readily available capacity elsewhere in the network. This is "switching restoral," essentially a traffic management capability.

When there is a cable cut, it is necessary to bring into service spare facilities called "protection" facilities. These are kept idle for just such situations, when a substitute route must be used to accommodate large amounts of traffic. Moving traffic onto spare facilities is called "facilities restoral."

The third type of restoral is to repair what is broken. For example, repairing a fiber-optic cable and allowing it to carry its normal load resolves the problem itself. The first two types of restoral are designed to help in the interim while the actual repair is taking place.

It is not necessary to have 100 percent restoral—that is, one circuit replaced for every circuit lost—to put a network back into service. By using a combination of switching and facilities restoral along with ongoing repair, enough capacity can be accessed to complete the great majority, if not all, customer calls in advance of the completed repair, which is known as "physical restoral."

SUMMARY

Although local and long-distance carriers build reliability into their networks at the design stage and monitor performance of the network on a continuing basis, there is always the chance that unforeseen problems will occur. Automated processes perform such functions as raise alarms, reroute traffic, activate redundant systems, perform diagnostics, isolate the cause of the problem, generate trouble tickets and work orders, dispatch repair technicians—all with the objective of returning primary facilities and systems to their original service configuration and level of performance. While many protection and restoral services are standard with voice and data services, others are extra cost options designed to meet the specific needs of customers.

See Also **Network Availability, Network Congestion, Network Latency**

Network Reliability and Interoperability Council

The Network Reliability and Interoperability Council (NRIC) is the successor to the Network Reliability Council (NRC) that was first organized by the FCC in January 1992. The Council was established following a series of major service outages in various local exchange and interexchange wireline telephone networks. Its job was to study the causes of the outages and develop recommendations to reduce their number and their effects on consumers.

The Council is composed of CEO-level representatives of about 35 carriers, equipment manufacturers, state regulators, and large and small consumers. Under its initial charter, the Council commissioned studies in the areas where it believed reliability concerns to be greatest—signaling (i.e., SS7), fiber cuts, switching systems, power failures, fires, 911 outages, and digital cross-connect systems. From time to time, the FCC has requested the assistance of the Council on other issues affecting the national telecommunications infrastructure.

For example, upon passage of the Telecommunications Act of 1996, the FCC asked its Network Reliability and Interoperability Council for recommendations on what should be done to implement Section 256 of the Act, which has two key provisions:

- To promote nondiscriminatory access by the broadest number of users and vendors of communications products and services to public telecommunications networks
- To ensure that users and information providers have the ability to seamlessly and transparently transmit and receive information between and across public telecommunications networks

At this writing, the NRIC is looking into the "Year 2000" (Y2K) problem by determining how it could adversely affect the reliability, interconnectivity and interoperability of public telecommunications networks. The Council is assessing the magnitude of risks, reviewing efforts taken to address those risks, and determining what additional steps should be taken to mitigate risks. In particular, the NRIC is making sure that appropriate internetwork testing and network monitoring have been arranged by the carriers.

Many of NRIC's recommendations relate to the voluntary standards processes that are used today to achieve key interoperability and reliability objectives in telecommunications. For example, it has recommended to the FCC that it monitor ANSI (American National Standards Institute) accredited and other open, consensus-based telecommunications standards developers to ensure that they support interoperability of national services and products.

SUMMARY

The NRIC is a federal advisory committee that reports to the FCC Chairman and to the FCC Defense Commissioner. It provides an opportunity for a broad array of players—including manufacturers, users, and others in the industry—to work with the Commission in developing industry-wide assessments, to coordinate testing and look for ways to share results, and to develop interconnecting contingency plans.

See Also **Reportable Incidents, Year 2000 Compliance**

Next-Generation Networks

Next-generation networks provide integrated voice-data services over an IP-based fiber-optic infrastructure. The services are aimed at businesses and consumers, and are more attractively priced than the comparable services of incumbent interexchange carriers, such as AT&T, MCI World-Com, and Sprint. Among the growing number of "nextgen" carriers are Frontier, Level 3 Communications, and Qwest LCI.

Nextgen carriers offer a variety of services, but they usually focus on those that have the most market potential. Examples include IP telephony, facsimile and VPNs. The use of IP for these and other services makes global distribution more efficient and economical than using traditional public switched telephone networks, while permitting their integration with legacy data and applications.

Importance of IP

IP is the underlying protocol for routing packets over the Internet and other TCP/IP-based networks. Within the context of next-generation networks, IP is important because it is non-proprietary, open and offers efficient, cost-effective ways to merge voice and data traffic on a common platform. Although there are other protocols that offer compelling advantages of their own—notably, ATM—none are capable of matching IP in terms of economy, efficiency and global reach.

Shared IP networks also are more efficient than circuit-switched PSTNs, which rely on dedicated connections that are set up between endpoints. These connections are idle much of the time. During a typical voice conversation, as much as 40 percent of the time is silence. In an IP network, packets are sent out onto the network only when there is voice to be conveyed. This frees the network to handle much more traffic over the available bandwidth. The packetization scheme allows the network to handle any mix of voice and data over the same links.

An IP-based infrastructure also meets the requirements for interoperability and integration, scalability, reliability, mediation, manageability, security, and global reach. Although the public switched telephone network (PSTN) comes close to meeting many of these requirements, an IP-based next-generation network takes these requirements to a higher level—and at only 20 percent of the cost of a circuit-switched network of comparable capacity.

Cost savings on network infrastructure is accrued in a number of ways. Since many nextgen service providers own the rights of way for their fiber routes, they save on permit fees. Since they are building their networks from scratch, they can deploy the latest Dense Wave Division Multiplexing (DWDM) equipment on new, higher-quality optical fiber. This combination results in far more network capacity than incumbent carriers have, and at a fraction of the cost to build.

Tremendous cost savings is achieved by using routers for moving traffic across the network, rather than the central office switches used by local telephone companies and interexchange carriers. Routers are much less complex than switches, yet can be as effective in handling real-time applications such as voice over an IP-based fiber optic network, especially one that is specifically designed and managed to minimize end-to-end delay.

SUMMARY

The service portfolios of nextgen carriers are still quite limited, compared to the incumbent interexchange carriers, such as AT&T, MCI WorldCom and Sprint. Many nextgen carriers resell excess fiber capacity that can be used for leased lines. Some offer frame relay, Internet access, Web hosting, VPNs, and IP fax. Others offer a different portfolio, which includes IP telephony. In addition to a broader range of services, the incumbent carriers have a strong track record when it comes to customer support. Nextgen service providers are still a relatively new element in the competitive telecommunications marketplace and customer support, which is taken for granted elsewhere, may not be available from the challengers for many years to come.

See Also **All-Optical Networks, Asynchronous Transfer Mode, Internet Protocol, Synchronous Optical Network, Virtual Private Networks, Voice-Data Integration**

Open Data-Link Interface

The Open Data-Link Interface (ODI) is a protocol-independent software interface for NetWare that provides similar functions to the Network Driver Interface Specification (NDIS) developed by Microsoft and 3Com for use on networked Windows computers.

Like NDIS, ODI operates at Layer 2 of the Open Systems Interconnection (OSI) reference model called Data-Link Control. It specifies how communication protocol programs—such as TCP/IP, IPX, AppleTalk and others—and network device drivers should communicate with each other (Figure 92). It enables multiple protocols to operate on the network simultaneously and provides the ability to install and support multiple types of network interface cards (NICs) in the same computer. As with NDIS, ODI consists of several discrete components:

MULTIPLE LINK INTERFACE The interface to which device drivers for the NIC are attached

LINK SUPPORT LAYER Provides a link for drivers, directing network traffic from the drivers to the proper protocol

MULTIPLE PROTOCOL INTERFACE Provides an interface for the connection of protocol stacks such as TCP/IP, IPX and AppleTalk.
When a packet arrives at a NIC, it is processed by the card's driver and passed to the Link Support Layer, where it is handed off to the appropriate protocol stack. The packet passes up through the protocol stack for higher-level processing before being transmitted over the network to its proper destination.

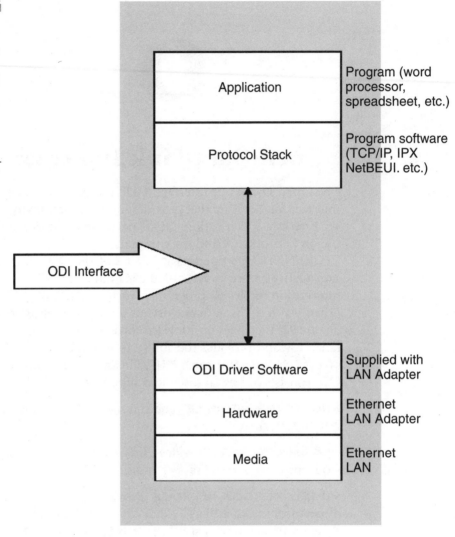

SUMMARY

Like NDIS, the NetWare ODI driver interface standardizes the development of NIC drivers so that vendors do not have to be concerned with writing separate drivers to work with each of the different network protocols.

See Also **Network Driver Interface Specification**

Outsourcing

Given the increasing complexity of today's information systems and communications networks, many companies are seeking ways to offload management responsibilities to those with more knowledge, experience and hands-on expertise than they alone can afford. Outsourcing arrangements allow companies to focus on core business issues rather than technology issues, so they can compete better in the global economy. The outsourcing firm typically provides an analysis of an organization's business objectives, application requirements, and current and future communications needs. Acting as the client's agent, the outsourcing firm coordinates the activities of equipment vendors and carriers to ensure efficient and timely installation and implementation.

Typical Services

Typically, outsourcing arrangements are open-ended, with the third-party firm providing services that are customized to meet each client's needs, including:

PROJECT MANAGEMENT Entails the coordination of many discrete activities, starting with the development of a customized project plan based on the client's organizational needs. For each ongoing task, crucial requirements are identified, lines of responsibility are drawn, and problem escalation procedures are defined.

INTEGRATION Entails unifying disparate computer systems and networks into a coherent, manageable utility. This typically involves the reconciliation of different physical connections and protocols. The outsourcing firm also ties in additional features and services offered through the public switched network. The objective is to provide compatibility and interoperability among different products and services, making access transparent to end users.

SITE ENGINEERING The outsourcing firm performs site survey coordination and preparation, ensuring that all power requirements, air conditioning, ventilation, and fire protection systems are properly installed and in working order before any new equipment and cabling is installed.

AGENT AUTHORITY Acting as the client's agent, the outsourcing firm interfaces with multiple suppliers and carriers to economically upgrade and/or expand systems and network.

STAGING SERVICES When an entire node must be added to the network or a new host must be brought into the data center, the outsourcing firm will stage all equipment for acceptance testing before bringing it online, thus minimizing potential disruption to normal business operations. When new lines are ordered from various carriers, the outsourcing firm will conduct the necessary performance testing before making them available to user traffic.

TROUBLE TICKET ADMINISTRATION In assuming responsibility for daily network operations, a key service performed by the outsourcing firm is trouble ticket processing, starting from alarm indications received at the network control center and ending with the faulty system or line being returned to service in proper working order.

VENDOR/CARRIER RELATIONS The outsourcing firm interfaces with all vendors and carriers for such things as procurement, installation, testing, upgrades and contract enforcement. Instead of having to manage multiple relationships, the client only needs to manage one: the outsourcing firm.

MAINTENANCE/REPAIR/REPLACEMENT Some outsourcing arrangements include maintenance, repair and replacement services. Not only does this arrangement eliminate the need for ongoing technical training, the company is also buffered from the effects of technical staff turnover. Repair and replacement services can increase the availability of systems and networks, while eliminating the cost of maintaining a spare parts inventory, test equipment, and asset tracking system.

DISASTER RECOVERY Contingency plans to address disaster situations may include numerous services customized to ensure maximum network availability and performance, such as risk assessment, the development of test-run guidelines, procedures for initiating the recovery process, and escalation procedures.

STRATEGIC PLANNING With experience drawn from a broad customer base, as well as its daily interactions with hardware vendors and carriers, the outsourcing firm can assist the client in assessing industry and technology trends, determining the impact of emerging services and products, and positioning the client to take advantage of new technologies.

EQUIPMENT LEASING An outsourcing arrangement can include financing options, including equipment leasing. Leasing can free up capital for other uses, and even cost-justify technology acquisitions that would normally prove too expensive to purchase. With new technology becoming available every 12 to 18 months, leasing can prevent the organization from becoming saddled with obsolete equipment.

TRAINING Outsourcing firms can fulfill the varied training requirements of users, including basic communications concepts, product-specific training, resource management, and help-desk operator training.

SUMMARY

While outsourcing promises numerous benefits, the arrangement can fail if it is not structured properly to avoid the company being locked into an inflexible contract. Many companies have found that the cost savings of outsourcing have not materialized. This can be attributed to inaccurate or overly optimistic projections of cost savings at the start. In addition, many companies have found that they must set up an infrastructure to manage the outsourcing agreement, which can greatly reduce the anticipated cost savings. Today, many companies enter into outsourcing arrangements not to save money, but to free themselves from technology concerns so they can refocus on core business issues such as productivity, competitiveness, and market expansion.

See Also **Managed Applications Services, Service Level Agreements**

P

Parallel Transmission

As the term implies, parallel transmission is a method of sending/receiving multiple bits simultaneously, with each bit carried over a separate channel or wire. If 8 bits are sent at a time, it will require 8 channels or wires, one for each data bit. To transfer data in this way, a separate channel or wire is used for the clock signal. This serves to inform the receiver when data is available. In addition, another channel or wire may be used by the receiver to acknowledge receipt of the data, indicating that it is ready to accept more data. Parallel transmission mainly takes place internally within computer equipment and over very short distances between computers and peripherals.

The most well-known parallel transmission application is printing, in which a cable connects from a computer's parallel port to a printer's parallel port. The cable's 25-pin D-shaped DB-25 connector plugs into the back of the computer, while the other end of the cable, which uses a 36-conductor Amphenol connector (also known as the Centronics connector), plugs into the back of the printer.

Using different connectors at each end of the cable is a result of the nonstandard way equipment was built in the early days of desktop computing. When IBM built its PC, it opted to not use the 36-conductor connector from the dominant printer manufacturer of the time, which was Centronics. Instead, IBM equipped its computers with the 25-pin D shell connector—the DB-25 connector. Over the years, printer manufacturers have stuck with the Centronics connectors, while PC manufacturers have standardized with DB-25 connectors. This is why a special adapter cable is needed to connect the two devices.

The parallel connection is fast and usually problem-free because there is no wait time for encoding and decoding the data. The standard parallel port typically sends data at 115,200 bits per second, while newer enhanced parallel ports are up to 100 times faster.

SUMMARY

Parallel transmission is a very simple method of moving data. It is good only for short distances, however, such as linking components within the computer, making a connection from the computer to a printer, or connecting two computers in the same room. Parallel transmission is obviously faster than serial, not only because more than one bit is sent at a time, but also because there is no need for data encoding/decoding at each end.

See Also **Serial Transmission**

Patch Panels

A patch panel is an assembly of pin locations and ports that provides a variety of cross-connect solutions for data and voice communications. Typically, network cables from around an office floor or building are brought into a wiring closet that contains one or more patch panels, where the cables are received and organized.

On a Category 5 patch panel, for example, the individual wires are punched down into numbered, color-coded slots in the back of the patch panel (technician side) with a special punch tool. When the wires are punched down properly, an electrical connection is made to a corresponding RJ45 port in the front (user side). Patch cable, with RJ45 connectors on each end, is used to connect the ports of patch panels and hubs. RJ45 ports and connectorized cables makes it easy to move equipment on the network simply by moving the patch cable between the panel and hub. If the hub has management capabilities, the changes can be implemented in software at a management station to create Virtual LANs (VLANs).

There are also fiber patch panels, used when multiple fiber cables are terminated for interconnection to active equipment or for transitioning from vertical to horizontal runs. The technician's side of the panel has facilities for securing the cable, storing fiber slack, as well as arranging and identifying the terminated fibers. The user's side of the panel has similar features for managing optical patch cords. Many panels include printed circuit boards with components that minimize cross-talk and insertion loss—thus improving performance at high data rates.

Regardless of the specific media types involved, patch panels are available for rack mounting or wall mounting. Rack mounting is the first choice, when practical, because it allows for better cable management and closer positioning of active equipment to the panel, such as hubs, switches, and routers. Wall mounting is usually chosen for low-density cable installations or when there is limited rack space.

SUMMARY

Any significant premises cabling installation runs the risk of chaos, with wiring closets becoming a tangled web of cables. Patch panels bring order to the mass of incoming cables by providing a method of organizing them in a way that facilitates network changes.
See Also **Hubs, Virtual LANs**

Peer-to-Peer Networks

In a peer-to-peer network, computers are linked together for resource sharing. If there are only two computers to link together, networking can be accomplished with a Category 5 cross-over cable that plugs into the RJ45 jack of the network interface card (NIC) on each computer. If the computers do not have NICs, they can be directly connected by either a serial or parallel cable. Once connected, the two computers function as if they were on a local-area network (LAN), and each computer can access the resources of the other. If three or more computers must be connected, a wiring hub is required.

Regardless of exactly how the computers are interconnected, each is an equal or "peer" and can share the files and peripherals of the others. For a small business using routine word-processing, spreadsheet, and accounting software, this type of network is the low-cost solution to sharing resources such as files, applications, and peripherals. Even an external modem can be shared to allow all networked computers to access the Internet at the same time.

Networking with Windows

Peer-to-peer networking is often implemented by computers running Windows 95/98 and NT. In addition to peer-to-peer network access, both provide network administration features and memory management facilities, support the same networking protocols—including TCP/IP for accessing intranets, virtual private networks (VPNs), and the public Internet—and provide options such as dial-up networking and fax routing.

One difference between the two operating systems is that in Windows 95/98 the networking configuration must be established manually, whereas in Windows NT the networking configuration is part of the initial program installation, on the assumption that NT will be used in a network. Although Windows 95/98 is good for peer-to-peer networking, Windows NT is more suitable for larger client/server networks.

Windows supports Ethernet, token-ring, and FDDI data-frame types. Ethernet is typically the least expensive network to implement. NICs can cost as little as $20 each, and a five-port hub can cost as little as $40. Category 5 cabling usually costs less than 50 cents per foot in 100-foot lengths, with the RJ45 connectors already attached at each end. Snap-together wall-plate kits, including a modular jack, cost about $8 each, eliminating the need for special tools to connect a cable's bare end to the modular jack that snaps into the rear of the wall plate.

Configuration Details

When setting up a peer-to-peer network with Windows 95/98, each computer must be configured individually. After installing a NIC and booting the computer, Windows 95/98 will recognize the new hardware and automatically install the appropriate network-card drivers. If the drivers are not already available on the system, Windows 95/98 will prompt for the manufacturer's disk containing the drivers, and they will be installed automatically (Figure 93).

Next, the client type must be selected. If a Microsoft peer-to-peer network is being created, add "Client for Microsoft Networks" as the primary network logon (refer back to Figure 93). Since networking's main advantage is resource sharing, it is important to enable the sharing of both printers and files. This is done by clicking on the "File and Print Sharing" button and choosing one or both of these capabilities (refer back to Figure 93). Through file and printer sharing, each workstation becomes a mini-server.

Identification and security are the next steps in the configuration process. From the "Identification" tab of the dialog box, select a unique name for the computer and the workgroup to which it belongs, as well as a brief description of the computer (Figure 94). This information will be seen by others when they use Network Neighborhood to browse the network.

Figure 93

To verify that the right drivers have been installed, users can open the Network Control Panel to check the list of installed components. In this case, a Linksys LNEPCI II Ethernet Adapter has been installed.

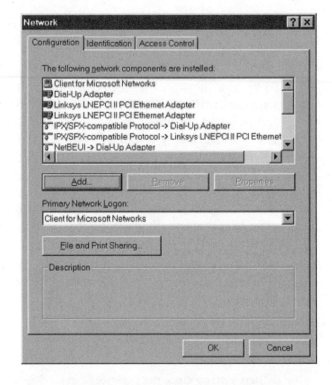

Figure 94

A unique name for the computer, the workgroup to which it belongs, and a brief description of the computer identify it to other users when they access Network Neighborhood to browse the network.

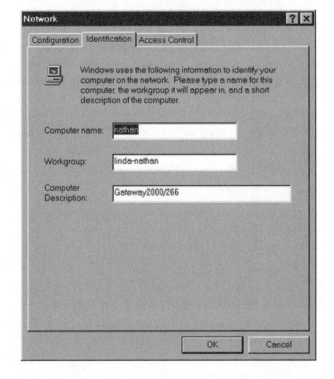

From the "Access Control" tab of the dialog box, select the security type. For a small peer-to-peer network, share-level access is adequate (Figure 95). It allows printers, drives, directories, and CD-ROM drives to be shared, and allows password access to be established for each of these resources. In addition, read-only access lets users view (not modify) a file or directory.

To allow a printer to be shared, for example, right-click on the printer icon in the Control Panel and select "Sharing" from the drop-down list (Figure 96). Next, click on the "Shared As" radio button and enter a unique name for the printer (Figure 97). This resource can also be password protected. When another computer tries to access the printer, the user will be prompted to enter a password. If a password is not necessary, the password field is left blank.

Another security option in the "Access Control" tab is user-level access, which limits resource access by user name. This function eliminates the need to remember passwords for each shared resource. Each user simply logs onto the network with a unique name and password; the network administrator governs who can do what on the network. However, this requires the computers to be part of a larger network with a central server—one running Windows NT Server—which maintains the access-

Figure 95

Choosing share-level access allows users to password-protect each shared resource.

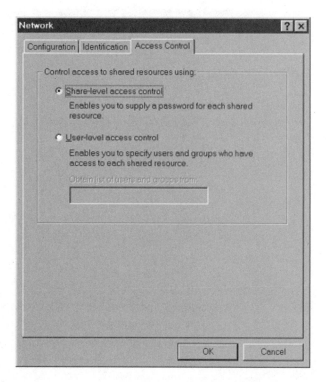

Figure 96
A printer can be
configured for
sharing.

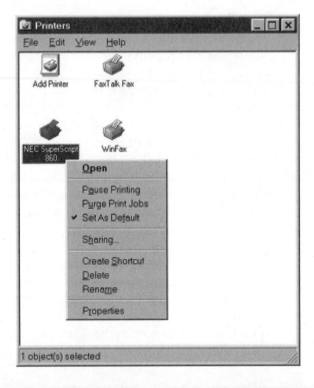

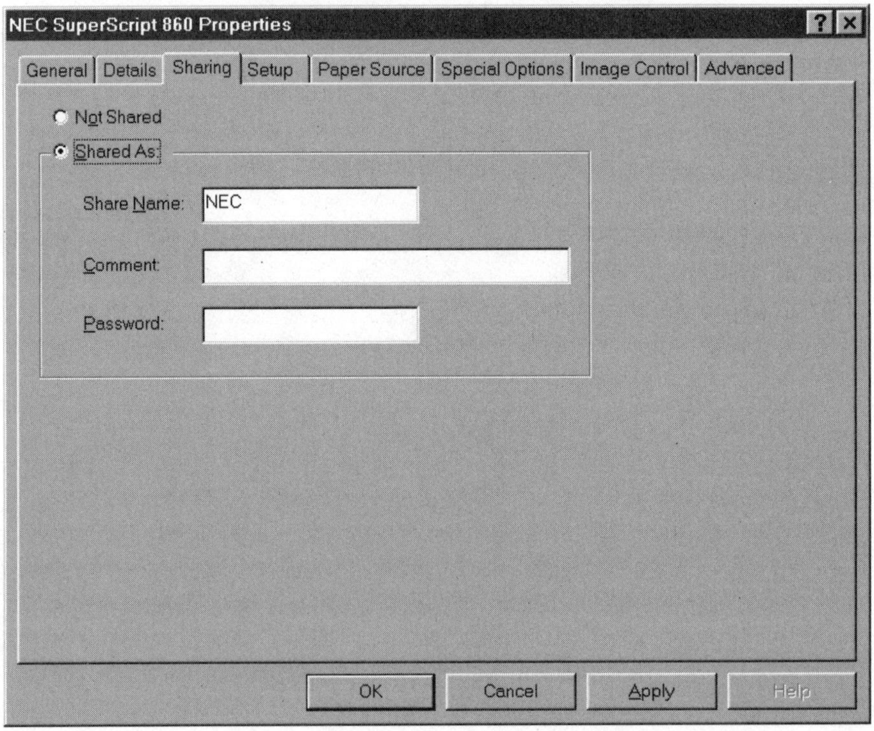

Figure 97
A printer can be
password-
protected if
necessary.

control list for the whole network. Since Windows 95/98 and Windows NT workstations support the same protocols, Windows 95/98 computers can participate in a Windows NT Server domain.

Peer services can be combined with standard client/server networking. For example, if a Windows 95/98 computer is a member of a Windows NT network and has a color printer to share, the resource "owner" can share that printer with other computers on the network. The server's access-control list determines who is eligible to share resources.

Once the networking infrastructure is in place, the NIC of each computer is individually connected to a hub with Category 5 cable. This cable has connectors on each end, which insert into the RJ45 jacks of the hub and NICs. For small networks, the hub will usually not be manageable with SNMP (Simple Network Management Protocol), so no additional software is installed. Once the computers are properly configured and connected to the hub, the network is operational.

SUMMARY

Peer-to-peer networking is an inexpensive way for companies and households to share resources among a small group of computers. This type of network provides most of the same functions as the traditional client/server network, including the ability to run network versions of popular software packages. Peer-to-peer networks also are easy to install. Under ideal conditions, the cards, software, hub, and cabling for five users will take only a few hours to set up.

Of course, there are other products besides Windows that implement peer-to-peer networks. One of the most popular is Artisoft's LANtastic, which includes software for Windows and DOS platforms, allowing users to mix and match the installation to fit changing needs. LANtastic allows hundreds of PCs to share resources on the same network. It includes drag-and-drop network administration and interfaces easily with Microsoft and Novell network operating systems.

See Also **Client-Server, Hubs**

Performance Baselining

Understanding the behavior of a properly functioning network can be of substantial value when trying to identify the cause of problems that may occur in the future. The only way to know a network's normal behavior is to analyze it while it is operating properly. Later, during troubleshooting, technicians and network managers can compare data from the properly functioning network with data gathered after conditions have begun to deteriorate. This comparison often points to the steps leading to a corrective solution. The process of compiling a network "snapshot" for this purpose is called performance baselining.

Information Requirements

The first step in baselining performance is to gather appropriate information from a properly functioning network. Much of this information may already exist; it is just a matter of finding it.

TOPOLOGICAL MAP   For example, many enterprise management systems can automatically discover devices on the network and create a topological map. This kind of information is necessary for knowing what components exist on the network and how they interact physically and logically. For WANs, this means locations, descriptions, and cable plant maps for equipment such as routers, bridges, and network access devices (LAN to WAN).

In addition, information on transmission media, physical interfaces (T1/E1, V-series, etc.), line speeds, encoding methods and framing, and access points to service provider equipment should be assembled. Although it is not always practical to map individual workstations in the LAN portions of the WAN, or to know exactly what routing occurs in the WAN cloud, understanding the general topology of the WAN can be useful in tracking down problems later.

WAN AND LAN PROTOCOLS   To fully understand how a network behaves, it is necessary to know what protocols are in use. Later, during the troubleshooting process, the presence of unexpected protocols may provide clues as to why network devices appear to be malfunctioning or why data transfer errors or failures are occurring.

LOGS   Some network problems begin to occur after new devices or applications are installed. The addition of new devices, for example, can cause network problems with a ripple effect throughout the network. A new end-user device with a duplicate IP address, for instance, could make it impossible for other network elements to communicate, or a badly configured router added to a network could produce congestion and connection problems. Other problems occur when new data communications are enabled or existing topologies and configurations are changed. A log of these activities can help pinpoint causes of network difficulty. In addition, previous network trouble—and its resolution—is sometimes recorded; this, too, can lead to faster problem identification and resolution.

STATISTICS Often, previously gathered data can provide valuable context for newly created baselines. Previously assembled baselines may also contain event and error statistics and examples of decoded traffic based on network location or time of day. These logs may have been gathered over long periods, yielding valuable information about the history of network performance.

USAGE PATTERNS A profile of users and their typical usage patterns can speed fault isolation. This entails having several types of information, including what kind of LAN traffic is carried over the WAN.

LAN TRAFFIC ON THE WAN With knowledge of what kind of LAN traffic to expect on the WAN, technicians and network managers will have a better idea of the analysis that might have to be performed later. In addition, knowing how LAN frames can be handled at end stations can help troubleshooters distinguish between WAN problems and end-station processing problems.

TRAFFIC CONTENT Knowing the WAN traffic type (voice, data, video, etc.) can help troubleshooters estimate when network traffic is most likely to be heavy, what level of transmit errors can be tolerated, and whether it makes sense to even use a protocol analyzer. For example, an analyzer may incorrectly report errored frames and corrupt data when attempting to process voice or video traffic based on data communication protocols.

PEAK USAGE Knowing when large data transfers will occur—such as scheduled file backups between LANs connected across the WAN—can help network managers predict and plan for network slowdowns and failures. It can also help technicians schedule repairs so the WAN performance is minimally impacted. Some of this information can be obtained from interviews with network administrators or key users. Other times, it must be gathered with network analysis tools.

HARD STATS AND DECODES After gathering information on topology, devices, protocols, and typical users of the WAN being baselined, "hard" statistics and examples of decoded network traffic should be gathered. Getting comprehensive baseline data may entail gathering data at regular intervals at numerous points throughout the network.

STATISTICS LOGS To understand usage trends and normal error levels over time, a statistics log must be created. Many protocol analyzers let technicians specify the period over which this kind of data is logged, the interval between log entries, and the type of statistics to log. The log file can then be exported to a spreadsheet or data-manipulation software for analysis.

FRAME OR PACKET DATA   To see details about typical WAN traffic, frame or packet data can be collected and saved to a file for later examination. Data collection can be done at specific periods during the day or week to find differences between peak and off-peak usage. Saved network traffic can also provide insight into device configurations for use later during routine upgrades or repairs.

TARGETED STATISTICS   Using configurable traffic filters and counters, selected blocks of data or statistics based on specific network events can be captured, which might include error-count thresholds, specific frame types, and in-channel alarms. A comprehensive collection of such data provides a benchmark for comparison if the network begins to malfunction. New protocols on the network, unexpected line and channel usage levels, and increases in normal errors and in-channel alarms can be isolated according to physical link location, helping narrow the search for the problem.

Applying the Baseline

If network performance and reliability problems occur, the information gathered during baselining can help identify the nature and source of the problem through comparison analysis and historical trends.

COMPARISON ANALYSIS   Baseline information is compared with current information to see network changes. For example, to isolate failing devices or connections, the number of errors recorded during baselining is compared to the current number of errors that occur over a similar time interval.

HISTORICAL TRENDS   Current network problems can result from subtly changing conditions that are detected only after examining a series of baselines gathered over time. For example, congestion problems may become apparent only as new users are added to a particular part of the network. Examining historical trends can help isolate these situations.

SUMMARY

Performance baselining provides a profile of normal network behavior, making it easier for technicians and network managers to identify devia-

tions so appropriate corrective action can be taken. This "snapshot" of the current network can also be used as the input data for subsequent performance modeling. For example, network administrators and operations managers can use the baseline data to conduct "what if" scenarios to assess the impact of proposed changes. A wide variety of changes can be evaluated, such as adding routers, WAN bandwidth or application workloads, and relocating user sites. During analysis, performance thresholds can be customized to highlight network conditions of interest. These capabilities enable users to plan and quantify the benefits of feature migrations, such as different routing protocols, and to make more accurate and cost-effective decisions regarding the location and timing of upgrades.
***See Also* LAN Analyzers, LAN Troubleshooting, Network Monitoring,. Network Probes, Performance Monitoring**

Personal-Area Networks

Personal-Area Networks (PANs) communicate information between portable devices that use the natural electrical conductivity of the human body to transmit electronic data. Developed by IBM, the technology enables information to be transferred via touch rather than issuing typed commands or pressing buttons.

The natural salinity of the human body makes it an excellent conductor of electrical current. PAN technology takes advantage of this conductivity by creating an external electric field that passes an imperceptibly small amount of current through the body, over which information can be carried.

The current uses only one-billionth of an ampere, which is lower than the natural currents already in the body. In fact, the electrical field created by running a comb through hair is more than 1,000 times greater than that being used by PAN technology. The speed at which the data is transmitted is 2400-bps modem. Theoretically, information could be transmitted at 4 Mbps using this method.

Applications

There are a number of applications for PAN technology. For example, the information on electronic business cards could be transferred between two people via a simple handshake. Or a computer might recognize an authorized user simply by that person touching the keyboard. This capability could yield significant benefits in the area of access control and data privacy.

The technology can be used to exchange information between personal information and communications devices carried by an individual, including cellular phones, pagers, personal digital assistants (PDAs), and smart cards. For example, upon receiving a page, the number could be automatically uploaded to the cellular phone, requiring the user to simply hit the send button. This automation increases accuracy and safety, especially in driving situations.

In addition, the technology could be used to automate and secure consumer business transactions. For example, a public phone equipped with PAN sensors would automatically identify the user, who would no longer have to input calling card number and PIN. This application significantly reduces fraud and makes calling easier and more convenient for users.

Sharing information in this manner increases the usefulness of personal information devices and provides users with features not possible with independent or isolated devices. Wiring all these devices together would be cumbersome and constrictive to the user. Infrared communications of information—used on TV remote controls—requires direct lines of sight to be effective, making it impractical for devices located inside wallets and pockets. Radio frequencies—such as those used with automated car locks—are susceptible to interference and eavesdropping. The near-field electrostatic coupling between the device transducer and the skin avoids these complications.

SUMMARY

PAN grew out of work conducted at the MIT Media Laboratory. Initial research was funded by the IBM Corporation, Hewlett-Packard, and Festo Didactic Corp. Although prototype PAN technology has been demonstrated by IBM since 1996, at this writing it is still in the research stages and no commercial products are yet available.
See Also **Bluetooth, Wireless LANs**

Phoneline Networking

Just as there is the need for high-speed data networks outside the home, there is a growing need for similar connectivity within the home. Businesses accomplish this by deploying LANs, but networks are not commonly deployed in the home due to the cost and complexity of installing the type of cabling required by traditional LANs. A specification developed by the Home Phoneline Networking Alliance (HomePNA), however, enables the creation of simple, high-speed, and cost-effective home networks using existing phone wiring.

Adding data to voice over existing phone wiring does not pose interference problems because different frequencies are used for each. Standard voice occupies the range from 20 Hz to 3.4 kHz in the U.S. (slightly higher internationally), while xDSL services, like ADSL, occupy the frequency range from 25 kHz to 1.1 MHz. Phoneline networking operates in a frequency range above 2 MHz. The frequencies used are far enough apart that the same wiring can support all three.

Home networking requires two types of network cards. One connects local devices to each other over the phone wiring. The other performs the same function, but also includes a modem for WAN access. PCs on home phone networks must run Windows 95, 98, or NT, or some other software that supports file sharing.

Once networked, the PCs could share a printer, as well as a dial-up connection for Internet or corporate network access. Users will also be able to work high-speed digital subscriber line (DSL) and cable modems into the mix. In fact, HomePNA lets the consumer choose the method of WAN access, which can also include ISDN and wireless services.

HPNA's schematic of a typical phoneline network shows home PCs with Internet access via V.90 modem, xDSL (digital subscriber line), ISDN, cable modem, or wireless connection. Also on the schematic are shared printers, cameras, scanners, a TV set-top box, and the Web, as well as ordinary phones (Figure 98).

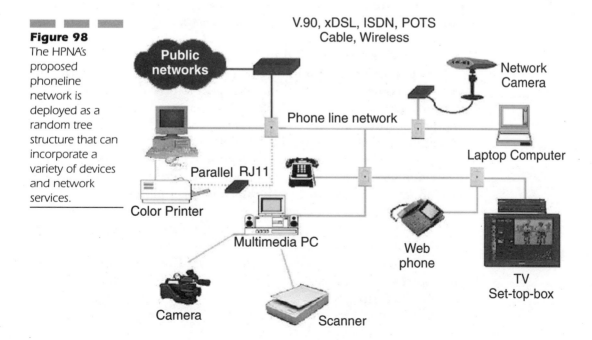

Figure 98
The HPNA's proposed phoneline network is deployed as a random tree structure that can incorporate a variety of devices and network services.

HomePNA Alternatives

IBM and Lucent Technologies, also members of the HomePNA, are taking the home network concept farther. IBM's Home Director Professional and Lucent's HomeStar Residential Wiring System interconnect proprietary home systems such as security, lighting, and heating and allow almost any intelligent device—PC and non-PC—to share data via a central server-like system. However, both companies' offerings require high-quality network cable in the home, in addition to the use of the existing phone line, making their solutions better suited for new buildings than for established homes. IBM, for example, is targeting the emerging intelligent home market. Its product will be sold to new home builders and be set up by certified installers.

Meanwhile, a vendor consortium called the Home Radio Frequency Working Group has issued the Shared Wireless Access Protocol (SWAP) specification for wireless technology within the home. Backed by Microsoft, Intel, Compaq, Hewlett-Packard, and IBM, the alliance is predicated on the notion that many users will not want to wire their homes to network their devices. Wireless networking is much more feasible and less costly, especially for areas such as garages, basements, and attics, which may be hard to reach with new telephone wire.

Still another standards effort for networks in the home comes from the Consumer Electronics Manufacturers Association (CEMA). Its standard for home networking divided into three parts, EIA-709 defines a common protocol (EIA-709.1) and transceivers for networking consumer products over existing power lines using narrowband signaling (EIA-709.2) and free topology twisted pair media (EIA-709.3).

EIA-709.1 defines a communication protocol for networked control systems in a home. The protocol provides peer-to-peer communication for networked control and is suitable for implementing both peer-to-peer and master-slave (controller) based systems. The protocol defines a rich set of features, which can be used to support simple on/off devices or complex devices. These features include acknowledged and unacknowledged messaging services, a full suite of network management services, authentication for message security, prioritization for important messages, and router-compatible addressing that can be used with off-the-shelf routers.

The protocol is media-independent and capable of supporting communications options that include twisted-pair, power line, radio frequency, infrared, coaxial cable, and fiber optics. Currently, power line and free topology twisted pair are defined in the EIA-709 standard.

EIA-709.2 defines physical communication over power lines inside and outside of homes over 120VAC to 240VAC wiring. The power line channel occupies the bandwidth from 125 kHz to 140 kHz and communicates at 5.65 Kbps.

EIA-709.3 supports free topology communication at 78.125 Kbps among devices in the control network. The standard defines a means for robust, field proven two-way communications among as many as 128 devices in a single twisted pair segment, and can be extended through the use of physical layer repeaters. Free-topology wiring supports any wiring topology, including bus, star, loop, or any combination of topologies.

SUMMARY

Telephone wiring in the home is designed to handle voice, but the new home phone network technology advocated by the Home Phone Networking Alliance allows the wiring to support data as well. Current home network technology supports 1 Mbps of data above the 3,000-Hz range used by voice, though vendors working within the Home Phone Networking Alliance hope to boost the LAN bandwidth to 10 Mbps. This means the home network soon will be capable of adding video to voice and data communications to provide videoconferencing, multimedia collaboration, distance learning, and on-demand entertainment.
See Also **Cable Modems, Digital Power Line, Digital Subscriber Line Technologies, Small Office/Home Office LANs, Wireless Home Networking**

Physical Network Management

Physical networks include LAN cabling and adapters, telephone wiring and connectors, WAN circuits and links, patch panels and cableways, and the diverse assemblage of hardware in between. Physical network management systems provide installers, technicians, service providers, and administrators with an easy way to store, retrieve, and manage information about cabling and hardware. Such products facilitate moves, adds, and changes. They also are used to resolve problems and support asset management, configuration management, help desk operations, and disaster recovery planning.

The Physical Network

A physical management solution documents and provides detailed information about the entire networking infrastructure, including the actual location of network devices, technical characteristics, and physical connectivity. All of this information is kept up-to-date and accessible via an online database, saving operators, technicians, and network managers time performing daily tasks.

Building a model of the physical infrastructure entails using tools to import existing data from portable testers, spreadsheets, and other databases, as well as floor plans from CAD systems. If the data is not in machine-readable format, modeling programs such as Cablesoft's Connectivity Routing and Infrastructure Modeling Program (CRIMP) contain numerous rapid data creation functions for building multiple records and making connections in bulk. Vendor-specific libraries may be available for importing graphical representations of network components into the modeling program.

Whether imported from another source or built using a component library, these drawings can then be linked to the site structure and displayed graphically at a workstation. The schematics show the physical connectivity and logical path of the equipment and cabling connections in the database. Specific information such as port or pair assignment, network addresses, and circuit connections are easily entered to match the level of detail required. Information can also be stored for multiple buildings or campuses, showing views by floor, zone, and equipment closet.

The physical network management system allows the network manager to place and connect devices graphically for an accurate picture of the network's design. Rule-based auto-configuration capabilities allow the most appropriate end-to-end circuit to be established. The software automatically traces the patches and cross-connections necessary to complete the circuit, easing design and change management. Complete documentation about the location and connectivity of every physical component is also provided to facilitate effective troubleshooting.

Rule-based move, add, and change functions allow an administrator to move and connect equipment through simple processes that do not require detailed knowledge of the network infrastructure or database. This is achieved by providing the network manager with the means—via a scripting language—to define a set of instructions for the physical network management system to follow whenever the operator performs a particular operation, such as moving a workstation.

Rules can be based on industry standards, or company or vendor requirements. They can also include error-checking, such as preventing a data line from being connected to a telephone jack or an Ethernet port from being connected to a token ring network. Once created, these rule-based processes can be listed on a pull-down menu so the operator can merely select an appropriate item and follow the prompts to complete the move or connection. If an error occurs during execution, the process will be rolled back to preserve data integrity.

Using these scripts, the system administrator can completely automate processes such as connecting and disconnecting equipment, creating new graphic instances in a drawing, changing object color, and invoking an action based on the result of a database query. When equipment is moved, its attributes move with it, automatically keeping the database up to date.

WAN Management

Some physical network management systems (NMS) provide the ability to model and manage wide-area circuits or links and to track bandwidth allocation. The associated tools provide the means to troubleshoot and isolate faults to reduce the costs associated with supporting enterprise networks. This capability also provides the means for documenting WAN equipment and topology.

Customer premises equipment (CPE) and wide-area nodes can be modeled, along with physical connectivity attributes such as cable type. In addition, other asset information can be incorporated into the relational database, including maintenance contract telephone numbers and equipment configurations. The level of detail is operator-definable.

Local- and wide-area assets can be incorporated into a single project. Typically, the router is used as the dividing line between LAN and WAN. A single piece of equipment can be displayed on multiple drawings, such as a floor plan and a wide-area closet on two different drawings, with a single entry in the relational database.

Circuit Modeling

Between the end pieces of equipment, one or more virtual circuits can be laid out, with or without the use of drawings. Sublevels of detail can also be modeled, with one circuit riding over another facility, such as a 64K service that is part of a 256K circuit, which in turn is actually made up of a series of T1 or T3 segments. Additional information can be tracked on the logical circuit, such as circuit identification, carrier, and contact name.

Once the physical, logical, and virtual circuits have been laid out, the relationship between them can be established by the physical NMS's routing system. This is a semi-automatic process in which the physical NMS graphically builds a circuit schematic based on connectivity information it has been given, then queries the operator at each decision point by presenting a list of possible paths before continuing on to the next hop. The schematic is dynamically updated as each segment is being routed, and a change log may also be activated to record the routing at the same time. Using this system, a typical WAN circuit may require only three or four decision points before it is completely routed.

The physical NMS also checks the available bandwidth for each circuit segment—both channelized and unchannelized—and either allocates the bandwidth required or informs the operator that the necessary bandwidth is unavailable. At a later time, additional submaps or facilities may be built if more detail is desired. Intermediate equipment can also be added to an existing circuit as required.

Once the WAN model is in place, a graphical or textual circuit trace can be automatically generated, bringing in components from one or more drawings. A circuit trace includes equipment, physical and virtual circuits, and port connectivity. On the circuit trace are text labels that identify the segment and equipment that makes up the circuit. These objects have intelligence and can be further queried for detail from the database, or the whole trace can be printed as part of a work order.

Bandwidth Allocation Tracking

Bandwidth allocation for circuits can also be modeled. Bandwidth is tracked by channels using the formula: channel bandwidth × the number of channels = the total capacity of the circuit. Circuits can be modeled using channelized or unchannelized bandwidth. Channelized bandwidth refers to the partitioning of the available bandwidth into separate channels. In the case of a T1 line, there can be 24 channels of 56/64 Kbps each. The total bandwidth of a T1 line is 1.544 Mbps. Taking out 8 bps for in-band management, the unchannelized bandwidth of a T1 line is 1.536 Mbps.

The channels of a circuit can be categorized as available, reserved, or used. Available channels are marked as "used" as the circuit is routed over them, or set to "reserved" using a channel allocation tool. Channels are made available again by unrouting the circuit from the facility.

From the channel allocation tool, a list of channels can be displayed for the selected circuit, along with their status. Used channels show the circuit that is routed on them. Any available channels can be set to "reserved" to mark them for future use. For a used or reserved channel, an owner (such as a department, group, cost center, project, or software application) can be assigned and optionally associated with a flat-rate cost for charge-backs.

Relational Database Management

Some physical network management systems provide links to industry-standard relational databases. Cambio Network's Command, for example, is a set of integrated applications implemented in a client/server architecture that provides enterprise-wide access to a centralized repository of network information. This repository resides within industry-standard database platforms supplied by Oracle, Sybase, or Informix. Command also features form and report building tools that enable users to customize methods of accessing and organizing the network information that has been recorded in the database. Regardless of product, asset and connectivity information are typically stored in multiple database tables, such as:

EQUIPMENT Defines equipment types, default connectors, equipment instances in the project, and operator-defined "info tables" that can be used for maintenance and inventory information

MEDIA Includes media types (such as cables, optical fiber, and microwave) and line information within the various media

MEDIA PATHS Include cableways, media segments, nodes, and routing information

CHANGE LOGS Used to define work orders and track tasks associated with changes to equipment, ports, cables, and locations

DATA/DRAWING INTEGRITY Handles communication between the database and drawings, tracking drawing locations, and maintaining system integrity

In addition, there may be a table that temporarily stores data for automatically populating drawings based on equipment and connectivity information supplied by an outside source.

The tables can be accessed from the physical NMS through SQL queries. The results of these queries can be used in a variety of ways, including displaying graphical information such as a circuit trace, maintaining a transaction log, and generating work orders and reports. In addition, operator-defined information that is stored in other databases, such as equipment costs and maintenance schedules, can be accessed from the system menu and maintained as part of a particular project.

The interactive interface to SQL enables the system operator to generate sophisticated database queries and display the results in a variety of ways. Depending on vendor, a set of predefined queries may be provided with the physical network management system. The predefined queries can be used as templates, allowing the operator to create an unlimited number of additional queries, using a combination of standard SQL functions and specialized system functions. The system functions include such things as changing object colors or performing a circuit trace on the results of the query. Frequently used SQL queries can even be named, saved to a file, and then displayed on the system's pull-down menus.

Standards

Several standards address the documentation and administration aspects of the physical network:

TIA/EIA 568 Commercial Building Telecommunications Wiring Standard

TIA/EIA 569 Commercial Building Standard for Telecommunications Pathways and Spaces

TIA/EIA 606 Administration Standard for the Telecommunications Infrastructure of Commercial Buildings

TIA/EIA 607 Ground and Bonding Requirements for Telecommunications in Commercial Buildings
A comprehensive physical network management product complies with the TIA/EIA 606 standard. Although it may not be appropriate for every environment, this standard offers a relatively complete framework for infrastructure documentation and administration. It addresses the record format for pathways and spaces, wiring, and grounding and bonding. It also specifies how the different records should be linked together in an administrative system. The following table summarizes the infrastructure elements:

	Record	Required Information	Required Linkages
Pathways and Spaces	Pathway	Identifier Type Fill Loading	Cable records Space records Pathway records Grounding records
	Space	Identifier Type	Pathway records Cable records Grounding records

	Record	Required Information	Required Linkages
Wiring	Cable	Identifier	Term. pos. records
		Type	Splice records
		Unterm. pair/ conductor nos.	Pathway records
		Damaged pair/ conductor nos.	Grounding records
		Available pair/ conductor nos.	
	Term. Hardware	Identifier	Term. pos. records
		Type	Space records
		Damaged position nos.	Grounding records
	Term. Position	Identifier	Cable records
		Type	Other term. pos. records
		User code	Term. hardware records
		Cable pair/ conductor nos.	Space records
	Splice	Identifier	Cable records
		Type	Space records
Grounding and Bonding	TMGB	Identifier	Bonding cond. records
		Busbar type	Space records
		Grounding cond. ID	
		Resistance to earth	
		Date measurement taken	
	Bonding Conductor	Identifier	Grounding busbar records
		Type	Pathway records
		Busbar identifier	
	TGB	Identifier	Bonding cond. records
		Type	Space records

Note: All records may also contain variable optional information and linkages to other records both within and outside the scope of this standard.

Acronyms: TMGB = Telecommunications Main Grounding Busbar and TGB = Telecommunications Grounding Busbar

For each infrastructure element (type of item), the standard specifies required identifiers, attributes, and linkages. An identifier is the unique name of a particular item, while an attribute is a property of a particular item. An identifier and its related attributes are a record. A list of records for a particular element is a table. A linkage represents a unique relationship between two items. The user is required to develop their own nomenclature (naming conventions) for identifiers and attributes. No length or format is specified in the standard.

SUMMARY

Physical network management systems provide a comprehensive view of the enterprise network, including hardware, software, connectors, data, and other important configuration details. It allows network and business support professionals to create, maintain and access a centralized model of the entire network, its components and their relationships. With API links among the various management platforms and third-party applications, collected information about the physical network can be shared among them. The result is faster resolution of network problems, reduced network downtime, more efficient diagnostic and troubleshooting activities, and cost savings.

***See Also* Cable Management, Inventory Management, Logical Network Management**

Ping

Ping is a function that allows users to check if a remote system on a network is currently up and running. Ping is often thought of as a UNIX utility, but it can also be run on PCs running in DOS mode to check the connectivity of clients in a peer-to-peer network or to check the connectivity of clients to a shared cable modem used for Internet access. On DOS and UNIX machines, the general command-line syntax for implementing ping is:

```
ping abc.com
```
or
```
ping 208.232.169.19
```

This will indicate whether the host at ABC Company is currently online. Either the domain name or the IP address may be used. Either way, the ping command sends one datagram per second and prints (or displays) one line of output for every echo response returned. No output is produced if there is no response. A count option can be used in the command-line syntax to specify the number of requests to be sent. Many implementations of ping also include an option that measures the roundtrip time of the sent packet in milliseconds (ms) as well as the packet loss between two hosts on the network. By pressing Ctrl-C on the keyboard, ping provides a brief statistical summary, as in the following example:

```
PING abc.com: 56 data bytes
64 bytes from 132.58.68.1: icmp_seq=0 ttl=251 time=66 ms
```

```
64 bytes from 132.58.68.1: icmp_seq=1 ttl=251 time=45 ms
64 bytes from 132.58.68.1: icmp_seq=2 ttl=251 time=46 ms
64 bytes from 132.58.68.1: icmp_seq=3 ttl=251 time=55 ms
64 bytes from 132.58.68.1: icmp_seq=4 ttl=251 time=48 ms
-- abc.com ping statistics --
5 packets transmitted, 5 packets received, 0% packet loss
round-trip min/avg/max = 45/52/66 ms
```

Using ping once or twice is generally enough to provide a reliable indication of a remote system's current state. Ping also can be used to continuously monitor the state of the connection. When using ping to isolate hardware or software problems, the local host is pinged first to verify that the local network is up and running before pinging hosts and gateways further away. For Windows (and Macintosh) machines, there are graphical utilities that implement ping (Figure 99). Among other things, they let the user specify:

- The ping data packet size
- The number of hosts to ping simultaneously
- The ping interval
- The amount of milliseconds to wait for echo reply
- Time to wait until next ping if the last ping succeeded

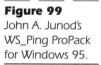

Figure 99
John A. Junod's
WS_Ping ProPack
for Windows 95.

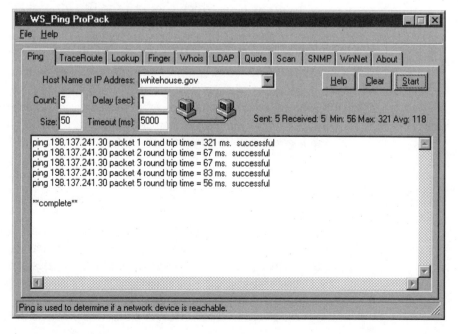

- Time to wait until next ping if the last ping failed
- The number of failed pings before the utility considers the host is down

Monitoring network activities is even easier if the ping utility is set to take a specific action whenever the remote host changes its state from up to down and vice versa. Among the actions the ping utility can implement in response to changes in network activity are playing audio messages and running custom programs.

For example, users can specify that a WAV file be played to indicate when a remote host crashes or recovers from a crash. The ping utility can also be configured to run a program as soon as the remote host recovers from a crash.

Major interconnect vendors bundle the ping capability with their network management systems. Bay Networks (a company of Nortel Networks), for example, offers the Bay Ping MIB, which allows network managers to set up ping sessions within the network. The ping MIB is a group of tables that stores the following information for one or more ping requests:

- General ping information, such as the IP address to ping—whether for trace routing or source routing—and the frequency of the ping
- Trace route data that shows the IP addresses the ping went through to reach its destination
- Source route data, which contains the IP addresses the ping is to go through instead of those in the routing table
- History data about previous pings to initiate at specific intervals

Sending a ping from the probe on the local segment to the default gateway provides response-time results, which can then be used to set color-coded baseline thresholds for various applications. Developing a graphical display makes it easier for network managers to tell at a glance if response time is good, bad or deteriorating from the norm.

SUMMARY ▪▪▪ ▪▪ ▪▪ ▪▪ ▪▪ ▪▪ ▪▪ ▪▪

Ping is a simple but useful troubleshooting tool for tracking down the source of a hardware or software failure on a network. Basically, when ping is run, a message is issued to one or more designated hosts. Various statistics are displayed which tell the user about the state of the remote host and the connection.

See Also **Performance Metrics, Network Monitoring**

Point of Presence

Point of presence (PoP) is a carrier facility that offers dial-up or leased-line access to its network for the purpose of providing services to customers. PoPs are configured with equipment and lines appropriate for the types of services that are offered from those locations. Examples of PoPs include:

- A local telephone company's central office

- A long-distance carrier's toll office

- An Internet service provider' (ISP) modem bank

- A national Internet backbone provider's metropolitan-area exchange (MAE) or Network Access Point (NAP)

- A cellular carrier's, paging company's, or wireless data service provider's base station transceivers

PoPs may be shared by several service providers. The NAPs on the major Internet backbones, for example, are equipped to handle large amounts of traffic to avoid bottlenecks on the Internet. The largest NAP, located in Chicago, is operated by Ameritech Advanced Data Services (AADS). The NAP has more than 40 companies, Internet service providers and universities connected to it via DS3 (45 Mbps), OC-3c (155 Mbps), and OC-12 (622 Mbps) Asynchronous Transfer Mode (ATM) connections.

Even company Web sites are called PoPs. When a company locates its Web server at an ISP's PoP to take advantage of its security and management services, the arrangement is sometimes referred to as a Virtual PoP.

SUMMARY

Points of Presence are the locations from which network services are provided. If a carrier does not have a local point of presence from which corporate customers can access the service, the carrier can route the traffic to the nearest PoP at little or no cost to the customer if the traffic volume makes the backhaul arrangement worthwhile. For consumers, the backhaul arrangement might be in the form of toll-free 800-number access to the nearest PoP.

See Also **Mail Gateways**

Policy-Based Management

Policy-based network management enables IT administrators to set policies that manage network resources and ensure that network bandwidth is appropriately allocated to users. Policies can determine who has access to certain resources and when those resources are available. For example, if an employee downloads a graphics-intensive file over the network during prime-time business hours, policy-based network management can assign a low priority and bandwidth for this type of transfer until off-peak hours. This capability ensures that crucial business is not disrupted.

If allowed to continue, congestion can cripple the network-centric business model in use at many organizations today. Chasing the problem with more bandwidth does not address the root problems of resource control and allocation. What network managers need is the ability to alleviate congestion right where it is created—at network access points. Policy-based management solutions make this level of control possible on a dynamic basis—all for the purpose of giving mission-critical applications the bandwidth they need without shutting out lower priority traffic.

In contrast to manually intensive management software products, policy management systems support centralized definitions of how the network should behave. They also enhance network operation by automating the distribution of those definitions to network components, such as switches, routers, and remote access servers. The result should be improved network quality through greater service standardization and automated operation.

Operation

Among the products that implement policy-based management is Check Point Software Technologies' FloodGate-1. Its real-time traffic-monitoring capabilities diagnose the source of network congestion to help the network manager set the appropriate policy and ensure the availability of critical services.

The process starts with all communications below the network layer being intercepted and screened by the software. The state-related information on each communication is then extracted and maintained in a table. This information is passed to the queuing engine, which enables organizations to define highly granular levels of traffic classification, queuing, and scheduling by dynamically controlling entire classes of traffic, not just individual connections. Using an enhanced, hierarchical Weighted Fair Queuing (WFQ) algorithm, the queuing engine intelligently and actively allocates available bandwidth for specified Internet services, users, groups, or designated network resources at an aggregate level.

The preemptive traffic scheduler within the engine ensures that high-priority inbound traffic is always given precedence over lower-priority data. However, in addition to bandwidth limits and guarantees, the queuing engine prioritizes traffic based on weighted priorities to ensure that even low-priority traffic receives some bandwidth on a consistent basis.

Policy Setting Mechanisms

Rules are established for traffic control via a combination of traffic classifications and bandwidth control criteria. Network managers can classify traffic on the basis of Internet service or application (e.g., HTTP, FTP, Telnet, Web), source, destination, group of users, groups of Internet services, Internet resource, and traffic direction—inbound or outbound.

Check Point's FloodGate-1 even allows network managers to classify traffic by a specific channel within a "push" application. Once traffic is accurately classified, FloodGate-1 applies one or more control criteria to actively manage bandwidth allocation. These criteria can be used alone or in concert to define appropriate bandwidth management policies. Control criteria categories include:

WEIGHTS Allocates bandwidth for users and Internet services based on designated merit or importance. The weight assigned to a particular class of traffic is proportionate to the weights of all other managed traffic.

GUARANTEES Provides guaranteed bandwidth for crucial applications or designated users and groups.

LIMITS Sets bandwidth restrictions for discretionary network services or user applications that are not time-sensitive.

Architecture

FloodGate-1 employs a three-tier client-server architecture that supports the ability to define and distribute an enterprisewide policy to all network access points. The software product consists of three components, all of which can run on a single platform or be distributed across multiple workstations and servers. These components include:

GRAPHICAL USER INTERFACE The policy editor, a Java-based application, can run on a variety of platforms and is used to create and modify the traffic management policy and define the network objects (Figure 100).

Figure 100
With Check Point's FloodGate-1, bandwidth management policies are defined with a GUI. The management policy consists of traffic rules that define bandwidth privileges for user-defined traffic classes.

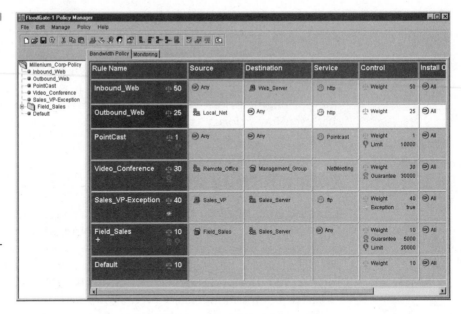

MANAGEMENT SERVER Controls and distributes the traffic policy to all policy modules.

POLICY MODULE Implements the traffic management policy at network access points and controls the flow of inbound and outbound traffic. Other policy-management products include 3Com's Policy Powered Network (PPN), Bay Networks' Optivity Policy, and Cisco Systems' CiscoAssure.

Standards

Key industry standards that support rule definition and policy implementation include the Lightweight Directory Access Protocol (LDAP), the Common Open Policy Service (COPS) protocols, and the Resource ReSerVation Protocol (RSVP).

LDAP is a standard for accessing network directory services that could take the place of multiple directories within a network, resulting in a more efficient and accurate directory scheme. Within the context of policy-based management, administrators can configure policies based on users and applications. This policy information is stored in an LDAP directory, where it can be accessed by the policy-management system to dynamically configure Quality of Service (QoS) on network devices. QoS parameters can specify dedicated bandwidth, controlled jitter and latency (required by some real-time and interactive traffic), and loss characteristics.

COPS provides secure policy information exchange between the policy servers and the software embedded in various network devices, while RSVP enables applications to dynamically request and reserve the network resources necessary to meet their specific QoS requirements.

Elements of LDAP, COPS, and RSVP are expected to be included in a future policy-management standard issued by the Internet Engineering Task Force (IETF).

SUMMARY

Policy-based management improves quality of service by eliminating the burst-and-delay effect inherent in most Internet traffic. Comprehensive logging capabilities provide detailed log information to show the allocation of bandwidth by user, application, and connection. Vendors offer hardware and software products for policy-based management. The hardware approach is faster, but the software approach makes upgrading easier. In addition, the software approach enables policy-based management to be incorporated into other network devices, such as firewalls, switches, and routers.

***See Also* Quality of Service, Service Level Agreements**

Port Replicators

A port replicator is a device that a laptop or notebook computer plugs into for the purpose of instantly connecting to full-size peripherals—including a keyboard, monitor, and printer (Figure 101).

For example, if a computer is used both at home and at work, a port replicator could be set up at both locations. Plugging the notebook computer into the port replicator at either location would give the user instant access to the desktop computer's peripherals, including its keyboard, monitor, and mouse—all without having to remove cables from the desktop computer and attach them to the notebook. Since all the peripherals plug into the back of the replicator, the notebook can be removed without unplugging all the cables.

In addition to serial and parallel ports, the port replicator typically has mouse and keyboard ports as well as a power port. Many port replicators also include a VGA port for an external monitor, so users are not limited to a small laptop screen, as well as a port for an external floppy disk drive. Some units also include an RJ11 modem jack, audio input/output ports, and a MIDI/joystick port for playing games. Newer port replicators now include a Universal Serial Bus (USB) port.

Most notebook computer manufacturers offer a port replicator as an additional option, but they work only with that vendor's products. Now

Figure 101
A port replicator is cabled to a desktop computer and peripherals. When users slide a laptop or notebook computer up the ramp, it locks into place and instantly connects to all the resources cabled to the port replicator.

there are "universal" port replicators that work with virtually any note-book computer. Mobility Electronics, for example, offers a universal port replicator called the EasiDock. The EasiDock's cable ends in a PC Card, which simply slides into the notebook's Type II PC Card slot.

SUMMARY

A port replicator is an easy and economical way to convert a laptop into a desktop computer. It allows laptop users to take advantage of full-size desktop computer components (such as a keyboard or monitor) without having to plug in each device separately. Port replicators are also called *docking stations.* Although the terms are often used interchangeably, a docking station is different in that it also provides slots for adding expansion boards and storage devices. One of the slots can be used for an Ethernet adapter, giving it a direct connection to the network via an RJ45 connector (UTP) or a BNC (coaxial) connector.
See Also **Coaxial Cable, Twisted Pair Cable, Universal Serial Bus**

Powerline Networking

Powerline networking means using a home's existing electrical wiring to interconnect PCs and printers. Creating the network is as simple as plugging special modules into electrical outlets throughout the home and then plugging a computer or printer into each module. Once connected, the modules recognize each other automatically. Built-in surge protectors prevent damage to the PCs and printers.

The first vendor to offer a powerline network kit is Intelogis, a spin-off company of Novell. It began shipping its PassPort Plug-In Network in mid-1998. It uses a home's existing electrical wiring to create an instant network that allows users to share a single Internet connection, printers, and files or play multiuser games between PCs.

At this writing, such a network is limited to providing 350-Kbps connections between devices, but improved technology can boost the connections to 2 Mbps and, later, to 10 Mbps and beyond, enabling powerlines to carry streaming audio and video.

Powerline networking is different from a digital powerline. The former is a product, while the latter is a value-added service being provided by utility companies to their customers.

SUMMARY

For those who live in apartments or temporary residences, a powerline network offers a low-cost alternative to wireless systems, installing Category 5 cable, or adding telephone extensions for phoneline networking. The powerline network can be easily dismantled and quickly set up again when it becomes necessary to move.
See Also **Digital Power Line, Phoneline Networking, Small Office/Home Office LANs, Wireless Home Networking**

Predictive Call Routing

Predictive call routing is an expert systems capability that uses sophisticated algorithms in an attempt to strike a balance between appropriate levels of customer care and call center operating efficiency.

A call-center facility is configured, staffed, and managed to handle a high volume of incoming calls, with multiple agents responding to calls. A traditional call center is typically equipped with an automatic call distributor (ACD) connected to the company's private branch exchange (PBX).[1] The function of the ACD is to take certain calls coming into the PBX and route them to a given extension or any free extension without operator intervention.

---

1. For a more in-depth discussion of ACDs and PBXs, see my book *Desktop Encyclopedia of Telecommunications*, published by McGraw-Hill (ISBN 0070444579). This 600-page reference can be conveniently ordered over the Web at Amazon.com.

Traditionally, incoming calls to an 800 number were held in queue at the ACD until the next available agent became free to answer the call. This "first-in/first-out" (FIFO) approach to handling customer calls led to a big problem: how to simultaneously provide individual customer care faster, leverage agent expertise and time more effectively, increase revenue, reduce operating expenses, and simplify call-center management.

Now, instead of using the traditional FIFO approach of queuing callers for the next available agent, a predictive call-routing algorithm is used that simultaneously looks at the needs of callers, their potential business value, and their ability to wait; it analyzes the skills of the agents and predicts how soon they will likely become available; and then it decides which agents should be matched to callers, regardless of their position in the queue. This approach shifts the call center to a model that more fairly balances the workload of the agents while matching them to the best prospect for the business.

When predictive call routing is used in conjunction with other expert system management capabilities, the ACD can automatically anticipate, control, and optimize the allocation of call-center resources to minimize caller wait times, reduce the number of abandoned calls, lower network costs, and balance workloads among agents.

Predictive call routing software makes real-time decisions based on a business's objectives for the various levels of care it wants to give customers, how it wants to optimize agent skills and time, its sales goals, and the expense it is willing to allocate to meet these objectives. Once the objectives are programmed into the software, the predictive call-routing software will consider all variables—the time a caller will likely wait, the media through which the call has come in (phone, Internet, fax, video, or e-mail), when the agent with the best skill set for the customer will become available, tiers of service, call volume, and sales goals—to determine the best and fairest use of the agent and the best call to take that will bring the greatest value to the business at that time.

SUMMARY

With customers ranging from individual consumers to large organizations, the challenge is to give each caller the precise level of individual care while providing an environment that balances agent skills and workloads against the needs of customers. The use of predictive call-routing technology enables a call center to be more customer-centric and takes the call center to a higher level of productivity and efficiency, making it even more of a strategic asset. It makes it easier for agents to give callers the type of service that generates customer loyalty and sales, but at a lower cost to the business.

See Also Load Balancing

Printer Management

The printer is the most commonly shared device on a network. Print server tools facilitate a large number of users by analyzing printer traffic and resource usage. This includes setting up multiple printers and, if necessary, multiple print servers. To optimize resources, print server tools analyze network, printer, and server speeds, all of which affect print spooling and despooling.

Before users can select a print destination, the administrator must enter information about each network printer into the printer catalog. The printer catalog identifies all the printers attached to the network, including specific information about each printer, such as server name, queue name, and print driver. Printer information is then saved in a database. Maintaining a printer catalog also allows the LAN administrator to restrict printer use based on criteria such as login ID and workstation ID. The printer catalog also is the maintenance tool for updating printer settings whenever they change. Once the catalog is established, the administrator has a single source for printer information. An icon can be assigned to each printer, facilitating printer selection among users.

Through the print manager catalog, LAN administrators can exercise full control over printer use. For example, a printer might be used solely for printing accounting forms, and the paper tray could be stocked with forms numbered sequentially. Submitting a print request for a document other than this form would disrupt the numbering system.

The LAN administrator can also manage printers to facilitate document distribution. For each report, a recipient profile can be created, which describes how it will be handled. A report can go to the sales department's printer, for example, or be delivered to specific users via electronic mail. E-mail is useful when the report requires further manipulation or integration into another file. Since various recipients may need only a few pages of a large report, the specific pages or portions of pages can be selected for print or electronic delivery via page selection profiles. The printer management tools track report delivery and maintain an audit trail of all reports.

Some vendors have Web-enabled their printer-management tools. For example, Hewlett Packard now provides a Web-based version of its JetAdmin printer management software. HP Web JetAdmin allows administrators to remotely install, manage, and troubleshoot network printers with a common Web browser, whether they are in the office or on the road.

Web JetAdmin keeps administrators informed of real-time printer status, diagnostics, and configuration information. Printers can be organized by logical groups and a specific printer or group of printers can be searched for based on a wide variety of criteria. Administrators can create a virtual office layout with dynamic custom maps, showing the location of each printer (Figure 102).

SUMMARY

As the most frequent source of trouble in the majority of networks, printers demand careful attention. Because many networks feature printers from a variety of vendors, it is important to find a printer-management tool that will support the broadest range of printers and server platforms. A tool that can be accessed with a Web browser gives LAN administrators more flexibility in performing various printer-management tasks.

***See Also* Network Printing**

Figure 102
HP Web JetAdmin's unique, custom mapping feature allows LAN administrators to set up maps to visually locate network printers and device groups.

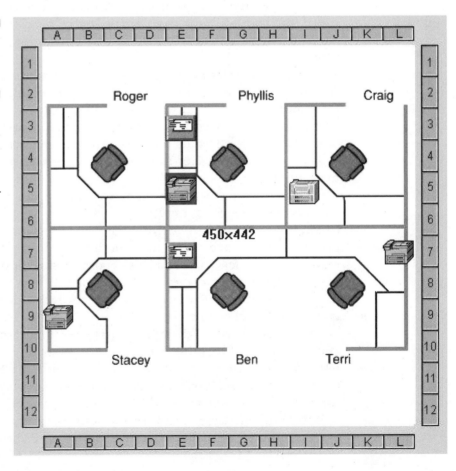

Proxy Servers

Proxy servers are used by companies and Internet service providers (ISPs) for caching frequently accessed files to conserve wide-area network bandwidth and reduce response times for clients. It also enables network administrators to maintain better control over the use of network resources by blocking access to specific servers by user or by document. Proxy servers often provide other functions as well, such as filtering certain types of Web content to prevent employees from wasting company time with non-work related Web browsing or downloading unauthorized software from the Internet which might contain viruses.

A proxy server can be deployed in a variety of ways. It can sit behind the firewall to facilitate access to the Internet and reduce response times. It can also protect information on the secure Web server behind the firewall and offer load balancing via caching. For companies that have several subnetworks, a proxy server deployed at each subnet can reduce traffic on the corporate backbone, eliminating the need for more bandwidth. In situations where remote offices are disconnected from the internal network, a proxy server can provide an inexpensive means for quickly replicating content. Outside the United States, where communications bandwidth is typically much more expensive, proxy servers are even more cost-effective for replicating content.

An ISP could deploy one proxy server at each point of presence (PoP) and cluster them at the Internet gateway to provide faster, more reliable service and reduce network congestion between the PoP and the central Internet gateway. Some ISPs deploy a proxy server only at their gateway to the Internet, which reduces traffic on their link to the Internet, but not on their own network from the PoP to the Internet gateway.

Caching

A proxy server typically supports HTTP, FTP, and Gopher for caching. It may also support the Secure Sockets Protocol (SSL) for the transmission of encrypted traffic, and SOCKS, which is a generic way of tunneling protocols (such as Telnet) that are not proxied. A proxy server uses sophisticated statistical analysis to store the documents most likely to be needed.

Among the many features of proxy servers is caching on-command, which enables an administrator to schedule batch updates to the cache. This includes the ability to preload documents or sites into the cache in anticipation of user demand and the ability to automatically refresh documents that already reside in the cache. Administrators can schedule batch updates to take place at regular intervals and off-peak hours so network bandwidth is not tied up caching documents during periods of heavy network use. Administrators can check the proxy access logs to determine whether frequently accessed sites are actually desirable for caching.

A proxy server may support the Cache Array Routing Protocol (CARP) and Internet Cache Protocol (ICP), which are proposed standards for distributed caching. CARP provides a mechanism for routing content requests among an array of proxy servers in a deterministic fashion. CARP enables load balancing, fault-tolerance, more efficient caching, and easier management for multiple proxy servers. ICP enables a proxy server to send queries to neighbor caches to determine whether they already have a document.

CARP is appropriate for a group of proxy servers serving the same audience of downstream clients or proxies, all under common administrative control. ICP is appropriate for proxies not under common administrative control that may be serving different clients.

Filtering

Network administrators can grant or limit access to network resources, including specific sites and documents, through the use of user name and password, IP address, host name, or domain name.

A proxy server allows administrators to ban access to particular sites using a list of URLs or wildcard patterns. For example, an administrator could use http://*.playboy.com/* to prevent access to all pages belonging to the Playboy site.

A proxy server can also filter based on content type, such as specific Multipurpose Internet Mail Extensions (MIME) types, and based on content, such as HTML tags. In addition, system administrators can implement their own security policies by stopping transmission of Java and JavaScript, and ActiveX components. Many proxy servers now include virus-scanning software to prevent damage to client data and applications.

Security

A proxy server can enhance firewall security in a variety of ways, such as by preventing external users on the Internet from being able to view the network structure and addresses. Blocking this information severely limits the chances of attack from hackers.

Reverse proxying is used to protect a Web server or database behind a firewall. A client connects to a proxy server (with an SSL session if necessary). Then the proxy server initiates a second connection from it to the Web server, from which it can retrieve data. All of this is transparent to the end user. Corporate data can remain behind the firewall and yet be accessible to the public as necessary. To provide additional security, the proxy server can be configured to speak only to the Web server's IP address and vice versa, and the firewall can be configured to allow HTTP traffic only between those two IP addresses.

A proxy server can also provide a circuit-level gateway for generic protocol support. For example, it may include an implementation of SOCKS version 5 for authenticated firewall traversal. SOCKSv5 is an open standard for facilitating traffic through the firewall at the circuit level. It provides generic protocol support for a variety of client platforms, including support for streaming media.

Analysis Tools

A proxy server automatically logs all requests using either the common or extended log-file format. The extended log-file format includes the referrer field and user agent. Administrators can also create their own log-file format by selecting which HTTP fields they would like to log. A built-in log analysis program includes reports such as total number of requests, total bytes transferred, most common URLs requested, most common IP addresses making requests, performance during peak periods, cache hit rates, and estimated response time reduction.

SUMMARY

For many companies and ISPs, a proxy server is a key element of their overall Internet gateway strategy because it improves the performance and security of communications across the TCP/IP-based Internet and private intranets. The proxy's disk-based caching feature minimizes use of the external network by eliminating recurrent retrievals of commonly accessed documents. This significantly improves interactive response time for locally attached clients. The resulting performance improvements provide a cost-effective alternative to purchasing additional network bandwidth. And since the cache is disk-based, it can be tuned to provide optimal performance based on network usage patterns.

See Also **Firewalls, Flow Management, Load Balancing, Network Caching**

Public Key Encryption

Historically, secure transmission of documents has relied on single key encryption, where both the sender and receiver used the same "key" to encrypt and decrypt documents. Often the key would simply be handed to the receiver for future use. In computer communications, the problem is how to arrange for the sender and receiver to agree on the secret key without anyone else possibly finding out what it is. If the parties are in separate physical locations, they must trust some form of transmission medium, such as a courier, phone system, or computer link to share the key. This immediately renders the technique vulnerable, because anyone who overhears or intercepts the key in transit can use it to read, modify, and forge all messages encrypted or authenticated using that key.

The concept of public key encryption (PKE) was developed in 1976 by Whitfield Diffie and Martin Hellman to solve the problems associated with private key encryption. In the Diffie-Hellman scheme, each participant has a pair of keys, one called the public key and the other called the private key. Each person's public key is published and freely accessible, while the private key is kept secret. Public keys are stored in databases (or key certificate authorities) that are available to everyone. Messages are encrypted using the public key of recipients, who decrypt it with a private key known only to them. In addition to protecting the privacy of personal email, PKE has important ramifications for electronic commerce over the Internet.

The advantage of public key encryption is that it is no longer necessary to trust a communications channel to be secure against eavesdropping. The only requirement is that public keys are associated with their users in a trusted or authenticated manner (e.g., in a trusted directory). Anyone can send a confidential message by using just public information, but the message can be decrypted only with the private key belonging to the intended recipient.

Public key encryption is used in a variety of commercial messaging products and is the basis for several electronic commerce standards, including the Secure Electronic Transaction (SET) payment protocol developed by MasterCard and Visa to approve credit-card transactions on the Internet.

Authentication Signatures

Public key encryption can be used not only for encryption but also for authentication, via digital signatures, which prove that a message really comes from its purported sender and that it has not been altered.

To "sign" a message, the sender carries out a computation using both his private key and the message itself. This creates the digital signature. This is attached to the message, which is then sent. To verify the signature, the receiver performs a computation that involves the message, the signature, and the sender's public key. If the same mathematical relation results, the signature is proven to be genuine; if not, the signature is fraudulent, indicating the message has been altered.

Authentication can be achieved by secret key systems, but as with the basic encryption process, it requires the sharing of a secret and sometimes the trust of a third party. Consequently, a sender can repudiate a previously authenticated message by claiming that the shared secret was somehow compromised by one of the parties sharing the secret.

For example, some secret key authentication systems use a central database that holds the secret keys of all users. A successful attack on the database would allow widespread forgery. Public key authentication does not suffer from this weakness, as all users are responsible for their private keys. This element of public key authentication is often called *nonrepudiation*.

PKE systems are not only for corporate and government use; there are effective solutions for personal use as well. One of the most popular programs that provide encryption and digital signatures is Pretty Good Privacy (PGP). This program, now available through Network Associates, is used to send secure files as email attachments. The company's PGP Desktop provides plug-ins for major e-mail products. Integration is performed through the tool bar or the clipboard. PGP is also integrated with Microsoft Explorer for Windows or Macintoshes, allowing the user to encrypt a file. Another component, PGP Disk, enables users to encrypt a hard drive.

Certificates

Certificates are digital documents attesting to the binding of a public key to an individual or other entity. This works in similar fashion to a notary seal in that it binds a user's identity to a signature in a manner that is verified by a trusted third party—the notary.

Certificates allow verification of the claim that a specific public key does in fact belong to a specific individual. Certificates help prevent someone from using a phony key to impersonate someone else. In some cases, it may be necessary to create a chain of certificates, each one certifying the previous one until the parties involved are confident of the identity in question.

A basic certificate contains a public key and a name. Typically, it also contains an expiration date, the name of the certifying authority that issued the certificate, a serial number, and perhaps other information. Most importantly, it contains the digital signature of the certificate issuer.

Corporations, government agencies, and universities issue their own certificates, which are signed and issued to clients by a trusted certificate server. A certificate granted to an individual is a signed recognition of the individual's identification and authenticity. It can be validated by checking the existing certificate against the public key of the certificate server and against the public key of the individual. There are public certificate issuers; Verisign, for example, is a public certificate issuer.

The most widely accepted format for certificates is defined by the ITU's X.509 international standard. Certificates can be read or written by any application complying with X.509. An X.509 certificate contains the following information:

- User name
- User organization
- Certificate start date
- Certificate end date
- User public key parameter
- Certificate authority name
- Certificate authority signature on certificate

Digital Envelopes

Sometimes it is necessary to protect a single message. In this case, a secret-key cryptography can be used. First, the sender and receiver must agree on a session key and each must have a copy. However, in accomplishing this task there is a risk the key will be intercepted during transmission. Public-key cryptography offers an effective solution to this problem within a framework called a digital envelope.

The digital envelope consists of a message that is encrypted using private-key cryptography, but which also contains the encrypted secret key. While digital envelopes usually use public-key cryptography to encrypt the secret key, this is not necessary. If the sender and receiver have an established secret key, they could use it to encrypt the secret key in the digital envelope.

Let's say that a sender uses secret-key cryptography for message encryption and public-key cryptography to transfer the message encryption key. The sender chooses a secret key and encrypts the message with it, then encrypts the secret key using the receiver's public key. The receiver gets both the encrypted secret key and the encrypted message. When the receiver wants to read the message, he or she does so by decrypting the secret key, using the private key, and then decrypting the message using the secret key.

This method can be used to encrypt just one message or for an extended communication. One of the nice features about this technique is that the sender and receiver may switch secret keys as frequently as they like. Switching keys often is beneficial because it is more difficult for an eavesdropper to find a key that is only used for a short period of time.

Not only do digital envelopes help solve the key management problem, but they increase performance, since secret-key cryptosystems are much faster than public-key cryptosystems.

Implementation

In many situations, public key encryption is not necessary. It is usually unnecessary for single users who merely want to keep personal files secure. They can do this with any secret key encryption algorithm using a personal password as the secret key. In general, public key cryptography is best suited for an open multiuser environment; hence its importance for electronic commerce on the Internet.

There are several issues to consider before choosing a specific type of key encryption system. The primary advantage of public key encryption is increased security and convenience, based on the fact that private keys never need to be transmitted or revealed to anyone other than their owner.

A serious disadvantage of public key encryption is that such systems are slower than secret-key cryptosystems. This slowness has prevented SET from being widely adopted by banks and other financial institutions for securing electronic transactions on the Internet.

The speed differential applies to hardware implementations of public key encryption systems as well. Even in software, private key encryption is generally at least 100 times faster than public, and when it is implemented in hardware, the difference can be magnified two to three times. Even though the gap has narrowed to an appreciable extent over the last few years, current public key encryption algorithms will never match the performance of those used for private key encryption. This problem may be solved in the future with new encryption algorithms based on elliptic-curve encryption that simply uses a mathematical technique that experts—including RSA—generally agree is much faster.

In addition, public key encryption systems require management and almost always require user support—both of which increase the cost of doing business. Furthermore, encryption is open to abuse, as when a disgruntled superuser decides to spice up his or her life by hacking into the security system to create a little database mayhem.

Despite the added costs, plus the potential risk of abuse, many security experts advocate biting the bullet and implementing a key cryptography system anyway. The consequences of not doing so are too compelling to ignore, especially since other security measures—such as passwords, authentication and login procedures, firewalls, and even physical defenses—are not 100-percent reliable.

Government restrictions on the use of strong encryption have serious ramifications for U.S. companies with international locations on their virtual private networks (VPNs). These restrictions mean that companies must first apply for an encryption export license from the U.S. Commerce Department and then get permission to use it from each country that requires it. France and Russia, for example, prohibit unregistered encryption use, while other countries, such as Saudi Arabia, simply ban encryption. Companies should not even think about getting export permission if their network traverses countries like Cuba, Iraq, or Libya. These are on the U.S. State Department's list of terrorist-sponsoring countries.

Cryptography products used strictly for authenticating users' identities, such as digital signatures, is usually not subject to U.S. export restriction or domestic-use rules. But Germany and Malaysia are among the few countries with rules pertaining to the provision of digital certificate services.

Tamperproof Security

The future of electronic commerce depends on the continued availability of strong security schemes. Many of the current e-commerce systems rely on public key encryption to protect credit card numbers and other sensitive information from falling into the wrong hands. However, there is no guarantee that such systems will remain tamperproof forever.

For over 20 years, the U.S. government's own Data Encryption Standard (DES) had been considered uncrackable because its 56-bit key makes possible 72 quadrillion (72,000,000,000,000,000) key combinations. Yet, even this formidable barrier has been breached many times. The latest game in the security industry is how fast the code can be broken. RSA Laboratory conducts the DES Challenge, a contest spearheaded by the Electronic Frontier Foundation (EFF), a non-profit civil liberties organization that deals with Internet privacy and security issues. The 1998 winners used a $250,000 supercomputer to break DES in less than three days.

Encryption systems that are essentially tamperproof already exist, but are difficult to find in many countries—including the U.S. The reason is that governments want the ability to easily crack encrypted messages and transactions for law enforcement and national security reasons. Accordingly, the U.S. government bans the export of software with more than 40 bits of encryption or, in some cases, 56 bits. The government does allow the sale of these products in the U.S. and Canada, but requires that a key escrow "trap door" be provided. The idea is to prevent individuals based outside of the U.S. and Canada from procuring extremely strong encryption schemes that the government cannot easily crack.

SUMMARY

Encryption should be the last line of defense against hackers who finally break into the corporate network and it should be a part of any strategy to safeguard the transmission of sensitive information, especially if it goes out over the wide open Internet. In many cases private key encryption provides effective security for personal information stored on a hard disk. When the data must be transferred to another machine over a transmission line or service, public key encryption is an effective and economical solution to protect it against eavesdroppers. Its slow operation eventually will be addressed by speedier encryption algorithms. Although the bit-length of the codes determines the strength of a particular security product, eventually governments will come to understand that it is futile to try and control the distribution of advanced encryption products, since they are already in the hands of criminals anyway.

See Also **Firewalls, LAN Security, Security Risk Assessment**

Q

Quality of Service

With networks carrying more bandwidth-intensive, real-time voice, video, and data, there is an increasing need for Quality of Service (QoS) solutions to provide better and more predictable network service so these applications can operate at peak performance. QoS solutions provide dedicated bandwidth, controlled jitter and latency, and improved loss characteristics. These goals are achieved by router software, which manages network congestion, shaping network traffic, using expensive wide-area links more efficiently, and setting traffic policies across the network that prioritize traffic to ensure that mission-critical applications get the service they require, while making sure that routine data gets adequate service as well.

Benefits

In addition to prioritizing traffic, QoS software provides the following benefits:

RESOURCE CONTROL Provides control over which resources are being used. For example, the bandwidth consumed over a backbone link by FTP transfers can be limited to give priority to remote database access.

RESOURCE EFFICIENCY Through the use of analysis and accounting tools, network managers can see what the network is being used for and verify that it is servicing the most important traffic.

TAILORED SERVICES The control and visibility provided by QoS enables network operators to offer different grades of service to their customers and charge for them accordingly.

APPLICATIONS COEXISTENCE QoS technologies make certain that the network is used efficiently by all applications. Mission-critical applications, time-sensitive multimedia, and voice will have the bandwidth and minimum delays they require, while other applications on the link will get service without interfering with priority traffic.

Service Levels

Essentially, QoS can provide three levels of "strictness" across a heterogeneous network from end to end:

BEST EFFORT Provides basic connectivity with no traffic handling guarantees. The Internet is an example of a best-effort service. Best effort is suitable for a wide range of networked applications such as general file transfers or e-mail.

DIFFERENTIATED On a statistical basis, some traffic is treated better than the rest, but there is no firm guarantee on faster handling, more bandwidth, or lower loss rate. Differentiated service can provide expedited handling appropriate for a wide class of applications, including lower delay for mission-critical and packet voice applications.

GUARANTEED Reserves network resources for specific traffic.
The "strictness" of QoS describes how tightly the service can be bound by specific bandwidth, delay, jitter, and loss characteristics. These characteristics can be offered to within tight tolerances on a terrestrial Time Division Multiplex (TDM) circuit, or for an ATM Variable Bit Rate Real-Time (VBR-rt), or Constant Bit Rate (CBR) service. However, they are much harder to bind on a typical Internet IP connection.

Congestion Management Tools

One way that network elements handle an overflow of arriving traffic is to use a queuing algorithm to sort the traffic, then prioritize it onto an output link. Router software offers several queuing tools.

FIFO QUEUING Basically, first-in/first-out (FIFO) queuing involves storing packets when the network is congested and forwarding them in order of arrival when congestion eases. As the default queuing algorithm, FIFO requires no configuration. However, it has several shortcomings. First, it makes no decision about packet priority; the order of arrival determines bandwidth, promptness, and buffer allocation. It also does not provide protection against ill-behaved applications (sources). Bursty sources can cause high delays in delivering time-sensitive application traffic, and potentially to network control and signaling messages. FIFO queuing was a necessary first step in controlling network traffic, but to-day's intelligent networks need more sophisticated algorithms. Today's router software implements queuing algorithms that avoid the shortcomings of FIFO queuing.

PRIORITY QUEUING   Priority Queuing (PQ) ensures that important traffic gets the fastest handling at each point where it is used. It gives strict priority to important traffic and can flexibly prioritize according to network protocol—IP, IPX or AppleTalk, for example—as well as incoming interface, packet size, and source/destination address.

With priority queuing, each packet is placed in one of four queues—high, medium, normal, or low—based on an assigned priority. Packets that are not classified by this priority-list mechanism default to the normal queue. During transmission, the algorithm gives higher-priority queues preferential treatment over low-priority queues. This is a simple and intuitive approach but can cause queuing delays that the higher-priority traffic might have experienced to be randomly transferred to the lower-priority traffic, increasing jitter on the lower-priority traffic. Higher-priority traffic can be rate limited to avoid this problem.

PQ is useful for making sure that mission-critical traffic traversing various WAN links gets priority treatment. For example, PQ may be used to ensure that important Oracle-based sales reporting data gets to its destination ahead of other less critical traffic. However, PQ uses static configuration and thus does not automatically adapt to changing network requirements.

CUSTOM QUEUING   Custom queuing handles traffic by assigning a specified amount of queue space to each class of packets and then servicing the queues in a round-robin fashion. This allows various applications to share the network, each having specific minimum bandwidth or latency requirements. CQ can be used to provide guaranteed bandwidth at a potential congestion point, assuring the specified traffic a fixed portion of available bandwidth and leaving the remaining bandwidth to other traffic. For example, encapsulated SNA requires a guaranteed minimum level of service. Half of available bandwidth can be reserved for SNA data, allowing the remaining half to be used by other protocols such as IP and IPX. Like priority queuing, custom queuing is statically configured and does not automatically adapt to changing network conditions.

WEIGHTED FAIR QUEUING   Weighted Fair Queuing (WFQ) is used in situations where consistent response time must be given to heavy and light network users alike without adding excessive bandwidth. WFQ is a flow-based queuing algorithm that schedules interactive traffic to the front of the queue to reduce response time, and fairly shares the remaining bandwidth between high bandwidth flows.

WFQ ensures that queues do not starve for bandwidth, and that traffic gets predictable service. Low-volume traffic streams—which comprise the majority of traffic—receive preferential service, transmitting their entire offered loads in a timely fashion. High-volume traffic streams share the remaining capacity proportionally between them.

WFQ is designed to minimize configuration effort and adapts automatically to changing network traffic conditions. In fact, WFQ does such a good job for most applications that it has been made the default queuing mode on most serial interfaces configured to run at or below T1/E1 speeds (1.544 Mbps/2.048 Mbps).

WFQ is efficient in that it will use whatever bandwidth is available to forward traffic from lower priority flows if no traffic from higher priority flows is present. This is different from Time Division Multiplexing (TDM) which simply carves up the available bandwidth and lets it go unused if no traffic is present for a particular traffic type.

The WFQ algorithm also addresses the problem of round-trip delay variability. If multiple high-volume conversations are active, their transfer rates and interarrival periods are made much more predictable. WFQ greatly enhances algorithms such as the SNA Logical Link Control (LLC) and the Transmission Control Protocol (TCP) congestion control and slow-start features. The result is more predictable throughput and response time for each active flow.

Congestion Avoidance

While congestion management techniques operate to control congestion after it occurs, congestion avoidance algorithms monitor traffic loads in an effort to anticipate and avoid congestion at common network bottlenecks. The primary congestion avoidance tool used in most routers, especially those of Cisco Systems, is Weighted Random Early Detection (WRED). This is actually a combination of two capabilities: Random Early Detection (RED) and IP Precedence.

The RED algorithms are designed to avoid congestion before it becomes a problem. RED works by monitoring traffic load at various points in the network and discarding packets if the congestion begins to increase. The source detects the dropped traffic and slows its transmission rate accordingly. RED is primarily designed to work with TCP in IP internetworks.

IP Precedence utilizes three bits in the IPv4 header's Precedence/Type of Service (ToS) field to specify a precedence for each packet in terms of 0 to 7, with 0 indicating normal and 7 providing the highest priority. ToS utilizes 4 bits to indicate a packet's delay, reliability, throughput, cost or security. The queuing technologies throughout the network can then use this information to provide the appropriate handling.

Combining the capabilities of the RED algorithm and IP Precedence provides for preferential traffic handling for higher-priority packets. It can selectively discard lower-priority traffic when the interface starts to get congested and provide differentiated performance characteristics for different classes of service.

QoS on IP Nets

As more applications converge on the use of IP as the primary networking protocol, it becomes increasingly necessary to implement a QoS solution to expedite traffic flow. However, true end-to-end QoS requires that every element in the network path—switch, router, firewall, host, clients, and others—take part in supporting QoS. This is fairly easy to do when an organization has control of the entire network end to end, as in a corporate intranet, for example, or a multi-company extranet. QoS is much more difficult to implement across different network domains, where there is no single management authority, as is the case with the public Internet.

Cisco Systems, which supplies 80 percent of the routers used on the Internet, takes advantage of the end-to-end nature of IP to address this situation. Its router software overlays Layer 2 technology-specific QoS signaling solutions with the Layer 3 IP QoS signaling methods of RSVP and IP Precedence. RSVP signals provide for Guaranteed QoS and IP Precedence signals provide for Differentiated QoS.

RSVP is an IETF Internet Standard (RFC 2205) protocol for allowing an application to dynamically reserve network bandwidth. RSVP enables applications to request a specific QoS for a data flow. Cisco's implementation also allows RSVP to be initiated within the network using configured proxy RSVP. This capability allows network managers to take advantage of the benefits of RSVP in the network, even for applications and hosts that have not been enabled with RSVP.

Hosts and routers use RSVP to deliver QoS requests to the routers along the paths of the data stream and to maintain route and host states to provide the requested service, usually bandwidth and latency. WFQ sets up the packet classification and scheduling required for the reserved flows and continues to handle non-reserved traffic by expediting interactive traffic and fairly sharing the remaining bandwidth between high-bandwidth flows. WRED applies its congestion avoidance on non-RSVP traffic flows.

SUMMARY

Just about any network can take advantage of QoS, whether it is a small corporate network, Internet service provider, or enterprise network. In addition, QoS can be applied to selected traffic running over various network technologies, including Asynchronous Transfer Mode (ATM), SONET, frame relay, and IP-routed networks that may use any or all of these underlying technologies. QoS can even be applied to Ethernet and token ring networks.

See Also **Latency, Network Congestion, Performance Metrics, Response Time, Service-Level Agreements**

R

Remote-Access Concentrators

A remote-access concentrator (RAC) is a highly scalable platform for providing the connection between the PSTN and data backbone networks, including the Internet. Such platforms are used by network service providers—including phone companies, Internet access providers, information service providers, and value-added network (VAN) providers—in environments where reliability, scalability, and fault tolerance are mandatory.

RACs are flexible in terms of configuration and are capable of integrally supporting channel banks, digital modems, and data network interfaces for TCP/IP and X.25. Such platforms can also be equipped to function as a wireless data communications platform, providing connectivity between PSTNs and Cellular Digital Packet Data (CDPD) networks based on the Internet Protocol (IP).

Today's most advanced RACs are completely software-defined, integrating the functions of T1/E1 and analog interfaces, ISDN and V.34 modems, and TCP/IP routing. This is accomplished with digital signal processor (DSP) cards that are software-configurable. Software-defined control gives service providers the means to remotely enhance or modify platforms at geographically dispersed sites from their network control center. This allows them to respond quickly to changing subscriber demands while eliminating costly, time-consuming trips to network sites by field engineers.

A single RAC can be scaled from several modem ports to several hundred modems or ISDN terminal adapters. Software-defined modems can serve a wide range of subscriber devices because the platform's integrated T1/E1 interface allows dynamic call routing based on Dialed Number Information available on the trunk. An integrated Data Network Gateway provides a TCP/IP interface supporting up to 96 user connections on a single plug-in card. With its high-density modem and IP connections, the remote-access concentrator gives service providers a high-performance, software-defined alternative to the hassle of managing and deploying discrete hardware components such as modem banks, terminal adapters, multiplexers, routers, and terminal servers.

Each board in a remote-access concentrator can be configured to collect various types of statistical information at a user-defined interval. T1/E1 performance, modem call statistics and call duration, X.25 packet and frame level statistics, and more are gathered in log files where they can be processed to create performance reports for an entire system and its individual components. The management system's real-time diagnostic capabilities monitor specific activity of system components. For example, T1/E1 monitors provide data such as alarm states, channel signaling states, and channel connection states. The SNMP agent provides extensive monitoring capabilities. The operator at an SNMP console can view each concentrator's overall operational state, each board's current state, and the state of individual modems. The agent also allows an operator at an SNMP console to monitor and control the management links between the network management system and each concentrator.

SUMMARY

Remote-access concentrators are high-density systems that consolidate all remote-access traffic from dial-in analog and ISDN ports as well as industry-standard WAN and remote connections. With the ever-increasing number of users accessing the Internet and corporate intranets, service providers are rapidly deploying points of presence (PoPs) that include RACs to reduce network complexity and costs. Through the use of DSP technology, features and interfaces can be defined in software, allowing service providers to offer customers access capabilities at a low cost per-port, while accommodating future growth as new features and connectivity standards emerge.

See Also **Statistical Time Division Multiplexers, Time Division Multiplexers**

Remote Login

Remote login (rlogin) client-server software was originally written by members of the University of California at Berkeley for the BSD UNIX operating system. Rlogin protocols are very simple compared to Telnet, with only elementary flow control and window-size exchanges passed between client and server. After the login is completed, rlogin does not use any protocol to affect the session. This means that whatever character is typed locally is sent to the other machine.

Although Telnet has generally replaced rlogin as the remote terminal of choice, rlogin is still used in many local environments because it allows users to access "trusted hosts" without a password. The "trusted host" concept means that if a user successfully logs into one host, all the other hosts assume the user will use the same user ID to log into them. Files control rlogin's assumption that users will want to use the same user ID for logging into multiple hosts. One of these files is /etc/hosts or /etc/hosts.equiv. A copy of this file is on every machine. It lists all the other machines it can "trust" to have matching user names. This configuration is typically used when all the machines are in the same workgroup, department, or organization.

Individual users on machines that are outside the local environment have a file called .rhosts, which is a private list of trusted machines. A user who logs in with rlogin from any of the machines on the list will not be asked for a user ID and password. Whenever a remote system does not recognize a user's ID and password—perhaps because it is not on the list—it will prompt the user for them, just as Telnet does.

SUMMARY

Telnet is used to log into all kinds of computers, whether they are UNIX machines or not. To remotely access another UNIX machine, however, rlogin is usually more convenient because it automates more of the process. With a few exceptions—such as escaping from rlogin—Telnet and rlogin are used in much the same way.

See Also Telnet

Repeaters

A repeater extends the effective distance of various networks by regenerating the signals. As such, the repeater operates at the lowest level of the OSI reference model—the physical layer (Figure 103).

Repeaters are used on wireless as well as wired networks to regenerate the signal. Signal regeneration is necessary because signal strength weakens with distance: the longer the path a signal must travel, the weaker it gets. This condition is known as signal attenuation. On a telephone call, a weak signal will cause low volume, interfering with the parties' ability to

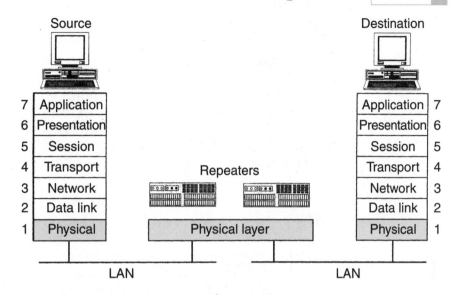

hear each other. In cellular networks, when a mobile user moves beyond the range of a cell site, the signal fades to the point of disconnecting the call. In the LAN environment, a weak signal can result in corrupt data, which can substantially reduce throughput by forcing retransmissions when errors are detected. When the signal level drops low enough, the chances of interference from external noise increase, rendering the signal unusable.

In the LAN environment, the protocols limit the use of repeaters. With Ethernet, for example, repeaters are usually required every 500 feet. The IEEE 802.3 design guidelines for Ethernet LANs specify that the span between the two farthest users, including the cable connecting each user to the LAN, cannot exceed 500 meters (about 1500 feet). Even with repeaters, the typical Ethernet LAN application requires that the entire path not exceed 1,500 meters end-to-end. This is the limit established for the proper operation of Ethernet's media access control mechanism, known as carrier sense multiple access with collision detection (CSMA/CD).

Repeaters are especially useful for extending a departmental network backbone. When the number of network users keep increasing, its cable length can easily exhaust the network constraints. To solve this problem, repeaters are used to join the network segments together. Each repeater has its own power supply and multiple ports. An auto-partition feature is provided for each individual port, which isolates a faulty cable from the rest of the network. To facilitate troubleshooting, repeaters have diagnostic LEDs that indicate power status, packet reception, collisions, port partitioning, and link status. When the fault is corrected, the repeater automatically reconnects the segment to the network.

SUMMARY

Aside from signal regeneration, repeaters can also be used to link different types of network media—fiber to coaxial cable, for example. Often LANs are interconnected in a campus environment by means of repeaters that form the LANs into connected network segments. The segments may employ different transmission media—thick or thin coaxial cable, twisted pair wiring, or optical fiber.

See Also **Bridges, Routers, Gateways**

Reportable Incidents

The Federal Communications Commission (FCC) describes a "reportable incident" as any network failure, caused by a hardware, software or procedural problem that blocks 90,000 or more calls and lasts more than 30 minutes. With this data, collected by the FCC, competitive carriers attempt to gain each other's customers.

A network failure is any time a call does not go through because of a hardware, software, or procedural error. Hardware errors are equipment problems, such as a power failure, equipment damaged by a hurricane or tornado, or perhaps a section of fiber-optic cable torn from the ground by a contractor's backhoe. A software error involves problems with the software that runs the network's complex signaling, switching and transmission equipment, while a procedural problem generally involves human error.

However, the FCC's definition of reportable incident is skewed in favor of smaller carriers. Based on the volume of calls handled by the various carriers, a relatively short failure on the AT&T network can block 90,000 calls, while significantly longer failures on other networks may never block 90,000 customer calls.

To put the number 90,000 in proper perspective, on an average business day, 90,000 calls represents less than 4/100ths of one percent of the total traffic on the AT&T network. However, other long-distance companies could crash a significant segment of their network or perhaps their entire operation for an extended period of time and still not block 90,000 calls.

A better gauge of network reliability is "defects per million." This measurement is a statistically valid record of how many calls per million did not go through the first time because of a network procedural, hardware, or software failure. Defects per million is not an average; it is an accurate accounting of network performance that is tallied by the day, as well as month to date and year to date.

For example, AT&T's defects-per-million performance in 1997 was 173, which means that of every one million calls placed on the AT&T network, only 173 did not go through the first time due to a network failure. That equals a network reliability rate of 99.98 percent for the entire year.

SUMMARY

Using the FCC definition of reportable incident is like trying to compare apples to oranges, especially when one network handles so much more traffic than the others. Any discussion of FCC reportable incidents does not provide a valid comparison of network reliability; however, reviewing the percentage of calls completed on the first attempt does provide a valid comparison.

See Also **Network Reliability and Interoperability Council**

Response Time

Response time is a measure of the total elapsed time between the start and completion of a task. On the Internet, for example, response time would be the time it takes after a user hits the Enter key and a requested Web page is displayed by the browser.

Response time on the Internet is often a subjective measure. If the user is only interested in reading the text of a Web page, and the page's image content is still loading, then response time would be the time it takes for the Web browser to display the text after the Enter key is hit.

Response time on the Internet is also variable, depending on factors such as network latency, the size of the requested Web page, and the load on the Web server. All of these factors can change at any given moment, causing response time to be inconsistent, even when the same task is performed repeatedly.

Response time can be measured in a variety of ways. For example, special probes installed on the network can look at the application protocol to measure application transactions times. Although these probes are not capable of measuring all transactions accurately, they do provide a good estimate of the response time users experience. There are some tools that are protocol-specific, such as TCP Acknowledge (ACK) Response probes, which measure the response times of TCP applications. Specifically, these probes measure the time it takes the end-system to acknowledge the receipt of TCP packets for a particular data stream. However, these types of probes are usually no more accurate than other types of probes.

A more accurate approach to measuring response time is to install a monitor on the end-user system. Response-time information captured here more accurately reflects what the user experiences. The disadvantage of this approach is that the monitoring software must be installed on each client.

SUMMARY

The accuracy of response-time measurements can be established over time through a process known as baselining. This involves taking measurements at times when the network is very busy and at times when it is less busy. If response time goes up at times when the network has less traffic and goes down when the network has more traffic—and these changes coincide with what the user experiences—then the measurement tool is performing reliably. This becomes the baseline against which all future measurements are compared. The accuracy of this baseline will diminish as the network changes: new equipment and lines may be added, for example. Or, more users may be added to the existing network. Obviously, corporations with private intranets will have more control over these factors than those who rely exclusively on the Internet, where changes often go unreported.
See Also **Latency, Ping, Quality of Service**

Routers

Routers are used to forward data packets through large mesh networks such as the global Internet or large corporate intranets. Routers are also used to segment a network with the goals of limiting broadcast traffic and providing security, control, and redundant paths. In addition to providing access to the wide-area network (WAN), routers can be equipped with firewall software to protect corporate resources from intruders.

A router is similar to a bridge in that both provide filtering and bridging functions across the network. But while bridges operate at the Data Link layer of the OSI reference model, routers operate at the Network layer (Figure 104), distinguishing among network layer protocols and making intelligent packet-forwarding decisions. Routers convert LAN protocols into wide-area network protocols such as TCP/IP, and perform the process in reverse at the remote location. They may be deployed in mesh as well as point-to-point networks and, in certain situations, can be used in combination with bridges.

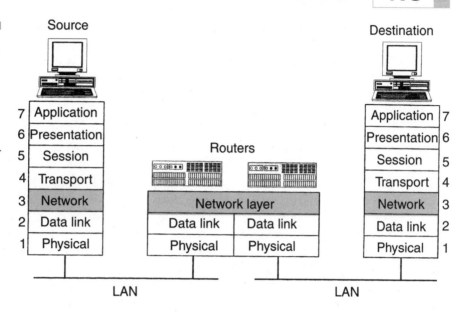

Figure 104
Routers operate at the Network layer (Layer 3) of the Open Systems Interconnection (OSI) reference model.

Although routers include the functionality of bridges, they differ from bridges in the following other ways: they generally offer more embedded intelligence and, consequently, more sophisticated network management and traffic control capabilities than bridges. Perhaps the most significant distinction between a router and a bridge is that a bridge delivers packets of data on a "best effort" basis, which may result in lost data unless the host computer protocol provides protection. By contrast, a router has the potential for flow control and more comprehensive error protection.

Types of Routing

There are two types of routing: static and dynamic. In static routing, the network manager configures the routing table to set fixed paths between two routers. Unless reconfigured, the paths on the network never change. Although a static router will recognize that a link has gone down and issue an alarm, it will not automatically reroute the traffic.

A dynamic router, on the other hand, reconfigures the routing table automatically, and recalculates the most efficient path in terms of load, line delay, or bandwidth. Some routers even balance the traffic load across multiple links. This allows the various links to better handle peak traffic conditions. And if one of the links goes down, the other link will take its traffic.

Routing Protocols

The routers within an autonomous system (AS) communicate with each other using interior gateway protocols of which there are two types: distance vector and link state. Distance vector protocols issue periodic broadcasts that propagate routing tables across the network. Such protocols, one of which is the Routing Information Protocol (RIP), are adequate for small, stable networks, but not for large, growing networks where the periodic broadcast of entire tables can consume an inordinate amount of network bandwidth. Link state protocols, such as Open Shortest Path First (OSPF), operate more efficiently. Instead of issuing periodic broadcasts of entire tables, they send routing information on a flash basis to reflect only the changes to network connections.

Although still supported by many vendors, RIP does not perform well in today's increasingly complex networks. As the network expands, routing updates grow larger under RIP and consume more bandwidth to route the information. When a link fails, the RIP update procedure slows route discovery, increases network traffic and bandwidth usage, and may cause temporary looping of data traffic. Also, RIP cannot calculate routes based on such factors as delay and bandwidth, and its line selection facility is capable of choosing only one path to each destination.

OSPF overcomes the limitations of RIP and even provides capabilities not found in RIP. The most important of the interior gateway protocols, OSPF is a link state routing algorithm in which each router knows the topology of the entire autonomous system. The table update procedure of OSPF entails each router on the AS transmitting a packet with a description of its local links to all other routers. The receipt of each packet is acknowledged and a routing table is built from the collected descriptions. If there is a network failure, then new topology information floods the AS, updating the routing tables.

There is a variation of OSPF specifically for handling multicast traffic. The Multicast Open Shortest Path First (MOSPF) protocol is used to distribute group membership information to all routers in the network so that each node can determine where to forward multicast packets. MOSPF is also well suited to campus deployment where global distribution of membership information is not problematic—particularly where OSPF is the unicast routing protocol.

Types of Routers

Multiprotocol nodal, or hub, routers are used for building highly meshed wide-area networks. In addition to allowing several protocols to share the same logical network, these devices pick the shortest path to the end node, balance the load across multiple physical links, reroute traffic around points of failure or congestion, and implement flow control in conjunction with the end nodes. They also provide the means to tie remote branch offices into the corporate backbone, which might use such WAN services as TCP/IP, T1, ISDN, Frame Relay and ATM. Some vendors also provide an interface for SMDS, although this service is less widely used.

Access routers are typically used at branch offices. These are usually fixed-configuration devices available in Ethernet and token-ring versions that support a limited number of protocols and physical interfaces. They provide connectivity to high-end multiprotocol routers, allowing large and small nodes to be managed as a single logical enterprise network. Although low-cost, plug-and-play bridges can meet the need for branch office connectivity, low-end routers can offer more intelligence and configuration flexibility at comparable cost.

Mid-range routers provide network connectivity between corporate locations in support of workgroups or the corporate intranet, for example. These routers can be standalone devices or packaged as modules that occupy slots in an intelligent wiring hub or LAN switch. In fact, this type of router is often used to provide connectivity between multiple wiring hubs or LAN switches over high-speed LAN backbones such as ATM, FDDI, and Fast Ethernet.

SUMMARY

Routers fulfill a vital role in implementing complex mesh networks such as the global Internet and large corporate intranets. They also have become an economical means of tying branch offices into the enterprise backbone network. Like other interconnection devices, routers are manageable via SNMP, as well as the proprietary management systems of vendors.

***See Also* Bridges, Gateways, Repeaters**

S

Security Risk Assessment

Security risk assessment is a proactive process that helps network managers and IT administrators identify and resolve security breaches before they are discovered and exploited by hackers and cause serious problems later. Using such tools also can prevent customer information from falling into the wrong hands, which can result in lawsuits, financial penalties, and damage to corporate image.

Appearance of SATAN

The first security risk assessment tool was created by computer-security researcher Dan Farmer and released in April 1995. Farmer achieved notoriety by selecting banking, credit union, government, newspaper, and Internet commerce sites as guinea pigs to see if these sites were safe from hacker attacks. He found that 60 percent of the sites probed with his newly developed tool, SATAN (Security Administrator Tool for Analyzing Networks), could be broken into or wiped out, indicating that many administrators were not too savvy about security.

SATAN scans IP nets and hosts for vulnerabilities. It recognizes common networking-related security problems and reports them. For each type or problem found, SATAN offers a tutorial that explains the problem and what its impact could be. The tutorial also explains what can be done about the problem: correct an error in a configuration file, install a bug fix from the vendor, use other means to restrict access, or simply disable service.

Freely available on the Internet, SATAN provides a good risk assessment and is fairly easy to use. However, it has been criticized for not having its problem database updated often enough. Meanwhile, more than a dozen vendors now offer this kind of risk assessment tool and the market for such tools has grown to about $60 million.

Commercial Tools

Among these tools is Netective developed by Netect Inc., a provider of network scanning and response software. Netective is a complete software solution that scans and detects networks for potential security holes, and offers the user patches or corrective actions to fix the breaches before they become a threat. Netective identifies and resolves security vulnerabilities at both the operating system level and the network level, protecting against both internal and external threats. It also monitors key system files for unauthorized changes and, by referencing a one-million-word dictionary, identifies vulnerable user passwords through a variety of password cracking techniques.

A detailed report provides IT administrators with a description of each vulnerability and corrective action as well as a ranking of vulnerabilities by the risk they pose to a site's security. Administrators are also presented with a high-level overview of the vulnerability and its solution with an option to link to a more detailed explanation and reference materials.

Employing an implementation model similar to anti-virus products, Netect provides ongoing security updates via the Internet to ensure that users are protected from the latest threats. Netect uses secure push technology to broadcast the vulnerability updates. Users are not required to reinstall the software in order to integrate the updates.

Axent Technologies, Inc. offers NetRecon, a tool that checks for vulnerabilities from various points within the network and reports them in real time. It runs on a Windows NT workstation and probes all networks and network resources, including UNIX servers, Windows NT servers, NetWare networks, Windows 95 and 3.x workstations, mid-range systems, mainframes, routers, gateways, Web servers, firewalls, and name servers.

NetRecon's employs a scanning technique that allows it to immediately display vulnerabilities as they are detected and quickly performs deeper probes. Multiple objectives are executed in parallel, checking each network and attached device for common vulnerabilities. The objectives feed each other information for maximum efficiency and effectiveness. For example, one objective looks for password information from an NIS server, another objective tries to crack passwords, while another looks for servers with rlogin (remote login) services to see if the cracked user passwords will provide access. The results from the first objective are loaded into the second and from the second into the third as soon as some results are available.

NetRecon's objectives are hierarchically organized, with the top objectives being the broadest. To start, the user generally clicks on the "Discover Network Vulnerabilities" objective to execute a scan. This automatically executes all of the necessary subordinate objectives needed to check the network for vulnerabilities.

NetRecon can scan a company's networks from the outside or the inside. Scanning from the outside shows how a network looks to an external attacker, while scanning from the inside reveals which parts of the network can be exploited internally or by an external attacker who has gained local network access.

Risk Assessment Service

A comprehensive risk assessment service is offered by the International Computer Security Association (ICSA). The organization's TruSecure is a package of security assurance services to help Internet-connected organizations assess their vulnerabilities. The service was developed in response to data ICSA compiled in a survey of 200 Internet-connected organizations, including small businesses, Fortune 500 companies and federal government agencies. According to the survey, 93 percent of the responding organizations had security flaws that left them open to malicious attacks, even though they had working firewalls.

Using a variety of home-grown, commercial, and underground hacking tools, ICSA performs a remote electronic assessment of an organization's network connections to make it more aware of what devices can be seen by an intruder. ICSA technicians can probe corporate Web servers, routers, and services such as FTP and Telnet.

To achieve TruSecure certification, ICSA clients undergo a six-step process. The first step is the testing and analysis of vulnerabilities. Next, a methodology of best practices is implemented to bring the network and systems up to security standards. Then ICSA conducts an electronic performance review, on-site audits, ICSA perimeter certification and periodic spot checks. The TrueSecure service starts at $39,000 per year.

One of the oldest security assessment and reporting services is offered by the Computer Emergency Response Team (CERT) Coordination Center, which is part of the Networked Systems Survivability program in the Software Engineering Institute, a federally-funded research and development center at Carnegie Mellon University. The program was set up in 1988 to study Internet security vulnerabilities, provide incident response services to sites that have been the victims of attack, publish a variety of security alerts, research security and survivability in wide-area networked computing, and develop information to help improve host security.

The Coordination Center issues CERT advisories—documents that provide information on how to obtain a patch or details of a workaround for a known computer security problem. The CERT Coordination Center works with vendors to produce a workaround or a patch for a problem, and does not publish vulnerability information until a workaround or a patch is available. A CERT advisory may also be a warning about ongoing attacks.

Vendor-initiated bulletins contain verbatim text from vendors about a security problem relating to their products. They include enough information for readers to determine whether the problem affects them, along with specific steps that can be taken to avoid problems. The goal of the CERT Coordination Center in creating these bulletins is to help the vendors' security information get wide distribution quickly.

SUMMARY

The problem of security is multi-faceted and there is no single solution that is capable of addressing all of the possible threats—from internal as well as external sources. Depending on the scale of the network and the kinds of applications that run over it, building a secure network often requires several products from different vendors. Used in combination, these products can provide a high degree of protection. However, maintaining security is an ongoing challenge. Networks continually undergo change. Once appropriate solutions are put into place, security managers should use security risk assessment tools to continually test their networks for new vulnerabilities.

See Also **Automated Intrusion Detection, Firewalls, LAN Security**

Serial Transmission

As the term implies, serial transmission is a method of sending/receiving data between Data Terminal Equipment (DTE) such as a computer and Data Communications Equipment (DCE) such as a modem—one bit at a time. A sequence of 7 or 8 bits is called a byte or character. The boundaries of a character are set by start and stop bits. It is the responsibility of the transmitter and receiver to keep track of when a character starts and ends as well as the sequence of the bits. The transmitter and receiver must transmit and receive at the same rate, which is expressed in bps, or bits per second.

In the PC environment, 7- or 8-bit characters are used. Seven bits are enough to encode all upper- and lowercase ASCII characters and symbols, which number 127. The eighth bit is used for the parity bit, which is used to check data integrity. If used, the parity bit is inserted between the last bit of a character and the first stop bit. There are several parity bit designations:

MARK   This parity bit is always set at a logical 1.

SPACE   This parity bit is always set at a logical 0.

EVEN   This parity bit is set to logical 1 by counting the number of bits in the character and determining if the result is even.

ODD   This parity bit is set to logical 1 if the result is odd.

The later two methods offer a means of detecting bit-level transmission errors. The use of parity bits is not mandatory. If not used, one bit in each frame is eliminated. This is often referred to as the nonparity bit frame.

SUMMARY

Of the two primary bit-oriented methods of data communication—serial and parallel—the former is the slowest. The RS-232 specification was developed by the Electronic Industries Association (EIA) in the 1960s to provide a standard serial interface between DTE and DCE. The standard evolved over the years and in 1969 the third revision, RS-232C, became the standard among PC makers. In 1987, a fourth revision was adopted—RS-232D, also known as EIA-232D. The serial port on most computers today is D-shaped and contains nine pins. Accordingly, it is called a DB-9 serial port. Each pin carries a different signal, with only nine signals required for data communication in the asynchronous mode.

***See Also* Parallel Transmission**

Servers

Servers are specially equipped computers that play a key role in the client/server environment. The client—typically a PC or workstation—provides the graphical interface, while the server provides the shared resources, such as applications for metered usage or a database for online analytical processing (OLAP). Servers can be dedicated to a specific task—such as facsimile, electronic mail or delivering Web content—or they can support multiple tasks and be partitioned according to various criteria to enforce security, improve performance, or simplify administration.

Servers can be configured to support an individual workgroup, a department, or an entire enterprise. They can be equipped with multiple processors to improve the performance of applications. Servers can even be linked together, or clustered, to appear as a single entity on the network to support mission-critical applications requiring high availability.

Types of Servers

There are many different types of servers, usually categorized by the task they perform:

FILE SERVERS   These act as shared repositories that store multimedia documents, images, engineering drawings, audio/video clips, and other large data objects. File servers are often used in a production environment where many clients need access to the same base files, which act as templates for the creation of new content.

DATABASE SERVERS   These process SQL (Structured Query Language) requests from clients, the results of which are returned over the network. Database servers provide the foundation for decision-support systems that require ad hoc queries and flexible reports.

TRANSACTION SERVERS   These enable clients to execute remote procedures that reside on a server with an SQL database engine. With a single request/reply message, the server executes a group of SQL statements, all of which either succeed or fail as a unit. This contrasts with a database server's approach of one request/reply message for each SQL statement. Grouped SQL statements are referred to as transactions.

GROUPWARE SERVERS   These enable enterprises to collaborate with the necessary level of privacy inside and outside the corporate boundary by enabling users to set up and manage private discussion groups and work projects. Lotus Notes/Domino is an example of such a system. Both can be used over corporate intranets and the public Internet, and have a Web-style interface (Figure 105).

WEB SERVERS   These return documents requested by clients over a corporate intranet or the public Internet. Each document is identified with its own Uniform Resource Locator (URL). Calling a document is as simple as clicking on a hypertext link from a client equipped with a Web browser. The request/response procedure is handled by a simple set of commands defined in the HyperText Transfer Protocol (HTTP).

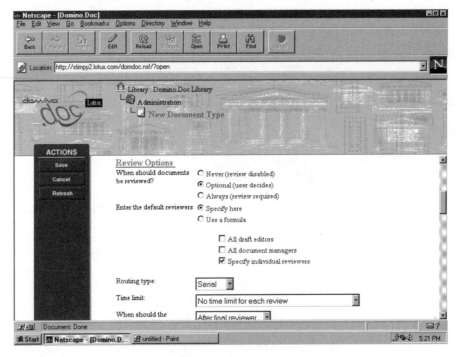

COMMERCE SERVERS   These resemble Web servers, but provide the means to buy products over the Internet, corporate intranets or multi-company extranets. Typically, the commerce server holds a catalog of items from which the buyer can choose by adding it to a "shopping cart." To complete the transaction, the buyer is presented with a list of selected items, along with the price, sales tax (if any) and shipping/handling cost. Payment is handled online via credit card.

MAIL SERVERS   These support messaging between clients over all kinds of networks. In the Internet environment, the two key e-mail protocols are the Post Office Protocol (POP) and the Simple Mail Transfer Protocol (SMTP), which are used for retrieving and sending e-mail, respectively. When a client establishes a connection to the server, POP3 is used to retrieve the new mail. Some mail servers support the newer Internet Message Access Protocol (IMAP) because it gives users more flexibility in accessing, filtering and storing messages than POP. SMTP is used for sending mail to the local server and moving it among other servers until it reaches its destination.

SPECIAL-PURPOSE SERVERS   These servers are dedicated to a specific task, such as network printing, in which case, it handles only client printing requests. The print server can reside anywhere on the network, and any user can send print jobs to any printer on the network. The print server also allows the organization to use fewer printers, and to use them more easily and across a wider network, which saves time for the network administrator and money for the organization. Other types of special purpose servers include CD-ROM "jukeboxes," fax servers, and proxy servers.

Server Management

As more business-critical applications rely on the availability of servers, the need to monitor configuration and performance increases because any unscheduled downtime can dramatically affect a company's productivity and profitability. There are management applications from numerous vendors that monitor all the vital areas of a server's operating system and hardware components to predict and prevent problems, and speed troubleshooting.

Such tools increase server availability by utilizing predictive failure alerts—temperature monitoring, disk problems or memory failure, for example. These tools also monitor and report possible server access security breaches by unauthorized users. In addition, they are used to configure the management console to generate an alert when resource usage values go over a specific limit. For example, when cache buffers reach a predefined threshold, the management system can warn the administrator before the server goes down. Or when disk capacity reaches an operator-defined warning level, the management system can warn the operator that additional disk space is needed.

Server management tools also include graphing capabilities to display usage rates such as the server log-in count throughout the day. At-a-glance utilities allow administrators to view information on one, a few or all servers on the network. Some management tools include a Web interface to enhance ease of use and expand accessibility to the management capabilities.

SUMMARY

Servers are an easy and economical way to share information, applications and other resources on a network. Instead of equipping every PC with its own storage, application software, modem/fax card and peripherals, one or more servers can be suitably equipped to service all the PCs on the network. This saves money on equipment and software, while simplifying administration.

See Also **Client/Server, Clustering, Fax Servers, Internet Servers, Proxy Servers**

Service-Level Agreements

Service-level agreements (SLAs) are contracts that specify the performance parameters within which a network service is provided. Although the contracts usually cover the services telecommunications carriers provide to corporate customers, they can also cover the services an information technology (IT) department provides to other business units within a company.

The SLA might define parameters such as the type of service, data rate, and what the expected performance level is to be in terms of delay, error rate, port availability, and network uptime. Response time to system repair and/or network restoral also can be incorporated into the SLA, as can financial penalties for noncompliance.

SLAs are available for just about any type of service, from traditional T-carrier services to frame relay. Internet service providers (ISPs) also offer SLAs for IP-based Virtual Private Networks (VPNs), intranets, and extranets. Some ISPs even guarantee levels of accessibility for their dial-up remote-access customers. IBM, for example, offers an SLA for its remote-access customers that guarantees a 95-percent success rate on dial-up connections to the IBM Global Network.

Although SLAs usually cover the services telecommunications carriers provide to corporate customers, they can also cover the services an IT department provides to other business units within the organization.

Service Provider SLAs

Among the growing number of service providers that offer SLAs is AT&T. The carrier offers SLAs at no extra charge in three different frame relay environments: domestic, international, and managed.

DOMESTIC The five SLAs for AT&T's domestic frame relay service include the following measures of network performance:

PROVISIONING If an agreed-on due date is missed for a port or PVC, recurring charges on the port or PVC are free for one month.

RESTORATION TIME If a customer reports a frame relay service outage (even if the problem is with local access) and it is not restored in four hours, recurring charges for the affected ports and PVCs are free for one month.

LATENCY If the customer reports a one-way delay from service interface to service interface (SI to SI) across the frame relay network of more than 60 milliseconds and AT&T can't fix the problem in 30 days, recurring charges for the affected PVC are free each month until repaired.

THROUGHPUT If 99.99 percent of the packets offered to the frame relay network within a PVC CIR (committed information rate) are not successfully transported through the network and AT&T can't fix the problem in 30 days, recurring charges on the PVC are free each month until repaired.

NETWORK AVAILABILITY If the customer's network is not available at least 99.99 percent of the time each month, the customer receives credits commensurate with network size.

AT&T also offers a suite of network management tools to support the SLAs. As part of its Customer Network Management (CNM) options, customers use two free Web-based tools: Order Manager and Ticket Manager. Using Order Manager, customers can enter service orders via the Internet and track their progress. With Ticket Manager, customers can enter trouble tickets related to performance metrics. CNM provides weekly and monthly reports on port and PVC usage, discards, network congestion, and other performance factors. CNM can also be used to create exception reports with definable thresholds.

INTERNATIONAL AT&T also offers SLAs for international customers, guaranteeing on-time provisioning and premise-to-premise availability of individual PVCs. The metric for on-time delivery is 100 percent, and there is no limit to the number of orders that can qualify for this credit. If the mutually agreed-to due date is not met, a one-time credit of $500 will be given per missed qualified order.

International network availability is measured from customer premise to customer premise. For example, if a 64-Kbps PVC experiences less than 99.8 percent availability, credits ranging from $25 to $525 will be applied. Credits are allocated based on PVC size and the amount of down time. Latency and throughput measures for international services are also available.

MANAGED Customers who use AT&T Managed Network Solutions (MNS) to manage their networks are covered by end-to-end SLAs, ensuring that AT&T will implement services on time, deliver predictable and reliable network performance, and provide consistent ongoing support. These comprehensive SLAs, available at no extra cost, apply both to AT&T's network transport services and to all related equipment up through the router on the customer's premises.

Individualized performance measures are defined for each customer based on specific network designs, including router, hardware, dial-backup, configuration and transport service designs. If service levels fall below the agreed-upon metrics, AT&T will credit customers for monthly charges and maintenance fees based on the terms outlined in each customer's contract.

Of course, AT&T is not the only service provider that offers SLAs. Others include Concentric Network, GTE Internetworking, IBM Global Network, Infonet Services Corp., MCI WorldCom, NaviSite Internet Services, Sprint, and UUNET. Eventually, most service providers will feel compelled to offer SLAs, or risk losing business to competitors.

Intracompany SLAs

An intracompany SLA describes the level of service required to support the various corporate applications. The metrics could be OnLine Transaction Processing (OLTP) response times, batch turnaround times for end-of-day reports, actual hours of system availability, and bandwidth availability. In essence, the SLA documents what a particular group of workers and managers need from IT to best fulfill their responsibilities to the company.

IT managers usually have access to online software tools to monitor system performance and resource consumption, giving them a general idea of how all systems on the network are behaving at any given moment. This information enables networks and systems to be managed effectively and provides the starting point for developing the SLA.

With information about current performance levels available and the expectations of business units and end users quantified, the IT department can write an effective SLA. At a minimum, this document should contain the following components:

BACKGROUND   This section should contain enough information to acquaint a nontechnical reader with the application, and to enable that person to understand current service levels and why they are important to the continued success of the business.

PARTIES   This section should identify the parties to the agreement, including the responsible party within IT and the responsible party within the business unit and/or application user group.

SERVICES   This section should quantify the volume of the service to be provided by the IT department. The application user group should be able to specify the average and peak rates, and the time of day they occur. The user may be provided with incentives to receive better service, or a reduced cost for service, if peak resource usage periods can be avoided.

TIMELINESS   This section should provide a qualitative measure of most applications to let end users know how fast they can expect to get their work accomplished. For OLTP applications, for example, the measure might be stated as "95 percent of transactions processed within two seconds." For more batch-oriented applications, the measure might be stated as "Reports to be delivered no later than 10:00 a.m. if input is available by 10:00 p.m. the previous evening."

AVAILABILITY This section should describe when the service will be available to the end users. The end users must be able to specify when they expect the system to be available in order to achieve their specified levels of work. IT must be able to account for both planned and unplanned system unavailability, and work these factors into an acceptable level of performance for end users.

LIMITATIONS This section should describe the limits of IT support during conditions of peak period demand, resource contention by other applications, and general overall application workload intensities. These limitations should be explicitly stated and agreed to by all parties to prevent finger pointing when problems arise.

COMPENSATION Ideally, a chargeback system should be implemented in which end users are charged for the resources they consume to provide the service they expect. This gives business units the incentive to apply management methods that optimize costs and performance. If this is impractical, the costs should still be identified and reported back to the business units to account for IT resources. The frequency and format of this information should also be described.

MEASUREMENT This section should describe the process by which actual service levels will be monitored and compared with the agreed upon service levels, as well as the frequency of monitoring. A brief description of the data collection and extrapolation processes should be included, and how users are to report problems to IT.

RENEGOTIATION This section should describe how and under what circumstances the SLA can be changed to reflect changes in the environment.

When the SLA is ready for implementation, the IT department must implement procedures to determine if service levels are being met. Additionally, IT needs to be able to forecast when the service levels can no longer be met due to growth or other external factors.

Standards

The effort toward standardizing SLAs has gone the farthest with respect to frame relay service. The Frame Relay Forum offers a set of common network service parameters that are described in its Service-Level Definition (SLD) implementation agreement. The SLD defines three metrics that should constitute the main elements of an SLA: delay, frame delivery rate, and connection availability. These metrics are the benchmarks by which network performance can be measured, whether the frame relay network is private or carrier provided.

DELAY   Delay metrics describe the time required to transport data from one end of the network to the other. Measuring delay involves three interdependent elements: access line speed, frame size, and wide-area network (WAN) delay. To be useful, measuring and reporting delay should be in the context of these elements.

Access line speed refers to the delay caused by the speed of the line from the user site to the frame relay network at both the local and remote ends of the network. Access line delay can contribute significantly to the overall delay of the network. For example, a 4000-byte packet might take approximately 500 ms to cross a 64-Kbps line. If the local and remote-access lines are 64 Kbps, the access lines alone could add nearly a second of delay.

Delay caused by the access line can be managed by increasing line speed or segmenting the data into smaller frames, which is usually handled by the router or frame relay access device (FRAD). Changing the frame size to 128 bytes reduces the access line delay to approximately 16 ms. Alternatively, increasing the line speed to T1 (1.536 Kbps), reduces the 4000-byte frame delay to a more manageable 20 ms.

The third element of delay, network delay, is difficult to manage in a public network scenario. However, measurement of delay across the WAN, separate from the delay imposed by the access line, can help the user pinpoint performance problems. Eliminating WAN delay will help the user focus on other causes of inadequate performance, such as configuration or application difficulties which account for as much as 70 percent of performance problems.

FRAME DELIVERY RATE   Frame relay networks typically categorize frames in two ways: below the Committed Information Rate (CIR) and above CIR. To provide a valid FDR, it must be determined if the measurement is for frames within CIR, in excess of CIR, or for the total number of frames presented to the network for delivery.

CONNECTION   Availability Connection availability measures the percentage of time the network connection is accessible to support the communications needs of the network. There are several elements to connection availability: overall availability, mean time to restore (MTTR) in the event a connection is lost, and mean time between service outages (MTBSO).

Overall availability refers to the total time the network connection is available, compared to the total measured time. If a network did not experience any services outages in a 30-day period, then its availability would be expressed as 100 percent. If the network is down for six hours, the availability would be 99.17 percent. Availability can be measured network-wide (all sites together) or individual measurements can be taken for each site and brought together to reflect a total network calculation.

Connection MTTR has a direct impact on availability because the longer it takes to restore a connection, the longer the service is unavailable. Most SLAs have specific measurements for MTTR. One method of reducing the impact of a service outage, usually caused by a failure in the local loop, is the use of ISDN as a backup service.

The degree of impact caused by service outages is more apparent if the time between outages is measured. Connection MTBSO measures the availability time between outages. Having four 6-hour outages in one day obviously has a greater impact than one 6-hour outage a week for a month. MTBSO gives the network manager the information needed to evaluate the other availability metrics.

SUMMARY ▪ ▪ ▪ ▪ ▪ ▪ ▪ ▪ ▪ ▪

To meet the challenges of the new competitive era, telecom service providers and ISPs are exploring new approaches to serving customers. A major step has been taken with service-level agreements that offer performance guarantees and credits to customers if various metrics are not sustained. In some cases, the customer need not even report the problem and provide documentation to support the claim. The carrier or ISP will report the problem to the customer and automatically apply appropriate credits to the invoice, as stipulated in the SLA.

At the same time, SLAs are becoming more important for ensuring the peak performance of enterprise networks. The purpose of the intracompany SLA is to specify, in mutually agreeable metrics, what the various end user groups can expect from IT in terms of resource availability and system response. SLAs also specify what IT can expect from end users in terms of system usage and cooperation in maintaining and refining the service levels over time. SLAs also provide a useful metric against which IT department performance can be measured. How well the IT department fulfilled its obligations, as spelled out in the SLA, can determine future staffing levels, budgets, raises, and bonuses.

See Also **Outsourcing, Performance Monitoring**

Short Message Service

Short Message Service (SMS) allows users to send and receive short text messages via a cellular phone or other portable appliances equipped with a display screen and a removable digital card, also known as a Subscriber Identity Module (SIM). SMS can be used to transfer messages of up to 160 characters between mobile telephone users, terminal users, or applications. Receipt notification allows mail boxes of all types—voice, fax, and text—to send an acknowledgment to a user's mobile telephone, indicating that the message has been delivered.

Callers can send text messages in two ways, by dialing a toll-free number to reach an operator who keys in the message and sends it or using a computer and the software package furnished by the service provider to create and send messages directly. Recipients are alerted to incoming messages by audible tones and/or visual indicators on their digital cell phones.

Types of Services

The Short Message Service originated with the digital cellular service standard known as the Global System for Mobile (GSM) telecommunications, but is now available with other technologies that implement cellular services and support a Subscriber Identity Module. The service itself is provided by the cellular network operator. There are two different types of SMS specified in the GSM standards: Point to Point (SMS/PP) and Cell Broadcast (SMS/CB).

SMS/PP   SMS/PP delivers short text messages from one subscriber to another. The network delivers the message from the sender to the phone with the number the sender sets when issuing the message. The sender does not know if or when the receiver receives the message.

Delivering a point-to-point SMS message from a mobile phone to another entails two separate tasks, sending the message from a mobile phone to a special facility in the network called Short Message Service-Service Center (SMS-SC) and then from the SMS-SC to the receiving mobile phone. The first part of the operation is called Mobile Originating (MO) SMS, while the latter operation is called Mobile Terminating (MT) SMS. Once the network has accepted the message, it can store it until it can be delivered to the receiver. The maximum storage time differs by network operator, but it can be specified by the sender with a special parameter that goes out with the message.

SMS/CB   Short Message Service Cell Broadcast (SMS/CB) functionality allows a number of unacknowledged general messages to be broadcast to all receivers within a particular region. Cell Broadcast (CB) messages are broadcast to one or more designated cells. CB service is typically used to deliver weather, traffic, advertising, and other local information. It can also be used to send messages to each member of one or more predefined groups.

The CB message comprises of 82 octets (93 characters). Up to 15 of these messages can be concatenated to form a macro message. To permit phones to selectively display or discard CB messages, the CB messages are assigned a message class that categorizes the type of information they contain and the language in which the message has been compiled.

Protocol Data Units

The SMS uses a connectionless protocol that makes use of six different Protocol Data Unit (PDU) types, the function and direction of which are described in the following table:

PDU Type	Function	Direction
SMS-DELIVER	Delivers a short message	SM-SC to Mobile phone
SMS-DELIVER-REPORT	Delivers a failure cause (if necessary)	Mobile phone to SM-SC
SMS-SUBMIT	Delivers a short message	Mobile phone to SM-SC
SMS-SUBMIT-REPORT	Delivers a failure cause (if necessary)	SM-SC to Mobile phone
SMS-STATUS-REPORT	Delivers a status report	SM-SC to Mobile phone
SMS-COMMAND	Delivers a command	Mobile phone to SM-SC

The primary function of SMS-DELIVER and SMS-SUBMIT is simply to deliver the actual message data and associated information between the mobile phone and the SM-SC. SMS-STATUS-REPORT carries information on whether or not the message was delivered to the actual receiver and when this happened. SMS-COMMAND contains a command to be executed to an earlier issued SMS-SUBMIT.

The PDUs are sent via GSM's control channels. During a call, SMS PDUs are sent through the Slow Associated Control Channel (SACCH), and otherwise through the Stand-alone Dedicated Control Channel (SDCCH). In addition to SMS, these control channels are used for the transmission of such important information as network quality, location update, and call establishment.

Added Functionality

In addition to handling text messages, the SMS platforms of various vendors enable subscribers to receive voice mail notification, digital pages, and informational services like stock quotes, sports scores, weather, and traffic bulletins.

For example, Motorola's SMS capability is enabled through the company's Message Register SMS Center and the cellular switch to provide a data-like information pipeline for enabling a variety of different applications. The Message Register resides on a fault-tolerant Tandem computer platform and integrates with other vendors' equipment via the Cellular Digital Messaging Protocol interface. Motorola's CDMA-based SMS platform supports time-stamping of messages, urgent message indication, and voice-mail message count. The SMS capability is also used with Motorola's Wireless Internet Service so cellular subscribers can access Internet e-mail and other informational services from their wireless phones. Motorola also offers an SMS-like service for narrowband AMPS (N-AMPS) cellular systems.

Another SMS platform is offered by Sema Group Telecoms. The company's Short Messaging Service Center, the SMS2000, provides additional SMS functions such as the ability to create distribution lists, multiple time-zone recognition, closed user-group configurations, and message compression.

The distribution list facility enables identical short messages to be delivered to lists of destinations. This capability is of particular benefit to corporate subscribers and information providers. Built into the distribution list facility are a number of flexible configuration options with which network operators can ensure that lists do not grow to unmanageable proportions or do not remain on the system indefinitely.

When a network operator's range extends over more than one time zone, the SMS2000 can be configured to convert the date/time of messages to that of the subscriber's local time zone. This facility eliminates any possibility of confusion such as the time of delivery being before the time of transmission, or a message appearing to take an inordinate length of time to deliver.

The General-Purpose Interface (GPI), provided as a module within the SMS2000, supports the creation of Closed User Groups (CUGs). Three different types of closed user groups (CUG) may be configured:

EXCLUSIVE Members can exchange short messages with each other, but not with anyone outside of their CUG.

OUTGOING BARRED Members can receive messages from anyone, but can send messages only to fellow members of the same CUG. Since the sender is billed for messages, this configuration option can help companies contain messaging costs.

INCOMING BARRED   Members can send to anyone, but can receive messages only from fellow members of the CUG. This configuration option may be of particular interest to companies concerned about the security of messages.

The GPI can support all three types of closed user groups simultaneously. In addition, subscribers can be members of more than one CUG simultaneously. Another significant facility provided by the SMS2000 is data compression. By compressing SMS messages, message throughput from mobile phone to the network can be greatly improved.

Subscriber Identity Module

The SIM provides personal mobility, so subscribers can have access to all services regardless of the terminal's location or the specific terminal used. By removing the SIM from one cellular phone and inserting it into another cellular phone, users can receive calls at that particular phone, make calls from that phone, or receive other subscription services. If the SIM contains frequently dialed numbers or a list of preferred roaming partners, they may be accessed from the new phone as well.

The SIM card is protected against unauthorized use by a password or personal identity number (PIN). When users insert the SIM into a telephone and switch it on, they are asked for the PIN to activate the service. (The PIN can also be stored in the cell phone for automatic entry whenever the phone is switched on.) Should users enter the wrong PIN three times in a row, the card will be blocked. If this happens, they must enter a Personal Unblocking Key (PUK) code to unblock the card. If this code is entered incorrectly ten times in a row, the card will become completely blocked and will have to be replaced.

An International Mobile Equipment Identity (IMEI) number uniquely identifies each Mobile Station. The SIM card contains an International Mobile Subscriber Identity (IMSI) number, identifying the subscriber, a secret key for authentication, and other user information. Since the IMEI and IMSI are independent, this arrangement provides users with a high degree of personal mobility.

The SIM comes in two form factors, credit card size (ISO format) or postage-stamp size (Plug-In format). Both sizes are offered together to fit any kind of cell phone (Figure 106). There is also a Micro SIM Adapter (MSA), which allows users to change from the Plug-In format SIM Card back into an ISO format card.

Figure 106
SIM issued to subscribers of Vodafone, the largest cellular service provider in the UK. Within the larger card is a detachable postage stamp sized SIM. Both use the same gold contact points.

Depending on how subscribers intend to use a GSM cell phone, they can usually choose an appropriate amount of memory for the SIM. Cards are available in capacities of 1, 3, 8, and 16MB. The latest generation of SIM cards offer 16KB of EEPROM memory, twice the amount of the previous generation of cards. With 16KB of storage space, up to 50 telephone numbers and up to 300 short messages can be stored on the card. These cards also make possible the use of EEPROM memory for loading new features and offering more services, such as:

- Remote access to airlines, banks, and retail chains for over-the-air purchases
- Enhanced user interface with proactive menus
- Dual service subscriptions, including Inmarsat and other satellite networks
- Prepayment functions implemented through disposable or reloadable cards

SIM cards also allow services to be individually tailored and updated over the air and activated without requiring users to locate a point-of-sale location in order to carry out the updating. SIM cards' remote control and modification possibilities allow carriers to offer their subscribers new, interactive services such as remote phonebook loading and remote recharging of prepaid SIMs. The cards can also contain company/private or parent/children subscriptions with separate PIN codes that can be changed over the air.

Management Software

There are two ways to send messages, from the telephone keypad (which is time-consuming) or by connecting the phone to a laptop or PC and using its larger keyboard for message entry. Nokia is among the vendors that offers a phone-to-computer cable and management software.

With Nokia's SMS Manager software, the company's line of GSM phones can be used as a personal phonebook to store names and phone numbers. Nokia SMS Manager is the software that enables subscribers to use their computers to enter and store this kind of information. With it, users can create several master phone directories with up to 2000 entries. There can be separate directories for work, travel, and leisure. This information can be uploaded to the phone's memory and can be updated from the PC whenever needed.

Nokia SMS Manager also allows users to store other kinds of information, such as mailing addresses and fax numbers, in the computer's master directory. In addition, users can create electronic business cards to send via SMS to other Nokia SMS Manager users. When received, these electronic business cards are automatically converted into new entries in the recipient's master directories.

SMS Manager allows a PC to be used to create, send, and receive short messages, and also supports all the usual Windows functions, including cut, copy, paste, clear, add, edit, delete, sort, find, and print.

Nokia SMS Manager is also useful for groups using Nokia GSM phones in the field. The software can be installed in an office PC with a Nokia GSM phone connected to it via the serial port. This enables quick transmission of individual or group messages via SMS to team members, much like with an e-mail system. Team members can also run Nokia SMS Manager on their notebook PCs to facilitate sending SMS messages back to the office.

SMS Manager also allows users to manage their phone's speed-dial memory and the SIM card's memory, where predefined user reply messages reside.

Enhanced SMS

There is an enhanced version of a Short Message Service called ESMS. The conventional SMS service often works only when the sender and recipient are on the same or similar network. For example, GSM-based SMS will work over DCS 1800 and PCS 1900 networks, which are basically GSM networks operating on different radio frequencies. ESMS, on the other hand, is a gateway to the Internet and to other networks. ESMS users can send messages to recipients on networks not normally compatible with their own and can send and receive e-mail to and from anyone who has an e-mail address on the Internet. This widens subscriber connectivity options considerably.

With conventional SMS, messages are limited to 160 characters. ESMS breaks longer messages into SMS-sized chunks for delivery via separate transmissions. Senders can limit the number of SMS chunks and even designate which parts of the message are sent. It is also possible to tell the system in which order to send the messages so they arrive in the right order on the recipient's phone screen.

Users do not require any special or additional equipment to take advantage of ESMS. All they need is a GSM or PCS/PCN mobile phone capable of sending and receiving SMS on a network that has been confirmed to work with ESMS.

ESMS also gives users the ability to forward e-mails that arrive on their computer to their mobile phone. If the computer can be set up to forward e-mail, then it can also be set up to pass mail to the subscriber's ESMS address just as it would to any other Internet address.

At the same time, subscribers can specify an address to which all ESMS mail is forwarded. For example, they might want to use ESMS to be notified of new mail. All mail could be sent to the ESMS address with just the sender and subject information being sent to the phone, with the full original message being forwarded by the ESMS network to another Internet address. This also has the advantage that subscribers need to publish only one e-mail address.

Like SMS, ESMS messages are stored on the network until they can be delivered or until enough time has elapsed that the network deletes undelivered messages automatically. If the recipient's phone is switched off or the subscriber is not in the coverage area, the network stores messages for a designated period—usually no more than 30 days. The networks will repeatedly try to send messages so when the recipient's phone is switched on or returns to coverage, queued messages will be delivered.

SUMMARY

SMS provides the means to send and receive short text messages via a digital cellular phone or other portable appliance. The main benefits of the service are that messages can reach the recipient even if he or she is already busy with a voice or data call. Recipients are alerted to incoming messages by audible tones and/or visual indicators on their digital cell phones. This service is available globally, including many countries in Europe and Asia where GSM-compliant wireless networks are in operation.
See Also **Electronic Mail, Fax Servers, Paging**

Simple Network Management Protocol

The Simple Network Management Protocol (SNMP) has been the management standard for multivendor TCP/IP-based networks since 1988. SNMP specifies a structure for formatting and transmitting messages between reporting devices and data- collection programs on the network. It runs on top of TCP/IP's datagram protocol—the User Datagram Protocol (UDP)—a connectionless-mode of transport. UDP is well suited to the brief request/response message exchanges characteristic of network management communications.

Architectural Components

Several components operate in the SNMP environment: managed devices, which run agents and contain Management Information Bases (MIBs), the management system, and SNMP itself.

MANAGED DEVICES   Managed devices are the specific pieces of hardware on the network such as bridges, routers, and gateways. Agents run on these managed devices to collect information about their performance or the performance of the link to which they are attached. The specific types of information that can be collected are defined in object-oriented MIBs.

An *object* refers to hardware, software, or a logical association such as a connection or virtual circuit. The *attributes* of an object might include things such as the number of packets sent, routing table entries, and protocol-specific variables for IP routing. For an Ethernet segment, for example, the following attribute is defined:

etherStatsPkts65to127Octets   This refers to the total number of packets (including error packets) received between 65 and 127 octets in length, inclusive—excluding framing bits but including Frame Check Sequence (FCS) octets. This type of information is defined in SNMP's Remote Monitoring MIB (RMON), specifically the RMON Ethernet Statistics Group.

The agent itself may be a passive monitoring device whose sole purpose is to read the network, or it may be an active device that performs other functions as well, such as bridging, routing, and switching. Devices that are non-SNMP compliant (i.e., do not have the capability of running an agent) must be linked to the management system via a proxy agent.

MANAGEMENT SYSTEM   The management system issues requests to agents and receives information that enables network administrators to comprehend the state of the network. The management system is a program that may run on one host or more than one host, each of which manages a particular subnetwork.

SNMP   SNMP's "get" request retrieves the values of specific objects from managed devices. The "get-next" request permits navigation of the MIB, enabling the next MIB object to be retrieved, relative to its current position. A "set" request is used to request a logically remote agent to alter the values of variables. In addition to these message types, there are "trap" messages, which are unsolicited messages conveyed from management agent to management stations.

Other commands are available that allow network managers to take specific actions to control a network. Some of these commands look like SNMP commands but are really vendor-specific implementations. For example, some vendors use a "stat" command to determine the status of network connections.

There are now SNMP management systems that offer graphical user interfaces. Castle Rock Computing Inc., for example, offers SNMPc for Windows NT. Besides its distributed architecture, SNMPc NT incorporates a variety of advanced management features, including real-time tabular and graphical displays, data export, event action filters, MIB import, device-specific GUIs, RMON support, custom menus, pager notification, and a variety of programming interfaces.

SNMPc NT supports multilevel hierarchical mapping. Each hierarchy can represent cities, buildings, or subnetworks. SNMPc NT can automatically lay out each mapped network as a tree, ring, or snaked bus topology (Figure 107). Each map object uses a device-specific or user-selected icon, and the object color indicates the device status. The user can start any device-specific application by double-clicking map icons.

SNMPc NT also provides multivendor graphical device views (Figure 108). The HubView and BitView scripting applications display a graphical image of multislot routers, bridges, and hubs. Through device-specific custom menus and mouse clicks, commands can be executed on the selected slot, port, or other graphical element. SNMPc includes device scripts for all devices that conform to standard MIBs and dozens of private MIB devices, including those created by Cisco, Bay Networks, and 3Com. Network managers can develop HubView or BitView scripts to graphically manage any SNMP device.

Figure 107
SNMPc NT provides a graphical display that supports multilevel hierarchical mapping.

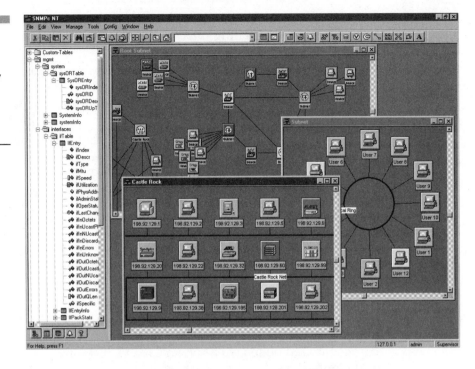

Figure 108
SNMPc NT provides multivendor graphical device views using HubView and BitView scripting applications to display a graphical image of multislot routers, bridges, and hubs.

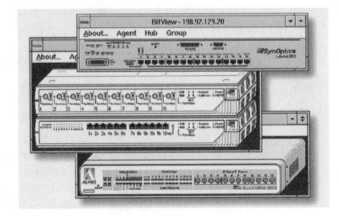

SNMPv2

In 1992, the Internet community recognized the need to enhance SNMP by adding security and management features, and make reporting more efficient. The result was a proposal for SNMP version 2 (SNMPv2). Among other things, SNMPv2 would also run over AppleTalk, IPX, and OSI transport layer software, as well as TCP/IP. But because the new version was not backward compatible with SNMP and agreement could not be reached in reconciling internal differences, the IETF working group on SNMPv2 was disbanded in 1996.

The idea of enhancing SNMP was picked up again in 1997, when a number of enhancement proposals were folded into SNMPv3. In April of 1998, all SNMPv3 specifications were submitted to the Internet Engineering Steering Group (IESG) for consideration as proposed standards. By the end of 1998, most of the development milestones for completion of SNMPv3 had been met. Meanwhile, there is a standard for Web-based network management in which the HyperText Transfer Protocol (HTTP) is used to convey management information between the managed devices and management system.

SUMMARY

SNMP's popularity stems from the fact that it works, it is reliable, and it is widely supported. The protocol itself is in the public domain. SNMP capabilities have been integrated into just about every conceivable device that is used on today's LANs and WANs, including carrier services such as frame relay.

See Also **Network Management Systems, Web-Based Network Management**

Single Point Sign-On

Single point sign-on, sometimes called global sign-on, is an administrative application that provides simplified, centralized access control to enterprise services. This capability enables a network administrator to control access to enterprise services by customizing individual desktops such that users can be assigned access to various resources on the enterprise network without having to remember different IDs and passwords for each resource. This enhances network security by simplifying access procedures, which have all too often become so complicated that the procedures and passwords are prominently posted on the sides of monitors so users do not forget them.

Single point sign-on provides controlled access to applications and services residing on a local disk, server, or host. Users merely access their desktops—or the sign-on program group—with one password. The desktop or window that appears provides users with a set of icons that invoke access to enterprise applications and services. User need only double-click on the icon. Single point sign-on does the work of controlling and managing all the procedures required to access and execute the applications, regardless of their location on the network. In addition to user IDs and passwords, this includes emulator selection, network navigation, and application subsystem selection.

All this takes place automatically, transparently, and securely. Single point sign-on even allows users to select their own passwords and then store remote logon information and passwords in encrypted form, so even network administrators do not know the passwords. Administrators can establish customized password aging policies of any length of time, setting up passwords to expire in a day, month, year—or never to expire.

Single point sign-on can be easily configured for individual end users and for groups of end users, in both cases providing access to only authorized enterprise applications and services. When administrators add services or applications for a group, all users in that group have instant access to them. This ensures the administrators' ability to consistently provide the information that end users need while preventing inappropriate access to information that should remain secure.

SUMMARY

Single point sign-on simplifies access to enterprise resources for authorized users without compromising security. This capability also simplifies the administrator's task in assigning access to resources. Once an application is set up, the assignment of resources is as simple as dragging and dropping them into place, from a library of resources to an icon representing the user's desktop.

See Also **LAN Security**

Small Computer System Interface

SCSI (Small Computer System Interface) is a local I/O bus that can be operated over a wide range of data rates. It traces its origins from the Shugart Associates System Interface (SASI), which was developed in 1979. Shugart and NCR presented SASI to the American National Standards Institute (ANSI) in 1981, where the X3T9.2 committee defined it as SCSI-1 in 1986. SCSI supports a wide variety of devices that can be connected to a host computer, including hard disks, tape drives, CD-ROM drives, printers, scanners, and communication devices.

SCSI allows multiple peripheral devices to be installed on the host system. The number of devices depends on the type of SCSI adapter that is installed on the host system. Most SCSI systems have an older narrow 8-bit SCSI host adapter, which has 8-bit data paths and allows for up to seven devices to be installed on the same channel. Newer wide host adapters have 16-bit-wide data paths and provide support for up to 15 devices.

SCSI-1

SCSI-1 is also referred to as *narrow SCSI*. This term refers to the 1-byte-wide data bus on a 50-pin parallel interface that is defined in the original ANSI standard X3.131-1986. The narrow bus consists of eight data lines with parity, a series of control lines, and the matching ground lines. It is capable of supporting up to seven devices and the host adapter. Data transfer rates of up to 4 MB/s can be achieved with cable lengths of up to six meters.

SCSI-2

This version of SCSI passes data through the SCSI bus along an 8-bit data path in 8-bit "computer words" and provides support for up to seven devices and the host adapter. This device uses a 50-pin connector, which connects via cable to the host adapter. Data transfer rates of up to 10 MB/s can be achieved with cable lengths of up to three meters.

Ultra SCSI 2

Ultra SCSI 2 is an enhancement of the SCSI 2 specification, which when properly set up allows for up to 20 MB/s transfer rates. Cable length requirements are very important or slower transfer rates will result. With four or less devices installed, the maximum cable length is three meters. With more than four devices, the maximum cable length 1.5 meters.

Fast Wide SCSI-2

This version of SCSI passes data through the SCSI bus along a 16-bit data path in 16-bit "computer words" and provides support for up to 15 devices and the host adapter. This device uses a 68-pin connector, which connects via cable to the host adapter. Data transfer rates of up to 20 MB/s can be achieved over a maximum cable length of three meters.

Ultra Wide SCSI-2

Ultra Wide SCSI 2 is an enhancement to the Fast Wide SCSI 2 specification, which when properly set up allows for up to 40 MB/s burst data transfer rates. Cable length requirements are very important or slower transfer rates will result. With four or less devices installed, the maximum cable length is three meters. With more than four devices, the maximum cable length is 1.5 meters.

Ultra2 SCSI

Ultra2 SCSI uses a technology called Low Voltage Differential (LVD), which allows for cable lengths of up to 25 meters (12 meters with 16 devices) and an 80 MB/s burst data transfer rate. This increased bandwidth means optimal performance for server environments where rapid response is required and random access and large queues are the norm. When using applications such as CAD and CAM, digital video and any RAID environment, the increased response time is immediately noticeable. Most of the devices that utilize this interface will utilize the 16-bit Wide SCSI format and will incorporate a high-density 68-pin interface.

Differential SCSI

Differential SCSI is a special version of SCSI that allows for cable lengths of up to 25 meters. To use this technology, a special "differential" host adapter is required that is not compatible with standard SCSI devices. Differential SCSI technology has a built-in dual data path and compares one line to its counterpart to interpret the data being read or written. All differential devices require external line termination, since the Differential SCSI specifications do not call for on-board termination.

SCSI Extenders

The distance limitations of SCSI can be overcome with devices called *extenders*, which transmit SCSI signals over coax or fiber-optic cable with little or no throughput degradation. Paralan Corp., for example, offer a SCSI to Coax Bus Extender that converts SCSI parallel signals to serial information and transmits them through one of two RG62A coax cables at a rate of over 100 Mbps to another extender that converts the information back to SCSI signals. The two extenders communicate in full duplex mode and provide transparent communications for the SCSI bus. The distance between the extenders can be as long as 300 feet. The company also offers a SCSI extender for fiber optic cable. The 62.5-micron graded index fiber-optic cable between two extenders can be over 1000 feet long.

SUMMARY

Although SCSI is installed in more than 90 percent of networks, native throughput, distance, and contention limitations have caused IT managers to examine alternative technologies such as Fibre Channel, which can push data through more devices at greater distances and at much higher speeds than conventional SCSI.

See Also **Fibre Channel, Fiber Distributed Data Interface, High-Performance Parallel Interface**

Small Office/Home Office LANs

The small office/home office (SOHO) is one of the fastest-growing segments of the networking market. A small office is one that has 2 to 50 users, while a home office typically has 2 to 5 users.[2] In addition to computers for one or more students, there might be one for a telecommuter, along with a docking station for a mobile professional's laptop computer. All of these computers can be networked together as peers to save money on printers and other resources, as well as the cost of Internet access.

The most popular network topology for the SOHO environment is 10/100BaseT Ethernet because it is relatively inexpensive and easy to set up and use. The same components used to build large enterprise networks are used to build the SOHO network. These components include cabling, media converters, network adapters, hubs, and network operating system. To access external networks a modem or router will be needed as well, depending on the type of connection desired.

Cabling

Although different media may be used in the SOHO environment—twisted pair, thin coax, and optical fiber—the most popular type of network cabling is Category 5 unshielded twisted pair (UTP) cable. Category

---

2. According to the market research firm Dataquest, more than 15 million homes in the U.S. have two or more PCs.

5 cabling is inexpensive, flexible, and adequate for short distances. This type of cable looks a lot like the thin coaxial cabling (10Base2) sometimes used on Ethernet LANs, but has eight wires inside. It can be used at a maximum distance of 328 feet. A modular plug at each end (RJ45) of the cable makes interconnecting the various network devices as simple as plugging them in.

Media Converters

All three types of media—twisted pair, thin coax, and optical fiber—can be used exclusively or together, depending on the type of network. Media converters are available that allow segments using different media to be linked together. Because media conversion is a physical layer process, it does not introduce significant delay on the network.

Network Adapters

A computer—whether configured as a client or server—is connected to the network with an adapter called a network interface card (NIC), which comes in internal and external versions. Typically, the NIC is installed inside the computer, plugging directly into one of the computer's internal expansion slots. Most older computers have 16-bit ISA slots, so a 16-bit NIC is needed. Today's newer, faster computers come with several 32-bit PCI slots. These PCs require 32-bit NICs to achieve the fastest networking speeds.

If a computer is going to be used with a Fast Ethernet network, it will need a network adapter that specifically supports 100 Mbps as well. Several NICs offer an "auto-sensing" capability, enabling them to determine the speed of the network and whether to run at 10 or 100 Mbps.

Some NICs support multiple types of media connectors. In addition to RJ45, they might also include a connector for the older BNC interface. If a PC lacks vacant expansion slots, you can use an external network adapter that plugs into the computer's parallel port. The adapter usually has LED indicators for items such as polarity, signal quality, link integrity, and power.

Portable computers use a PC Card (formerly known as PCMCIA) network adapter. This credit-card sized adapter communicates with the portable using special software drivers called "socket and card services." Another set of drivers called "network drivers" enable the PC Card to communicate with the network at large. Some client operating systems, such as Windows 95, provide the sets of drivers that enable Plug-and-Play (PnP) operation; otherwise they must be installed by the user from the vendor's installation disk. Some PC Card adapters include an integral 56-Kbps modem that can be used for remote dialup connection to the SOHO LAN.

Hubs

A hub is basically a wiring box with a row of jacks into which the cables from NIC-equipped computers and other devices are plugged, providing the means to interconnect them. Most small-office hubs have 4 to 16 jacks, but some may have more to accommodate the proliferation of equipment now found in the SOHO environment, including printers, a cable modem, a CD-ROM jukebox, a file server, and other types of devices.

Some switching hubs allow both 10- and 100-Mbps networking hardware to be used on the same network. An auto-sensing feature associated with each port determines the speed of the NIC at the other end of the cable and transparently bridges the two speeds together.

To accommodate growth, "stackable" hubs allow additional units to be joined together with an ordinary Category 5 cable that plugs into a special uplink port on each hub. Stackable hubs appear as a single hub to the network, regardless of how many are connected together. Since data does not pass through the hub's regular RJ45 ports, it is not slowed down by error correction and filtering.

In a SOHO environment where just a few computers are networked together, users can get by with a hub, some network adapters, and 10BaseT cables. If significant growth is anticipated in the future, a stackable 10/100 Ethernet hub is the better choice. It provides tighter integration and maximum throughput, and the means to scale up to virtually any number of nodes if necessary without external repeaters or switches.

Hubs also come in managed and unmanaged versions. Hubs that are not manageable are easy to set up and maintain. They offer LED indicators that show the presence of send and receive traffic on each port, collisions, and hub usage and power status. However, unmanaged hubs do not offer the configuration options and administrative features of managed hubs.

When the number of PCs reaches 20, it becomes more economical over the long term to have a management system in place that can facilitate daily administration and provide information for troubleshooting purposes. A management system offers ways to configure individual ports. For example, address filtering can be applied to ports to limit access to certain network resources. A management system can also provide performance information about each port to aid troubleshooting and fault isolation. Most hubs of this type support the Simple Network Management Protocol (SNMP) and the associated Remote Monitoring (RMON) standard, which enables more advanced network monitoring and analysis.

Network Operating System

Every computer attached to a network must be equipped with a network operating system (NOS) to monitor and control the flow of information between the interconnected devices. Network operating systems are either peer-to-peer or client/server. Examples of peer-to-peer NOSs commonly used by small businesses are Windows 95, Artisoft LANtastic, and NetWare Lite. These NOSs are useful for sharing applications, data, printers, and other resources across PCs that are interconnected by a hub. Examples of client/server NOSs are Windows NT and NetWare, which are used by larger organizations whose users require fast network access to a variety of business applications.

Peer-to-Peer

A simple peer-to-peer network can be built very inexpensively with Category 5 cabling and a 10/100 Ethernet switching hub. After the networking hardware has been installed, a peer-to-peer network software package must be installed in all the PCs. If the PCs come with Windows 95, the basic protocols for peer-to-peer networking are probably already installed in each system, and it is simply a matter of configuring them through the operating system's control panel. If a different NOS is preferred, such as Artisoft LANtastic or NetWare Lite, it can be installed separately.

Most NOSs allow each peer-to-peer user to determine which resources are available to other users. Specific hard and floppy disk drives and directories, files, printers, and other resources can be attached or detached from the network via software. Access to each resource can be controlled in a variety of ways. For example, when configuring Windows 95 for peer-to-peer networking, a password can be applied to each shared resource. Alternatively, specific users and groups can be granted access to each shared resource.

When one user's disk has been configured to be "sharable," it will appear as a new drive to the other users. Because drives can be easily shared between peer-to-peer PCs, applications need to be installed on only one computer instead of all computers. If the company relies on a spreadsheet application, for example, it can be installed on one user's computer and be accessible to other computers whenever it is not in use. If the spreadsheet program does not need to be used continuously by all potential users, then it makes more sense to share one copy rather than buy one for each machine. Of course, sharing applications over a peer-to-peer (or client/server) network might require a network license from the vendor, or the company risks a penalty for copyright infringement.

Client/Server

In a client/server environment such as Windows NT or Novell NetWare, files are stored on a centralized, high-speed, file-server PC that is made available to client PCs. Network access speeds are usually faster than those found on peer-to-peer networks. Virtually all network services, such as printing and electronic mail, are routed through the file server, which allows networking tasks to be tracked. Inefficient network segments can be reworked to make them faster, and users' activities can be closely monitored. Public data and applications are stored on the file server, where they are run from client PC locations, which makes upgrading software a simple task; network administrators can simply upgrade the applications stored on the file server, rather than having to physically upgrade each client PC.

In a client/server network, client PCs are subordinate to the file server. The clients' primary applications and files are stored in a common location. File servers are often set up so all users on the network can access their "own" directories, along with a range of "public" directories where applications are stored. If two clients want to communicate with each other, they must go through the file server to do it. A message from one client to another is first sent to the file server, where it is then routed to its destination. A small business with 100 or more client PCs might find that a server-centric network is the best way to meet the needs of all users.

Take a simple task such as printing, for example. Instead of equipping each desktop or workgroup with its own printer, it is more economical for many users to share a few high-speed printers. In client/server networks, network printing is usually handled by a print server, a small box with at least two connectors: one for a printer, and another that attaches directly to the network cabling. Some print servers have more than two ports; they may, for example, support two, three, or four printers simultaneously. When users send print jobs, they travel over the network cabling to the file server, where they are stored. When the print server senses that jobs are waiting, it moves them from the file server to its attached printer. When the jobs are finished, the print server returns a result message to the file server, indicating that the printing process is complete.

Routers and Modems

Data communication between remote sites requires a router or modem. A modem is used for dial-up access at speeds of up to 56 Kbps. A router is usually used on a dedicated leased line, which is always available. All kinds of hybrid devices now offer multiple functions, offering users more flexibility in making connections.

A router can be a stand-alone device or it can come in the form of a module that plugs into a managed hub. Routers operate at layer 3 of the OSI reference model (the network layer). Basically, they convert LAN protocols into wide-area packet network protocols such as TCP/IP, and perform the process in reverse at the remote location.

ISDN-based routers provide high-quality dial-up connections to a local Internet service provider (ISP) so traffic can be carried economically between far-flung locations over the public TCP/IP-based Internet. Some products come with scripts that detect the specific type of ISDN connection the company has before setting up the TCP/IP information on the network. (Seventy types of ISDN circuit configurations are available, each with an order code.) Some routers even offer firewall software to protect the corporate network from intruders who might attempt to enter through the Internet.

A stand-alone ISDN access router comes with an Ethernet port for connection to the hub. It also has an ISDN port for connecting to the company's network termination (NT) point, where the carrier has a digital line for ISDN service. Some ISDN routers also have "plain old telephone service" (POTS) ports, which also allows the unit to be used for faxing. In some cases, two phone lines capable of supporting 56-Kbps modems can be aggregated to achieve bandwidths close to ISDN's 128.8 Kbps. Depending on the software package ordered with the router, the device's WAN port can be configured for use over leased lines, ISDN, frame relay, switched 56K service, SMDS, and X.25.

Getting Started

One of the easiest ways to set up a LAN in the SOHO environment is to buy a starter kit. This is a convenient, affordable solution that comes complete with the necessary hardware and cabling needed to create a two-node network. A typical kit includes a four- or five-port stackable hub, two auto-sensing Ethernet 10/100 NICs, and two Category 5 cables. These kits are ready to run with all major network operating systems.

Another way for a small business to get started with a LAN is to buy products from a local value-added reseller (VAR), who will install and configure the network to meet specific needs. Usually, VARs specialize in providing products from a single vendor and its authorized partners, so the offerings are tightly integrated and have a record of proven performance. Resellers and integrators bring a lot of value to the small-business customer, including pre- and postsales consulting, installation, and application customization.

SUMMARY

Just a few years ago, building a corporate network was a daunting task undertaken only by seasoned professionals. Ordinary folks who could not afford to spend lavishly on technical assistance either had to struggle along without a network or try to put the pieces together themselves, often on a trial-and-error basis, until they got it right. Recognizing the growing importance of this market, many interconnect vendors have started to design products that are easy to install and use. In some cases, the equipment is either ready to use out of the box or self-configuring after installation. In the few instances where manual procedures are still necessary to get the equipment configured properly, a graphical interface guides users through the process.

See Also **Client-Server, Hubs, Media Converters, Modems, Network Interface Cards, Network Operating Systems, Peer-to-Peer Networks, Phoneline Networking**

SNA over IP

Data communications was once a relatively simple function, conducted in a host-centric environment and primarily under the auspices of IBM's Systems Network Architecture (SNA). After more than 20 years, SNA is still a stable and highly reliable architecture and, despite the trend toward distributed computing (including the advent of new architectures such as client-server), the mainframe is still valued for its ability to handle mission-critical applications. Additional advantages include accounting, security, and management tools, which enable organizations to closely monitor all aspects of performance and to contain costs. The sheer financial investment in legacy hardware and software—estimated to exceed $1 trillion worldwide—provides ample incentive to protect and leverage these assets.

Companies faced with the need to support a growing base of telecommuters, small branch offices, and far-flung international locations are looking at ways to efficiently and economically extend the reach of SNA without inflicting performance penalties on end users. A popular solution is to run SNA traffic over private TCP/IP intranets using a process called data link switching (DLSw).

DLSw was developed by IBM in 1992 as a way to let users transport SNA traffic over TCP/IP. Because DLSw entails a fair amount of processing, it imposes a considerable resource burden on networking nodes, typically routers. Nevertheless, DLSw eliminates session timeouts and offers predictable response time.

In DLSw, SNA frames are encapsulated within TCP/IP packets. While the encapsulation process is not too burdensome from a network-processing point of view, it is the additional, higher-level processing of DLSw that makes it worthwhile. For example, DLSw uses a form of spoofing to prevent host-issued SNA polls from continually being sent over the underlying TCP/IP network and possibly causing congestion. A DLSw router at the host end intercepts and responds to these host polls, while another DLSw router at the remote end conducts its own polling exchange directly with a communicating SNA station.

Another task performed by DLSw is maintaining SNA-session information across the TCP/IP router network. This involves additional processing on all the intermediate DLSw routers in the TCP/IP network—not just the periphery routers that directly interface to the SNA host and workstation.

Switch-to-Switch Protocol

The DLSw standard describes the Switch-to-Switch Protocol (SSP) used between routers (called data-link switches) to establish DLSw peer connections, locate resources, forward data, handle flow control, and perform error recovery. RFC 1795 requires that data-link connections are terminated at the peer routers; that is, the data-link connections are locally acknowledged.

By locally terminating data-link control connections, the DLSw standard eliminates the requirement for link-layer acknowledgments and keep-alive messages to flow across the WAN. In addition, because link-layer frames are acknowledged locally, link-layer timeouts should not occur. It is the responsibility of the DLSw routers to multiplex the traffic of multiple data-link controls to the appropriate TCP pipe and transport the data reliably across an IP backbone.

Before any end-system communication can occur over DLSw, the following processes must take place:

- A peer connection is established.
- A description of capabilities is exchanged between the routers.
- A circuit is established between a pair of end-systems.
- Flow control is applied to regulate the flow of data between peers.

Establish Peer Connections

Before routers can switch SNA traffic, they must establish two TCP connections between them. The standard allows one of these TCP connections to be dropped if it is not required. Additional TCP connections can be made to allow for different levels of priority.

Exchange Capabilities

After the TCP connections are established, the routers exchange their capabilities. Capabilities include the DLSw version number, initial pacing windows (receive window size), NetBIOS support, a list of supported link Service Access Points (SAPs), and the number of TCP sessions supported. Media Access Control (MAC) address lists and NetBIOS name lists can also be exchanged at this time and, if desired, a DLSw partner can specify that it does not want to receive certain types of search frames. It is possible to configure the MAC addresses and NetBIOS names of all resources that will use DLSw and thereby avoid any broadcasts. After the capabilities exchange, the DLSw partners are ready to establish circuits between SNA or NetBIOS end systems.

Establish Circuit

Circuit establishment between a pair of end systems includes locating the target resource (based on its destination MAC address or NetBIOS name) and setting up data-link control connections between each end-system and its data-link switch (local router). SNA and NetBIOS are handled differently. SNA devices on a local-area network (LAN) find other SNA devices by sending an explorer frame with the MAC address of the target SNA device. When a DLSw router receives an explorer frame, the router sends a "can you reach" frame to each of the DLSw partners. If one of its DLSw partners can reach the specified MAC address, the partner replies with an "I can reach" frame. The specific sequence includes a canureach ex (explorer) to find the resource and a canureach cs (circuit setup) that triggers the peering routers to establish a circuit.

At this point, the DLSw partners establish a circuit that consists of three connections: the two data-link control connections between each router and the locally attached SNA end system, and the TCP connection between the DLSw partners. This circuit is uniquely identified by the source and destination circuit IDs, which are carried in all steady state data frames in lieu of data-link control addresses such as MAC addresses. Each circuit ID is defined by the destination and source MAC addresses, destination and source link SAPs, and a data-link control port ID. The circuit concept simplifies management and is important in error processing and cleanup. Once the circuit is established, information frames can flow over the circuit.

Most DLSw implementations cache information learned as part of the explorer processing so that subsequent searches for the same resource do not result in the sending of additional explorer frames.

Flow Control

The DLSw standard specifies flow control on a per-circuit basis and calls for two independent, unidirectional-circuit, flow-control mechanisms. Flow control is handled by a windowing mechanism that can dynamically adapt to buffer availability, TCP transmit queue depth, and end-station flow-control mechanisms. Windows can be incremented, decremented, halved, or reset to zero. The granted units (the number of units that the sender has permission to send) are incremented with a flow-control indication from the receiver (similar to classic SNA session-level pacing). Flow-control indicators can be one of the following types:

- Repeat-Increment granted units by the current window size
- Increment-Increment the window size by one and increment granted units by the new window size
- Decrement-Decrement window size by one and increment granted units by the new window size
- Reset-Decrease window to zero and set granted units to zero to stop all transmission in one direction until an increment flow-control indicator is sent
- Half-Cut the current window size in half and increment granted units by the new window size

Flow-control indicators and flow-control acknowledgments can be piggybacked on information frames or can be sent as independent flow-control messages, but reset indicators are always sent as independent messages.

SUMMARY

Throughout their development histories, SNA and TCP/IP have moved along separate paths. Today, with trends as diverse as corporate downsizing, distributed computing, and participation in the competitive global economy, companies are virtually forced to consolidate the two types of networks to achieve greater efficiencies and economies. Given the vast installed base of SNA systems and the ubiquity of TCP/IP networks, the movement toward integration makes sense, especially with the advent of such techniques as DLSw.

In 1997, IBM introduced a better way to encapsulate SNA in IP than Data Link Switching. The new technology, called Enterprise Extender, can reroute network traffic without disrupting the SNA session if a node fails. It can also control network congestion, and it uses fewer mainframe processor cycles than Data Link Switching.

Enterprise Extender uses the High-Performance Routing (HPR) function of IBM's Advanced Peer-to-Peer Networking (APPN), the next generation of SNA. It runs on the user datagram protocol UDP), on top of IP. This way, the software can deliver some of the benefits of APPN to SNA users while allowing them to migrate to an IP network.

See Also Routers

Software-Defined Radio

As its name implies, a software-defined radio system uses software to perform demodulation. The radio system can easily change to receive a different modulation scheme simply by using a different software routine to demodulate. This architecture offers several compelling advantages over first-generation hardware-based radio systems, which are built to receive a specific modulation scheme.

The concept of software-defined radio originated with the military where it was originally used for electronic warfare applications. Now the cellular/wireless industries in the U.S. and Europe have begun work to adapt the technology to commercial communications services in the hope of realizing its long-term economic benefit. If all goes according to plan, future radio services will provide seamless access across cordless telephone, wireless local loop, PCS, mobile cellular and satellite modes of communication, including integrated data and paging.

Generations of Radio Systems

First-generation hardware-based radio systems are built to receive a specific modulation scheme. A handset is built to work over a specific type of analog network or a specific type of digital network. The handset works on one network or the other, but not both, and certainly cannot cross between analog and digital domains.

Second-generation radio systems are also hardware-based. Miniaturization enables two sets of components to be packaged into a single, compact handset. This enables the unit to operate in dual mode—for example, switching between AMPS or TDMA modulation as necessary. Such handsets are implemented using "snap-in" components; two existing chipsets—one for AMPS and one for TDMA, for example—are used together. Building such handsets typically costs only 25 to 50 percent more than a single-mode handset, but offers network operators and users far more flexibility.

Handsets that work across four or more modes/bands entail far more complexity and processing power and call for a different architecture altogether. The architecture is based in software and programmable digital signal processors (DSP). This architecture is referred to as software-defined radio, or just software radio. It represents the third generation of radio systems.

As new technologies are placed onto existing networks and wireless standards become more fragmented—particularly in the U.S.—the need for a single radio unit that can operate in different modes and bands becomes more urgent. A software radio handset could, for example, operate in a GSM-based PCS network, a legacy AMPS network, and a future satellite mobile network.

Operation

As noted, a software radio is one in which channel modulation waveforms are defined in software. Waveforms are generated as sampled digital signals, converted from digital to analog via a wideband Digital to Analog Converter (DAC), and then upconverted from an Intermediate Frequency (IF) to the desired Radio Frequency (RF).

In similar fashion, the receiver employs a wideband Analog to Digital Converter (ADC) that captures all of the channels of the software radio node. The receiver then extracts, downconverts, and demodulates the channel waveform using the software loaded on a general-purpose processor.

Multimode/Multiband

As competing technologies for wireless networks emerged in the early 1990s, it became apparent that subscribers would have to make a choice. The newer digital technologies offered more advanced features, but coverage would be spotty for some years to come. The older analog technologies offered wider coverage, but did not support the advanced features. A compromise was offered in the form of wireless multimode/multiband systems that let subscribers have the best of both worlds.

At the same time, wireless multimode/multiband systems allow operators to economically grow their networks to support new services where the demand is highest. With multimode/multiband handsets, subscribers can access new digital services as they become available, while retaining the capability to communicate over existing analog networks. The wireless system gives users access to digital channels wherever digital service is available, while providing a transparent handoff when users roam between cells alternately served by various digital and analog technologies. As long as subscribers stay within cells served by advanced digital technologies, they will continue to enjoy the advantages provided by these technologies. When they reach a cell that is supported by analog technology, they will have access only to the features supported by that technology. The intelligent roaming capability of multimode/multiband systems automatically chooses the best system for the subscriber to use at any given time.

Third-generation radio systems are agile in frequency and extend this flexibility even further by supporting more modes and bands. It is important to remember, however, that software radio systems may never catch up to encompass all the modes and bands that are available today and which may become available in the future. Users will always be confronted by choices. Making the right choice will depend on calling patterns, the features associated with the different technologies and standards, and the type of systems in use at international locations visited most frequently.

Multimode and multiband handsets have been available from several manufacturers since 1995. These handsets support more than one technology for its mode of operation and more than one frequency band.

An example of a multimode wireless system is one that supports both Advanced Mobile Phone Standard (AMPS) and Narrowband AMPS. Narrowband AMPS is a system-overlay technology that offers enhanced digital-like features, such as Digital Messaging Service, to phones operating in a traditional analog-based AMPS network. Among the vendors offering dual-mode AMPS/N-AMPS handsets is Nokia, the world's second largest manufacturer of cellular phones.

An example of a multiband wireless system is one that supports both GSM at 900 MHz and GSM at 1800 MHz in Europe. Among the vendors offering dual-band GSM handsets is Motorola. The company's International 8800 Cellular Telephone allows GSM 1800 subscribers to roam on either their home or GSM 900 networks (where roaming agreements are in place), using a single cellular telephone.

Of course, handsets can be both multimode and multiband. Ericsson, for example, offers dual-band/dual-mode handsets that support communication over both 800 MHz AMPS/D-AMPS and 1900 MHz D-AMPS networks. Subscribers on a D-AMPS 1900 channel can handoff both to/from a D-AMPS channel on 800 MHz as well as to/from an analog AMPS channel.

Multimode and multiband wireless systems allow operators to expand their networks to support new services where they are needed most, expanding to full coverage at a pace that makes economic sense. From the subscribers' perspective, multimode and multiband wireless systems allow them to take advantage of new digital services that are initially deployed in large cities, while still being able to communicate in areas served by the older analog technologies.

With its multimode capabilities, the wireless system preferentially selects a digital channel wherever digital service is available. If the subscriber roams out of the cell served by digital technology—from one served by CDMA to one served by AMPS, for example—a handoff occurs transparently. As long as subscribers stay within CDMA cells, they will continue to enjoy the advantages the technology provides, such as better voice quality and soft handoff, which virtually eliminates dropped calls. When subscribers reach a cell that supports only AMPS, voice quality diminishes and the chances for dropped calls increases.

However, these multimode/multiband handsets are not software-programmable. They rely instead on packaging dual sets of hardware in the same handset. Miniaturization of the various components makes this both practical and economical, but this approach has its limitations when the number of modes and frequencies that must be supported goes beyond two or three. Beyond that point, a totally new approach is required that relies more on programmable components.

Standards

The Software-Defined Radio Forum (formerly the Modular Multifunction Information Transfer System, or MMITS Forum) is an industry consortium dedicated to open architecture Plug-and-Play digital and software radios. Created in March 1996 with an initial focus on SPEAKeasy, MMITS has grown to embrace civil and commercial markets with a global outreach including the U.S., Europe and Asian suppliers of software and digital radio technology.

SPEAKeasy   Originally designed for military applications to bridge the wireless gap of diverse tactical radio systems in use by different armed services, SPEAKeasy can also meet the diverse requirements of the global wireless industry.

SPEAKeasy is a Department of Defense and industry program initiated to develop a software-programmable radio operating from 2 MHz to 2 GHz, employing waveforms selected from memory, downloaded from floppy disk, or reprogrammed over the air. This programmability allows a SPEAKeasy radio to become interoperable with whatever radio system it encounters. This becomes advantageous when it is not practical or cost effective to have as many radios as would be needed to communicate on every desired waveform or system.

SPEAKeasy is being hyped as the "PC of the communications world." It brings data to the aircraft cockpit, networking to the battlefield, and rapid modernization to all communication platforms. The result is a new level of flexibility that ultimately ensures important tactical and strategic advantages on the battlefield of the future.

For military applications, the benefits of SPEAKeasy are summarized as follows:

- Tactical radio systems for all armed forces
- Voice and data communications to all aircraft
- Voice and data communications onto the battlefield

For civilian use, the applications of SPEAKeasy include:

- Civil government emergency communications
- Law enforcement radio communications
- Public safety
- Global wireless communications

SPEAKeasy is not only designed to implement a totally open architecture, offering interoperability and full programmability, but it offers secure communications as well—all combined into a very compact, lightweight configuration. Future capabilities can be easily incorporated by upgrading internal components with current state-of-the-art technology.

FIRST

Flexible Integrated Radio Systems Technology (FIRST) is a European collaborative research project investigating the new technology of Intelligent Multimode Terminals (IMTs), and in doing so embraces the concept of software radio.

The consortium brings together a wide range of expertise from the telecommunication and related industries, from network operators and base-station manufacturers to university research departments and high-tech R&D companies. The goal of the project is for the telecommunications industry to make a significant contribution to the adoption of IMTs and software radio technology, and influence the emerging Universal Mobile Telecommunication System (UMTS) standard in Europe.

The UMTS effort combines key elements of Time Division Multiple Access (TDMA) and Code Division Multiple Access (CDMA) technologies, with an integrated satellite component to deliver wideband multimedia capabilities over mobile communications networks. UMTS makes possible a wide variety of mobile services ranging from messaging to speech, data, and video communications; Internet and intranet access; and high bit-rate communication up to 2 Mbps. As such, UMTS is expected to take mobile communications well beyond the current range of wireline and wireless telephony, providing a platform that will be ready for implementation and operation by the year 2002.

SUMMARY

Software radio architectures not only reduce the complexity and expense of serving a diverse customer base, but they simplify the integration and management of rapidly emerging standards. With software-based radio systems, access points, cell sites, and wireless data network hubs can be reprogrammed to meet changing standards requirements rather than replacing them or maintaining them in parallel with a newer infrastructure.

From the perspective of users, the same hardware would continue to be used—only the software gets upgraded. This could signal the end of outdated cellular telephones. Consumers will be able to upgrade their phones with new applications—much like purchasing new programs to add new capabilities to their computers.

Although the benefits are clear, commercial software-defined radio systems are still a few years away. Until they become available, users will have to make do with the current generation of multimode/multiband handsets.

See Also **Third-Generation Wireless Networks**

StarLAN

AT&T originally developed StarLAN to satisfy the need for a low-cost, easy-to-install, local-area network that would offer more configuration flexibility than token ring and more availability than Ethernet. StarLAN was offered in two versions—1 Mbps over unshielded twisted pair (UTP) wire and 10 Mbps over UTP or optical fiber. Cable segments were joined at a SmartHUB.

While the early version of StarLAN was proprietary, later versions offered compliance with IEEE 802.3 standards, making them interoperable with Ethernet and token ring. StarLAN never approached the market share of Ethernet and token ring. Its survival was further jeopardized by

AT&T's failure to become a significant player in the computer market with the acquisition of NCR.

In 1991, AT&T and NCR merged under the name AT&T Global Information Solutions, where responsibility for StarLAN resided until 1996. That year, AT&T Global Information Solutions changed its name back to NCR Corp. in anticipation of being spun off to AT&T shareholders as an independent, publicly traded company. Around that time, NCR discontinued the StarLAN product line. Used StarLAN components are still available from a few third-party sources.

See Also ARCnet, Ethernet, Token Ring

Statistical Time Division Multiplexers

Statistical Time Division Multiplexers (STDMs) are used to combine traffic from different sources over a higher-speed line. The technique is statistical in that the input/output devices are not assigned their own channels, as with Time Division Multiplexers (TDMs). If an input device has nothing to send at a particular time, another input device can use the channel. This uses the available bandwidth more efficiently. An STDM can be purchased as a stand-alone device or it can be an add-on feature to a TDM, providing service over one or more assigned channels.

Operation

STDM operation is similar to that of TDMs; the high-speed side appears very much like a TDM high-speed side, while the low-speed side is quite different. The STDM allocates high-speed channel capacity based on demand from the devices connected to the low-speed side. This allocation by demand (or contention) provides more efficient use of the available capacity on the high-speed line. In the variable-allocation scheme of an STDM frame, the time slots do not occur in a fixed sequence.

An STDM increases high-speed line usage by supporting input channels whose combined data rates would exceed the maximum rate supported by the high-speed port. When any given channel is idle (not sending or receiving data), input from another active channel is used in the time slot instead. The STDM has the option of turning off the flow of data from a sender if there is insufficient line capacity and then turning the flow back on when the capacity becomes available.

Features

TDMs and STDMs share many of the same operational and management features. Among the features found in STDMs is data compression; like TDMs, STDMs support techniques for compressing data so they can actually transmit fewer bits per character. Data compression shrinks the time slot for the STDM and allows it to transmit more time slots per frame.

STDMs also provide error detection and correction. While TDMs detect and flag errors, STDMs can correct them. The sending STDM stores each transmitted data frame and waits for the receiving STDM or computer to acknowledge receipt of the frame. A positive acknowledgment (ACK) or negative acknowledgment (NAK) is returned. If an ACK is received, the STDM discards the stored frame and continues sending the next frame. If a NAK is returned, the STDM retransmits the questionable frame and any subsequent frames. The process is repeated until the problematic frame is accepted, or a frame retransmission counter reaches a predetermined number of attempts and activates an alarm. For applications such as asynchronous data transmission, where error detection is not performed as part of the protocol, STDM error detection and correction is a valuable feature. However, for protocols such as IBM's binary synchronous communication (BSC), which contains its own error-detection algorithm, STDM-performed error control adds additional delays and redundancy that may not be appropriate for the application.

Several throughput enhancements are available for STDMs which can be added later when needs change. These features include:

PER-CHANNEL COMPRESSION In which each channel has its own compression table

FAST-PACKET TECHNOLOGY Which increases throughput by sending part of a frame before the entire frame is built

DATA PRIORITIZATION Inserts shorter interactive frames between larger variable run-length frames

RUN-LENGTH COMPRESSION Removes redundant characters from the transmission to improve performance

Despite the continuous price reductions on leased lines, businesses are always looking to cut the cost of telecommunications. One of the most effective ways to do this is through the deployment of TDMs and STDMs that increase the bandwidth usage of leased lines. Such lines are billed at a flat monthly rate by the carriers, no matter how little or how much traffic is actually carried over them. Not only can the business save money by using multiplexers, but the cost of the devices themselves can be recovered in a matter of a few months out of the money saved. Using different levels of voice and/or data compression, the channel capacity of a leased line can be easily doubled or quadrupled to save even more money, enabling the business to recover the cost of the equipment even faster.

SUMMARY

While TDMs are better suited to carrying time-sensitive applications, each input/output device requires a dedicated channel, which may be wasted if the attached devices have no traffic to send. STDMs excel at handling routine data that is not time sensitive, while enabling many more devices than channels to share the available bandwidth, thereby ensuring its most efficient use of the line.

See Also **Time-Division Multiplexing, Wavelength Division Multiplexing**

Storage-Area Networks

Customer demand for increased performance, availability, and manageability of storage, combined with the emergence of Fibre Channel technology, are driving the convergence of storage and networking architectures. To harness the full capabilities and performance of storage hardware and connectivity, a new network-based storage topology is beginning to emerge—the Storage Area Network (SAN). In providing any-to-any connectivity for storage resources on a dedicated high-speed network, the SAN offloads storage traffic from daily network operations while establishing a direct connection between storage elements and servers (Figure 109).

SAN Concepts

Essentially, a SAN is a specialized network that enables fast, reliable access among servers and external or independent storage resources. In a SAN, a storage device is not the exclusive property of any one server. Rather, storage devices are shared among all networked servers as peer resources. Just as a LAN can be used to connect clients to servers, a SAN can be used to connect servers to storage, servers to each other, and storage to storage.

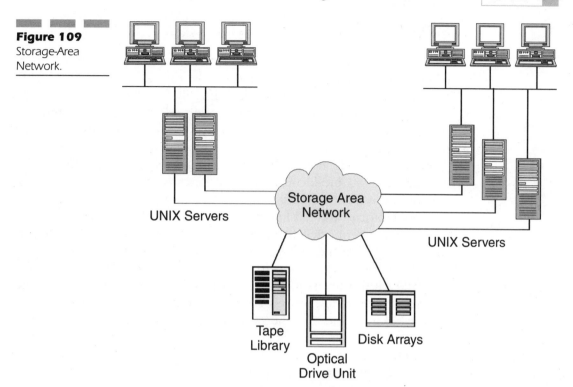

Figure 109
Storage-Area
Network.

SANs provide an open, extensible platform for storage access in data intensive environments like those used for video editing, prepress, OLTP, data warehousing, storage management, and server clustering applications.

SANs offer a number of benefits. The redundancy that is an inherent part of SAN architectures makes high availability more cost-effective for a wider variety of application environments. The pluggable nature of SAN resources—storage, nodes, and clients—enables much easier scalability while preserving ubiquitous data access. And with storage centralized, more efficient management of the data for tasks such as optimization, reconfiguration, and backup/restore become much easier.

SANs are particularly useful for backups. Previously, there were two choices; either a tape drive had to be installed on every server and someone went around changing the tapes, or a backup server was created and the data moved across the network, which consumed bandwidth. Performing backup over the LAN can be excruciatingly disruptive—and slow. A daily backup can suddenly introduce gigabytes of data into the normal LAN traffic. With SANs, organizations can have the best of both worlds: high-speed backups in a centralized location.

SAN Origins

SANs have existed for years in the mainframe environment in the form of Enterprise Systems Connection (ESCON). In mid-range environments, the high-speed data connection was primarily SCSI (Small Computer System Interface)—a point-to-point connection, which is severely limited in terms of the number of connected devices it can support as well as the distance between devices.

An alternative to network-attached storage was developed in 1997 by Michael Peterson, president of Strategic Research (Santa Barbara, CA). He believed network-attached storage was too limiting because it relied on network protocols and did not guarantee delivery. He proposed that SANs could be interconnected using network protocols such as Ethernet, and the storage devices themselves could be linked via nonnetwork protocols.

In a traditional storage environment, a server controls the storage devices and administers requests and backup. With a SAN, instead of being involved in the storage process, the server simply monitors it. By optimizing the box at the head of the SAN to do only file transfers, users are able to get much higher transfer rates, such as 100 Mbps via Fibre Channel. Traditional SCSI connections offer transfer rates of only 40 Mbps—80 Mbps with the newer Ultra2 SCSI.

Using Fibre Channel as the connection between storage devices also increases distance options. While traditional SCSI allows only a 25-meter distance (about 82 feet) between machines and Ultra2 SCSI allows only a 12-meter distance (about 40 feet), Fibre Channel supports spans of 10 kilometers (about 6.2 miles). SCSI can connect only up to 16 devices, whereas Fibre Channel can link as many as 126. By combining LAN networking models with the core building blocks of server performance and mass storage capacity, SAN eliminates the bandwidth bottlenecks and scalability limitations imposed by previous SCSI bus-based architectures.

SAN Features

The following are key features of storage-area networks:

- Storage and archival traffic are routed over a separate network, offloading the majority of data traffic from the enterprise network.
- Data transfers are fast with Fibre Channel—up to 100 Mbps with a single loop configuration and up to 200 Mbps with a dual-loop configuration.
- A shared data storage pool can be easily accessed by remote workstations and servers.

- The SAN can be easily expanded to a virtually unlimited size with hubs or switches.

- Nodes on the SAN can be easily added or removed with minimal disruption to the active network.

- For totally redundant operation, the SAN can be easily configured to support mission-critical applications.

A key feature of SANs is zoning. This is the division of a SAN into subnets that provide different levels of connectivity between specific hosts and devices on the network. In effect, routing tables are used to control access of hosts to devices. This gives IT managers the flexibility to support the needs of different groups and technologies without compromising data security. Zoning can be performed by cooperative consent of the hosts or can be enforced at the switch level. In the former case, hosts are responsible for communicating with the switch to determine if they have the right to access a device.

There are several ways to enforce zoning. With hard zoning, which delivers the highest level of security, IT managers program zone assignments into the flash memory of the hub. This ensures that there can be absolutely no data traffic between zones.

Virtual zoning provides additional flexibility because it is set at the individual port level. Individual ports can be members of more than one virtual zone, so groups can have access to more than one set of data on the SAN.

Broadcast zoning can be used to restrict the scope of broadcasts. For example, IP ARP (Address Resolution Protocol) broadcasts can be kept from SCSI ports on the switch. These IP broadcasts can otherwise cause storage devices to crash.

Technology Mix

As the SAN concept has evolved, it has moved beyond association with any single technology. In fact, just as LANs and WANs use a diverse mix of technologies, so can SANs. This mix can include FDDI, ATM and IBM's Serial Storage Architecture (SSA), as well as Fibre Channel. SAN architectures also allow for the use of a number of underlying protocols, including TCP/IP and all the variants of SCSI.

Instead of dedicating a specific kind of storage to one or more servers, a SAN allows different kinds of storage—mainframe disk, tape, and RAID—to be shared by different kinds of servers, such as Windows NT, UNIX, and OS/390. With this shared capacity, organizations can acquire, deploy and use storage devices more efficiently and cost-effectively.

SANs also let users with heterogeneous storage platforms use all available storage resources. This means that within a SAN users can back up or archive data from different servers to the same storage system. They can also allow stored information to be accessed by all servers, create and store a mirror image of data as it is created, and share data between different environments.

With a SAN, there is no need for a physically separate network to handle storage and archival traffic. This is because the SAN can function as a virtual subnet that operates on a shared network infrastructure. For this to work, however, different priorities or classes of service must be established. Fortunately, both Fibre Channel and ATM provide the means to set different classes of service.

Although early implementations of SANs have been local- or campus-based, there is no technological reason why they cannot be extended much farther over the WAN. As WAN technologies such as SONET and ATM mature, and especially as class-of-service capabilities improve, the SAN can be extended over a much wider area—perhaps globally in the future.

SANs also promise easier and less expensive network administration. Today, administrative functions are labor-intensive and time-consuming, and IT organizations typically have to replicate management tools across multiple server environments. With a SAN, only one set of tools is needed, which eliminates the need for replication and associated costs.

SAN Components

Several components are required to implement a SAN. A Fibre Channel adapter is installed in each server. These are connected via the server's PCI bus to the server's operating system and applications. Because Fibre Channel's transport-level protocol wraps easily around SCSI frames, the adapter appears to be a SCSI device. The adapters are connected to a single Fibre Channel hub, running over fiber-optic cable or copper coaxial cable. Category 5 cable, the high-end twisted pair rated for Fast Ethernet and 155-Mbps ATM, can also be used.

A LAN-free backup architecture may include some type of automated tape library that attaches to the hub via Fibre Channel. This machine typically includes a mechanism capable of feeding data to multiple tape drives and may be bundled with a front-end Fibre Channel controller. Existing SCSI-based tape drives can be used also through the addition of a Fibre Channel-to-SCSI bridge.

Storage management software running in the servers performs contention management by communicating with other servers via a control protocol to synchronize access to the tape library. The control protocol maintains a master index and uses data maps and time stamps to establish server-to-hub connections. Currently, control protocols are specific to the software vendors. Eventually, the storage industry will likely standardize on one of the several protocols now in proposal status before the Storage Network Industry Association.

From the hub, a standard Fibre Channel protocol, Fibre Channel-Arbitrated Loop (FC-AL), functions similarly to token ring to ensure collision-free data transfers to the storage devices. The hub also contains an embedded SNMP agent for reporting to network management software.

Role of Hubs

Much like Ethernet hubs in LAN environments, Fibre Channel hubs provide fault tolerance in SAN environments. On a Fibre Channel-Arbitrated Loop, each node acts as a repeater for all other nodes on the loop, so if one node goes down, the entire loop goes down. For this reason, hubs are an essential source of fault isolation in Fibre Channel SANs. The hub's port bypass functionality automatically bypasses a problem port and avoids most faults. Stations can be powered off or added to the loop without serious loop effects. Storage management software is used to mediate contention and synchronize data—activities necessary for moving backup data from multiple servers to multiple storage devices. Hubs also support the popular physical star cabling topology for more convenient wiring and cable management.

To achieve full redundancy in a Fibre Channel SAN, two fully independent, redundant loops must be cabled. This scheme provides two independent paths for data with fully redundant hardware. Most disk drives and disk arrays targeted for high-availability environments have dual ports specifically for the purpose. Wiring each loop through a hub provides higher availability port bypass functionality to each of the loops.

Many organizations need multiple levels of hubs. Hubs can be cascaded up to the Fibre Channel-Arbitrated Loop limit of 126 nodes (127 nodes with an FL or switch port). Normally, the distance limitation between Fibre Channel hubs is three kilometers. However, Hewlett-Packard offers technology that extends the distance between hubs to 10 kilometers, allowing organizations to link servers situated on either side of a campus, or even spanning a metropolitan area.

Management

The tools needed to manage a Fibre Channel fabric are available through the familiar SNMP (Simple Network Management Protocol) interface. The draft proposal for a MIB (Management Information Base) has been circulated at the Internet Engineering Task Force (IETF). New vendor-specific MIBs will emerge as products are developed with new management features. Hubs and other central networking hardware provide a natural point for network management.

Among the companies currently offering SAN management solutions is Legato Systems, whose Enterprise Storage Management Architecture (ESMA) enables customers to take advantage of the benefits offered by SAN topologies for backup and restore. One element of ESMA is the Legato NetWorker Storage Node, which enables large, business-critical servers to be backed up directly to tape, under the control of a central Legato NetWorker server. This results in the ability to off-load backup traffic from the local-area network and move it to the storage-area network, taking advantage of the bandwidth offered by Fibre Channel.

Another ESMA component is Legator SmartMedia. The any-to-any connectivity of SANs enables tape libraries to be connected to multiple servers and Legato SmartMedia manages the sharing of media and devices between them. Legato SmartMedia enables drive and library sharing, managing application requests for media from a central location. This centralized management of distributed libraries results in increased manageability and reduced operating costs.

SUMMARY

The move to Storage-Area Networks over the next few years will provide a new level of scalability to system administrators and allow a much greater degree of flexibility than the traditional attached storage paradigm. Fibre Channel technology provides the basic foundation of this shift, allowing enterprises to start implementing a solid foundation today to support the environments of the future. However, SCSI has not yet exhausted its potential as a storage connectivity option. SCSI is installed in more than 90 percent of networks, and the latest SCSI variant—Ultra160/m SCSI—may even pose a challenge to Fibre Channel. In addition to supporting data transfers at up to 160 Mbps, Ultra160/m SCSI offers intelligent data management. Products incorporating Ultra160/m intelligently test and manage the storage network so that the maximum reliable data transfer rate is used. If the cabling, backplanes, and terminators can support the desired speed, the data flows at full throttle. If not, the hardware will be smart enough to gracefully negotiate a lower transfer rate automatically. The result will be more system autonomy and less IT manager involvement.

See Also **Fibre Channel, Small Computer System Interface**

Switched LANs

A recent trend in tying together network resources is to interconnect the various network segments using switches rather than hubs or routers. The easiest way to improve the performance of shared LANs is to add port or segment switching in which LAN segments are assigned to new ports instantly, allowing bottlenecks to be eliminated through the reassignment of very active LAN nodes.

For example, several dozen workstations running network bandwidth-intensive applications such as imaging, video editing, and computer-aided design on the same Ethernet segment can produce a serious bottleneck. LAN switching can segment an overcrowded, shared-bandwidth workgroup into multiple virtual LANs in which each user or group of users can access 10-Mbps Ethernet or 16-Mbps token-ring bandwidth. The solution is cost-effective and improves network performance for each user.

Several types of switches are now available for Ethernet LANs. In addition to the basic 10BaseT switch, there are auto-sensing 10/100-Mbps switches and Gigabit Ethernet switches. There are also switches for Token Ring. However, the hottest trend is the use of Layer 3 switches, which improve the performance of segmented Ethernet and Token Ring LANs without having to use routers. Aside from cost savings, the benefits are improved availability and throughput. Sometimes Layer 3 capabilities come with or can be added to the other types of switches.

Since existing investments in network topology, cabling, adapter cards, and operating systems are retained, the costs associated with implementing a switched LAN solution come mainly from the purchase of the new switch itself and any additional management tools that may be required.

Layer 3 Switches

Layer 3 switches deserve some explanation because they are not really switches. With all the pressure to differentiate their products, some vendors confuse buyers by applying the term to products that range from bridges to routers. Like ordinary LAN switches, Layer 3 switches make forwarding decisions based on the packet's destination address. But they also have the intelligence to make decisions based on information stored inside the packet, just like routers. So these "Layer 3 switches" really are not switches at all, but routers with faster frame-forwarding capabilities.

One reason why Layer 3 devices operate so fast is that they usually support only IP. When other protocols are supported, such as Novell's IPX, Apple's AppleTalk, or IBM's APPN, the packet forwarding rate is the same as traditional routers or even slower. When added to the network, Layer 3 devices can make routers more efficient by off-loading the IP routing task. Since IP is the most often used protocol, off-loading IP from the routers can provide a fairly substantial performance gain.

Adding the Layer 3 capability to a network does not always require buying a whole new system. Some vendors, such as Cisco, offer add-ons to their existing LAN switches that transform them into Layer 3 systems.

Workgroup Switches

Although there are feature-rich enterprise-level switches, most LAN switches are implemented at the department or workgroup level. Prices, performance, and features vary widely. Some devices are basic, no-frills switches, while others come with virtual LAN support, protocol filtering, SNMP management, high-speed uplinks, and remote monitoring (RMON) support.

For example, 3Com's SuperStack II Switch 1100 is priced at $65 per port. It has 12 10BaseT ports and features two built-in auto-sensing 10/100 Fast Ethernet ports. A matrix port enables high-speed connection to other SuperStack II units. The stacked switches can be managed as a single entity and share a single IP address. The switch automatically provides full-duplex/half-duplex capability on all ports to boost bandwidth for servers and power users. It also implements advanced policy-based management across the network and supports Fast IP, IGMP snooping, IEEE 802.1p prioritization, and IEEE 802.1q standards-based VLANs. The company offers a 24-port unit priced at about $100 per port.

There are 10/100BaseT switches that fall into the same price range as 10BaseT switches. D-Link Systems, for example, offers a 24-port, 10/100-Mbps, dual-speed switch priced at $80 per port. The Layer 2 switch has 5.5G bps of total bandwidth, supports SNMP and RMON, and is targeted for workgroup or departmental deployments.

Usually, Token Ring networks carry a higher price tag than Ethernet. Although prices for Token Ring workgroup switches are dropping, the lowest price of $169 per port from Madge Networks is still twice the per-port price of an Ethernet switch.

In the 12- to 24-port range, 100BaseT switches tend to be stackable, allowing the system to scale to as many as eight units. Many are of fixed configuration and do not allow for the addition of more ports, except by adding an entire system. When designing switched networks, using systems of differing port capacities may be appropriate. However, two eight-port switches do not necessarily equate to one 16-port switch. With two switches, there is usually a bigger bottleneck when packets must travel between ports on different switches.

For network managers unsure of which technology their organizations might migrate to in the future—ATM or Gigabit Ethernet—dual-speed Ethernet stackable switches now offer high-speed uplink modules for ATM or Gigabit Ethernet backbone connections. This kind of switch is a risk-free buy for companies not sure which high-speed LAN technology they will choose but who will likely need to upgrade to ATM or Gigabit Ethernet in the future.

Management Tools

Although current shared-LAN tools handle visibility of the network and the bandwidth usage of hub segments in a switched environment, visibility of the switched ports is lost. This leaves network managers unenlightened about the source of the network traffic. The difference in managing a shared versus a switched network is that in a shared environment, all the tools and agents assume a single, shared interface into the hub. In a switched environment, every port behaves as an individual LAN. This behavior requires an individual interface to every port, in addition to an interface into the box.

External network management tools such as LAN analyzers, port probes, and monitors have become standard equipment for network managers. But they can be expensive; the hardware alone could cost between $3,000 and $5,000, and the RMON agent software could cost as much as $10,000. In addition, a comprehensive data collection and analysis tool such as a network sniffer costs about $25,000. Alternatively, monitoring only one LAN segment at a time—on a 24-port switch, for example—requires a lot of time and effort to gather all the necessary information.

Some vendors use the switch's internal capability to create a roving RMON probe that allows statistics to be gathered on port segments at specified intervals. Management information can be gathered from multiple ports with a single probe without having to incur the expense of putting RMON on every port. Although the roving RMON probe is a simpler and more economical solution than the manual network sniffer, it still does not provide a global view of the network.

Some vendors offer enhanced network configuration and management features, which are aimed at LAN managers scrambling to optimize growing networks. Bay Networks, for example, offers enhancements for its BayStack Ethernet switches that include the ability to configure port-based VLANs, port mirroring, Web-based management, and unlimited media access control (MAC) addressing. Network management support for these switches is provided through integration with Bay Networks' Optivity network management software. Other vendors, including 3Com, offer many of the same management features, plus broadcast traffic control, RMON on every port, and policy-defined management.

Web-based management tools facilitate installation, configuration, control, and troubleshooting of network switches. Using a familiar Web interface, such tools enable LAN managers to administer switches from any networked PC. The interface presents graphical information the LAN administrator can interpret quickly instead of having to read lines of data. Using a Web browser instead of a dedicated management console or a terminal at the switch itself saves a LAN manager considerable time and support costs. Many vendors offer Web-based management tools free, even making them available for download from their Web site.

SUMMARY

LAN switches are available in high-end, multislot matrix systems, which typically provide aggregate switching capacities in excess of 1 Gbps and support multiple high-speed technologies besides 100BaseT. These switches, usually with built-in Layer 3 capability, can accommodate virtually any type of LAN interface. Vendors offer modules that can interconnect 100BaseT, 10BaseT, 100VG-AnyLAN, FDDI, and ATM networks.

There is a wide variety in the price and throughput performance of LAN switches of all kinds. The performance claims of vendors do not always hold up under rigorous testing by independent sources. Given the disparity in claimed versus actual performance—and all other things being equal, such as management and flow- and traffic-control features—the high prices charged by some vendors needs to be negotiated down before committing to a purchase, or the buyer needs to look elsewhere.
See Also **Hubs, Routers**

Synchronous Communication

Synchronous communication relies on the presence of a clocking system at both ends of the transmission. These clocks must be synchronized at the beginning of the session so that the timing of the transmission—not the use of start and stop bits, as in asynchronous communication—defines where data begins and ends.

When data is transmitted in synchronous fashion, a unique 8-bit pattern is used to define the start of the data stream (Figure 110). This special bit pattern is embedded in the digital signal to assist in maintaining the timing between the sender and receiver.

Synchronous communication is used over digital lines and can achieve much higher data rates than asynchronous communication, which is typically used over analog telephone lines. For example, the synchronous method of communication is used on ISDN lines because it handles data more efficiently than the typical modem's asynchronous technique.

While asynchronous communication sends small blocks of data with many control bits for error correction, synchronous techniques use large blocks of data with control bits only at the start and end of the transmission. The higher quality of digital lines permits minimal error checking. A modem used over an analog line does not handle synchronous communication well because noise on the line disrupts synchronization.

An example of synchronous communication is Ethernet for local-area networking. On the wide-area network, ISDN is an example of a service that is based on synchronous communication.

SUMMARY

Synchronous communication is usually much more efficient in its use of bandwidth than asynchronous communication because more data can be transmitted without the use of control and error bits. Another advantage of synchronous communication is that the frame structure allows for easy handling of control information that may be required by data link layer protocols. There is a natural position—usually at the start of the frame—for any special codes needed by the communication protocol.

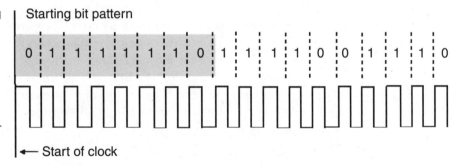

Figure 110

In synchronous communication, a unique 8-bit pattern indicates the start of the clock.

Starting bit pattern

0 1 1 1 1 1 1 0 1 1 1 1 0 0 1 1 1 0

← Start of clock

For example, the High-Level Data Link Control (HDLC) ensures that data passed up to the next layer has been received exactly as transmitted—error free, without loss, and in the correct order. HDLC also performs flow control, which ensures that data is transmitted only as fast as the receiver can receive it.

See Also **Asynchronous Communication, Bisynchronous Communication, Isochronous Communication**

Synchronous Data Link Control

Synchronous Data Link Control (SDLC) was introduced by IBM in 1973 and is the preferred link level protocol for its Systems Network Architecture (SNA) networks. It was intended to replace the older Binary Synchronous Communication (BSC) protocol developed in 1965 for wide-area connections between IBM equipment. It is equivalent to the High-Level Data Link Control (HDLC) developed by the International Organization for Standardization (ISO) and adapted by many non-IBM vendors.

Like HDLC, SDLC ensures that data passed up to the next layer has been received as transmitted—error free, without loss, and in the correct order. However, SDLC is not a peer-to-peer protocol like HDLC and it is used only on leased lines where the connections are permanent. In a point-to-point configuration, there is one primary station that controls all communications, and one secondary station. In a point-to-multipoint configuration, there is one primary station, but multiple secondary stations arranged as drops along the line.

The primary station can be a mainframe or mid-range central computer, or a communications controller that acts as a concentrator for a number of local terminals. It operates in full-duplex mode, while the secondary stations operate in half-duplex mode. The primary station is aware of the transmission status of the secondary stations at all times. The drops can be in different locations. A mainframe in New York, for example, may support a multidrop line with controllers connected to drops in offices in Atlanta, Chicago, Dallas, and Los Angeles.

SDLC uses the same frame format as HDLC (Figure 111). The variable-length frame is bounded by two 8-bit flags, each containing the binary value of 01111110. The 8-bit address field of each SDLC frame always identifies the secondary station on the line. When the primary station invites a secondary station to send data (i.e., polling), it identifies the station being polled. Each secondary station sees all transmissions from the primary, but only responds to frames with its own address. In a point-to-multipoint configuration, up to 254 secondary station addresses

Figure 111

An SDLC frame with flag delimiters at each end.

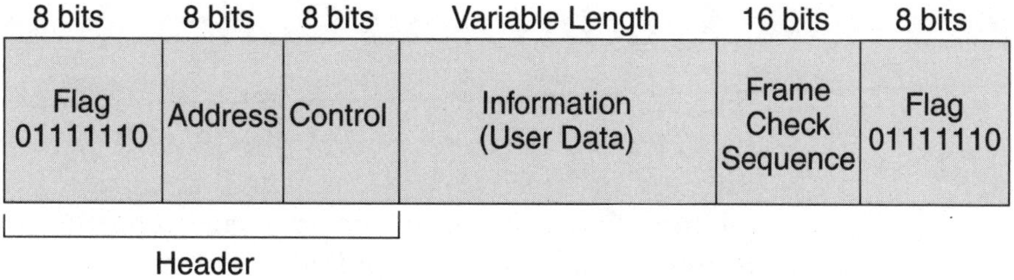

8 bits	8 bits	8 bits	Variable Length	16 bits	8 bits
Flag 01111110	Address	Control	Information (User Data)	Frame Check Sequence	Flag 01111110

Header

are possible, with one additional used for testing and another for broadcasting information from the primary station to all secondary stations. The SDLC frame's 8-bit control field is used to indicate whether the frame contains application-specific data, supervisory data, or command data.

The variable-length information field contains application-specific data, in other words the user's data. The content of this field is always in multiples of 8 bits. This field is optional, however, since control fields containing unnumbered commands do not transmit application data in the frames.

The 16-bit frame check sequence (FCS) field is used to verify the accuracy of the data. As in HDLC, the FCS is the result of a mathematical computation performed on the frame at its source. The same computation is performed at the receive side of the link. If the answer does not agree with the value on the FCS field, this means some bits in the frame have been altered in transmission, in which case the frame is discarded. A window of up to seven frames can be sent from either side before acknowledgement is required. Acknowledgement of received frames is encoded in the control field of data frames, so that if data is flowing in both directions, no additional frames are needed for frame acknowledgement.

SUMMARY

SDLC is a full-duplex protocol, which enables the primary and secondary stations to send data to each other at the same time. Since SDLC is a bit-oriented protocol, it is insensitive to code, which may be industry-standard ASCII or IBM's EBCDIC. SDLC is much more efficient than the older BSC. With the former, the acknowledgment for the data is usually sent with the data itself, while in the latter, the acknowledgment is a separate transmission.

See Also **High-Level Data Link Control**

Synchronous Optical Network

Synchronous Optical Network (SONET) is an industry standard for high-speed transmission over optical fiber. SONET networks offer many advantages for large corporations as well as carriers, including equipment interoperability between the two environments, virtually unlimited bandwidth, fault recovery, and network management.

The importance of SONET has been underscored in recent years with the rapid increase of data traffic on the public networks, fueled by Internet usage. Data now accounts for about 50 percent of the traffic on the telephone networks in North America. Today, all the major long distance carriers, RBOCs, national Internet service providers, and next generation carriers—and their competitors—have SONET networks in place. Of course, many large corporations as well as state and federal government agencies have deployed SONET as well.

Transmission Rates

The SONET standard specifies a hierarchy of rates and formats for optical transmission, ranging from 51.84 Mbps to 13.271 Gbps. The following table summarizes the standard optical carrier (OC) transmission rates:

OC Level	Line Rate
OC-1	51.84 Mbps
OC-3	155.520 Mbps
OC-9	466.560 Mbps
OC-12	622.080 Mbps
OC-18	933.120 Mbps
OC-24	1.244 Gbps
OC-36	1.866 Gbps
OC-48	2.488 Gbps
OC-192	9.95 Gbps
OC-256	13.271 Gbps

Network Elements

SONET network infrastructures consist of various types of specialized equipment including ADM, BDCS, WDCS, digital loop carriers, regenerators, and SONET CPE (customer premises equipment).

ADD-DROP MULTIPLEXER The ADM provides an interface between network signals and SONET signals. It is a single-stage multiplexer/demultiplexer that converts DS-n signals into OC-n signals. The ADM can be used in terminal sites and intermediate (add-drop) sites; at an add-drop site, it can drop lower-rate signals down or pull lower-rate signals up into the higher-rate OC-n signal.

BROADBAND DIGITAL CROSS-CONNECT The BDCS interfaces various SONET signals and legacy DS3s; it accesses the STS-1 signals, and switches at this level. It is the synchronous equivalent of the DS3 digital cross-connect; however, the BDCS accepts optical signals and allows overhead to be maintained for integrated Operations, Administration, Maintenance, and Provisioning (OAM&P). Most asynchronous systems prevent overhead from being passed from signal to signal, but the BDCS makes two-way cross-connections at the DS3, STS-1, and STS-c (concatenated) levels. It is typically used as a SONET hub that grooms STS-1s for broadband restoration purposes, or for routing traffic.

WIDEBAND DIGITAL CROSS-CONNECT WDCS is a digital cross-connect that terminates SONET and DS3 signals, and maintains the basic functionality of a virtual tributary (VT) and/or DS1-level cross-connects. It is the optical equivalent of the DS3/DS1 digital cross-connect and accepts optical-carrier signals as well as DS1s and DS3s. In a WDCS, switching is done at the VT, DS1, or DS0 level. Because SONET is synchronous, low-speed tributaries are visible in VT-based systems and directly accessible within the STS-1 signal; this allows tributaries to be extracted and inserted without demultiplexing. Finally, the WDCS cross-connects constituent DS1s between DS3 terminations, and between DS3 and DS1 terminations.

DIGITAL-LOOP CARRIER Like the DS1 digital-loop carrier, this network element accepts and distributes SONET optical-level signals. Digital-loop carriers allow the network to transport services that require large amounts of bandwidth. The integrated overhead capability of the digital-loop carrier allows surveillance, control, and provisioning from the central office.

REGENERATOR A SONET regenerator drives a transmitter with output from a receiver, and stretches transmission distances far beyond what is normally possible over a single length of fiber.

PROTOCOL STACK The SONET transmission protocol consists of four layers: photonic, section, line, and path.

PHOTONIC LAYER The photonic layer is the electrical and optical interface for the transport of information bits across the physical medium. Its primary function is to convert STS-N electrical signals into OC-N optical signals. This layer performs functions associated with the bit rate, optical-pulse shape, power, and wavelength; it uses no overhead.

SECTION LAYER The section layer deals with the transport of the STS-N frame across the optical cable, and performs a function similar to the data-link layer (layer 2) of bit-oriented protocols such as high-level data-link control (HDLC) and synchronous data-link control (SDLC). This layer establishes frame synchronicity and the maintenance signal; functions include framing, scrambling, error monitoring, and orderwire communications.

LINE LAYER The line layer provides the synchronization, multiplexing, and automatic protection switching (APS) for the path layer. Primarily concerned with the reliable transport of the path layer payload (voice, data, or video) and overhead, it allows automatic switching to another circuit if the quality of the primary circuit drops below a specified threshold. Overhead includes line-error monitoring, maintenance, protection switching, and express orderwire.

PATH LAYER The path layer maps services such as DS3, FDDI, and ATM into the SONET payload format. This layer provides end-to-end communications, signal labeling, path maintenance, and control, and is accessible only through terminating equipment. A SONET ADM accesses the path layer overhead; a cross-connect system that performs section and line-layer processing does not require access to the path layer overhead.

Channelization

Channelized interfaces provide network configuration flexibility and contribute to lower telecommunications costs. For example, channelized T1 delivers bandwidth in economical 56/64K bps DS0 units, each of which can be used for voice or data and routed to different locations within the network or aggregated as needed to support specific applications. Likewise, channelized DS3 delivers economical 1.544 Mbps DS1 units, which can be routed separately or aggregated as needed. Channelized interfaces also apply to the SONET world—channelized OC-48, OC-12, and OC-3 can all be subchannelized down through DS3 speeds. OC-48, for example, can be channelized as follows:

- Four OC-12 tributaries, all configured for IP (Packet over SONET) framing, or all configured for ATM framing
- Four OC-12 tributaries, with two configured for IP (Packet over SONET) framing, and the other two for ATM framing

- Two OC-12, with one configured for IP framing, and the other for ATM framing; 8 OC-3, with four configured for IP framing, and four configured for ATM framing

- Two OC-12, with one configured for IP framing, and the other for ATM framing; 6 OC-3, with three configured for IP framing, and three configured for ATM framing; 6 DS3, with three configured for IP framing, and three configured for ATM framing

- Forty-eight DS3s, with 24 configured for IP framing, and 24 for ATM framing

Multiservice channelized SONET is implemented on a single OC-48 line card, which provides IP packet and ATM cell encapsulation to support business data and Internet services from the same hardware platform. One vendor that offers such products is Argon Networks. The company's GigaPacket Node (GPN) is a native IP router as well as a native ATM switch in a single modular platform, which enables any port to be configured for either Packet over SONET (PoS) or ATM service. The configurations are implemented through keyboard commands at service time, rather than permanently assigned at network build out time.

Survivable Networking

Network failures come in the form of hard failures such as blown circuit packs and soft failures such as degraded optics. SONET offers several ways to recover from both types of failures, including:

AUTOMATIC PROTECTION SWITCHING If a transmission system detects a failure on a working facility, it switches to a standby facility to recover the traffic. One-to-one and one-to-n protection switching are provided.

BIDIRECTIONAL LINE SWITCHING If two fiber pairs exist between each recoverable node and a fiber facility fails, the node preceding the break loops the signal back toward the originating node, where it travels different fiber pairs to its destination.

UNIDIRECTIONAL PATH SWITCHING If one fiber pair is between each node and a signal is transmitted in two different paths around the ring, the network determines and uses the best path at the receiving end. If a fiber facility fails, the destination node switches traffic to the alternate receive path. The switchover occurs in as little as 50 milliseconds.

SYSTEM REDUNDANCY All SONET network elements use circuit pack redundancy, such as crucial optic cards with 1×1 backup cards. Other service cards are backed up in a $1 \times n$ scheme. So for four DS1 mapper cards, there will be a spare on "standby" for redundancy.

SONET's embedded control channels enable end-to-end performance tracking and the identification of elements that cause errors. With this capability, carriers can guarantee transmission performance, and users can verify it without going offline to implement test procedures. These capabilities allow problem identification prior to service disruption. Combined with the self-healing capabilities, SONET's diagnostic capabilities ensure that properly configured networks experience virtually no downtime.

IP vs. ATM over SONET

Organizations of all types and sizes have an insatiable appetite for more bandwidth. Because most carriers are having difficulty meeting bandwidth demand as corporate data-network usage continues to increase, they are turning to bandwidth-multiplying technologies such as Wave Division Multiplexing (WDM) and Dense Wave Division Multiplexing (DWDM), as well as SONET. Often these technologies are used together as an integrated optical networking system.

The industry is now focusing on the fact that LANs and WANs are now merging, as both are now used to carry voice, data, and video traffic—a theme initiated by the Internet under the rubric of "voice-data convergence." ATM is conducive to switching connections in WANs, but it is still too expensive to implement on LANs.

IP has become prevalent throughout both the Internet and in LANs. Therefore, many experts believe that IP traffic should be the primary protocol. There will be continued debate on IP vs. ATM for building transparent networks. However, both can run on top of SONET in any combination of channels over the same optical fiber. This gives companies the flexibility they need in building large-scale transparent networks capable of supporting any application and bandwidth requirement.

The choice of SONET as the transmission facility—rather than relying exclusively on WDM, DWDM, and emerging proprietary technologies—would give organizations the bandwidth they need now and position them for future growth in virtually any protocol, application, and technology direction they choose to go.

Network planners should evaluate broadband equipment in terms of its migration path to SONET; vendors should be evaluated on the extent to which they plan to implement path/overhead signaling for network management, user-to-network signaling, network synchronization, partitioning, channelization that supports both IP and ATM, remote management and diagnostics, and bandwidth on demand. Compliance with standards and participation in interoperability testing should also be considered.

Standards

SONET standards were developed by the Alliance for Telecommunications Industry Solutions (ATIS), formerly known as the Exchange Carriers Standards Association (ECSA). These standards were published and distributed by the American National Standards Institute (ANSI). Bellcore, the research and development arm of the RBOCs, was involved in the development of SONET standards from the beginning, and continues to issue technical specifications that ensure standards compliance. ATIS sponsors the SONET Interoperability Forum (SIF)—a membership organization comprised of equipment vendors, service providers and end users—which drives the industry to implement interoperable SONET products and services based on open industry and international standards.

SUMMARY

Data transfer on existing phone lines is now doubling in size annually, with the portion of data transferred on lines via the Internet now more than quadrupling each year. Optical internetworking technologies such as WDM and DWDM, coupled with SONET applications and legacy asynchronous multiplexing, could result in as much as a 100-fold increase in available bandwidth. Greater bandwidth translates into easier and more efficient transfer of data—enabling, for example, end users of the Internet to receive information more quickly and take advantage of real-time applications such as IP telephony, videoconferencing, and collaborative computing.

See Also **All-Optical Networks, Next-Generation Networks**

Systems Integration

An integrator ties together disparate systems and applications so they work in a seamless fashion. This entails reconciling physical connections and overcoming problems related to incompatible protocols. The systems integrator uses its hardware and software expertise to customize the necessary interfaces. The objective is to provide compatibility and interoperability among different vendors' products at a price the customer can afford.

The systems integrator not only provides the integration of various vendors' hardware and software, it also can tie in additional features and services offered through the public switched network. To do this, the systems integrator draws upon its experience in information systems, telephony and data communications. Added value is provided through strong project management skills and accumulated experience with customer requirements in a variety of operating environments.

Integration Drivers

In highly competitive markets, companies constantly strive to add value. This often means partnering with other firms to address a broader range of customer needs. Other times, it means merging with or acquiring firms that offer complementary products or services. In either case, the number of new technologies that must be accommodated and the number of legacy systems that must communicate increases—both of which pose formidable technical and management challenges.

For example, the spiraling cost of health care delivery in the U.S. is driving the need to maximize use of advanced information systems and networking technologies to reduce overhead expenses and stay competitive. Rising health care costs have led to dramatic changes in the way health care is delivered, administered, and paid for. The growth of managed care has unleashed a wave of mergers and acquisitions, and introduced new players who bring with them new information systems and networks, which must be integrated with the existing infrastructure.

Systems integration brings together hardware, software, telecommunication services, and people to solve diverse business problems. It can include designing and implementing new applications, providing ongoing hardware platform support, enhancing and expanding networks, training users, and managing maintenance. The objective is to unify an organization's activities, information technologies, and people to meet clinical and business goals.

Services

A number of discrete services are typically provided by systems integration firms, including:

DESIGN AND DEVELOPMENT   Includes such activities as network design, facilities engineering, equipment installation and customization, and acceptance testing

CONSULTING   Includes needs analysis, business planning, systems/network architecture, technology assessment, feasibility studies, RFP development, vendor evaluation and product selection, quality assurance, security auditing, disaster recovery planning, and project management

SYSTEMS IMPLEMENTATION Procurement, documentation, configuration management, contract management, and program management

SOLUTION INTEGRATION Staging, integrating, and testing the solution prior to actual deployment

CABLING AND MEDIA TESTING Covers all forms of transport media including wiring, cabling, fiber optic, and wireless

FACILITIES MANAGEMENT Operations, technical support, hot-line services, move and change management, and trouble ticket administration

NETWORK MANAGEMENT Network optimization, remote monitoring and diagnostics, network restoral, technician dispatch, and carrier and vendor relations

In the United States and Canada alone, about 6,000 companies claim to perform systems integration. Today's systems integrators include firms such as Electronic Data Systems (EDS), which rose to prominence by servicing the data-center environment; computer system vendors that specialize in their own product lines; and value-added resellers (VARs) with roots in the UNIX or Windows environment. Carriers, both local exchange and interexchange, also provide systems integration services. Carriers are especially useful when the project involves computer telephony integration.

Of course, large companies often have the resources within their IT departments to handle systems integration internally. Sometimes a company's IT staff may become expert in a particular kind of systems integration and be spun off as a separate profit center, offering their services to other companies.

Selection Criteria

When selecting a systems integrator, the candidates should be evaluated for competency in the following areas, as appropriate:

INTEGRATION COMPETENCY Large-scale systems integration services; project management for large or complex systems; strategic information technology planning and consulting; data conversion and technology migration; large database architecture, design, and operation; and data security, confidentiality, and privacy issues

ENTERPRISE COMPETENCY Strategic planning; organizational restructuring/alignment; process reengineering, CPI, and Total Quality Management; systems feasibility studies; systems adoption and transition; information management; and technology assessment and planning

PROJECT MANAGEMENT COMPETENCY Demonstrated leadership/organizational skills; flexible teaming arrangements, including subcontracting; proven methodologies with published project schedules and plans; ability to complete projects within budget and time constraints; procedures for cost control and conflict resolution; configuration management experience; ability to define common project processes; and experience with changing and evolving technologies

ARCHITECTURE COMPETENCY Open-systems and vendor-neutral approaches; pure integration focus, independent of hardware sales or transaction fee services; experience with and processes for trade studies and life cycles analysis, network topologies, and protocols design; and all phases of systems design including top requirements definition, baseline concept definition, systems specifications and documentation, testing (checkout and system verification), troubleshooting and problem resolution, and system turnover and training

DESIGN COMPETENCY   Client/server-based systems of central and distributed databases; legacy data conversion; interprogram and intraprogram interface definition; procedures for security, access control, and data protection; e-mail implementation (data packaging and routing); EDI, electronic funds transfer and data transfer definitions; data exchange standards; and proven methodologies for testing and validation

COMMUNICATIONS COMPETENCY   Connectivity/interfaces; telephony (microwaves, wires, fibers); wireless systems; switches and routers; facilities definition; multimedia technology; and videoconferencing

TECHNICAL COMPETENCY Data conversion tools; high-speed search and data extraction; authentication and fraud control solutions; biometric processing for identification and verification; Internet, intranet, and Web technologies; electronic document and workflow processing; automatic coding of medical documents; diagnostic image analysis and processing; distributed database solutions; telecommunications, networking and gateways; and repositories, data warehousing, and data marts

OPERATIONAL COMPETENCY Product improvement, update support, and version control; user support including user groups, bulletin boards, newsletters, and help desks; problem resolution; transaction processing as well as data repository/warehousing/data marts; network and security administration; and data security and integrity

Beyond all this, candidates should be able to show that their resources have been successfully deployed in previous projects of similar nature and scope to the one at hand.

SUMMARY ▮ ▮ ▮ ▮ ▮ ▮ ▮ ▮ ▮ ▮

Today's companies are looking to integrators for business solutions that improve their competitiveness in particular markets. While some companies have the resources to design and install complex systems and networks, others are turning to outside firms to handle the process. An alternative to the do-it-yourself approach or hiring a systems integrator is to share integration responsibilities with a contractor. This can save as much as 30 percent of integration costs. However, projects that are co-managed must be carefully defined, particularly the responsibilities of each party, so there are no gray areas regarding expectations and accountability. *See Also* Network Integration

Systems Network Architecture

Systems Network Architecture (SNA) refers to the protocol used by IBM mainframes. Introduced in 1974, the original hierarchical version of SNA has evolved to support applications in the distributed computing environment and to work with other networking technologies. Legacy SNA networks are hierarchically structured with mainframes at the top of the hierarchy (Figure 112).

IBM's mainframe-based VTAM (Virtual Telecommunications Access Method) software controls the operation of SNA networks and provides the interface between mainframe-based applications and the network via a communications controller. SNA networks include at least one VTAM host. The backbone of SNA networks is made up of interconnected communications controllers, which have dedicated networking processors that run IBM's Network Control Program (NCP) software. These controllers route traffic over the campus- and wide-area backbones of SNA networks.

Peripheral Nodes are the terminals, which provide users and applications with access to the SNA network. They can also be PCs equipped with terminal emulation software. In either case, the nodes support only local management functions and do not perform any intermediate routing. Most Peripheral Nodes allow users to communicate only with applications residing on a VTAM host, but they can be allowed to support peer-to-peer communications with users on other Peripheral Nodes.

Functional Layers

The protocols and services defined by SNA are described by a seven-layer model (Figure 113), which is similar in scope to the OSI reference model

Figure 112
Traditional centralized, hierarchical SNA environment.

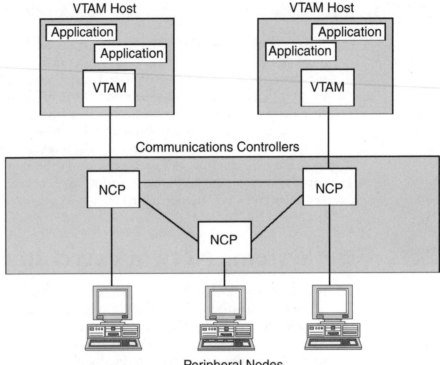

and is the same layering model defined by the newer version of SNA called Advanced Peer-to-Peer Network (APPN).

The lower three layers—Physical Control, Data Link Control, and Path Control—describe the networking services and protocols that provide basic message forwarding in SNA networks. These layers are collectively called the Path Control Network Function.

The Physical Control and Data Link Control layers provide connections between adjacent nodes in an SNA network. The data links are LAN and WAN facilities over which the nodes communicate with each another. The physical and data link layers defined by industry-standard LANs, such as Ethernet, token ring, and FDDI are supported. For wide-area networking, SNA supports physical interface standards such as RS-449 and RS-232. SNA uses Synchronous Data Link Control (SDLC) protocols for communications over dedicated wide area links. SDLC is compatible with the international standard High-level Data Link Control (HDLC) protocol. Industry-standard packet-switching interfaces including X.25, frame relay and ATM are also supported.

SNA also supports IBM's local channel interfaces for connecting mainframes to local nodes. Parallel bus and tag channel as well as Enterprise Systems Connection (ESCON) technologies are supported by SNA.

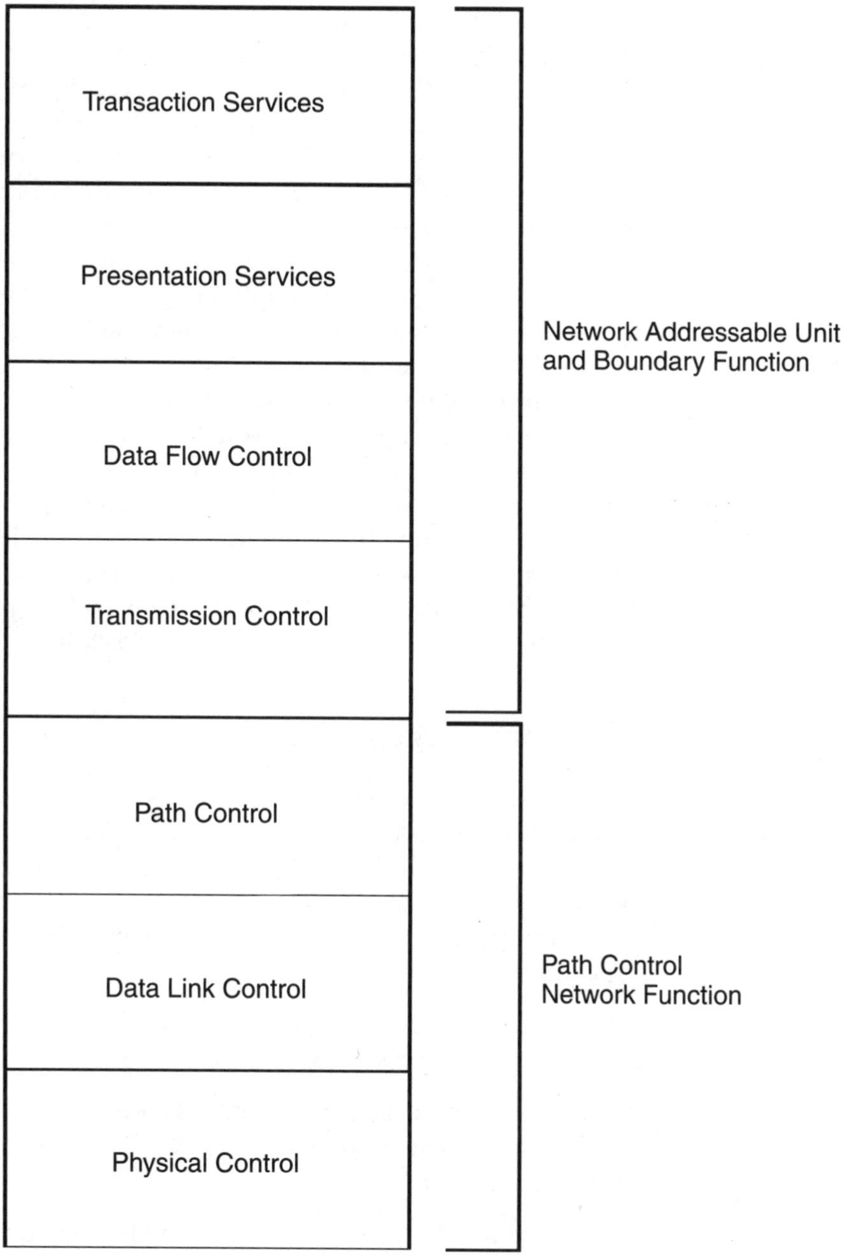

Figure 113
The seven-layer
SNA protocol
model.

Transaction Services
Presentation Services
Data Flow Control
Transmission Control
Path Control
Data Link Control
Physical Control

Network Addressable Unit
and Boundary Function

Path Control
Network Function

SNA requires that its data links provide reliable message delivery. While some other types of networking protocols, such as TCP/IP, can operate over either reliable or unreliable data links, SNA always requires reliable data links. The data link protocols used in SNA networks also require that link-level acknowledgments to polls be received within a fixed time interval—usually several seconds. This can present problems when dedicated LAN and WAN data links are replaced by networks that cannot provide fixed transit delays for messages. This situation can occur when SNA data is sent over packet-switching networks or when SNA data is tunneled through TCP/IP internets. (This situation is ameliorated by APPN's High-Performance Routing protocol discussed below.)

The Path Control layer is responsible for routing messages hop-to-hop across an SNA network. Although Subarea and APPN networks use different addressing and routing techniques, the level of service provided to the upper layers is similar.

Within APPN networks, two different Path Control routing technologies are used. Intermediate Session Routing (ISR) is the original APPN routing protocol and is included in the base set of protocols, which are part of every APPN product. An optional routing protocol, called High performance Routing (HPR), is available for APPN. HPR provides better packet forwarding performance than ISR and adds capabilities such as nondisruptive session rerouting and a flow control protocol that is optimized for operation over high bandwidth data links.

Network Accessible Units

The SNA software components that support end-to-end communications are called Network Accessible Units (NAUs). The top four layers of SNA are implemented within NAUs. The protocols implemented within the four layers are end-to-end protocols that are designed to support communications between a single pair of NAUs. Subarea SNA networks have three types of NAUs:

LOGICAL UNITS (LUs) Software that resides in the various devices that make up an SNA network, which support end-user communications

SYSTEM SERVICES CONTROL POINTS (SSCPs) Central control points of SNA networks, which provide network management and user interconnection functions

PHYSICAL UNITS (PUs) Represent the actual devices on the network, which are managed by SSCPs

Architecture

Each type of node has an architectural designation in addition to its commonly used name. VTAM Hosts are called Type 5 Nodes, Communications Controllers are Type 4 Nodes, and Peripheral Nodes are designated as either Type 2.0 or Type 2.1 Nodes. Type 2.0 nodes support only hierarchical communications with mainframe-based applications, while Type 2.1 Nodes also support peer-to-peer communications. The SNA node types are also commonly called PU Types—an older designation.

Logical Units represent the end users of SNA networks. Examples of end users include applications running on VTAM Hosts or Peripheral Nodes, or the displays and printers used by interactive terminal users. End users communicate with one another via LU-to-LU sessions.

SNA defines several categories of LUs, called logical unit types. Each LU type defines a subset of end-to-end SNA protocols that are used to support communications between a specific category of end users. Among the more commonly used LU types are:

LU TYPE 0 Supports industry-specific terminals used in the financial and retail industries

LU TYPE 1 Supports communications with character-oriented printers

LU TYPE 2 Supports communications with interactive 3270 display stations

LU TYPE 3 Supports communications with 3270 printers

LU TYPE 6.2 Current SNA program-to-program communications protocols, also called Advanced Program-to-Program Communications (APPC)

System Services Control Points and Physical Units support network management functions. SSCPs are part of VTAM and always reside on SNA Hosts. SSCPs operating under the control of host-based management software, usually IBM's NetView, manage all of the resources of an SNA network.

Each of the managed nodes in an SNA network contains a PU, which is responsible for the management of resources of their local nodes and they are controlled by an SSCP. SSCPs and PUs interact over SSCP-PU sessions.

Some types of LUs, called SSCP-dependent LUs, enter into sessions with their SSCPs. These SSCP-LU sessions are used to activate and deactivate LU-LU sessions between dependent LUs. Other types of LUs, called SSCP-independent LUs, can start and stop LU-LU sessions without the intervention of an SSCP. LU Type 6.2 is the only LU type that can function independently.

Users access SNA networks primarily through 3270 terminals or systems providing 3270 emulation. When users initiate an LU-LU session they can specify the class of service (CoS) required for optimal applications performance. CoS defines the required route characteristics including bandwidth, security, and availability needed to satisfy the user's performance requirements.

APPN

APPN is the latest generation SNA technology for linking devices without requiring the use of a mainframe. Specifically, it is IBM's proprietary SNA routing scheme for client/server computing in multiprotocol environments. As such, it is part of IBM's LU 6.2 architecture (i.e., APPC), which facilitates communications between programs running on different platforms.

APPN routes SNA traffic natively across PC LANs. Large IBM shops that must prioritize and route traffic in a peer-to-peer fashion are good candidates for APPN. HPR, used exclusively in the SNA environment, enhances SNA routing and adds to APPN by further prioritizing and routing SNA traffic around failed or congested links. It is used in situations where bandwidth is critical, especially in packet-switched networks.

Included in the APPN architecture are Automatic Network Routing (ANR) and Rapid Transport Protocol (RTP) features. These features route data around network failures and provide performance advantages, closing the gap with TCP/IP. ANR provides end-to-end routing over APPN networks, eliminating the intermediate routing functions of early APPN implementations, while RTP provides flow control and error recovery. To these features, HPR adds a very advanced feature called Adaptive Rate Based (ARB) congestion prevention.

ARB uses three inputs to determine the sending rate for data. As data is sent into the network, the rate at which it is sent is monitored. At the destination node, that rate is also monitored and reported back to the originating node. The third input is the allowed sending rate. Together, these inputs determine the optimal throughput rate, which minimizes the potential for packet discards to alleviate congestion.

By enabling peer-to-peer communications among all network devices, APPN helps SNA users connect to LAN networks and more effectively create and use client-server applications. APPN supports multiple protocols, including TCP/IP, and allows applications to be independent of the transport protocols that deliver them.

APPN's other benefits include allowing information routing without a host, tracking network topology, and simplifying network configuration and changes. For users still supporting 3270 applications, APPN can address dependent LU protocols as well as the newer LU 6.2 sessions, which protects a company's investment in applications relying on older LU protocols.

SUMMARY

The original, hierarchical nature of SNA networking, called subarea SNA, is still widely used to support large-scale, mission-critical applications, such as credit card authorization systems, automatic teller machine networks, and airline reservation systems. However, in response to the networking requirements of new decentralized applications, IBM came up with Advanced Peer-to-Peer Networking, a decentralized, mainframe-independent version of SNA.

See Also **Communications Processors, SNA over IP, Synchronous Data Link Control**

T

T-Carrier Facilities

T-carrier is a type of digital transmission system employed over copper, optical fiber, or microwave to achieve various channel capacities for the support of voice and data. The most popular T-carrier facility is T1, which is implemented by a system of copper wire cables, signal regenerators, and switches that provides a transmission rate of up to 1.544 Mbps using digital signal level 1 (DS1). In Europe, the UK, Mexico, and other countries that abide by International Telecommunication Union (ITU) standards, the equivalent facility is E1, which provides a transmission rate of 2.048 Mbps.

T-carrier had it origins in the 1960s. It was first used by telephone companies as the means of aggregating multiple voice channels onto a single high-speed digital backbone facility between central office switches. The most widely deployed T-carrier facility is T1, which has been commercially available since 1983.

Digital Signal Hierarchy

To achieve the DS1 transmission rate, selected cable pairs with digital signal regenerators (repeaters) are spaced approximately 6,000 feet apart. This combination yields a transmission rate of 1.544 Mbps. By halving the distance between the span-line repeaters, the transmission rate can be doubled to 3.152 Mbps, which is called DS1C. Adding more sophisticated electronics and/or multiplexing steps makes higher transmission rates possible, creating a range of digital signal levels, as follows:

North American Digital Signal Hierarchy

Signal Level	Bit Rate	Channels	Carrier System	Typical Medium
DS0	64 Kbps	1	—	Copper wire
DS1	1.544 Mbps	24	T1	Copper wire
DS1C	3.152 Mbps	48	T1C	Copper wire
DS2	6.312 Mbps	96	T2	Copper wire
DS3	44.736 Mbps	672	T3	Microwave/fiber
DS4	274.176 Mbps	4032	T4	Microwave/fiber

International (ITU) Digital Signal Hierarchy

Signal Level	Bit Rate	Channels	Carrier System	Typical Medium
0	64 Kbps	1	—	Copper wire
1	2.048 Mbps	30	E1	Copper wire
2	8.448 Mbps	120	E2	Copper wire
3	34.368 Mbps	480	E3	Microwave/fiber
4	44.736 Mbps	672	E4	Microwave/fiber
5	565.148 Mbps	7680	E5	Microwave/fiber

For example, a DS3 signal is achieved in a two-step multiplexing process whereby DS2 signals are created from multiple DS1 signals in an intermediary step. DS1C is not commonly used, except in highly customized private networks where the distances between repeaters is very short, such as between floors of an office building or between buildings in a campus environment. Some channel banks and multiplexers support DS2 to provide 96 voice channels over a single T-carrier facility. DS4 is used mostly by carriers for trunking between central offices.

Quality Objectives

The quality of T-carrier facilities is determined by two criteria: performance and service availability. The performance objective refers to the percentage of seconds per day when there are no bit errors on a circuit. The service availability objective refers to the percentage of time a circuit is functioning at full capability during a three-month period. If these objectives are not met, the carrier issues credits to its users. Each carrier has its own quality objectives for T-carrier services, which are based on circuit length.

The quality objectives for AT&T's Fractional T1 service, for example, is nine errored seconds per day, which translates into 99.99-percent error-free seconds per day, four severely errored seconds per day, and 99.96-percent service availability per year. According to AT&T, "severely errored" means that 96 percent of all frames transmitted in a second have at least one error.

A related measure of performance is "failed seconds," defined by AT&T as the time starting after 10 consecutive severely errored seconds and ending when there have been 10 consecutive seconds that are not severely errored. Channel Service Unit/Data Service Unit (CSU/DSU) at each end of the circuits collect and issue reports on this type of information.

T1 Lines

As noted, the most popular T-carrier facility is T1, which is a digital line or service providing a DS1 transmission rate of up to 1.544 Mbps. T1 lines are used for economical and efficient voice and data transport over the wide-area network. The available bandwidth is divided into 24 channels operating at 64 Kbps each, plus an 8-Kbps channel for basic supervision and control. Voice is sampled and digitized via Pulse Code Modulation (PCM).

Economy is achieved by consolidating multiple lower-speed voice and data channels over the higher-speed T1 line. This is more cost-effective than dedicating a separate lower-speed line to each terminal device. The economics are such that only five to eight analog lines are needed to cost-justify the move to T1.

Efficiency is obtained by compressing voice and data to make room for even more channels over the available bandwidth. Individual channels can also be dropped or inserted at various destinations along the line's route through the use of an add-drop multiplexer (ADM) or digital cross-connect system (DCS). Network management information can be embedded in each channel for enhanced levels of supervision and control.

A T1 multiplexer usually provides the means for companies to realize the full benefits of T1 lines, but channel banks offer a low-cost alternative. The difference between the two devices is that T1 multiplexers offer higher line capacity, support more types of interfaces, and provide more network management features than channel banks.

D4 Framing

T1 multiplexers and channel banks transmit voice and data in D4 frames. These frames are bounded by bits that perform two functions: they identify the beginning of each frame and help locate the channel-carrying signaling information.

D4 frames consist of 193 bits, which equates to 24 channels of 8 bits each, plus a single framing bit. Each frame contains framing bits or signaling bits in the 193rd position, which permits management of the facility itself. This is done by robbing the least significant bit from the data stream, which alternatively carries information or signaling data. Another bit is used to mark the start of a frame. Twelve D4 frames comprise a superframe.

Extended Superframe Format

Extended Superframe Format (ESF) is an enhancement to T-carrier, which specifies methods for error monitoring, reporting and diagnostics. Use of ESF allows technicians to maintain and test the T1 line while it is in service, and often fix minor troubles before they adversely affect service. ESF extends the normal 12-frame superframe structure of the D4 format to 24 frames. By doubling the number of bits available, more diagnostic functionality also becomes available.

Of the 8-Kbps bandwidth (repetition rate of 193rd bit or framing bit) allocated for basic supervision and control, 2 Kbps are used for framing, 2 Kbps for Cyclic Redundancy Checking (CRC-6), and 4 Kbps for the Facilities Data Link (FDL). With CRC, the entire circuit may be segmented so that it can be monitored for errors, without disrupting normal data traffic. In this manner, performance statistics can be generated to monitor T1 circuit quality. Via FDL, performance report messages are relayed to the customer's equipment, usually a Channel Service Unit (CSU), at one-second intervals. Alarms can also use the FDL, but performance report messages always have priority.

ESF diagnostic information is collected by the CSU at each end of the T1 line for both carrier and user access. CSUs gather statistics on such things as clock synchronization errors and framing errors, as well as errored seconds, severely errored seconds, failed seconds, and bipolar violations. A supervisory terminal connected to the CSU displays this information, furnishing a record of circuit performance.

Originally, the CSU compiled performance statistics every 15 minutes. This information would be kept updated for a full 24 hours so that a complete one-day history could be accessed by the carrier. The carrier would have to poll each CSU to retrieve the collected data and clear its storage register. By equipping the CSU with dual registers—one for the carrier and one for the user—both carrier and user alike have full access to the performance history.

Today, the CSU is not required to store performance data for 24 hours. Also, the CSU no longer responds to polled requests from carriers, but simply transmits ESF performance messages every second.

ESF also allows end-to-end performance data and sectionalized alarms to be collected in real-time. This allows the carrier's technical staff to narrow down problems between carrier access points and on interoffice channels, and to find out the direction the error is occurring.

Monitoring Services

Carriers offer optional services that monitor private T1/T3 networks, providing customers with protection against unexpected service impairments. Such services include:

- On-site network monitoring
- Configuration monitoring
- Fault (alarm) monitoring
- T1 performance monitoring
- Trouble ticket management
- Onsite supervisory workstation

End-to-end performance monitoring is implemented using ESF to collect error data from the circuits. The carrier will usually report on three ESF parameters for a T1 circuit: errored seconds, severely errored seconds, and failed seconds. This data is updated on an hourly basis, then compared to the circuit error thresholds set by the carrier. When recorded ESF errors exceed 80 percent of the carrier's preset 24-hour thresholds, an alarm is triggered.

When problems arise, both visual and audio indicators alert the customer of the event. Color-coded changes on the workstation indicate the severity of the alarms. At the same time, alarm text is automatically output to the workstation's printer. At the workstation, the customer can view detailed circuit data such as circuit number, alarm type, start date and time of alarm, direction for the alarm (whether "from" or "to" transmission direction), plus the circuit segment where the alarm occurred— either an interoffice circuit or access circuit. The customer can view or print historical alarm monitoring information later, using an alarm record that is maintained on each circuit on a rolling, 31-day basis.

Meanwhile, the carrier's technical staff acts immediately to resolve the alarm conditions. Any changes are reflected graphically at the supervisory workstation, where the customer can stay informed of T1/T3 performance and the carrier's problem resolution process. In addition, the customer has access to the trouble ticket manager to electronically open, close and track trouble tickets. The customer can even use this tool to contest tickets closed by the carrier.

SUMMARY

T-carrier underlies just about every type of carrier facility available today, including T1 and T3, and their fractional derivatives—dedicated or switched. These facilities, in turn, support such services as frame relay and ISDN, and provide access to SMDS, ATM, and virtual private networks. Through multiplexing and compression techniques, companies can subdivide T-carrier facilities to achieve greater bandwidth efficiency and cost savings.

See Also **Channel Service Units, Microwave, Fiber-Optic Technology**

T-Carrier Services

T-carrier facilities provide the network platform for several switched dedicated and packet-switched services. The switched dedicated services usually rely on the Public Switched Telephone Network (PSTN) for transport, but the channels are dedicated to a specific customer. A variety of data services are available, many of which run over separate networks.

Regardless of the type of network or the type of network, T-carrier often is used for the underlying transport medium. Even though fiber optic networks are increasingly used to consolidate traffic of all types over high-speed backbones, the traffic often enters or leaves in the T-carrier format used by PBXs, channel banks, multiplexers, cross-connect systems, and other voice-oriented systems. In addition, network devices that are data-oriented, such as routers, also have T1 interfaces.

Switched Dedicated Services

Among the switched dedicated services are N×64 Kbps, Multirate ISDN, Switched 384 Kbps, and Switched 1,536 Kbps.

N×64-KBPS SERVICE N×64-Kbps Service entails the dynamic aggregation of multiple 64-Kbps channels into a high-speed switched data pipe. A network access device—which could be a T1 multiplexer, an inverse multiplexer, a PBX, or a video coder/decoder (codec)—equipped with ISDN PRI is used to implement this service. The device would transmit a message with the telephone number of the remote site and the number of 64-Kbps channels needed to support the application. The message would then be transmitted over the PRI link's D signaling channel to the carrier's Class 5 central office switch. The switch establishes the number of 64-Kbps channels needed to support the application. Channels could be added or dropped from the switched link as needed. With $N \times 64$-Kbps service, the channels would almost always take different paths through the carrier's network and be reassembled at the receiving end.

MULTIRATE ISDN   A related type of service, called Multirate ISDN, enables a user to establish on-demand, switched digital links at speeds ranging from 64 Kbps to 1.544 Mbps. An ISDN-equipped network access device such as a T1 multiplexer, an inverse multiplexer, or a PBX transmits an ISDN Q931 message that contains the telephone number of the remote site and the number of 64-Kbps channels needed to support the particular application. The message is transmitted over the PRI link's D channel to a carrier switch. The switch then establishes a single, contiguous pool of bandwidth across the user's carrier network and signals the network access device that the link has been established. The user's network device can also instruct the central office switch to add or delete channels if needed. However, the existing link must first be torn down. By contrast, some inverse multiplexers can add bandwidth during a transmission.

SWITCHED 384-KBPS SERVICE   As its name implies, Switched 384-Kbps Service provides a channel that is equivalent to six contiguous 64-Kbps channels or DS0s. Throughout the network, the 384-Kbps channel is switched as a single channel from its point of origination to its destination. Since it is a switched service, the bandwidth is called up only when needed. With each call, the traffic may take a different path within the network to its destination.

SWITCHED 1,536-KBPS SERVICE   Switched 1,536-Kbps Service is equivalent to 24 contiguous 64-Kbps channels or DS0s. Throughout the network, the entire 1,536-Kbps of bandwidth is switched as a single channel from its point of origination to its destination. Since it is a switched service, the bandwidth is called up only when needed. With each call, the traffic may take a different path within the network to its destination.

Advanced Packet Services

T-carrier also provides the foundation for several innovative advanced packet services, including frame relay, SMDS, and ATM.

FRAME RELAY   Frame-relay backbones are based on T-carrier links. Frame relay is a streamlined version of X.25 that achieves throughput rates that are orders of magnitude greater than conventional X.25. Frame relay achieves high throughput by eliminating 66% of the overhead functions, including error correction, that are traditionally carried out at intervening nodes by X.25.

Frame relay offers two types of connections: permanent virtual circuits (PVCs) and switched virtual circuits (SVCs). A PVC is a predefined path between end points. An SVC provides a temporary connection that is set up for the duration of the call in accordance with instructions embedded within the frame itself. This type of connection is very useful when only occasional network access is required, since network resources can be allocated to other users after the call is completed. Where groups of user devices create a continual stream of information, as in LAN interconnection, PVCs are the preferred type of connection.

SMDS   Access to SMDS is provided by T-carrier links. SMDS itself is a high-speed, packet-switched service aimed at the LAN interconnection needs of corporate locations within the same metropolitan area. SMDS is a shared service and employs a dual counter-rotating fiber-ring architecture. The two rings transmit data in opposite directions so that if one ring fails, the other is capable of handling the traffic and getting it to its proper destination, thus circumventing the fault.

CPE can access an SMDS switch using either twisted-pair wiring (at the T1 rate) or optical fiber (at the T3 rate). Each customer has private access to an SMDS switch for up 16 devices per access link. The devices are connected in a bus arrangement, just like an Ethernet LAN. SMDS access may be customized to suit the individual bandwidth needs of subscribers. By means of access classes, limits can be enforced on the level of sustained information transfer and on the burstiness of the transfer. In the case of T3, the access classes are 4, 10, 16, 25, and 34 Mbps. For T3 access paths, an "ingress access class" may be applied to the information flow from the CPE to the carrier's switch, and an "egress access class" applied to information flowing from the switch to the CPE. Both types of access classes may be selected by the subscriber. For T1 access paths, the same 1.536 Mbps access class is applied to both directions of information flow. In addition, several local carriers enable their customers to access their SMDS networks at speeds of 56/64 Kbps or lower.

ATM   ATM switching, or cell relay, is designed as a general-purpose broadband switching method for voice, data, image, and video traffic. With ATM, data is switched at the packet level—53-byte cells—rather than at the bit or byte level, thus minimizing internodal processing requirements. The benefits of ATM include high speed, low latency, and increased network availability through the automatic and guaranteed assignment of network bandwidth, which supports isochronous traffic such as voice and video, as well as real-time data traffic for online, interactive applications. ATM can run over T-carrier facilities and some multiplexer vendors offer T1/E1 ATM User-Network Interfaces (UNIs) to allow users to access ATM services without requiring investments in broadband network infrastructures.

SUMMARY

The main driving force in the growth of T-carrier services is the Internet. Services based on T-carrier scale fairly easily to accommodate user needs, as demonstrated by the availability of fractional services and the ability of carriers to aggregate multiple T-carrier facilities to deliver additional bandwidth. For example, Sprint offers a service that aggregates four or eight T1 lines to provide, respectively, 6- or 12-Mbps dedicated links. MCI WorldCom offers multiple T1 access at rates of 3, 4.5, and 10 Mbps. UUNET Technologies offers a Double T1 service with a 3-Mbps access speed. These services are aimed at corporations that have saturated their T1 links into the Internet, but still do not need a 45-Mbps T3 connection. Other big attractions driving T-carrier demand are the ease with which the line termination equipment can be managed and the ability to consolidate traffic—both voice and data—from multiple remote sites onto a single access line into a company.

See Also **T-Carrier Facilities**

Technology Procurement Alternatives

There are three methods of paying for equipment and systems: purchasing, leasing, and getting an installment loan. Each has its advantages and disadvantages, depending on the financial objectives of the company.

Purchasing

The attraction of paying cash is that it costs nothing extra; no interest is paid, as with a loan, so the cost of the item from a financial perspective is the purchase price. However, there are several reasons why cash payment may not be a good idea, aside from the obvious reason that the company may have no cash to spare.

The first reason not to pay cash for a purchase is that the money to be taken from cash reserves could probably be used to finance other urgent activities, and the interest rate to be paid when those activities require financing may be higher, if financing can be obtained at all. Sometimes a set of T1 multiplexers, for example, or other major equipment can be purchased based on a very low interest rate relative to prevailing commercial rates. To pay cash for equipment when it could instead be financed at 7 percent, then pay 11 percent for money six months later, is unsound financing. If financing is at an attractive rate, it is probably better to use that rate and finance the system, unless company cash reserves are so high that future borrowing for any reason will not be required.

Another reason not to pay cash is the issue of taxes and cash flow. If a company has accumulated a profit in cash and can use it to pay for a major network upgrade or expansion, for example, the cost cannot be deducted in a single year; it must be depreciated over a period of five or more years. The company will thus be placed in the position of owing tax on its profits, less first-year depreciation (and other operating expenses), and possibly not having enough cash to pay that tax.

A final reason for not using reserves to finance a large purchase concerns the issue of credit worthiness. Often a company with little financial history can borrow for a collateralized purchase such as a network, but it cannot borrow readily for intangibles such as ordinary operating expenses. Other times, an unexpected setback will affect the credit standing of the firm. If all or most of a firm's reserves have been depleted by a major purchase, it may be impossible to secure quick loans to meet new expenses and the company may falter.

Installment Loan

For organizations that elect to purchase equipment, but cannot or choose not to pay cash, loan financing will be required. These arrangements are often difficult to interpret and compare, so users should review the cost of each alternative carefully.

If a company is not able or willing to pay cash for the system, the alternative is some form of deferred payment. The purpose of these payments is to stagger the purchase effect across a longer period. This improves the cash flow in a given year, but a substantial price to be paid is the interest; whether the equipment is leased or installment purchased, interest must be paid. This may result in a conflict of accounting goals—is cash flow or long-term cost, including interest, the overriding factor?

The question of financial priorities must be answered early in the equipment acquisition process. A purchase financed over a longer period, providing that the interest premium for that longer term is not unreasonable, will have a lower net cost per year, taking tax effects of depreciation into consideration.

If the company is in a critical position with cash flow, longer-term financing and a minimum down payment are the major priorities. The cash flow resulting from a PBX purchase will be the annual payments for the system less the product of the company's marginal tax rate times the annual depreciation. If the system's payments are $5,000 per month on a $200,000 note and first-year depreciation is 20 percent, the company will pay $60,000 in the first year and have a tax deduction of $40,000. If the marginal rate is 30 percent, that deduction will be worth $12,000, so the net cash flow is negative at $48,000. Longer terms will reduce the payments, but not proportionally due to increased interest.

Where cash flow is not a major concern, financing should be undertaken to minimize the interest payment to be made. Interest charges can be reduced by increasing the down payment, by reducing the term of the loan, and by shopping for the best loan rates. Each of the ways users can finance a system will affect user mobility in these areas.

Leasing

Assuming the decision has been made to lease rather than purchase equipment, it is important to differentiate between the two types of leases available because each is treated differently for tax purposes. One type of lease is the operating lease, in which the leasing company retains equipment ownership. At the end of the lease, the lessee may purchase the equipment at its fair market value. The other kind of lease is the capital lease, in which the lessee can retain the equipment for a nominal fee, which can be as low as one dollar.

OPERATING LEASE With the operating or tax-oriented lease, monthly payments are expensed, that is, subtracted from the company's pretax earnings. With a capital or nontax-oriented lease, the amount of the lease is counted as debt and must appear on the balance sheet. In other words, the capital lease is treated as just another form of purchase financing and, therefore, only the interest is tax deductible.

A true operating lease must meet the following criteria, issued by the Financial Accounting Standards Board (FASB), some of which effectively limit the maximum term of the lease:

- The term of the lease must not exceed 80 percent of the projected useful life of the equipment. The equipment's "useful life" begins on the effective date of the lease agreement. The lease term includes any extensions or renewals at a preset fixed rental.

- The equipment's estimated residual value in constant dollars (with no consideration for inflation or deflation) at the expiration of the lease must equal a minimum of 20 percent of its value at the time the lease was signed.

- Neither the lessee nor any related party is allowed to buy the equipment at a price lower than fair market value at the time of purchase.

- The lessee and related party are also prohibited from paying, or guaranteeing, any part of the price of the leased equipment. The lease, therefore, must be 100-percent financed.

- The leased equipment must not fall into the category of "limited use" property, that is, equipment that would be useless to anyone except the lessee and related parties at the end of the lease.

With the operating lease, the rate of cash outflow is always balanced to a degree by the rate of tax recovery. With a purchase, the depreciation allowed in a given year may have no connection with the amount of money the buyer actually paid out in installment payments.

CAPITAL LEASE   For an agreement to qualify as a capital lease, it must meet one of the following FASB criteria:

- The lessor transfers ownership to the lessee at the end of the lease term.

- The lease contains an option to buy the equipment at a price below the residual value.

- The lease term is equal to 75 percent or more of the economic life of the property. (This does not apply to used equipment leased at the end of its economic life.)

- The present value of the minimum lease rental payments equals or exceeds 90% of the equipment's fair market value.

From these criteria, it becomes quite clear that capital leases are not set up for tax purposes. Such leases are given the same treatment as installment loans; that is, only the interest portion of the fixed monthly payment can be deducted as a business expense. However, the lessee may take deductions for depreciation as if the transaction were an outright purchase. For this reason, the monthly payments are usually higher than they would be for a true operating lease.

Depending on the amount of the lease rental payments and the financial objectives of the lessee, the cost of the equipment may be amortized faster through tax deductible rentals than through depreciation and after-tax cash flow.

With new technology becoming available every 12 to 18 months, leasing can prevent the user from becoming saddled with obsolete equipment. This means that the potential for losses when replacing equipment that has not been fully depreciated can be minimized by leasing rather than purchasing. With rapid advancements in technology and consequent shortened product life cycles, it is becoming more difficult to sell used equipment. Leasing eliminates such problems.

Leasing can also minimize maintenance and repair costs. Because the lessor has a stake in keeping the equipment functioning properly, it usually offers on-site repair and the immediate replacement of defective components and subsystems. In extreme cases, the lessor may even swap out the entire system for a properly functioning unit. Although contracts vary, maintenance and repair services that are bundled into the lease can eliminate the hidden costs associated with an outright purchase.

Finally, leasing usually allows more flexibility in customizing contract terms and conditions than normal purchasing arrangements. This is simply because there are no set rates and contracts when leasing. Unlike many purchase agreements, each lease is negotiated on an individual basis. The items typically negotiated in a lease are the equipment specifications, schedule for upgrades, maintenance and repair services, and training.

Another negotiable item has to do with the end-of-lease options, which can include signing another lease for the same equipment, signing another lease for more advanced equipment, or buying the equipment. Many lessors will allow customers to end a lease ahead of schedule without penalty if the customer agrees to a new lease on upgraded equipment.

Of course, leasing has its downside. Although leasing can be an alternative source of financing that does not appear on the corporate balance sheet, the cost of a conventional lease arrangement generally exceeds that of outright purchase. Excluding the time value of money and equipment maintenance costs, the simple lease versus purchase break-even point can be determined by the formula: $n = p/l$, where p is the purchase cost, l is the monthly lease cost, and n is the number of months needed to break even. Thus, if equipment costs \$10,000, and the lease costs \$250 per month, the break-even point is 40 months. This means that owning equipment is preferable if it is expected to last more than 40 months.

As in any financial transaction, there may be hidden costs associated with the lease. If the lease rate seems very attractive relative to that offered by other leasing companies, a red flag should go up. Hidden charges may be embedded in the lease agreement that would allow the leasing company to recapture lost dollars. These hidden charges can include shipping and installation costs, higher than normal maintenance charges, consulting fees, or even a balloon payment at the end of the lease term.

The lessee may even be required to provide special insurance on the equipment. Some lessors even require the lessee to buy maintenance services from a third party to keep the equipment in proper working order over the life of the lease agreement. The lessor may also impose restrictions, such as where the equipment can be moved, who can service it, and what environmental controls must be in place at the installation site.

SUMMARY

The choice of procuring technology through purchasing, installment loan, or leasing arrangement depends on the financial condition and financial objectives of the company. Each procurement method has its advantages and disadvantages, which must be weighed in the decision. Often, the choice will be influenced by how frequently the company believes it must upgrade its systems and networks to stay competitive. For some companies, frequent changes might favor a leasing arrangement, whereby the lessor takes responsibility for keeping the system or network up to date with new technology.

See Also **Financial Management of Technology, Technology Migration Planning**

Technology Transition Planning

Most organizations recognize the need to transition to advanced technologies and understand the relationship between successful technology implementation and competitive advantage. However, the transition process involves a fair amount of risk, mainly because new technologies are not easy to integrate with legacy environments that cannot merely be thrown away.

For many organizations, planning technology transitions is a continuous process that closely parallels the drive to stay competitive in global markets. Assessing business requirements, evaluating emerging information technologies, and incorporating them into the existing infrastructure is an endless, complex exercise aimed at pushing the price/performance curve to stay ahead of the competition.

All of these issues can be effectively addressed by implementing a transition methodology that starts with a participative approach to planning and ends with the development of a timeline against which the project's progress is measured.

Participative Planning

The best way to defuse emotional and political time bombs that may jeopardize the success of implementing a new technology is to include all affected employees in the planning process. The planning process should be participative and start with the articulation of the organizational goals that the move to the new technology is intended to achieve, outlining anticipated costs and benefits. This stage of the planning process is also intended to address the most critical concern of the participants: "How will I be affected?" Once the organizational goals are known, these become the new parameters within which the participants can influence their futures.

Department managers, too, may feel threatened. They might view the change in terms of having to give up something: resources (in the form of budget, staff, power, and prestige) and control of various operations. These are very real concerns in this era of corporate downsizing. Perhaps as important, they see themselves as having to sacrifice an operating philosophy in which they have invested considerable time and effort to construct and maintain throughout much of their tenure. To suddenly put all this aside for something entirely new may be greeted with little or no enthusiasm, or worse—a lack of cooperation, which can spoil the best transition plans.

This participative approach not only facilitates cooperation; it has the effect of spreading ownership of the solution among all participants. Instead of a solution dictated by top management, which often engenders resistance through emotional responses and political maneuvering, the participative approach provides the most affected people with a stake in the outcome of the project: with success, comes the rewards associated with a stable work environment and shared vision of the future; with failure, comes the liabilities associated with a tumultuous work environment and uncertainty about the future. Although participative planning takes more time, its effects are often more immediate and long lasting than imposed solutions, which are frequently resisted and short-lived.

Another benefit of the participative approach to planning is that it gives all parties concerned a chance to buy into the new system or network and to recommend improvements that can benefit the entire organization. Consider including in the planning process IS and department managers as well as representatives from the various business units.

In some cases, as with technologies that transcend organizational boundaries (i.e., electronic procurement, inventory, distribution, and payment systems), suppliers, customers, and strategic partners should even be included in the planning process—at their own corporate locations, if possible. This is particularly important for planning and implementing extranets—TCP/IP networks that span multiple companies. It is also necessary for planning and implementing systems sharing arrangements between organizations as part of an emergency restoration plan.

Education

A crucial aspect of preparing personnel for change is managing the fear that change invokes. Fear of change has long been noted as a reason for delays in all phases of a new project. Compounding fear of change is the fact that many IT professionals have witnessed a short-lived resolve for certain decisions and remain hesitant to embrace new technologies for fear of wasting their efforts. In other cases, introductions of new technologies occur so rapidly that many IT personnel find it difficult to envision and comprehend practical applications. The hesitation to embrace new technologies is counterproductive and is often the result of obsolete management policies where change is perceived as an error-correction process.

Organizations must develop a strategy for dealing with the technology transition period. During this period, the organization will be transformed from one with a set of well-developed, yet increasingly outdated skills, into an organization with a set of newly acquired skills, ready to redefine its future. A key element of any strategy for dealing with the technology transition period is training and education. This can be achieved in either of two ways.

One is to work with an experienced consulting team to create a transition solution, which provides internal IT staff with the skills needed for success. This process includes assessing the impact of the proposed solution on the organization, performing a training needs analysis, and developing an appropriate education approach. Follow-up analysis can be performed at a later date to determine the effectiveness of training and education, and to identify new needs.

A second way to manage the transition period is to blend the skills of internal IT staff with the focused expertise of a third-party integration team through selective outsourcing. This coordinated, comprehensive approach ensures that an organization's internal staff, from executives to front-line professionals, all develop the attitudes and skills necessary to use and support the new technology.

Effective training not only provides a shared vision of how new systems and networks will strengthen the organization, it also allows faster development of the skills needed for the transition. Training also reduces resistance to change by removing doubts about the new technology. It builds confidence among staff, so they can make a valuable contribution in the new environment. Investing in training also goes a long way toward demonstrating a company's commitment to success.

An education program might include instruction on how to perform traditional data center functions in a client/server environment and explain the similarities and differences between mainframe and UNIX systems. Such instruction can be provided by an outside consulting firm experienced in providing client/server-related seminars, workshops, and courses. To identify skills gaps, the firm should be capable of performing a thorough training needs analysis. An important early step involves preparing both business managers and IT staff for change by building a common awareness and understanding of open systems and client/server and their impact on business strategy and goals. The objective is to align business and IT. This step raises the learning levels of employees, creates support for the technology vision, facilitates organizational change, and increases effective follow-up after implementation.

Solution Design

The solution design of the transition plan requires a broad understanding of leading-edge technology. The components will include hardware and software, networks, applications and administration tools. The process involves matching technologies to both the previously identified business and information goals to the needs of the proposed solution approach. A target architecture will include standards, technologies, products, and processes.

A crucial step in this process is to determine which platform is best for the organization's applications and networking requirements, and which operating system standards will achieve maximum interoperability and portability. If applications will be replaced or reengineered for open, client/server computing, for example, it will be necessary to develop an application data model. This will be useful in formulating target application systems and the physical system design.

Evaluating Alternatives

Objectively evaluating alternative IT strategies is an important element to achieving business goals. Alternatives should be evaluated and compared to determine the appropriate model that best meets the company's needs. These evaluations will help to formulate solution alternatives for implementing a transition. Transition recommendations ultimately may include rehosting, replacing, redesigning, or even outsourcing the existing IT environment.

With the completion of these evaluations, the next step is to develop a conceptual transition model. This model should be based on the transition recommendations for alternative solution approaches. It should include a target architecture, diagrams of the transition approach, and a preliminary business case. With all the alternatives laid out, a risk assessment can be performed.

Risk Assessment

While change is constant and provides an opportunity to add greater value to the business, change also brings risk. Therefore, it is important to identify any risk elements of the transition plan—the potential risks in the transition of the staff and organization. Once the probability and impact of these risks are evaluated, a contingency plan can then be developed to offset each identified risk.

While limiting transition costs is an effective risk-management method, a number of others may be applied:

- Create a transition plan that can adapt over time
- Take an evolutionary instead of revolutionary approach to change
- Work with suppliers, consultants and service organizations experienced in implementing technology transitions for a broad customer base

An evolutionary approach is an effective method for managing risk. When moving from a mainframe to a client/server alternative, for example, the evolutionary approach often translates into a plan detailing the coexistence of mainframe and client/server technologies during a transition period. In some cases, the mainframe will continue to operate as a database server well into the transition period and beyond.

Another risk management strategy is to select a number of mainframe applications that could be simply "rehosted" or transferred to open systems platforms. Remaining essentially unchanged, the application would operate on a more cost-effective platform and become more widely accessible. The conversion of mainframe programs written in COBOL, for example, to an open systems platform is relatively easy: numerous firms specialize in performing such conversions. In addition to minimizing risk, rehosting applications can provide cost benefits, which can then be extended to include networks.

For example, the majority of organizations considering a move to open networking have made a significant investment in the SNA infrastructure. A transition needs to be made in carefully planned stages to minimize the risk and cost of upgrading the network and allowing the organization to continue leveraging its existing investments as long as possible. Organizations should therefore consider transitioning their networks in stages.

In stage one, as new workgroups or subnets are added to the network, they should be based on TCP/IP, the most commonly used set of protocols for internetworking. These workgroups can communicate with the established SNA network using SNA gateway products running on open systems servers.

In stage two, implemented over time, the legacy systems on the established SNA network will be replaced with open systems alternatives. The new systems can use TCP/IP-based communication among themselves, and emulate SNA devices when dealing with remaining legacy systems. IBM 3270 terminals can remain on the network with the use of emulation and interface software, either indefinitely or until they are phased out.

In stage three, at the point when a significant number of nodes on the network are open systems communicating via TCP/IP, the SNA network backbone itself can be transitioned to a TCP/IP network using a cost-effective mix of dedicated lines and switched technology. This move will likely improve communications among the various systems, reducing the need for emulation software. Any remaining SNA-dependent mainframes can be migrated to TCP/IP-based communications by installing TCP/IP interfaces on the legacy systems.

After these transition stages have been achieved, what remains will be an open, flexible network positioned for cost savings and the cost-effective integration of future technologies.

Project Timeline

The final step of the transition methodology is to create a timeline to identify milestones for various phases of the project. Activities should be prioritized according to overall requirements, application strategy, and solution availability. It also is useful to include an estimate of the time required for each component. In creating this timeline, it is helpful to work with external consultants and third parties as well as internal staff. Once the timeline is developed and each party signs off on it, the timeline is distributed and becomes the baseline against which progress is measured.

SUMMARY

Many of the architectures announced, deployed, and implemented during the early 1990s are no longer effective. The solutions of several years ago worked well for the needs of the time, but today's solutions must meet expanded business requirements and be capable of growing with the corporation as it seeks to respond to fast-changing customer demands and markets. Organizations seeking to remain competitive are driving the transition from proprietary, centralized, often mainframe-based architectures to an open, distributed, scalable client/server architecture. Yet that transition carries with it inherent complexities and risks. By focusing not just on technology, but also on the IT processes and people skills, organizations can build flexible enterprise-wide networks which provide strategic advantages in increasingly competitive global markets.

See Also **Financial Management of Technology, Technology Procurement Alternatives**

Telnet

Telnet, short for telecommunications network, is part of the TCP/IP suite of protocols, which is used to access a remote host on the Internet. Telnet was developed to provide a means for two machines to directly access each other without having to use a dedicated port. Although a dedicated port is an easy method of access, it is also processing intensive. When connected to another machine, the user's computer has to translate terminal codes between the two. Although less of an issue today, in the early days of TCP/IP such translations imposed a heavy load on the CPUs. Telnet solved this problem by placing all of the terminal codes within the Telnet protocol itself. This enabled Telnet to determine and set terminal parameters automatically. When Telnet establishes a connection, the machines on each end agree on the parameters governing information exchange, thereby reducing the burden on their CPUs.

When connected via TCP, the user's machine behaves as if it were a directly attached terminal. The remote host need not run UNIX, but must support the Telnet protocol. Finding the host only requires that the user know its name, after which a user ID and password may be required for access. Although Telnet client software is available with a user-friendly graphical user interface, once access is granted to the remote host, menus and text-based input must be relied upon for navigation and command execution. The Telnet command supports both character-at-a-time and line-by-line data entry.

The hosts that support Telnet may contain such things as software archives, library catalogs, specialized databases, games and other programs. They may also provide access to other Internet resources such as e-mail, FTP, and Gopher sites. Some of the more than 10,000 public Telnet sites even support Internet Relay Chat (IRC), an interactive method of communicating with other online users. Lists of publicly available Telnet sites can be obtained by searching the Web using the keywords *telnet sites*.

Another common use of remote host access via Telnet is for program initialization. For example, a Web site developer may want to use the Perl language for processing HTML form input. Once the Perl program is written, it must be placed in a cgi-bin directory at the Web server (i.e., the "host"). This can be done via the File Transfer Protocol (FTP). But before the program can start processing forms data, its permissions must be set. The way to do this is by accessing the host directly via Telnet, getting into the cgi-bin directory, and issuing a command such as:

chmod a+rx filename.pl

Essentially, chmod changes the permissions for a file, allowing filename.pl to read (r) and execute (x) for everyone (a+). Filename.pl is the name assigned for the particular program, and .pl is the extension indicating that it is a Perl program.

Starting a Session

A Telnet session is started with either the name or the IP address of the remote host. Within a Web browser, this information is entered as:

telnet://lib.dartmouth.edu

The plain-text name of the remote host is usable only if the system can translate the name into an IP address using the Domain Name System (DNS). In this case, the Telnet session is with the Dartmouth College Library Online System, which contains such items as the World Factbook, the MLA Bibliography, the full text of 33 Shakespeare plays, the King James version of the Bible, and a portion of the MEDLINE database.

When the connection is established, the remote host may request a user ID and password. If the host is configured for public access, users can usually gain entry by typing guest if the system asks for a password. Any commands the user issues must be those of the remote host. Hosts that are configured for Telnet access usually have menus to aid navigation and provide help on the commands that are supported.

A Web browser can be configured to automatically open a Telnet client (Figure 114) when a Telnet host name or IP address is typed on the command line normally used to enter the addresses for Web sites.

Terminal Emulation Mode

Several software vendors offer emulation programs that are based on Telnet. One of these emulation programs is TN3270, which lets workstations and PCs communicate with a mainframe host via TCP/IP's Telnet terminal emulation mode. By using a terminal emulation mode that is native to

Figure 114
A Web browser, such as Netscape Navigator, can be configured to automatically start a Telnet client, such as InterSoft International's NetTerm.

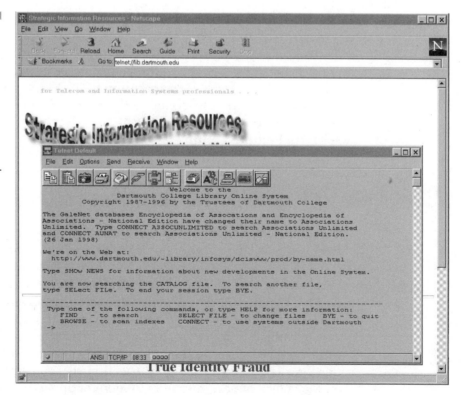

TCP/IP, TN3270 users can communicate with a mainframe across a TCP/IP network without requiring a boundary router to convert the protocols. This approach is inexpensive and easy to implement, since Telnet software is widely installed and available from a variety of vendors.

TN3270 has some limitations, however. For example, it does not offer the reliability and integrity of native SNA (Systems Network Architecture). For example, TN3270 cannot support the SNA traffic prioritization schemes often required in mission-critical SNA applications. And because it uses TCP/IP, TN3270 does not offer the data integrity of methods that use SNA on an end-to-end basis. A newer version of the standard called TN3270 Extended (TN3270E) offers additional SNA capabilities, such as 3270 printing, but it also does not support APPC (Advanced Program to Program Communications), which is at the heart of IBM's next-generation SNA applications.

Finally, some TN3270 configurations require a dedicated terminal server to convert 3270 data streams into TCP/IP and back again. Users can avoid this problem by implementing TCP/IP on the mainframe or front-end communications server as well as on the desktop machines.

SUMMARY

Although it originated as a means of interconnecting UNIX machines via TCP, Telnet has become a general method of logging into remote machines—whether or not they are running UNIX. A number of Windows programs and Web browsers are available that simplify the use of Telnet with a graphical user interface.

See Also **Remote Login**

Thin Client Architectures

The thin-client computing model originated with Oracle Corp. in 1995. In this model, applications are deployed, managed, supported, and executed on a server. This allows organizations to deploy low-cost client devices on the desktop and, in the process, overcome such application deployment challenges of management, access, performance, and security.

The clients themselves come in three types: network computers, Windows terminals, and netPCs. Each type of client is suited to a particular corporate need. Since all three solutions rely on servers for their applications, realizing the benefits of the thin-client environment depends on the ability of administrators to monitor application usage, fine-tune server performance, and sometimes take control of the clients for support, diagnostics, and training.

By centralizing management of applications and using the network to deliver a GUI to desktop devices, organizations can reduce the total cost of ownership (TCO). Saving on the total cost of ownership and reducing complexity are the primary motivations for companies installing thin-client hardware. They dramatically reduce application deployment time by updating clients automatically whenever an application is installed on the server, making it easy to provide the latest applications to all users. The thin client architecture improves management and security by enabling systems administrators to control an entire network from the server. This eliminates the need for end users to install software or make changes to existing software, creating an endless array of software bugs, glitches, freezes, and crashes.

Applications of Thin Clients

Businesses that have embraced thin clients are using them for a variety of applications. Most thin clients are used to access an office suite such as Microsoft Office, but some are used to run mission-critical applications, such as accounting, transaction-processing, and order-entry applications. To a lesser extent, thin clients are also running engineering, Enterprise Resource Planning (ERP), and medical applications.

Users of thin-clients are usually task-oriented and prefer to do their work without being distracted by technology issues. These are front-line professionals, such as doctors in HMOs, accountants, engineers, and salespeople of big-ticket items, such as industrial equipment and real estate. Thin clients are also used for back-office operations supported by clerical and administrative staff, low-level salespeople, and workers on the shop floor.

Operation of Thin Clients

The operation of thin clients is fairly simple; they are dependent on servers for boot-up, applications, processing, and storage. Since some thin clients may not have a hard drive, the server provides booting service to the network computers when they are turned on. The server can be a suitably equipped PC, a RISC-based workstation, a mid-range host such as the IBM AS/400, or even a mainframe. The server typically connects to the LAN with an Ethernet or token-ring adapter and supports TCP/IP for WAN connections to the public Internet or a private intranet.

Since all applications reside on the server, they must be installed only once—not hundreds or thousands of times at individual desktops. Periodic updates to applications are conducted on the server. This ensures that every network computer uses the same version of the application every time it is accessed.

Network computers can access both Java and Windows applications on the server, as well as various terminal emulations for access to legacy data. Users accessing Java applications do so through a Java-enabled Web browser, which also gives them access to applications on the Internet or intranet. For access to Windows applications, the server must run a multiuser implementation of Windows NT, such as WinCenter from Network Computing Devices, Inc. The server's operating system may also include terminal support for 3270, 5250, and X-Windows servers.

Role of Java

To one degree or another, Java plays a role in all of these thin-client architectures. Java was designed by Sun Microsystems to provide a cleaner, simpler language that could be processed faster and more efficiently than C or C++ on nearly any microprocessor.[1] Whereas C or C++ source code is optimized for a particular model of processor, Java source code is compiled into a universal format. It writes for a virtual machine in the form of simple binary instructions. Compiled byte-code is executed by a Java run-time interpreter, performing all the usual activities of a real processor, but within a safe, virtual environment instead of a particular computer platform.

Much of the Java applications development at major corporations hinges on the Web because the Internet is increasingly being considered the foundation for network computing, providing an economical way to access corporate information from remote locations and mobile computers. A Java-enabled Web browser is the interface for accessing these applications. This arrangement minimizes the number of copies of application software and streamlines software maintenance tasks.

With the advent of thin clients and the increasing reliance on network servers, the concept of the "thin server" has emerged. This is a dedicated, special-purpose server optimized for supporting thin clients. In addition to supporting a narrow range of network applications, the thin server supports localized services to reduce network traffic congestion and pro-

---

1. For a more in-depth discussion of Java, see my book *Desktop Encyclopedia of the Internet*, published by Artech House (ISBN: 0-89006-729-5). This 600-page reference can be conveniently ordered through the Web at Amazon.com.

vides fast access to routine applets used for many database and spreadsheet programs.

Unlike traditional server-based systems that require an investment in separate hardware and software, thin servers come complete with hardware and software at a fraction of the cost. They usually come ready to set up, install, and configure out of the box. Typically, all users need to do is plug in an Ethernet cable and set the IP address. A thin server usually comes with a browser GUI, a Java-based management application, and an embedded HyperText Transfer Protocol (HTTP)-compliant operating system.

Types of Thin Client

As noted, there are three types of thin clients: network computers, netPCs, and Windows terminals.

NETWORK COMPUTERS   Network computers emphasize access to server-based applications and data via Web browsers. They combine high-speed Internet and network application performance with Java capabilities. The management software runs on the server and is used to administer the network computers and to set up their boot parameters. Some network computers can also run Windows applications via multi-user implementations of Windows NT on a PC server. Others, such as those available from IBM, can also access 3270 and 5250 terminal applications and work with applications on AIX and UNIX servers using X-Windows server support.

NetPCS   Unlike network computers, netPCs are designed to run programs locally. This type of thin client comes in many different configurations. Of the three types of thin clients, the netPC is the one that most closely resembles a traditional PC. Among the first vendors to offer netPCs were Compaq and Hewlett-Packard. Both companies offer netPCs that are compliant with the DMI 2.0 specification for desktop management.

WINDOWS TERMINALS   Windows terminals provide local and remote access (e.g., Internet and intranet) to Windows, Java, Web browser, and host-based applications. These machines are suited for retail, call center, financial, transaction processing, manufacturing, and office sites. In the home office, they can access a remote server using standard modems. Resident VT220, VT100 and VT52 ANSI terminal emulators make the Windows terminals ideal as character terminal replacements, ready for integration with Windows NT. Among the most prominent vendors of Windows terminals is Wyse Technologies.

DMI Compliance

Since thin clients are network-oriented, there must be a way to incorporate them into the overall management scheme. This is necessary because organizations usually have a mix of thin clients and fully configured desktop systems. Many vendors have extended and enhanced the manageability of thin clients by making them compliant with the Desktop Management Interface (DMI), specifically DMI 2.0, issued by the Desktop Management Task Force (DMTF). DMI 2.0 provides a standard way to access management data about a system, which is a key element in reducing the total cost of ownership. Version 2.0 also includes support for Intel's Wired for Management (WfM) specification.

Wired for Management

The WfM baseline specification, developed by Intel, facilitates the central management of Intel processor-based systems over a network. WfM-enabled computers contain up to four key capabilities to support advanced systems management:

INSTRUMENTATION This is a technology that enables asset management and remote systems monitoring. It gathers information from components of the system, disk drive, and network interface, as well as I/O cards and other add-ins. The values provided by the instrumentation are described by a file called the Management Information Format (MIF) file, which is accessible through the DMI 2.0 Service Provider.

NETWORK SERVICE BOOT Also known as Preboot Execution Environment (PXE), this capability includes platform agents that allow an operating system and/or applications to be loaded on a system from a remote server. This capability is implemented as an option ROM on the NIC, or in system ROM for the LAN on motherboard products.

REMOTE WAKE-UP This is the ability to bring a system from a reduced-power state, including a full shut-down of all system functions, to a fully-powered state in which all management interfaces are available. This capability enables off-hours, remote, and automated maintenance and software upgrades. It is implemented via Magic Packet support in the NIC or LAN on motherboard (LOM) and wake-up circuitry, such as Wake-on-LAN on the system baseboard.

POWER MANAGEMENT This capability allows a PC to function in a reduced-power state, but still be "awakened" and quickly become fully operational when needed. The WfM Baseline Specification recommends that desktop systems comply with the Advanced Configuration and Power Interface (ACPI), and requires such compliance for mobile systems. ACPI compliance means that an ACPI-compliant BIOS must be resident in the system's non-volatile memory, the platform must include an ACPI-compliant chipset, and all peripherals, such as the network interface, must support ACPI.

These capabilities are implemented via technologies that reside in the computer's hardware, nonvolatile memory, disk drive, and network interface card (NIC).

SUMMARY

The promise of thin clients is that they allow organizations to more quickly realize value from the applications and data required to run their businesses, they receive the greatest return on computing investment, and they accommodate both current and future enterprise computing needs. This does not mean thin clients will replace PCs. The two are really complementary architectures that can be centrally managed. In some cases, it may even be difficult to distinguish between the two—the line of demarcation seems to be quite fluid. Overall, thin clients should provide the economies and efficiencies organizations are looking for, but these advantages will primarily come from among the user population that is task-oriented.

See Also Client-Server, Network Booting

Third-Generation Wireless Networks

The ability to deliver multimedia services over cellular networks requires more capacity, robustness, and flexibility than narrowband technologies can provide. Accordingly, the wireless industry is focusing on wideband versions of Code Division Multiple Access (CDMA) and Time Division Multiple Access (TDMA) technologies to implement third-generation (3G) networks capable of delivering multimedia services and Internet content.

Competing Standards

There are several competing WCDMA standards in North America, including wideband cdmaOne, WIMS (Wireless Multimedia and Messaging

Services) W-CDMA, W-CDMA/NA (W-CDMA North America), and TDMA (IS-136). Most wireless operators have chosen from among these technologies for building and enhancing their cellular networks, and all have been submitted to the International Telecommunication Union (ITU) for proposed inclusion into its IMT-2000 "family of systems" concept for globally interconnected and interoperable 3G networks.

The ITU, a part of the United Nations, is working to establish wireless standards and specifications that will allow global roaming, including high-speed data and Internet access, full-motion video, and other sophisticated multimedia services. Under the ITU's IMT-2000 initiative, the new global information infrastructure will be supported by both satellite and terrestrial systems, serving fixed and mobile users in public and private networks.

Wideband cdmaOne technology was submitted to the ITU by the CDMA Development Group (CDG) as cdmaOne-2000. The WIMS W-CDMA technology was submitted to the ITU by a group of vendors headed by AT&T Wireless Laboratories, Hughes Network Systems, and InterDigital Communications Corporation. The North American GSM Alliance, a group of 12 U.S. and one Canadian digital wireless PCS carriers, submitted the WCDMA/NA technology to the ITU.

In September 1998, the W-CDMA/NA and WIMS W-CDMA proposals were merged into what is referred to as the enhanced W-CDMA/NA proposal. The new proposal offers enhanced data capabilities, such as enabling packet data to be delivered to up to ten times as many users. Supporters of the enhanced W-CDMA/NA declined to unify their proposal with wideband cdmaOne, claiming that the necessary changes would cause a significant degradation in system capacity and performance, affect additional capabilities, and probably raise the price of services to customers.

Wideband cdmaOne

Wideband cdmaOne uses a CDMA air-interface based on the existing TIA/EIA-95-B standard to provide wireline quality voice service and high-speed data services, ranging from 144 Kbps for mobile users to 2 Mbps for stationary users. It will fully support both packet- and circuit-switched communications such as Internet browsing and landline telephone services, respectively.

CDMA (TIA/EIA-95-B) is a spread-spectrum approach to digital transmission. Each conversation is digitized and then tagged with a code. The mobile phone is then instructed to decipher only the particular code it is tuned to, enabling it to pluck the right conversation off the air. This is similar in process to an English-speaking person being able to pick out the conversation of the only English-speaking person in a room of international visitors.

Spread-spectrum technology entails tagging groups of bits from digitized speech with a unique code that is associated with a particular call on the network. Groups of bits from one cellular call are pseudo-randomly combined in a multiplexing process with those from other calls and transmitted across a broader band of spectrum—1.25 MHz—and then reassembled in the right order to complete the conversation.

Spreading is achieved by applying a pseudo-noise code (also called a *chip code*) to the data bits, which increases the overall data rate and expands the amount of bandwidth used. Since each call has been tagged with a unique code, a specific conversation can be identified when the spread-spectrum signal is recombined (despread) at the receiving end. However, for a given channel within a cell, the aggregate signal for all other conversations will be perceived as interference.

For the spread-spectrum system to operate properly, the receiver must acquire the correct phase position of the incoming signal. Acquisition is accomplished by a search of as many phase positions as necessary until one is found that results in a large correlation between the phase of the incoming signal and the phase of the locally generated spreading sequence at the receiver. The receiver must also continually track that phase position so lock loss will not occur. The processes of acquisition and tracking are performed by the synchronization subsystem of the receiver. These functions work together to ensure that incoming spread signals can be properly "despread."

As noted, advanced services require more capacity, robustness, and flexibility than narrowband technologies can provide. CDG and its members have issued a specification of the 64-Kbps data rate service, and commercial deployment of this service is ongoing. The 64-Kbps data capability will provide high-speed Internet access in a mobile environment, a capability that cannot be matched by other narrowband digital technologies, including CDMA.

The CDG believes that mobile data rates up to 114 Kbps and fixed peak rates beyond 1.5 Mbps are within reach before the end of the decade using today's cdmaOne and that these capabilities can be provided without degrading the system's voice transmission capabilities or requiring additional spectrum. In other words, cdmaOne is expected to double capacity and provide a 1.5-Mbps data rate capability—all within the existing 1.25-MHz channel structure. At the same time, cdmaOne will continue to support existing TIA/EIA-95-B services, including speech coders, packet data services, circuit data services, fax services, Short Messaging Services (SMS), and Over the Air Activation and Provisioning.

The cdmaOne extends support for multiple simultaneous services far beyond the services in TIA/EIA-95-B by providing much higher data rates and a sophisticated multimedia quality of service (QoS) control capability to support multiple voice/packet data/circuit data connections with differing performance requirements. For example, one channel can carry circuit data with low bit error rate (BER) and low latency transmission requirements, while another channel carries packet data that can tolerate a much higher BER and relatively unconstrained latency.

The QoS negotiation procedures provide a service that is functionally equivalent to B-ISDN Q2931 procedures. This provides for ease of implementing transparent multimedia call service via a gateway to ATM/B-ISDN networks (e.g., landline ATM networks). Additionally, cdmaOne packet data services (IP) supports QoS negotiation upper layer protocols such as the Resource ReSerVation Protocols (RSVP) that perform end to end service negotiation procedures to provide multimedia call support.

Enhanced W-CDMA/NA

The enhanced W-CDMA/NA proposal is based on the Global System for Mobile (GSM) telecommunications (formerly known as Groupe Spéciale Mobile, for the group that developed it) standard. Since GSM is based in digital technology, it allows synchronous and asynchronous data to be transported as a bearer service to or from an ISDN terminal. The data rates supported by GSM are 300, 600, 1200, 2400, and 9600 bps. Data can use either the transparent service, which has a fixed delay but no guarantee of data integrity, or a nontransparent service, which guarantees data integrity through an Automatic Repeat Request (ARQ) mechanism, but with variable delay.

Supplementary services are provided on top of bearer services, and include such features as caller identification, call forwarding, call waiting, and multiparty conversations. There is also a lockout feature, which prevents the dialing of certain types of calls, such as international calls.

GSM uses a combination of Time- and Frequency-Division Multiple Access (TDMA/FDMA) to divide up the bandwidth among as many users as possible. The FDMA part involves the division by frequency of the total 25-MHz bandwidth into 124 carrier frequencies of 200 kHz. One or more carrier frequencies are then assigned to each base station. Each of these carrier frequencies is then divided in time, using a TDMA scheme, into eight time slots. One time slot is used for transmission by the mobile device and one for reception. They are separated in time so that the mobile unit does not receive and transmit at the same time.

Within the framework of TDMA, two types of channels are provided: traffic channels and control channels. Traffic channels carry voice and data between users, while the control channels carry information that is used by the network for supervision and management.

The third generation of GSM will be an evolution and extension of current GSM systems and services currently available, optimized for high-speed packet data-rate applications, including high-speed wireless Internet services, video on demand and other data-related applications. Specifically, third-generation GSM adds the use of CDMA multiplexing at the radio interface level.

As noted, Enhanced W-CDMA/NA represents a merger of WIMS (Wireless Multimedia and Messaging Services) W-CDMA and W-CDMA/NA (W-CDMA North America). The merger results in a technology that allows the acquisition of packet data within 10 milliseconds, which is much faster than other proposed 3G technologies.

The W-CDMA/NA proposal incorporates two key WIMS elements into its technology: use of multiple parallel orthogonal codes for higher data rates, and a pilot/header structure that enables very rapid packet acquisition and release. This improves data performance and throughput to address the growing marketplace and the demanding requirements of multimedia and Internet based services.

SUMMARY

Broadband wireless technologies are being developed not only with an eye toward building third-generation networks capable of providing enhanced multimedia services to North American customers, but with the goal of meeting the requirements established by the ITU for its global IMT-2000 initiative. Both the wideband cdmaOne and enhanced W-CDMA/NA technology proposals have been developed to meet these twin objectives. When formally accepted by the ITU, wideband cdmaOne and enhanced W-CDMA/NA will be members of IMT-2000's "family of systems." This means they will globally interconnect and interoperate with other systems in the family.

Most users of second-generation network services will not have to worry about migration issues because the onus is on the network operators to make these changes as painless as possible for existing users. Sometimes this will entail only a software upgrade that can be implemented over the airwaves by dialing a special number. For legacy terminals requiring changes to hardware, the network provider will have every incentive to offer an attractive trade-up policy.

See Also **Wireless LANs**

Time Division Multiplexers

Time Division Multiplexers (TDMs) combine voice, data, and video traffic from various input channels so they can be transmitted over a single higher-speed digital link, such as a T1 line. At the other end, another device separates the individual lower-speed channels, sending the traffic to the appropriate terminals. TDMs enable businesses to reduce telecommunications costs by making the most efficient use of a leased line's available bandwidth. Being billed at a flat monthly rate is enough of an incentive to load the line with as much traffic as possible. Using different levels of voice and/or data compression, the channel capacity of a leased line can easily be doubled or quadrupled to save even more money.

Although used mostly on private networks, TDMs can interface with a Public Switched Telephone Network (PSTN) as well, and route traffic between them. For example, if a private T1 line degrades or fails, the TDM can be configured to automatically switch traffic to an ISDN link until the private line is restored to service. Then the TDM takes the traffic off the ISDN link and puts it back onto the private line, where the costs are lower.

Operation

With TDM, each input device is assigned its own time slot, or channel, into which data or digitized voice is placed for transport over a high-speed link. When a T1 line is used, there are 24 channels, each of which operates at 64 Kbps. The link carries the channels from the transmitting multiplexer to the receiving multiplexer, where they are separated out and sent on to assigned output devices. If an input device has nothing to send, the assigned channel is left empty.

The TDM manages access to the high-speed line and cyclically scans (or polls) the terminal lines, extracts bits or characters, and interleaves them into the assigned time slots (e.g., frames) for output to the high-speed line. The multiplexer includes channel cards for each low-speed channel and its associated device, a scanner/distributor, and common equipment for the high-speed line. Low-speed channel cards handle the data and control signals for the terminal devices. They also provide storage capacity through registers that provide bit or character buffering for placing or receiving data from the time slots in the high-speed data stream.

The TDM's scanner/distributor scans and integrates information received from low-speed devices into the message frame for transmission on the high-speed line, and also distributes data received from the high-speed line to the appropriate terminals at the other end.

On the transmit side, the TDM samples data from each terminal input channel and integrates it into a message frame for transmission over the high-speed line. Message frames consist of time slots, and each time-slot position is allocated to a specific terminal. Interleaving is the technique that multiplexers use to format data from multiple devices for aggregate transmission over the link (Figure 115).

The common equipment provides logical functions that are used to multiplex and demultiplex incoming and outgoing signals. It contains the necessary logic to communicate with both the low-speed and high-speed devices. It also generates data, control, and clock signals, which ensure that the time slots are perfectly synchronized at both ends of the link.

When digital lines are used on the network side, the TDM is connected to a CSU/DSU (channel service unit/digital service unit), a required network interface for carrier-provided digital facilities, which is typically provided as a plug-in card by most multiplexer vendors. The CSU is positioned at the front end of a circuit to equalize the received signal, filter both the transmitted and received waveforms, and interact with the carrier's test facilities. The DSU element transforms the encoded waveform from alternate mark inversion (AMI) to a standard business equipment interface, such as RS-232 or V.35. It also performs data regeneration, control signaling, synchronous sampling, and timing.

Figure 115

Data from multiple input sources is interleaved by the time-division multiplexer for transmission over the high-speed link. Note the empty time slot. If a device has nothing to send, this amount of bandwidth goes unused.

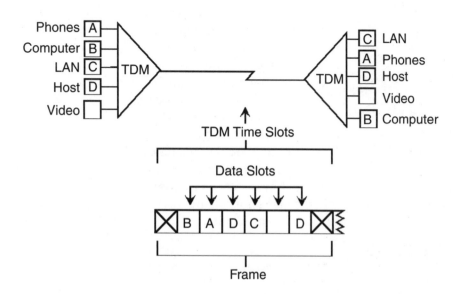

SUMMARY

Several types of multiplexing are in use today. On private leased lines, the dominant technologies are TDM and statistical time division multiplexing (STDM). Each lends itself to particular types of applications. TDMs are used when most of the traffic must have predictable and consistent throughput delay for near real-time operation, such as voice, videoconferencing, and multimedia applications. However, when an input device has nothing to send, its assigned channel goes empty and is wasted. With STDM, if a device has nothing to send, the channel it would have used is taken by a device that does have something to send. If all channels are busy, input devices wait in queue until a channel becomes available. STDMs are used in situations where efficient bandwidth usage is valued and the applications are not bothered by delay.

***See Also* Statistical Time Division Multiplexers, Wavelength Division Multiplexing**

Token Ring

A 4-Mbps token-ring architecture for local-area networks was introduced by IBM in 1985 in response to the commercial availability of Ethernet, which was developed jointly by Digital Equipment, Intel, and Xerox. When Ethernet was introduced, IBM did not endorse it, mainly because its equipment would not work in that environment. Later in 1989, 16-Mbps token ring became available.

Token-ring LANs operate at rates of 4 or 16 Mbps. The ring is essentially a closed loop, although various wiring configurations that employ a multistation access unit (MAU)[2] and patch panel may cause it to resemble a star topology (Figure 116). In addition, today's intelligent wiring hubs and token-ring switches can be used to create dedicated pipes between rings and provide switched connectivity between users on different rings.

The cable distance of a 4-Mbps token ring is limited to 1600 feet between stations, while the cable distance of a 16-Mbps token ring is 800 feet. Because each node acts as a repeater in that data packets and the token are regenerated at their original signal strength, token-ring networks are not as limited by distance as are bus-type networks. Like its nearest rival, Ethernet, today's token-ring networks normally use twisted pair wiring, shielded or unshielded.

---

2. MAUs are nonintelligent concentrators that can be used as the basis for implementing token-ring LANs.

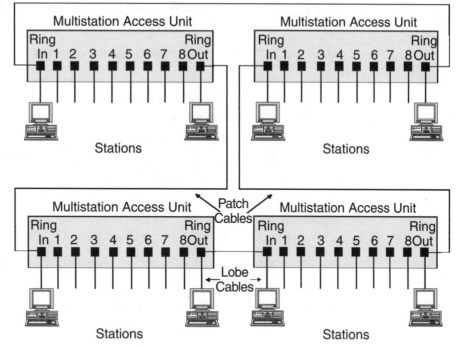

Figure 116
Token-ring stations
are directly
connected to
MAUs, which can
be wired together
to form one large
ring. Patch cables
connect MAUs to
adjacent units.
Lobe cables
connect MAUs to
stations. MAUs
include bypass
relays for
removing stations
from the ring.

Frame Format

The frame size used on 4-Mbps token rings is 4048 bytes, while the frame
size used on 16-Mbps token rings is 16192 bytes. The IEEE 802.5 standard
defines two data formats—tokens and frames (Figure 117). The token,
three octets in length, is the means by which the right to access the
medium is passed from one station to another. The frame format of token

Figure 117
Format of IEEE
802.5 token and
frame.

| Token | SD | AC | ED | | | | | | |

| Frame | SD | AC | FC | DA | SA | Data | FCS | ED | FS |

AC   Access control
DA   Destination address
ED   Ending delimiter
FC   Frame control
FCS Frame check sequence
FS   Frame status
SA   Source address
SD   Starting delimiter

ring differs only slightly from that of Ethernet. The following fields are specified for IEEE 802.5 token-ring frames:

START DELIMITER (SD)   Indicates the start of the frame.

ACCESS CONTROL (AC)   Contains information about the priority of the frame and a need to reserve future tokens, which other stations will grant if they have a lower priority.

FRAME CONTROL (FC)   Defines the type of frame, either Media Access Control (MAC) information or information for an end station. If it is a MAC frame, all stations on the ring read the information. If it contains information, such as user data, it is read only by the destination station.

DESTINATION ADDRESS (DA)   Contains the address of the station that is to receive the frame. The frame can be addressed to all stations on the ring.

SOURCE ADDRESS (SA)   Contains the address of the station that sent the frame.

DATA   Contains the data "payload." If it is is a MAC frame, this field may contain additional control information.

FRAME CHECK SEQUENCE (FCS)   Contains error-checking information to ensure the integrity of the frame to the recipient.

END DELIMITER (ED   Indicates the end of the frame.

FRAME STATUS (FS)   Indicates whether one or more stations on the ring recognized the frame, whether the frame was copied, and whether the destination station is available.

Operation

A token is circulated around the ring, giving each station in sequence a chance to put information on the network. The station seizes the token, replacing it with an information frame. Only the addressee can claim the message. At the completion of the information transfer, the station reinserts the token on the ring. A token-holding timer controls the maximum amount of time a station can occupy the network before passing the token to the next station.

A variation of this token-passing scheme allows devices to send data only during specified time intervals. The ability to determine the time interval between messages is a major advantage over the contention-based access method used by Ethernet. This time-slot approach can support voice transmission and videoconferencing, since latency is controllable.

To protect the token ring from potential disaster, one terminal is typically designated as the control station. This terminal supervises network operations and does important housecleaning chores, such as reinserting lost tokens, taking extra tokens off the network, and disposing of "lost" packets. To guard against the failure of the control station, every station is equipped with control circuitry so that the first station detecting the failure of the control station assumes responsibility for network supervision.

Dedicated Token Ring

When operating in full-duplex mode—the capability to send and receive simultaneously—the total throughput of each station is increased from 16 to 32 Mbps. Under the IEEE 802.5r standard for DTR, end stations and concentrators that operate in this mode adhere to the token-passing access protocol, enabling them to coexist with conventional half-duplex token-ring equipment.

The DTR concentrator consists of C-Ports and a data transfer unit (DTU). The C-Ports provide the basic connectivity from the device to token-ring stations, traditional concentrators or other DTR concentrators. The DTU is the switching fabric that connects the C-Ports within a DTR concentrator. In addition, DTR concentrators can be linked to each other over a LAN or WAN via data transfer services such as ATM.

SUMMARY

For over 15 years, token ring has provided a stable network platform for handling a variety of business applications. Yet, as business applications have grown in complexity and bandwidth consumption, token ring has not kept pace the way Ethernet has in terms of allowing users to migrate to higher speeds. While new versions of Ethernet have increased throughput 10 times with Fast Ethernet and 100 times with Gigabit Ethernet, token ring's DTR standard only doubled throughput to 32 Mbps. Faced with the prospect of losing market share to Ethernet, token-ring vendors formed the High-Speed Token-Ring Alliance (HSTRA) in 1997. A year later, the alliance issued a specification for High-Speed Token Ring (HSTR), which offers 100 Mbps, preserves the native token-ring architecture, and provides a migration path to a future gigabit token-ring standard.

See Also **ARCnet, Ethernet, High-Speed Token Ring, Multistation Access Units, StarLAN**

Transceivers

A transmitter-receiver (transceiver) connects a computer, printer, or other device to a local-area network (Figure 118). The transceiver can be integrated on the Network Interface Card (NIC) or it can be a separate device that connects to the NIC with a drop cable. In the latter case, the transceiver cable and connectors form the Attachment Unit Interface (AUI) and the transceiver is the Medium Attachment Unit (MAU). In the Ethernet environment, the MAU has four basic functions:

TRANSMIT   Transmits serial data onto the medium

RECEIVE   Receives serial transmission and passes these signals to the attached station

COLLISION DETECTION   Detects the presence of simultaneous signals on the network and alerts the station

JABBER FUNCTION   Automatically interrupts the transmit function to inhibit abnormally long data stream output

The MAU consists of the Physical Medium Attachment (PMA), which provides the functions and two connectors. On the network side, the MAU attaches to the Medium-Dependent Interface (MDI). The specific interface depends on the type of media used. For example, 10BaseT (twisted pair) uses a RJ-45 connector and 10Base2 (thin coax) uses a BNC connector (Figure 119), while the older 10Base5 (thick coax) implements a special "vampire" tap that pierces the coaxial cable and makes contact with both the center conductor and shield.

Figure 118
Transceiver
architecture.

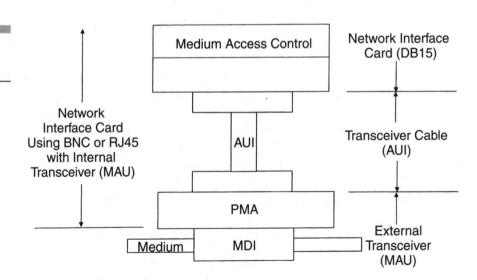

Figure 119
This four-port transceiver from Allied Telesyn has a BNC connector (top) that is used for connecting the unit to a 10Base2 (thin coax) Ethernet LAN.

Status Indicators

Transceivers offer several indicators to inform users of performance status at any given time:

TRANSMIT Indicates packets are being transmitted onto the media

RECEIVE Indicates packets are being received from the media

SQE Indicates a Signal Quality Error (SQE) test signal is present

COLLISION Indicates a collision has occurred
A user-selectable switch is provided, which allows network managers to choose between enabling or disabling the SQE test function. This feature permits the transceiver to be used with repeaters that cannot support the test function.

SUMMARY

Transceivers are available in a variety of configurations to support different LAN types and media. There are transceivers for all versions of Ethernet, as well as Fiber Distributed Data Interface (FDDI) and Asynchronous Transfer Mode (ATM) networks. There are transceivers for coaxial cable (thick and thin), twisted pair wiring (shielded and unshielded), optical fiber (single- and multimode), and wireless (spread spectrum and infrared).
See Also **Media Converters, Network Interface Cards**

Transmission Control Protocol

Transmission Control Protocol (TCP) is a key component of the TCP/IP suite of Internet protocols. The function of TCP is to forward data delivered by IP to the appropriate process at the receiving host. Among other things, TCP defines the procedures for breaking up the data stream into packets and reassembling them in the proper order to reconstruct the original data stream at the receiving end. Since the packets typically take different routes to their destination, they arrive at different times and out of sequence. All packets are temporarily stored until the missing packets arrive so they can be put in the correct order. If a packet arrives damaged, it is simply discarded and another one is resent.

To accomplish these and other tasks, TCP breaks the messages or data stream down into a manageable size and adds a header to form a packet. The packet's header (Figure 120) consists of:

SOURCE PORT ADDRESS (16 BITS) / DESTINATION PORT ADDRESS (16 BITS)
The source and destination ports correspond to the calling and called TCP applications. The port number is usually assigned by TCP whenever an application makes a connection. There are well-known ports associated with standard services such as Telnet, FTP, and SMTP.

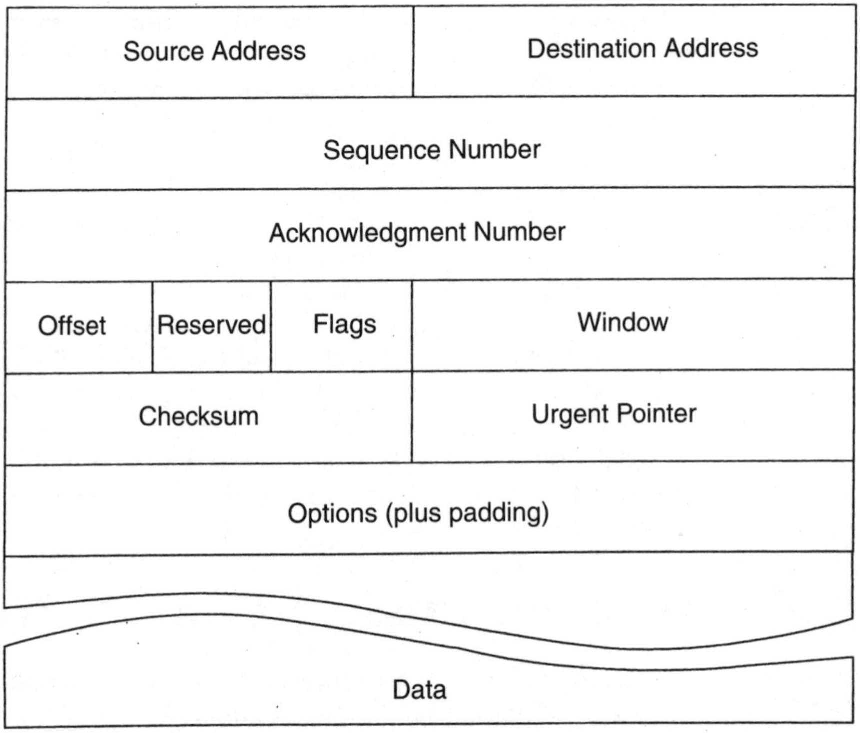

Figure 120
TCP packet header.

Source Address			Destination Address
Sequence Number			
Acknowledgment Number			
Offset	Reserved	Flags	Window
Checksum			Urgent Pointer
Options (plus padding)			
Data			

SEQUENCE NUMBER (32 BITS) Each packet is assigned a unique sequence number that lets the receiving device reassemble the packets in sequence to form the original data stream.

ACKNOWLEDGMENT NUMBER (32 BITS) The acknowledgment number indicates the identifier or sequence number of the next expected packet. Its value is used to acknowledge all packets transmitted in the data stream up to that point. If a packet is lost or corrupted, the receiver will not "acknowledge" that particular packet. This negative acknowledgment triggers a retransmission of the missing or corrupted packet.

OFFSET (4 BITS) The offset field indicates the number of 32-bit words in the TCP header. This is required because the TCP header may vary in length, according to the options that are selected.

RESERVED (6 BITS) This field is not currently used, but may accommodate some future enhancement of TCP.

FLAGS (6 BITS) The flags field serves to indicate the initiation or termination of a TCP session, reset a TCP connection, or indicate the desired type of service.

WINDOW (16 BITS) The window field, also called the receive window size, indicates the number of 8-bit bytes that the host is prepared to receive on a TCP connection. This provides precise flow control.

CHECKSUM (16 BITS) The checksum determines whether the received packet has been corrupted in any way during transmission.

URGENT POINTER (16 BITS) The urgent pointer indicates the location in the TCP byte stream where urgent data ends.

OPTIONS (0 OR MORE 32-BIT WORDS) The options field is typically used by TCP software at one host to communicate with TCP software at the other end of the connection. It passes information such as the maximum TCP segment size that the remote machine is willing to receive.

SUMMARY

The Transmission Control Protocol is implemented in the end hosts and offers assured delivery of Internet Protocol (IP) packets. TCP is needed for this purpose because the function of IP is only to route data from the source to the destination—there is no extra functionality to guarantee its delivery. TCP makes sure data is delivered in a timely fashion, that the packets arrive in sequence, and without error. Services such as e-mail, file transfer, terminal login, and World Wide Web run over TCP connections and depend on TCP for their reliability.

See Also Internet Protocol, User Datagram Protocol

Transmission Modes

Three modes of transmission are supported on today's networks: simplex (in which data is transmitted in only one direction), half-duplex (in which data can go in either direction, but only one way at a time), and full-duplex (in which data can travel in both directions simultaneously).

Simplex

Some familiar forms of simplex communication include over-the-air radio and TV broadcast. In the corporate environment, some video systems are used in simplex mode, such as when a stockholder meeting is multicasted to PCs over the Internet. In this case, users can subscribe to a channel or enroll in the multicast session by filling out a form posted on the company's Web page. At the time of the multicast, only users who have specifically signed up for it will receive the feed. Another example of simplex communication is the delivery of ticker tape messages on Web pages, which display stock quotes, news, sports scores, and advertisements.

Half-Duplex

A familiar type of half-duplex communication is citizens band (CB) radio in which communication is possible in both directions, but the users must wait their turn. When done speaking, they usually end by saying "over." This alerts the other party to start talking. Early sound cards for PCs were half-duplex, enabling users to converse over the Internet by taking turns. International calls carried over the Public Switched Network (PSTN) are half-duplex.

Full-Duplex

Full-duplex transmission can be accomplished with a four-wire line, coaxial cable, or radio wave transmission system. A familiar type of full-duplex communication is the ordinary local or domestic long-distance call over the PSTN. Private T1 lines are also full-duplex because, like the public network, they are inherently four-wire lines. Most CATV systems are half-duplex, but they are rapidly being upgraded for full-duplex so cable operators can offer real-time interactive services through television set-top boxes as well as telephone service in competition with local telephone companies. Cellular phone networks are an example of a radio wave transmission system that supports full-duplex communication.

Asymmetric Duplex

There is another mode of data communication, called asymmetric duplex. This mode allows data transmission in both directions—half-duplex or full-duplex—but at very different rates. For example, old teletex systems allowed 1200 bits per second in one direction and 75 bits per second in the reverse direction. This was done because a lot of data was sent to a terminal but the terminal only made short queries to the host.

Asymmetric duplex is also used in some digital subscriber line (DSL) technologies. With Asymmetric Digital Subscriber Line (ADSL), for example, data travels downstream (in the direction of the computer) at 1.5 to 8 Mbps and upstream (in the direction of the network) at 64 to 640 Kbps.

SUMMARY

The nature of the application determines what mode of communication—simplex, half-duplex, or full-duplex—is used. Generally, any real-time, interactive application will use the full-duplex mode of communication. Interactive applications that are not time-sensitive will use the half-duplex mode of communication. Applications that are not interactive and which are not time-sensitive will use the simplex mode of communication.

See Also Digital Subscriber Line, Multicast Transmission

Transparent LAN Service

Transparent LAN services (TLS) are high-speed, LAN interconnection services that hide the complexity associated with WAN technology, design, and management from corporate users who do not want to deal with installing frame relay or Asynchronous Transfer Mode (ATM) equipment or configuring virtual circuits and ports.

With a TLS, a service provider interconnects the corporation's LANs in such a way that they appear to be interconnected by a LAN segment, making them "transparent." Geographically separated users can then communicate with one another and access remote servers as easily as if they were all at the same location.

Service Rationale

TLS is intended for companies without staff members who are familiar with frame relay and do not have the resources to hire someone with this kind of expertise. When subscribing to a basic frame relay service, the company is responsible for designing, planning, and optimizing the network. When subscribing to a TLS, the company simply tells its service provider what LAN segments from each site must be interconnected. The service provider then interconnects the sites. From the company's perspective, the WAN becomes as easy to manage as a LAN. Since the company only needs to work with familiar LAN technology, no in-house WAN expertise is required.

Companies that do not want to purchase or manage WAN access equipment may find a TLS appealing as well. With most frame relay services—except managed frame relay services—the subscribing company is responsible for testing, installing, and implementing WAN equipment to interface with the service provider's network. With TLS, the WAN access equipment is often included in the service price and the service provider is responsible for managing and maintaining the equipment.

If a company want its WAN to operate at the same speed as its LANs, TLS comes in various speeds, which makes this possible. The most common speeds are 10 and 16 Mbps, but there are also services at 100 Mbps. If low-speed frame relay is not adequate and DS3 (45-Mbps) frame relay is too fast, transparent LAN service may be a good choice. Since transmission across the WAN occurs at native LAN speeds, bottlenecks between the LAN and WAN are eliminated.

Implementation Methods

The service provider decides what type of transmission method it will use for the service. There are three available options:

ASYNCHRONOUS TRANSFER MODE   LAN packets are encapsulated or subdivided into 53-byte ATM cells and transported over a Synchronous Optical Network (SONET).

FRAME RELAY   LAN packets are converted into frame relay frames and transmitted over SONET.

DARK FIBER The service provider uses time division multiplexing (TDM) to transmit LAN packets across the fiber.

ATM over SONET is the only TLS backbone alternative that enables providers to offer service-level agreements (SLAs) for performance. The uniformity of ATM cells and the detailed information contained in the ATM cell header allow for predictable latency and throughput, which allows the precise control of traffic necessary to meet performance guarantees.

Fiber Distributed Data Interface (FDDI) rings were used in some early TLS offerings, but because of its distance limitations and lack of scalability, carriers generally are not considering FDDI as an infrastructure for new TLS offerings.

Typical Service

One TLS provider is TCG CERFnet, which offers OmniLAN (Figure 121). This connectivity solution supports LAN protocols at their native rate, including Ethernet, Fast Ethernet, token ring, and FDDI. Bandwidth can be procured on an incremental basis. The service is fully managed by TCG CERFnet and includes customer premises equipment in the form of a LAN/ATM concentrator.

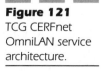

Figure 121
TCG CERFnet
OmniLAN service
architecture.

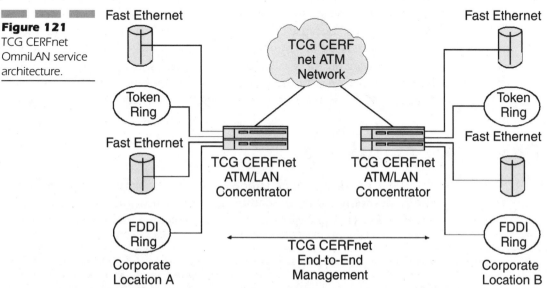

The OmniLAN services leverage ATM technology to provide LAN connectivity solutions to customers not ready to deploy enterprise ATM networks of their own. TCG CERFnet's customer premises equipment converts LAN traffic into ATM cells, which are then transported across its ATM/SONET backbone. The cells are reassembled at the receiving location, providing a seamless interconnection to corporate users. In addition, because ATM supports multiple logical connections over the same physical port, the need for a company to purchase and maintain the costly equipment associated with separate private line circuits and interfaces was eliminated.

TCG CERFnet proactively monitors and manages the service 24 hours per day and seven days per week via its Network Operation Centers (NOCs). Using a set of third-party and custom network management tools, TCG CERFnet provides:

- Fault management
- Trouble isolation
- Remote diagnostics
- Configuration management
- Performance/utilization reporting

SUMMARY

Transparent LAN services connect corporate LANs together at native speeds in a point-to-point, point-to-multipoint, or mesh network configuration using the service provider's ATM backbone. TLS is considered to be "transparent" because end users do not need to know how the carrier delivers the service or about the infrastructure used to support it. End users simply hand the service provider a LAN segment from each site to be interconnected, and the service provider does the rest.

***See Also* Managed Applications Services, Service-Level Agreements**

Trunking

Trunking provides higher bandwidth capacity between a LAN switch and application server by combining the bandwidth of several lower-speed links.[3] This aggregation scheme is implemented in hardware and/or

---

3. A similar concept, known as *inverse multiplexing*, exists in the WAN environment. For a more in-depth discussion of inverse multiplexing, see my book *Desktop Encyclopedia of Telecommunications*, published by McGraw-Hill (ISBN: 0-0704-4457-9). This 600-page reference can be conveniently ordered through the Web at Amazon.com.

software, so the bandwidth of four 100-Mbps Fast Ethernet links can be aggregated to quadruple performance to 400 Mbps, for example (Figure 122). This saves organizations from having to move up to Gigabit Ethernet and buy a new hub or switch, pull new cable, and worry about distance limitations. All that is really needed is specialized trunking software and possibly some cards to aggregate several switch-to-server links into one bigger logical pipe.

Any combination of ports can be trunked and the scheme can work over multiple paths in addition to point-to-point connections. The goal is to deliver flexible, dedicated throughput for bandwidth-hungry applications such as multimedia and file sharing. The trunking software can even perform load balancing and apply fault tolerance to crucial connections. By balancing traffic across various switch ports and server interfaces, performance is exploited fully. And if a port goes down, it is automatically removed from the trunking group and its traffic redistributed across the remaining ports.

Load Balancing

Typically, traffic loads are balanced through a round-robin algorithm that shunts packets through consecutive ports: packet 1 goes to port A; packet 2, to port B; packet 3, to port C; and so on. This ensures that every port transmits the same number of packets. While this scheme can cause

Figure 122
Trunking helps relieve switch-to-server bottlenecks by logically aggregating individual physical connections into a single high-speed pipe.

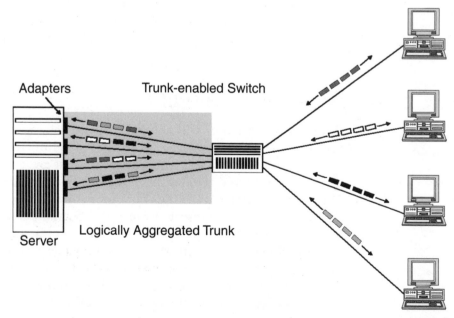

Adapters   Trunk-enabled Switch

Logically Aggregated Trunk

Server

packets to arrive out of order, it is usually not a significant problem because buffers are used at both ends of the connection to allow time for sequencing out-of-order packets and smoothing traffic flow. Since the routes are close to equal 95 percent of the time, packets do not have a chance to stray too far out of order.

Some vendors have improved on the round-robin approach by giving it enough intelligence to adapt to the load. The algorithm makes a best guess as to which port should be used. Unlike straight round-robin, it can compensate for excess traffic on one adapter and adjust itself to favor the least loaded card, thus optimizing traffic flow.

Server Processing

To minimize the processing burden on the server in processing packets and moving them to the right trunking group, some vendors offer cards with on-board processors to run the trunking software. These products still rely on the server, since adapter drivers put some load on the server's operating system (OS), but this is the case with conventional cards as well. In most cases, the trunking software will only work with the card from the same vendor. The trunking software may not even work with older cards from the same vendor.

The software differs among vendors in the maximum number of links that can be aggregated in a trunking group. Some vendors permit only four Fast Ethernet connections, for a total theoretical throughput of 800 Mbps in full duplex mode. Others can aggregate a dozen 100-Mbps links for a claimed capacity of 1.2 Gbps in half duplex mode or 2.4 Gbps in full duplex mode. Still other vendors allow up to 16 or 32 trunked connections.

Types of Trunking

Trunked performance will also be influenced by whether the software, adapters, and attached switches are capable of symmetrical or asymmetrical trunking. Symmetrical trunking means that every link is full duplex; packets are sent and received by every port within the trunking group. With asymmetrical trunking, however, while all of the ports in the trunking group can transmit packets, only one port can receive. Thus, the server may be able to pump out 400 Mbps but accept only 100 Mbps. Some products are asymmetrical or symmetrical, while others let network managers choose between symmetrical or asymmetrical trunking.

To monitor traffic on the trunked links, most vendors allow SNMP traps to be set on their trunked connections. Performance statistics can be viewed from a central console. Some vendors go a step further by adding RMON (remote monitoring) to their aggregated links.

Standards

At this writing, there is no standard for LAN trunking. The IEEE 802.3ad Link Aggregation Task Force, however, is studying trunking for both adapters and switches. A finished specification should be ready by mid-year 2000. Although prestandard trunking products typically work together, they do not perform automatic negotiation, a feature that enables adapters and switch ports to perform a handshake and then automatically set up trunks. This must currently be done manually.

SUMMARY

Trunking offers an effective way to increase the bandwidth capacity of switch-to-server links in order to improve the performance of mission-critical applications, while providing the means to leverage existing investments in LAN hardware. The trunking products of different vendors vary widely in terms of throughput performance, the types of trunking they support, the maximum number of links that can be aggregated, and in network management features.

See Also **Load Balancing, Fault Tolerance**

Twisted Pair Cable

The most commonly installed cabling for local-area networks (LANs) are Unshielded Twisted Pair (UTP) and Shielded Twisted Pair (STP) in Ethernet, Token Ring, and FDDI copper environments. Of these, UTP is most often used, even for high-speed networks of 100 Mbps and emerging gigabit-per-second networks.

Both types of cable look like ordinary telephone wire, except they have eight wires inside instead of four. Only four of the wires are actually used, but it is a different combination of wires that is used for each type of network. The eight wires are color-coded for easy identification when attaching RJ45 connectors (Figure 123). The RJ45 connectors have eight conductive pins that the wires make contact with during the installation process. The connectors, in turn, plug into a port on the Network Interface Card (NIC) of a computer, hub, or switch. When properly installed, these components comprise a complete circuit over which data can be sent and received between networked devices.

Figure 123
RJ45 connector.

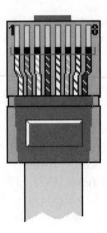

Color Codes

The eight wires are bundled together into what is known as a Category 5 cable (Figure 124) for which there are two wiring standards, EIA/TIA 568A and 568B (also known as AT&T 258A), and consequently two different wiring color codes. The following table provides the color codes for the eight wires inside UTP and STP cable, and identifies which ones are used for each type of network using both standards:

Pin	EIA/TIA 568A	AT&T 258A or EIA/TIA 568B	Ethernet 10Base T	Token Ring	FDDI
1	White/Green	White/Orange	X		X
2	Green/White	Orange/White	X		X
3	White/Orange	White/Green	X	X	
4	Blue/White	Blue/White		X	
5	White/Blue	White/Blue		X	
6	Orange/White	Green/White	X	X	
7	White/Brown	White/Brown			X
8	Brown/White	Brown/White			X

Note: The color codes are in the format of wire/stripe, where "wire" is the solid color.

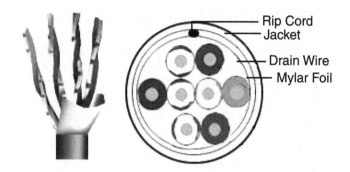

Figure 124
Category 5 cable, showing the four twisted pairs (left) and how the cable is constructed (right).

Ethernet Connections

A segment of Category 5 cable is inserted between each device's NIC and a hub. Each cable segment cannot exceed 325 feet in length. Because the cables from all of the network devices converge at a common point—the hub—an Ethernet network forms a star configuration. A hub is basically a box with a row of RJ45 jacks. Most hubs have 5, 8, 12, 16 or 24 jacks, but some have more. Most hubs also have an uplink port, a special jack that allows the hub to be connected to other hubs to expand the network to accommodate more computers. The uplink port can also be used to link to a cable modem, allowing users to share the same Internet access connection. The uplink port can also be used to connect the hub to a high-speed backbone.

The following table shows the functions of the wires at each end of the connection for a straight-through cable, in conformance to the EIA/TIA 568B (AT&T 258A) standard:

Pin	Hub Port	NIC Port
1	RX+	TX+
2	RC−	TX−
3	TX+	RX+
4	Not used	Not used
5	Not used	Not used
6	TX−	RX−
7	Not used	Not used
8	Not used	Not used

Token-Ring Connections

The following table shows the functions of the wires at each end of the token-ring connection for a straight-through cable, in conformance to the EIA/TIA 568B (AT&T 258A) standard:

Pin	Ring-In Port	Ring-Out Port
1	Not used	Not used
2	Not used	Not used
3	TX–	RX–
4	RX+	TX+
5	RX–	TX–
6	TX+	RX+
7	Not used	Not used
8	Not used	Not used

FDDI Connections

The following table shows the functions of the wires at each end of the FDDI connection for a straight-through cable, in conformance to the EIA/TIA 568B (AT&T 258A) standard:

Pin	Hub Port	NIC Port
1	TX+	TX+
2	TX–	TX–
3	Not used	Not used
4	Not used	Not used
5	Not used	Not used
6	Not used	Not used
7	RX+	RX+
8	RX–	RX–

SUMMARY

Although there are other grades of cable, Category 5 is installed on 85 percent of networks because it supports transmission rates of 10, 100, and

1000 Mbps. Installing lower-grade cable means risking time-consuming and expensive cable installations to keep up with new standards for higher data rates. Category 5 cabling can be run under floors, around office dividers, and over dropped ceilings—even snaked through walls. However, for maximum performance, this type of cabling—notably unshielded twisted pair—should be kept away from power outlets, florescent lighting fixtures, uninterruptable power supplies, and other sources of strong electromagnetic interference. Cable should be cut as near to their required lengths as possible, since coiling up the excess can also cause interference as well.

See Also **Cable Management, Cable Testing, Coaxial Cabling, Fiber-Optic Technology, Network Interface Card**

U

Unified Messaging

Unified messaging integrates voice, e-mail, fax, and pager messages into a single in-box that can be accessed from any location via a telephone, cell phone, PC, or hand-held computer. In some cases, users can have their e-mail messages read to them through text-to-speech translation software.

There is no standard way to implement the technology, so products differ widely and vendors have different ways to integrate their offerings with existing messaging systems. Among the vendors offering unified messaging products is Nortel. The company's CallPilot is a Windows NT-based adjunct server to a PBX that stores voice and fax messages in a single mailbox. Using a LAN-side connection, it then pops header information for these messages onto desktop e-mail interfaces.

CallPilot lets users manage their messages in several ways. They can view all messages at once, quickly prioritizing each message based on who it is from, when the message was sent, and what the message status is. Then they can reply, fax, or call the senders back with the click of a button. Users can also speak predefined spoken commands from their offices or remote locations to compose, send, delete, skip, and play back messages. This speech-recognition feature also allows users to access their messages from a cell phone without having to press any buttons.

The system itself comes in two hardware options for the adjunct server: an eight-port mid-range Pentium-class server that provides 30 hours of message storage, or a rack-mounted server. An alternative to the adjunct hardware is to install the CallPilot system on a Meridian extension shelf as a single-board Pentium server. Interfaces are available to competitors' PBXs as well.

System Management

The notable features of unified messaging systems that simplify overall management and reduce cost of ownership include:

SYSTEM MANAGEMENT INTERFACE Usually a familiar, easy-to-use Windows application that can administer any number of unified messaging systems from a single location anywhere on a corporate intranet or the public Internet.

APPLICATION BUILDER A graphical user interface for creating and distributing custom voice and fax applications such as automated attendants, voice menus, and fax-back menus.

REPORTER A Windows-based application that provides an extensive set of management reports and monitors the system for "suspicious" activity, such as an excessive number of failed log-on attempts.

CONTEXT-SENSITIVE HELP Minimizes the need to use documentation for implementing system management features.

MONITORING Enables SNMP (Simple Network Management Protocol) and delivers alarms and alerts to any compliant enterprise management application.

Services

An alternative to purchasing a unified messaging system is subscribing to a service that offers similar capabilities. MCI Worldcom, for example, offers a unified messaging service called networkMCI Contact, which gives subscribers a single telephone number to receive calls, e-mail, faxes, and pages. Users can access messages via a Web browser, an e-mail client, or a telephone. In addition, subscribers can specify preferences, such as where live phone calls or messages should be routed at specific times or under certain conditions.

Subscribers can also link their e-mail and Internet access, and thus program the unified messaging service through an MCI Worldcom Web site. If users choose only to link their voice calls, faxes, and paging services, then they can set up their service through MCI Worldcom's voice response system.

StarTouch International offers a unified massaging service called the Electronic Secretarial Administrator (ESA), which offers more than 95 features to subscribers, including voice, fax, and e-mail integration; eight-way domestic and international conferencing; call screening; fax and voice-on-demand; fax broadcasting; and domestic/international direct-dial capability. ESA can be managed entirely from the Web. By logging on to the STI Web site, subscribers can listen to their voice mail, review faxes, build broadcast lists, look over their billing statement, delete old voice and fax messages, initiate international callback, and set up conference calls.

Internet Call Centers

Another facet of unified messaging is the Internet-enabled call center. In this type of operation, ordinary 800 calls are handled as usual by agents, and customer records are pulled up from a database using Computer-Telephony Integration (CTI) technology to enhance the agent response. When Internet connectivity is added to the call center, Web visitors can communicate with call center agents for product information by choosing a preferred method of communication, including an IP voice call, text chat, e-mail, or fax—or they can request a callback on the public switched telephone network. To assist further, agents can escort customers through appropriate Web pages, mark the page to highlight important items, and even help customers fill out portions of the order form.

Lucent Technologies is one of many vendors that have integrated multimedia messaging into existing call centers. In addition to using CTI, the company's CentreVu Internet Solutions includes an Internet Telephony Gateway (ITG) that downloads a Java call control applet to customer desktops during Web-site visits. The applet provides customers with the means to talk with a service representative or initiate a text chat session (Figure 125).

Lucent also offers optional Message Care software that route customer faxes to an electronic Internet call center mailbox via an e-mail server's fax interface, where the fax messages can be accessed and routed just like Internet calls or e-mail. With this application, agents do not have to stop what they are doing to periodically check their e-mail box. They can handle both phone calls and electronic messages as they arrive at the call center. Since the Message Care software automates this process, agents spend less time handling each message, which significantly cuts the labor cost per message.

SUMMARY ▪ ▪ ▪ ▪ ▪ ▪ ▪ ▪ ▪

Today's businesses rely on several methods of communication, including electronic mail, facsimile, Web pages, and voice messages. Traditionally, there have been separate systems for each, forcing users to check all of them throughout the day by opening a different application or using a

Figure 125
The Web page of Micron Electronics, for example, issues Lucent's Java call-control applet to customers on request, allowing them to communicate via voice or text to a company representative.

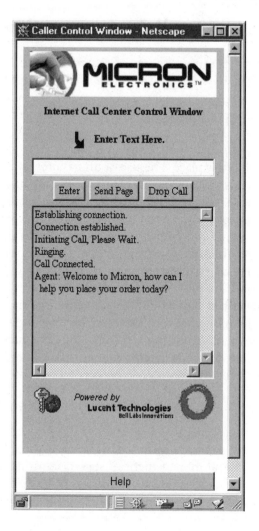

different device, slowing productivity. As the number of messages grows, so does the need for an improved way of managing them using a common set of management tools from a single terminal device. This makes users more productive, regardless of where they happen to be on a daily basis. The Internet-enabled call center treats all forms of communication as if they were telephone calls. This ensures that all customers receive prompt service, regardless of which method of communication they choose.

See Also **Computer-Telephony Integration, Electronic Mail, Fax Servers, Short Message Service**

Uninterruptible Power Supplies

One of the most common causes of system and network downtime is power-related. On a system, power surges often corrupt data, while power outages shut down sessions, wiping out data that has not yet been written to disk. On a network, a power failure that affects a server, hub, switch, or router will stop traffic flow among users, applications, and databases.

The best way to deal with power problems is to prevent them from happening in the first place. The use of an uninterruptible power supply (UPS) provides this measure of protection. Via an internal battery that stays constantly charged, a UPS provides enough standby power to prevent intermittent problems from doing damage and permits the orderly closing of one or more systems during prolonged outages. Some UPSs even change over to other power sources, such as diesel-powered generators, to keep crucial systems running until primary power can be restored.

While most central sites have UPSs, many remote sites typically do not, usually as a cost-saving measure. However, battery backup for remote sites can be very inexpensive, costing less than $100 per system, which is cheap compared to the cost of indeterminate network downtime and trying to recreate lost data that may not have been saved to disk.

Features

The management software included with UPSs provides graceful unattended shutdown and management of servers running multiple network operating systems such as Novell, Windows NT, OS/2, SCO UNIX, and UNIXWare. Some UPSs also support the Simple Network Management Protocol (SNMP), which lets network managers monitor battery backup from a central management console. Via SNMP, for instance, every remote UPS can be instructed to test itself once a week and report back if the test fails. Other important features commonly included in today's UPSs are:

PHONE-LINE PROTECTION Prevents "back-door" surge damage to a computer's modem, network interface card, power supply, and motherboard

SITE-WIRING FAULT INDICATOR Identifies potentially dangerous wiring problems before a system is plugged into the power source

SURGE SUPPRESSION Prevents catastrophic hardware damage caused by power surges and spikes, which extends system life

NOISE FILTERING Prevents electrical noise from affecting computer operation or introducing "glitches" into data files

BATTERY MANAGEMENT Monitors battery performance and maximizes battery life through automatic self-tests and reporting

STATUS INDICATORS Displays such conditions as on battery, low battery, and battery recharging

ADVANCED WARNING DIAGNOSTICS Indicate battery overload and replace battery conditions

AUDIBLE ALARMS Alert the user when the system is overloaded, has reached the low battery level, or needs battery replacement.

HOT-SWAP CAPABILITY Allows users to replace batteries by swapping them out without powering down the connected load.

Management and Diagnostics

Power failures can occur at any time—at night, on weekends, or while the system administrator is out of the building. In such cases, the UPS works with the power management and diagnostic software to provide automatic unattended system shutdown. The software automatically stores data and shuts down the affected system during extended power outages before UPS batteries are drained.

The software also provides important power management and diagnostic features, such as UPS status testing and remote UPS management. The former automatically monitors the health of the UPS, ensuring that problems such as weakened batteries are detected before they impact uptime. The latter allows UPS parameters to be configured, servers to be rebooted, and power problems diagnosed remotely.

American Power Conversion (APC), for example, offers a management system for its line of UPSs called PowerChute PLUS (Figure 126), which notifies the network administrator of such things as utility voltage, UPS load, and run time of each UPS. It also features an event log that the administrator can use to diagnose problems and their severity. If a minor voltage fluctuation triggered the alert, and the fluctuations do not threaten hardware, the administrator can adjust the remote UPS's voltage tolerance window to prevent unnecessary system shutdowns.

Sizing the UPS

UPSs come in many configurations. At the low end, certain UPSs are suitable for protecting LAN nodes, stand-alone computers, and point-of-sale (POS) terminals. In the middle range, UPSs are suitable for protecting heavily configured LAN nodes, workstations, and peer-to-peer servers.

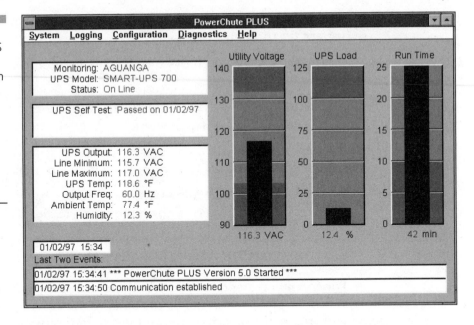

High-end UPSs can protect multiple systems, minicomputers, LAN hubs, and telecommunications equipment.

To determine the appropriate size of UPS for a given environment, network administrators should make a list of all crucial equipment that requires power protection, including monitors, terminals, and external hard drives.

Next, the total load must be determined. The VA (volts-amps) ratings will help determine what size UPS is needed to protect the equipment. These numbers are usually found on a sticker on the back or bottom of the equipment and are multiplied together (i.e., volts × amperes) to arrive at the VA rating. If the components are measured in watts, then the number of watts is multiplied by 1.4 (watts × 1.4) to arrive at the VA rating.

The VAs of each piece of equipment are then added to find the total VA. The total VA is matched against the various UPS models offered by the vendor, which are priced accordingly. To take into consideration any future expansion needs, it is often a good idea to oversize the UPS. Another factor that typically goes into UPS sizing (and pricing) is the amount of runtime a particular system offers. This is the amount of time supported by the UPS under full load or half load while primary power is unavailable. Vendors usually offer a range of UPS models that provide backup power for only a few minutes to those that can provide backup power for several hours.

SUMMARY ▪ ▪ ▪ ▪ ▪ ▪ ▪ ▪

The overwhelming majority of power problems likely to hit computers come from undervoltages. Instantaneous power from a UPS prevents data loss caused by these brownouts, as well as longer outages that occur less frequently from such catastrophic events as cable cuts, transformer failures, or lightening strikes at power distribution stations. With the increasing reliance on computerized data, the use of uninterruptible power supplies should be a key element in the disaster recovery plan of any workgroup, department, and organization.

***See Also* Power Conditioning**

Universal Serial Bus

The Universal Serial Bus (USB) is designed to replace the various serial and parallel port connectors with one standardized plug and port combination that provides a data transfer rate of 12 Mbps. To add new capabilities to a PC, users just plug in new USB-compliant peripherals, daisy-chaining them off the PC's USB port. Alternatively, a USB hub can be used to branch out to the various peripherals. Either way, the configuration process is automatic; the PC detects the peripheral and configures the necessary software.

Automatic configuration also eliminates the need to open the machine to add cards, set DIP switches, or resolve conflicts with interrupt requests (IRQ). No conflicts occur because each USB device uses the interrupt that is reserved for it. With USB's "hot-swapping" feature, users do not even have to shut down and restart their PCs to attach or remove peripherals. Devices can be added or removed as necessary, making USB especially suited for users who want to share peripherals between desktop and laptop computers.

Device Connections

USB also allows many peripherals to be connected to a computer at one time. With USB, up to 127 devices can be daisy-chained together from the same computer port. USB also sends power through its connections. USB distributes electrical power to peripherals by allowing a PC to automatically sense the amount of power required and deliver it to each device.

The 500 milliamps (mA) of current is adequate to run many scanners, digital cameras, mice, or keyboards. Most available USB peripherals are small and make modest electrical demands, so they can draw their power from the USB connection. But the power from a single port can diminish as more devices are added. The solution is to use a powered USB hub, which provides additional current—enough to supply either four or seven additional USB ports. This USB feature eliminates those bulky power supply boxes that come with many peripherals.

USB connections allow data to flow both ways between a PC and peripherals. This means the PC can be used to control peripherals in new and creative ways. For example, a PC can automatically manage a telephone call center to maintain voice, fax, and data mailboxes; screen and forward calls; and even deliver a variety of selected outgoing messages.

USB also provides benefits for multimedia-equipped PCs. USB allows streaming of digital data directly from the sound source to the subwoofer, where digital-to-audio conversion takes place. By taking the digital-to-audio conversion out of the PC box, the audio signals avoid internal PC distortion activities (e.g., CD drive, fans, and RF signals). As a result, the audio reproduction is virtually distortion-free. USB also allows PCs to tune a USB-compliant speaker system to match the acoustics of the listening environment.

System Requirements

Computers with Windows 95 must run the SR2.1 release to tap into USB. Windows 98 has built-in USB support, but Windows NT 4.0 does not. Windows 2000 (the former NT 5.0) has full USB support. USB-ready computers come with one or two rectangular USB ports alongside the usual parallel and serial ports. New laptops usually have only one USB port. If users experience problems when daisy-chaining multiple devices off the same USB port, they can increase the number of USB ports by using an inexpensive USB hub. These usually come with four USB ports. To use the hub, however, it must be plugged into a computer that already has a USB port.

Older Windows 95 PCs lacking USB ports can be upgraded with the addition of a PCI-based card. These cards typically provide two or four full-powered USB ports. The card installation is plug-and-play, meaning that Windows 95 will automatically recognize it and install the necessary updates and drivers.

Older peripherals can be connected to USB ports with cables or adapters. For example, a parallel-to-USB cable or adapter can be used to connect a printer to a USB port, freeing the parallel port for use by other peripherals such as an Iomega ZIP drive.

SUMMARY

A standard called FireWire has been available longer than USB, but USB's less costly implementation pushed it ahead of FireWire in terms of vendor support. Now the two are viewed as complementary technologies. Systems equipped with both USB and FireWire are available, with USB handling the lower-speed peripherals—such as input, modem communications, imaging, audio, and telephony devices—and FireWire handling devices that demand greater bandwidth, such as video peripherals and mass storage units. Standard serial and parallel ports will continue to be supported long into the future to support the vast installed base of older peripherals.

See Also FireWire, Small Computer System Interface

UNIX

UNIX was originally developed by AT&T Bell Laboratories in the mid-1960s. One of the schools that received an early copy of UNIX was the University of California at Berkeley. With government funding, students added many features to UNIX, which began to take on an identity all its own and came to be known as BSD (Berkeley Software Distribution) UNIX. Today, companies such as Digital Equipment Corp. (now owned by Compaq Computer Corp.), Hewlett-Packard Co., IBM Corp., and Sun Microsystems use a version of BSD UNIX on their workstations.

The UNIX operating system is used by most Web servers today. The three most compelling features of UNIX are its portability, its support of multiple tasks and multiple users, and its integral TCP/IP networking capabilities.

The operating system's portability comes from the fact that it is written in the C programming language instead of assembly language. In the 1960s, this was a breakthrough in operating systems because it meant UNIX and UNIX-based applications could be moved from one type of computer to another, even if they had widely divergent architectures. Operating systems written in assembly language, on the other hand, were limited to running on the computers of specific manufacturers.

Unlike DOS, UNIX was designed as a multitasking operating system, which allowed users to perform more than one operation at a time. These tasks or processes can be run in the background so users can continue with other activities. For example, a user could sort a file or make a calculation in the background and at the same time edit or print a file, or check e-mail. Of course, UNIX is no longer the only multitasking operating system; Microsoft Windows and IBM's OS/2 also support multitasking.

However, unlike Windows and OS/2, UNIX is a multiuser operating system. Many users can have simultaneous access to the same machine. The multiuser capability allows groups of people to work easily together, sharing files and utilities. Of course, to operate effectively, every multiuser system must have a system administrator to establish appropriate procedures. Among other things, the system administrator adds and removes user accounts, backs up and restores files, changes the system configuration, manages security, and takes care of troubleshooting and performance tuning.

TCP/IP networking capabilities are built into the UNIX operating system, including applications such as FTP, SMTP, and Telnet, which have been developed and refined over the lifetime of UNIX. Today, many of these tools have become standards in non-UNIX systems as well, including Windows and Macintosh. Even the more esoteric applications, such as Ping, are available as shareware in Windows and Macintosh versions.

In fact, UNIX and TCP/IP are very much related. Much of the development for each was conducted at the University of California at Berkeley. In 1983, UCB released a version of UNIX that incorporated TCP/IP as an integral part of the operating system. This made it easier for UNIX systems to be networked over the growing ARPANET funded and operated by the Department of Defense. Until it was dismantled in 1990, ARPANET served mainly the military, defense contractors, and academic researchers. Of course, TCP/IP is the foundation of the worldwide Internet, which has been undergoing increasing commercialization since the early 1990s.

SUMMARY ■ ■ ■ ■ ■ ■ ■ ■

UNIX is a powerful operating system that contains a rich set of features, including integral TCP/IP networking. Much of the migration of UNIX functionality to Windows and Macintosh platforms is directly related to the growing popularity of the TCP/IP-based Internet. For its part, TCP/IP has evolved to become one of the most mature of all networking environments; in fact, TCP/IP and UNIX have provided a model of multivendor interconnectivity that really works. While UNIX's value comes from its ability to run on different computer platforms, TCP/IP's value comes from its ability to provide connectivity among different platforms, ranging from microcomputers to minicomputers to mainframes.

See Also **Transmission Control Protocol/Internet Protocol, Windows NT**

User Datagram Protocol

User Datagram Protocol (UDP) is a low-overhead message exchange service. While the Transmission Control Protocol (TCP) offers assured delivery of Internet Protocol (IP) packets, it does so at the price of more overhead. UDP, on the other hand, functions with minimum overhead; it merely passes individual query-response messages to IP for transmission on a best-efforts basis. Since IP is not reliable, there is no guarantee of delivery. If a response is not received within a reasonable time, it is up to the application to reissue the query. If duplicate queries arrive at the server, the server recognizes the duplicate and discards it.

Despite its low reliability, UDP is very useful for certain types of applications, such as quick database lookups. For example, the Domain Name System (DNS) on the Internet consists of a set of distributed databases that provide a service that translates between system names and their IP addresses. For simple messaging between applications and these network resources, UDP is adequate.

Header Format

The UDP header consists of the following fields (Figure 127):

SOURCE PORT (16 BITS) This field identifies the source port number.

DESTINATION PORT (16 BITS) This field identifies the destination port number.

LENGTH (16 BITS) This field indicates the total length of the UDP header and data portion of the message.

Figure 127
The UDP header.

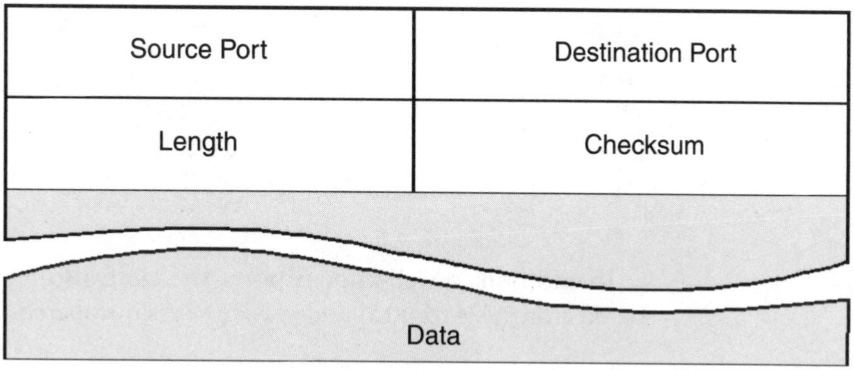

Source Port	Destination Port
Length	Checksum
Data	

CHECKSUM (16 BITS)   This field validates the contents of a UDP message, and is optional. If it is not computed for the request, it can still be included in the response.

Port Assignments

Applications using UDP communicate through a specified numbered port that can support multiple virtual connections called *sockets*. A socket is an IP address and port, and a pair of sockets (source and destination) forms a connection. One socket can be involved in multiple connections.

Registered, "well-known" ports are numbered from 0 to 1023. Telnet, for example, always uses port 23 for communications, while FTP uses port 21. These ports are assigned by the Internet Assigned Numbers Authority (IANA) and on most systems can be used only by system (or root) processes or by programs executed by privileged users. Other examples of UDP well-known ports are:

Service	Port	Description
Users	11	Shows all users on a remote system
Quote	17	Returns a "quote of the day"
Mail	25	Used for electronic mail via SMTP
Domain Name Server	53	Translates system names and their IP addresses
BOOTpc	68	Client port used to receive configuration information
TFTP	69	Trivial File Transfer Protocol used for initializing diskless workstations
World Wide Web	80	Provides access to the Web via the HyperText Transfer Protocol (HTTP)
snagas	108	Provides access to an SNA Gateway Access Server
nntp	119	Provides access to a newsgroup via the Network News Transfer Protocol (NNTP)
SNMP	161	Used to receive network management queries via the Simple Network Management Protocol

In addition to well-known ports, there are also registered ports numbered from 1024 to 49151 and private ports numbered from 49152 to 65535.

SUMMARY

While UDP is not a very reliable form of communication, it is adequate for simple messaging between applications, such as the query-response transactions that take place between clients and servers within the DNS system on the Internet. The reason these lookups are so fast is that UDP is not burdened with overhead functions; it merely passes individual messages to IP for transmission on a best-efforts basis.

See Also **Internetwork Packet Exchange**

Video over Frame Relay

Companies want to use frame relay for video because it is the most widely deployed data transmission service in the world. It is also less expensive than private leased lines, ISDN, or ATM for the bursty kinds of data companies send most of the time. But video requires a continuous data stream in order to deliver images without jitter and distortion.

Today, virtually every router can deliver prioritization at the frame level, and every major carrier offers a service-level agreement (SLA) for frame relay that guarantees minimum availability, minimum latency, and other performance guarantees. These service levels can be monitored by carrier-provided or third-party management tools. In providing performance guarantees, video can be safely passed over frame relay networks with a high degree of reliability.

Another way network administrators can ensure picture quality is by reducing packet size. Frame relay allows the payload portion of packets to be adjusted to carry larger or smaller amounts of information. Since distortion is directly proportional to the amount of information dropped by a packet, reducing packet size reduces the potential for distortion.

Frame relay's most compelling feature for video is its cost advantage, which can be dramatic, especially when compared to leased lines or switched ISDN. Three hours of a 384-Kbps interconnection over ISDN might cost about $2,000, versus about $800 for a 384-Kbps frame relay link.

SUMMARY

Frame relay has proven itself to be highly reliable and cost-effective for a variety of applications. In less than a decade it has become a risk-free choice, especially for organizations that maintain remote sites and are seeking a flexible, reliable, high-speed WAN solution to provide data con-

640

nectivity between offices. Even applications very sensitive to delay, such as video, can be run reliably over frame relay links.
See Also **Frame Relay, Voice over Frame Relay**

Videoconferencing Systems

Videoconferencing has become an efficient and cost-effective means of bringing people together over vast distances. The technology enables people in diverse locations to participate in meetings, consultation, and continuing education—without resorting to expensive, time-consuming travel.

Applications

Videoconferencing systems have been deployed in a variety of applications, including distance learning, corporate workgroup collaboration, and telemedicine. One noteworthy application of videoconferencing comes from the healthcare industry. With videoconferencing technology, healthcare organizations can reduce the cost of doing business with remote offices, make valuable clinical resources available to patients too far away or sick to travel, strengthen relationships with the referral base by enabling remote diagnosis and consulting, reach out to new populations for revenue generation, and take advantage of resource-efficient education programs. The technology not only enables participants miles apart to talk face-to-face but also allows them to share notes, charts, and other materials—even work on spreadsheets, databases, and other computer applications simultaneously.

Types of Systems

Videoconference systems come in several configurations, from individual desktop installations to portable, mid-range roll-about units to full-scale conference room set-ups. There are also table-top group videoconferencing systems, with a footprint about the size of a notebook, which can interconnect up to four sites. Videophones are also available that are suited for personal, one-to-one communication.

ROOM-BASED SYSTEMS   Room-based systems usually entail the use of one or more large screens in a dedicated meeting room equipped with environmental controls. The system components—screens, cameras, microphones, and auxiliary equipment—can be permanently installed since they will not be moved to another room or building. These systems provide high-quality video and synchronized audio. Prices for room-based systems start at around $70,000, but can go much higher as more sophisticated equipment and features are added.

MID-RANGE SYSTEMS   If an organization does not rely extensively on videoconferencing but considers it an important capability to have when the occasion arises, a mid-range or roll-about videoconferencing system is a viable solution. Typically, these portable systems use one screen and no more than two cameras and three microphones. Prices range from $15,000 to $50,000, depending on options.

DESKTOP SYSTEMS   Desktop videoconferencing is becoming popular because it allows organizations to leverage existing assets—making videoconferencing just another application running on the desktop. When equipped for videoconferencing, desktop machines can be used for video mail over LANs as well as videoconferencing over WANs. The ability to use widely available Ethernet networks makes videoconferencing technology more accessible, less costly, and easy to deploy. Data sharing can be accomplished through an optional whiteboard capability. Fully equipped desktop videoconferencing systems are available in the range of $1200 to $5000 per unit, depending on options. At the high end, the vendor provides PCs already configured for videoconferencing, eliminating the need and difficulty in installing and configuring add-in components.

Of course, anyone with the proper expertise can add videoconferencing capability to an existing PC with appropriate hardware and software. The following components are necessary:

- Modem or ISDN terminal adapter to establish a dial-up connection to another similarly equipped computer

- Microphone for collecting the speech signals of the user

- Sound card for digitally processing speech

- Camera for collecting the video image of users

- Videoconferencing software to compress the digitized speech and video components and packetize them for transmission on the network

- Speakers or headset for listening to the audio portion of the videoconference

TABLETOP SYSTEMS   An economical alternative to mid-range videoconferencing systems for small workgroups is a tabletop system. The unit, including camera, is typically no bigger than a laptop computer, yet provides high-quality audio and video over LAN or ISDN connections. The unit accepts video input from a VCR or external document cameras. When not being used for videoconferencing among up to four locations, the unit can function as a stand-alone speakerphone or presentation system.

VIDEOPHONES Videophones are used for one-on-one communication. This type of equipment satisfies the desire for impulse videoconferencing. The videophone unit includes a small screen, built-in camera, video coder/decoder (codec), audio system, and keypad. The handset lets the unit work as an ordinary phone as well as a videoconferencing system. Prices start at about $1000 for models that work over ordinary phone lines.

Features

Videoconferencing systems are continually being enhanced with new features and add-on devices, including:

AUTOMATED CONNECTION SETUP Automatically determines the service provider identification numbers (SPIDs) and switch type for quick ISDN installation and provides auto-IP address configuration for quick setup on a LAN.

VOICE TRACKING CAMERA Automatically locks onto the voice of the person speaking, while ignoring background noise.

ANTICIPATING USER INTERFACE When the remote control is picked up, the picture-in-picture (PIP) windows and various help screens appear.

EMBEDDED WEB CAPABILITY AND ETHERNET HUB Facilitates remote system management and diagnostics, software upgrades, and presentation via the Internet or corporate intranet.

INTEGRATED PRESENTATION SYSTEM Presentations, such as those created in Microsoft PowerPoint, can be displayed and presented to the remote site while simultaneously being presented to other remote users connected via the Internet/intranet.

GRAPHICS TABLET Captures and saves images from a variety of sources so they can be organized into presentations and shared with other conference participants.

ELECTRONIC WHITEBOARD Used primarily with dual monitor systems, displays meeting notes on one monitor while the presenter's image remains visible on the other monitor.

CD-ROM INTEGRATION   Allows multimedia content from CD-ROM to be shared with conference participants.

ONE-BUTTON ADDRESS BOOK DIALING   Permits easy setup of point-to-point or multipoint calls.

Multipoint Control Units

Videoconferencing among more than two locations requires a multipoint control unit (MCU), a switch that connects video signals among all locations, enabling participants to see each other, converse, and work simultaneously on the same document or view the same graphic. The multipoint conference is set up and controlled from a management console connected to the MCU.

The MCU makes it relatively easy to set up and manage conference calls between multiple sites. Among the features that facilitate multipoint conferencing are the following:

MEET-ME   Enables participants to enter a conference by dialing an assigned number at a prearranged time.

DIAL-OUT   Automatically dials out to other locations and add them to the conference at prearranged times and dates.

AUDIO ADD-ON   Allows participants to hear or speak to others who do not have video equipment (or compatible video equipment) at their location.

TONE NOTIFICATION   Provides special tones to alert participants when a person is joining or leaving the conference, and when the conference is about to end.

DYNAMIC RESIZING AND TONE EXTENSION   Allows locations to be added or deleted, and the duration of a bandwidth reservation to be extended, during a conference without the session having to be restarted.

INTEGRATED SCHEDULING   Permits video conferences to be set up and scheduled days, months, or a year in advance using an integral calendar or scheduler application. The MCU automatically reserves the required bandwidth, configures itself at the designated time, and dials out to participating sites to establish the conference.

The MCU also provides the means to precisely control the video conference in terms of who is seeing what at any given time. Some of the advanced conference control features of MCUs are:

VOICE-ACTIVATED SWITCHING Allows all participants to see the person speaking, while the speaker sees the last person who spoke.

CONTRIBUTOR MODE Works with voice-activated switching to allow a single presenter to be shown exclusively on a conference.

CHAIR CONTROL Allows a person to request or relinquish control, choose the broadcaster, and drop a site or the conference.

PRESENTATION OR LECTURE MODE Allows a speaker to make a presentation and question participants in several locations. Participants can see the presenter at all times, but the presenter sees whoever is speaking.

MODERATOR CONTROL Allows a moderator to select which person or site appears onscreen at any given time.

ROLL CALL Allows a conference moderator to switch to each participant for the purpose of introducing them to others or to screen the conference for security purposes.

SUBCONFERENCING Allows the conference operator to transfer participants into and out of separate, private conferences associated with the meeting, without having them disconnect and reconnect.

BROADCAST WITH AUTOMATIC SCAN Allows participants to see the presenter at all times. To gauge audience reaction, the speaker sees participants in each location on a timed, predetermined basis.

Depending on the choice of MCU, a number of network connectivity options are available. Generally, the MCU can be connected to the network either directly via T1 leased lines or ISDN PRI trunks, or indirectly through a digital PBX. Some MCUs can be connected to the network using dual 56-Kbps lines or ISDN BRI. Others connect videoconferencing systems over the WAN using any mix of private and carrier-provided facilities, or any mix of switched services regardless of the carrier (Figure 128). This capability offers the most flexibility in setting up multipoint conferences.

For users who do not have high-speed links, some MCUs include an inverse multiplexing capability that combines multiple 56/64-Kbps channels into a single, higher-speed, 384-Kbps channel on a demand basis, thus improving video quality. Most MCUs use the inverse multiplexing method standardized by the Bandwidth on Demand Interoperability Group (BONDING).

While most MCUs are designed for use on WANs, some MCUs are available for LANs. These devices are useful for providing multipoint desktop videoconferencing over local networks within a campus environment or among many floors in a high-rise building.

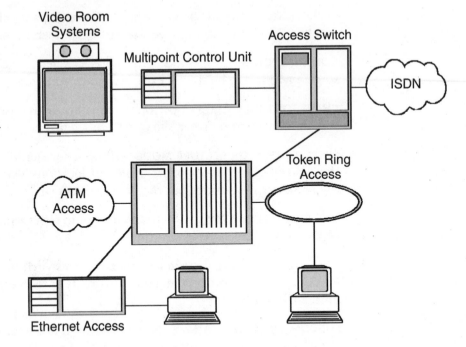

Figure 128
A typical video
conferencing
arrangement
implemented by a
multipoint control
unit (MCU).

Standards

Recommendation H.320 is a set of videoconferencing standards developed by the ITU's Telecommunications Study Group 15. H.320 is the umbrella standard that defines the operating modes and transmission speeds for videoconferencing system codecs, including the procedures for call setup, call tear-down, and conference control. The codecs that comply with H.320 are interoperable with those of different manufacturers, delivering a common level of performance.

The H.320 videoconferencing standard includes associated specifications that define how the videoconferencing products of different vendors interoperate. Among the key H.320 standards are:

H.322 A standard for LAN-based videoconferencing with guaranteed bandwidth.

H.323 A standard for LAN-based videoconferencing with nonguaranteed bandwidth (non-isochronous), such as Ethernet or token ring. H.323 specifies G.711 as the mandatory speech codec standard.

H.324 A standard set for videoconferencing over high-speed modem connections using standard telephone lines. H.324 specifies G.723 as the mandatory speech codec standard.

Some videoconference systems include integral gateways that convert an H.320 signal to an H.323 signal and vice versa, enabling two users to video-conference even if one has an ISDN connection and the other an Internet connection.

Another important component of H.320 is the H.261 video compression specification, which defines how digital information is coded and decoded. H.261 also allows the signals to be transmitted at a variety of data rates, from 64 Kbps to 2.048 Mbps in increments of 64 Kbps. H.261 also defines two resolutions. One is the Common Intermediate Format (CIF), a format usually used in high-end room systems, which provides the highest resolution at 352 x 288 pixels. The other is the Quarter Common Intermediate Format (QCIF), a format used by most desktop video-conferencing systems and videophones, which provides lower resolution at 176 x 144 pixels.

Another standard, H.230, describes the signals used by conferencing systems and MCUs to communicate during a conference. These signals enable conferencing systems and the MCU(s) to exchange instructions and status information during the initiation of a conference and while the conference is in progress. A related standard, Recommendation H.243, defines the basic MCU procedures for establishing and controlling communication between three or more videoconferencing systems using digital channels up to 2 Mbps.

There is also a set of ITU recommendations that standardize audio compression for videoconferencing equipment. The three key standards in this area are:

G.711 Defines the requirements for 64-Kbps audio. This is the least compressed and offers the highest quality audio.

G.722 Defines 2:1 audio compression at 32 Kbps.

G.728 Defines 4:1 audio compression at 16 Kbps.
The reason compression is important is that it squeezes the audio component of a videoconference into a smaller increment of bandwidth, freeing more of the available bandwidth for the video component. This results in higher-quality video, without appreciably diminishing audio.

SUMMARY

Videoconferencing is finally taking its place as a strategically significant corporate communications tool. The ultimate low-cost and ubiquitous method of videoconferencing may be the Internet. Currently, image quality—and its synchronization with an audio component—over the Internet is not the same as what can be achieved over high-speed LANs and digital WAN services like ISDN. However, corporations with well-managed intranets or subscriptions to a carrier-managed Virtual Private Network (VPN) service can overcome the poor performance of the public Internet to experience high-quality videoconferences at nearly 30 frames per second. *See Also* **Audioconferencing Systems**

Virtual LANs

The bandwidth requirements of shared media communications networks, such as Ethernet, have increased dramatically in recent years as applications have grown in sophistication to take advantage of the speed and power of today's computer systems and the size and complexity of today's networks. To meet increased bandwidth requirements, changes have been made in Ethernet signaling to attain gigabit-per-second speed. In addition, new strategies have been developed to ease transmission bottlenecks and facilitate network management. One of these strategies is that of the virtual LAN, or VLAN.

Under the VLAN concept, all Ethernet clients connected through a switch can be organized into logical subgroups. This logical grouping of Ethernet clients can be used by the switch to reduce—on a per-port basis—the number of outgoing packets that are queued for transmission and also ease the task of network management.

In the case of an Ethernet switch that does not support or recognize VLAN subgroupings, any broadcast traffic received by any of its ports must be forwarded to all other ports within the switch. Because of this, large amounts of broadcast traffic can cause forwarding buffers within a switch to become overloaded, reducing its switching efficiency and in extreme cases, causing packet loss.

By creating VLANs to logically group clients within a network, a VLAN-aware switch can use this logical grouping to selectively filter broadcast traffic based upon its VLAN identification number. This results in broadcast traffic being forwarded only between members of the same VLAN, thereby reducing the amount of packet forwarding within the switch. This can substantially increase a switch's performance, which is of benefit to all users.

Network Management

VLANs can also ease network management. Without VLANs, network administrators would typically put all workstations within a workgroup on the same physical LAN segment. Modifying workgroups requires physically changing the wiring connection of a workstation from one LAN segment to another.

One of the biggest advantages of VLANs is that when a computer is physically moved to another location, it can stay on the same VLAN without any hardware reconfiguration. The network administrator can also use software to add and delete workstations from a workgroup. Network management software keeps track of relating the virtual picture of the local-area network with the actual physical picture. Even the combining of workgroups into a single larger workgroup can be software controlled.

Another aspect of network management is to ensure security between workgroups. Without the creation of VLANs, broadcast traffic will be seen by all devices interconnected through hubs, bridges, or switches. VLANs can be used to improve this situation since broadcasts can be restricted to only those devices within a specific subgroup—that is, all devices with the same VLAN identification number.

A VLAN may be locally configured and monitored through UNIX commands, a user-friendly Windows interface or remotely through a Web browser. A graphical user interface has obvious advantages in that VLANs can be created by simply defining collections of member ports selected from a directory of switches and switch ports provided in a VLAN management window.

Once the selections are made, the management application automatically sends the appropriate SNMP information to the list of affected switches and assigns the designated ports to that VLAN. Members can be attached to VLANs by dragging a switch port to a defined VLAN name. Groups of members can then be moved from one VLAN to another by dragging the group name, or any member of the group, and dropping it on the desired VLAN name.

VLANs can also be managed through a Web browser, enabling them to be configured and troubleshooted from any computer equipped with a standard Web browser, freeing the user from having to be at an SNMP-based management platform, such as Hewlett-Packard is Co.'s OpenView or IBM's NetView/6000.

Special monitoring tools provide a comprehensive view into VLAN traffic—not only between switches, but on all VLAN member ports within a switch—enabling greater control, security, and management of network resources.

By applying a RMON2 probe to a specific VLAN, for example, network managers can monitor protocols and applications for fault isolation, real-time troubleshooting, and effective capacity planning. Among other things, RMON2 provides traffic statistics, Host, Matrix, and TopN tables for both the network and application layers. This information is important for determining how traffic patterns are evolving, and for ensuring that users and resources are placed in the correct subgroup to optimize performance and reduce costs.

Standards

IEEE 802.1Q provides the industry standard for VLAN interoperability. The standard determines the VLAN frame format, membership rules and management procedures, which enable the switches of different vendors to participate in multivendor VLANs. Some vendors have built their switches so they can be upgraded via software to comply with the specification. Other switch vendors cannot change their switching code, so their products will likely require a more expensive hardware upgrade.

SUMMARY

VLANs address the uncontrolled proliferation of broadcast and multicast traffic on large networks. The benefits of VLANs come down to increasing the efficiency, speed and manageability of networks. VLANs are configured through software rather than hardware, which makes them extremely flexible. Graphical management tools and Web accessibility make setting up and changing VLANs an easier and faster process than having to physically configure the switch to organize users into appropriate subgroups.
See Also **Switched LANs**

Virtual Private Networks

Carrier-provided networks that appear and function like private networks are referred to as "virtual private networks," or VPNs. Originally, VPNs provided voice services and were an economical alternative to private leased lines. Today, VPNs are increasingly data oriented, enabling companies to link geographically separated LANs, or remote users and LANs, over the public Internet.

VPNs for Voice

AT&T introduced the first VPN service in 1985. Its voice-oriented Software-Defined Network (SDN) was offered as an inexpensive alternative to private lines. Since then, VPNs have added more functionality and expanded globally. Today, the big three carriers—AT&T (Software-Defined Network), MCI WorldCom (Vnet), and Sprint (VPN Service)—each offer virtual private networks. In the case of AT&T, various services—including data and cellular calls—can be combined under a single service umbrella, expanding opportunities for cost savings within a single discount plan.

VPNs allow users to create their own private networks by drawing on the intelligence embedded in the carrier's network. This "intelligence" is actually derived from software programs and configuration information that reside in various switch points and databases throughout the network. All services and features are defined in software, allowing users to control how their networks are configured. Without carrier involvement, you can reconfigure an entire network by changing a few parameters in a network database from an on-premises management station.

For example, a feature called *flexible routing* enables network managers to reroute calls to alternate corporate locations when a local node experiences a temporary outage or peak-hour traffic congestion. This feature can also be used to extend customer service business hours across multiple time zones.

The ubiquity and user-friendliness of the World Wide Web has prompted long-distance carriers to offer customers a view into their VPNs. Among the long-distance carriers that offer Web-based management tools are AT&T, MCI WorldCom, and Sprint. AT&T, for example, offers a suite of Web-based management utilities under the name Customer Direct.

One of these utilities is Network Call Attempt Monitor, with which customers can access their network data at any time from any Internet-connected PC that runs a JavaScript-enabled Web browser. Another utility lets customers submit trouble tickets over the Web—for both voice and data networks. This greatly speeds the response time to problems on customer networks without tying up help desk phone lines. AT&T provides storage for customers' network data and operates the servers where customer network information is hosted. Security of customer data is provided through the use of individual user login names and passwords, as well as through data encryption. Customers of AT&T's network management services do not have to pay extra to access this service over the Internet.

VPNs for Data

Although the first VPNs were voice-oriented, they could also handle low-speed data. But in early 1997, a wholly new trend emerged in which private data was routed between corporate locations worldwide over carrier-provided IP networks. The carrier is responsible for network management, security and quality of service issues. In many cases, service-level guarantees are available from the carriers, which provide users with credits for poor performance. These IP-based VPNs are now among the fastest growing category of data service.

IP-based VPNs are an increasingly popular option for enhancing communication between corporate locations over the Internet. They can also be used for electronic commerce and to make enterprise applications available to customers and strategic partners worldwide. Basically, this type of data service lets business users carve out their own IP-based WANs within the carrier's high-speed Internet backbone. The major carriers provide service-level guarantees to overcome concerns about latency and other quality-of-service issues traditionally associated with the public Internet.

Implementation

Several protocols are used to implement IP-based VPNs: Point-to-Point Tunneling Protocol (PPTP), Layer 2 Forwarding (L2F) protocol, Layer 2 Tunneling Protocol (L2TP), and IP Security (IPSec) protocol.

PPTP supports flow control and multiprotocol tunneling between two points (Figure 129). L2F also supports multiple protocols, but can be used to create tunnels between multiple locations. Both protocols are vendor-driven: PPTP is backed by Microsoft and 3Com, among others, and L2F is backed by Cisco Systems, Nortel Networks, and Shiva. Since PPTP and L2F are not interoperable, L2TP is under consideration as an IETF standard. L2TP is a combination of PPTP and L2F, and offers the advantage of interoperability. The drawback to all three tunneling protocols, however, is their lack of integrated authentication, encryption, and integrity features.

IETF's IPSec, on the other hand, provides for packet-by-packet authentication, encryption, and integrity. Authentication positively identifies the sender. Encryption allows only the intended receiver to read the data. Integrity guarantees that no third party has tampered with the packet stream. These security functions must be applied to every IP packet because Layer 3 protocols such as IP are stateless; that is, there is no way to be sure whether a packet is really associated with a particular connection. Higher-layer protocols such as TCP are stateful, but connection information can be easily duplicated or "spoofed" by knowledgeable hackers. The key limitation of IPSec is that it can only carry IP packets, whereas the other three protocols support IPX and AppleTalk, as well as IP.

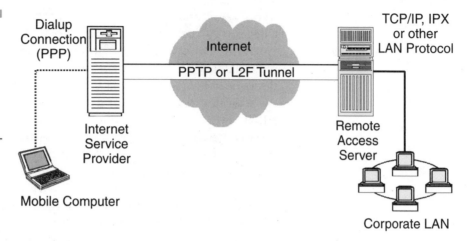

Figure 129
Tunneling offers
the means of
transporting
multiprotocol
traffic in a private
session over the
Internet.

Carrier Services

The major carriers offer some compelling features with their IP-based VPN offerings. To encourage customers to consider their services, carriers also are making promises about trouble response time, network uptime, and dial port availability. The overriding concern of corporate managers, however, is end-to-end latency. If the VPN cannot get the packets through, then it is of little importance that the network is available 100 percent of the time.

Accordingly, some service providers offer a latency guarantee. Uunet, for example, guarantees 150 milliseconds for latency and 99.9-percent network availability. The credit is 25 percent of the customer's daily bill if Uunet fails to meet one criteria, and 50 percent of the daily bill if it fails to meet both criteria.

AT&T WorldNet offers dedicated and dial-up VPN service. Dedicated VPN service is provided on AT&T's frame relay network and allows customers to create closed user groups (CUGs) for intranets and extranets, while the dial-up VPN service provides remote users with access to corporate LANs. AT&T enhances the reliability of its VPN with secure nodes and PoPs (points of presence), a self-restoring FASTAR system that minimizes the impact of cable cuts, redundant router configurations, redundant ATM switching architecture, and alternate access paths for Internet-bound packets. AT&T offers one free day of service if the VPN goes down for more than 15 minutes. Network reliability for all Internet connections is 99.7 percent, which is slightly less than the 99.97-percent reliability of the public network, but high enough to help convince organizations to place daily business applications on the Internet.

Because AT&T owns and controls all the IP backbone network equipment and facilities, it can exercise absolute control over its backbone, a key factor in ensuring reliability. AT&T maintains a physically secure Network Operations Center (NOC), as well as an identically equipped facility in a remote location. There, systems and software have been tuned to enable the NOC staff to detect, isolate, and fix any network troubles in a proactive fashion.

The IP backbone network is managed by both in-band and out-of-band network monitoring systems, allowing NOC technicians to monitor the real-time status of all network elements at all times: in-band via the network and out-of-band via secure dial-up. Although SNMP is the primary network management protocol used in monitoring the IP backbone network, customized alarm correlation and filtering software are also used to allow quick detection of network alarms, along with custom-built tools to monitor routing tables, routing protocols, and physical and logical circuits.

For security, AT&T uses RADIUS (Remote Authentication Dial-In User Service) servers and Novell directory services technology to validate the authenticity of users dialing into the network via the Challenge Handshake Authentication Protocol (CHAP). Packet filters are used to prevent source address spoofing, which blocks outsiders from entering its network to access closed user groups and client Web servers. All points of presence, modem pools, and authentication servers are in protected buildings.

SUMMARY

Whether voice- or data-oriented, VPNs combine the advantages of both private facilities and public services, drawing on the intelligence embedded in the carrier's network. With services and features defined in software and implemented via out-of-band signaling methods, users can configure and monitor their virtual networks from on-premises terminals and management systems. These capabilities, combined with service-level agreements, make VPNs attractive for data as well as voice—for regional, national, and international corporate locations—and portend success for VPNs long into the future.

See Also **Service-Level Agreements, Virtual LANs**

Voice Compression

To transmit voice over any type of digital network, speech must be sampled and encoded into bits. This is done with a coder. At the other end of the line, the bits are decoded to recreate the speech to its original form. The quality of the signal largely depends on the number of bits used to encode it. If an infinite number of bits were used, they could represent the signal exactly as it had been transmitted over the line. The problem with this approach is that bandwidth is limited, whether a voice conversation occurs over the public switched telephone network (PSTN) or a private corporate network of leased lines. Furthermore, the encoding/decoding process must be highly efficient, or a significant delay will result.

Digitizing Voice

Voice compression begins after the waveforms are digitized through a standard process called *pulse code modulation* (PCM), which is described in the ITU's G.711 specification. This process involves assigning a level to each sample at every 1/8000 second. Only eight bits are sent to encode each sample, so only 256 different levels may be encoded. This produces a rate of 64 Kbps, which is the basic channel within the T-carrier signal hierarchy. Digital services that rely on T-carrier technology, such as ISDN and Switched 56K, provide 56/64-Kbps bearer channels that are created using PCM.

The Adaptive Differential Pulse Code Modulation (ADPCM) compression technique, described under ITU standard G.726, uses four bits and a standard prediction algorithm. Since only four bits are used, the bandwidth of the compressed signal is only 32 Kbps, for a compression ratio of 2:1. This is an extremely simple compression algorithm, so there is not much delay in transmission. It also provides very good quality. ADPCM is often used on private T1 leased lines to double the number of available voice channels. This saves the organization from the expense of adding another T1 line to handle increases in voice traffic.

Other types of services, however, have bandwidth constraints that are so severe that voice sampling and encoding must be done at much lower bit rates. Examples of these services are digital cellular and IP telephony services. The challenge is to meet the bandwidth limitations of each type of service, but without sacrificing speech quality.

International Standards

Today's digital cellular and IP telephony services require highly efficient use of channel resources and storage space, while maintaining high-quality speech that is virtually indistinguishable from conventional phone service provided over the PSTN. A set of international standards are available that guarantee various levels of speech quality and facilitate interoperability between standards-compliant products.

The international umbrella standard under which the audio standards fit is H.323, which is recommended by the International Telecommunication Union (ITU). H.323 defines how audio/visual conferencing data is transmitted across networks. The audio component is addressed by several G.7xx voice codec specifications for toll-quality voice. For voice-over-IP (VoIP) applications, for example, the most commonly supported codec specifications are:

G.711 Describes the requirements for a codec using PCM of voice frequencies at 64 Kbps, providing toll quality voice on managed IP networks that have sufficient available bandwidth.

G.723.1 Describes the requirements for a dual-rate speech codec for multimedia communications (e.g., videoconferencing) transmitting at 5.3 and 6.3 Kbps. This codec provides near toll quality voice on managed IP networks.[4]

G.729A Describes the requirements for a low-complexity codec that transmits digitized and compressed voice at 8 Kbps. This codec provides toll quality voice on managed IP networks.

The specific codec to be used is negotiated on a call-by-call basis between IP/PSTN gateways using the H.245 control protocol. Among other things, the H.245 protocol provides for capability exchange, enabling the gateways to implement the same codec at the time the call is placed. The gateways may be configured to implement a specific codec at the time the call is established, based on predefined criteria, such as:

USE G.711 ONLY In this case, the G.711 codec will be used for all calls.

USE G.729 (A) ONLY In this case, the G.729 (A) codec will be used for all calls.

USE HIGHEST COMMON BIT RATE CODEC In this case, the codec that will provide the best voice quality is selected.

USE LOWEST COMMON BIT RATE CODEC In this case, the codec that will provide the lowest packet bandwidth requirement is selected.

---

4. The mean opinion score (MOS) used to rate the quality of speech codecs measures toll quality voice as having a top score of 4.0. With G.723.1, voice quality is rated at 3.98, which is only 2 percent less than that of analog telephone.

This capability exchange feature provides carriers and Internet service providers (ISPs) with the flexibility to offer different quality voice services at different price points. It also allows corporate customers to specify a preferred proprietary codec to support voice or a voice-enabled application through a corporate intranet or virtual private network (VPN).

Voice over IP

With more voice conversations and voice-enabled applications running over IP networks, there is a need for efficient low-bit-rate codecs to reduce the bandwidth required to deliver high quality speech. With a voice-enabled learning application, for example, voice could be compressed to 5.3 Kbps using an off-the-shelf codec that adheres to the G.723.1 specification. However, getting voice down to less than 3 Kbps so the application can be accessed over 28-Kbps dial-up connections—making it available to more potential users, but without compromising quality—requires more than a good low-bit-rate voice codec.

Even the best codec can break down in the face of packet loss, network jitter, and other difficult-to-predict conditions. Accordingly, there are companies that specialize in developing technologies designed specifically to address these issues. Voxware, for example, offers OEMs (original equipment manufacturers) a software development kit (SDK) that works with its codecs to reduce latency, improve overall speech and conversation quality, and dynamically adjust to real-time network conditions. The key features of the toolkit include:

ADAPTIVE FRAME LOSS CONCEALMENT (AFLC) This algorithm, closely linked to the Voxware codec, synthesizes missing speech data by extracting more information from the speech stream for accurate reconstruction. This capability is exploited in the AFLC algorithm, enabling the SDK to be tolerant of higher packet delay and loss conditions. This, in turn, allows systems built with Voxware SDKs to use much smaller jitter buffers, thereby greatly reducing latency.

AUTOMATIC JITTER BUFFER MANAGEMENT (AJBM) Automatically adjusts the size of the system's jitter buffer. The AJBM system collects packet arrival data, analyzes it, and makes jitter buffer control decisions based on another algorithm. Since the jitter buffer is perhaps the most crucial element of a VoIP system, this capability greatly improves the overall quality of the system's sound. By minimizing the jitter buffer size, conversations will demonstrate greatly reduced latency without sacrificing the intelligibility of individual words.

AUTOMATIC TRANSMISSION CONTROL Establishes a communication path between the receiver (decoder side of the transmission) and the transmitter (encoder side of the transmission). Since the receiver understands the condition of the network—by evaluating packet delays and other IP network performance measures—it provides instructions to the transmitter/encoder about how the data should be packetized and shipped over the IP network. This capability works in conjunction with the Voxware codecs to actually control the bit rate of the data stream sent over the network.

SUMMARY

The goal of voice transfer over any kind of network is to be able to send as much information about the voice as possible, providing good reproduction, with as few bits as possible. The voice must first be digitized, then compressed and transmitted. The usual first step is to nonuniformly quantize the signal using PCM. Compression involves various other techniques that use more complex algorithms to represent the signal as a set of parameters, which takes fewer bits to encode. Due to the compression techniques used, perfect quality is never achieved. Using the best algorithms, however, the quality may be extremely close to telephone quality. The best algorithms are not necessarily those that are internationally accepted as standards. Some proprietary algorithms perform better that the standards, but at the expense of interoperability with other vendors' products. Fortunately, many of these products can default to a standard compression method when communication occurs between the systems of different vendors.

See Also **Data Compression**

Voice and Data Integration

Voice-data integration entails the transmission of multiple voice and data channels over a single digital transport with the objective of eliminating separate networks and/or making more efficient use of available bandwidth to reduce recurring costs.

Parallel Networks

Most companies today use two different networks with different infrastructures to handle their voice and data communications needs. Typically, a digital PBX provides full-featured voice communication and switches calls to external parties over the public switched telephone network (PSTN). The PBXs at different company sites may be linked over leased lines, interconnected through ISDN permanent virtual connections (PVCs) or through a carrier's virtual private network (VPN) service at discounted rates based on call volume.

Regardless of the method of interconnection, "on-net" users throughout a company have access to the sophisticated features provided by the company's PBXs, such as conference calls, voice mail, and universal inhouse numbers. At the same time, limiting intracompany calls to a PBX network is an effective means of containing telecommunications costs, since most calls do not go out over the more expensive PSTN.

Data communication, on the other hand, usually takes place over a completely independent network, which links a variety of devices including PCs, servers, and hosts over a local-area network installed on the company's premises. Wide-area connections between LANs are achieved in various ways, including public X.25 networks and, increasingly, frame relay networks, while leased lines are also used where there is a high volume of data traffic or where time-critical sessions are involved, as in SNA terminal-to-host communication.

WAN charges are made up of volume-sensitive parts (for X.25, frame relay), usage-sensitive parts (based on call duration and distance of voice calls), and flat-rate basic subscription charges (for the use of PVCs or leased lines). Clearly, integrating voice and data networks can improve the usage of WAN links and result in considerable reductions in the monthly costs for these lines.

Another reason organizations are considering voice-data integration concerns the rise of multimedia applications. Since these applications consist of synchronized voice, data, and video components, effective deployment is dependent on networks that are capable of handling this type of traffic. For these reasons, more organizations are showing interest in the various options for integrating voice and data.

Integration Solutions

Over the years, different network technologies and services have been used to integrate voice and data, including private lines, frame relay, IP, and ATM. Each option has strengths and weaknesses that must be evaluated based on factors such as traffic volume, the applications involved, the expected level of performance by end users, and management complexity. Of course, cost also plays a key role in the decision—both the up front cost of implementation and the anticipated timeframe for return on investment.

Circuit-Switched PSTN

The circuit-switched PSTN dominates as the ultimate example of ubiquity and global connectivity. Over the past two decades, these voice networks have experienced significant growth in the transport of images, primarily fax messages. The advent of voice network interface modems has also led to the rapid growth of data transmissions over these same voice networks. E-mail, file transfers, and bulletin board downloads are routinely done over voice networks. The Internet is also accessed over voice networks, so much so that the telephone companies have complained to the FCC that their local central office switch ports are tied up for long periods, which greatly increases the rate of unsuccessful call attempts among telephone users.

PSTNs, having been designed for voice, have only a limited capability to handle integrated voice and data. Although interoffice trunks are digital, most local lines are not. A modem (modulator-demodulator) converts the digital signals generated by a computer into analog signals suitable for transmission over dial-up telephone lines. Another modem, located at the receiving end of the transmission, converts the analog signals back into digital form for manipulation by the data recipient.

Modems are often required to transfer files, access bulletin boards and Web sites, send electronic mail, and connect to host computers with various terminal emulation software. They support multimedia traffic to the extent that voice, data, and video are conveyed as packets within the data stream. But because the analog local loop is susceptible to a variety of impairments that can affect data, the transmission speed is limited to about 56 Kbps.

When downloading the contents of a multimedia Web page, for example, all the information will not arrive at the same time. Some text might arrive first and then stop until some bit-intensive images load. A video component will usually be grainy and jerky, while the audio component may be clipped and not be well synchronized with the video component.

ISDN

Where digital local lines are available, ISDN can be used to achieve higher and more reliable transmission than modems can currently provide. ISDN was the first PSTN service promoted as the means to integrate voice and data. With ISDN, the available bearer channels could be used for voice, data, or video on call-by-call basis for multimedia applications such as videoconferencing, distance learning, and telemedicine. With a network device called an *inverse multiplexer*, additional ISDN channels can be called up as needed or removed when no longer required.

Multirate ISDN service lets organizations select appropriate increments of bandwidth on a per-call basis rather than leasing more expensive, high-speed, dedicated circuits. The service, known as Multirate ISDN, Nx64 Kbps, and SWF-DS1 (Switched Fractional DS1), is more economical for customers because they use only the bandwidth they need. Speeds, in increments of 64 Kbps, are available between 128 Kbps and 1.536 Mbps. Multirate ISDN eliminates the need for a separate inverse multiplexer for building the required amount of circuit bandwidth, although inverse multiplexers support other types of circuits as well, providing users with additional flexibility.

At the individual level, a PC equipped with an ISDN/modem card can support up to two analog devices, such as a telephone and a fax machine. Users can talk on the phone and transmit data from the desktop simultaneously—over the same ISDN line. Users can also access multimedia content on the Web using the combined bandwidth of the two bearer channels. When a channel is needed for a phone call or fax, a channel is taken from the Internet connection. When the call or fax is finished, the bandwidth is reassigned to the Internet connection. For most PC users, however, ISDN is still prohibitively expensive, not available, or too complicated to configure.

Today, telephone companies claim they can reach about 95 percent of their serving areas with ISDN. Despite this impressive figure, the phone companies generally have not been able to extend ISDN to customers located on the perimeter of the local loop, that is, distances beyond 18,000 feet of a serving office. Due to changing demographics over the past decade, there are now almost as many people located around the perimeter of the local loop and beyond as there are within the local loop itself.

Telephone companies are implementing a variety of local loop enhancement technologies that greatly increase the bandwidth capacity of the vast installed base of existing twisted pair copper wiring. Of the dozen or so digital subscriber line technologies available, Asymmetric Digital Subscriber Line (ADSL) is the most promising because it is the only one that includes a voice channel as well as information channels.

ADSL

Asymmetric Digital Subscriber Line is an economical local loop upgrade technology that uses existing twisted pair telephone lines, converting them into access paths for voice, multimedia, and high-speed data communications. The electronics at both ends compensate for line impairments, increasing the reliability of high-speed transmissions.

ADSL offers more than 6 Mbps of downstream bandwidth (to the user) and as much as 640 Kbps upstream bandwidth (to the network). Such rates expand the existing access capacity by a factor of 50 or more without the need for new cabling.

ADSL creates three independent information channels suitable for any combination of services, including voice, ISDN, video-on-demand programming, and interactive gaming—a high-speed downstream channel, a medium-speed transfer (duplex) channel, and a voice channel. By filtering the voice channel away from the digital channels, telephone service is guaranteed, even if the broadband connections fail.

The problem with all DSL technologies is that they are not yet widely available; the implementations are not interoperable; and items commonly used in the local loop, such as load coils and bridge taps, degrade high-speed data.

CATV Networks

Community Antenna Television (CATV) companies are mounting a major challenge to the PSTN with the deployment of cable modems. Initially providing up to 10 Mbps, the likes of TCI, Cox, and Comcast envision capturing the consumer, home office, and small business markets for both voice and image (local and long distance) and Internet access (data and image). The investment in the rollout is substantial, but it is potentially a lower-cost solution than the necessary PSTN technology upgrades.

A cable modem contains both a regular phone modem for sending data over a phone line on the upstream and a demodulator for receiving data on the downstream. The unit has connectors for a telephone line, coaxial cable, and Ethernet cable. The phone line connection is used for dial-up access into the Internet. The cable connection is used for downloading the requested data to the demodulator at speeds of up to 10 Mbps. The Ethernet connection provides the link between the cable modem and the computer's 10/100BaseT network interface card, so the downloaded data can get from the cable modem to the computer.

Instead of equipping each PC with its own cable modem card, an external cable modem can be connected to a 10/100BaseT hub, providing up to 32 computers with simultaneous access to the Internet. Hubs also provide the means to implement an internal peer-to-peer LAN, allowing all computers to share files and resources, and engage in messaging.

In addition to the Internet, cable modems can be used to access corporate virtual private networks (VPNs). Software is available that allows users to restrict access and offers Web filtering and protection against viruses.

In addition, most vendors, including Motorola and Nortel, include some form of encryption or authentication. Cable modems represent an economical and effective way to integrate voice and data. They will eventually offer other access options.

Private Line Facilities

Private line facilities (dedicated or leased lines) differ from switched lines in that each user has the sole use of a preestablished circuit. Multiplexers at each end of the circuit(s) combine voice, data, and video traffic from various input channels so they can be transmitted over a single higher-speed digital link—usually a T1 line. At the other end, another device separates the individual lower-speed channels, sending the traffic to the appropriate terminals. Multiplexers enable businesses to reduce telecommunications costs by making the most efficient use of the leased line's available bandwidth. Since the line is billed at a flat monthly rate, this is incentive enough to load it with as much traffic of any type as possible. Using different levels of voice and/or data compression, the channel capacity of a leased line can easily be doubled or quadrupled to save even more money.

Several types of multiplexing are in common use today. On private leased lines, the dominant technologies are time division multiplexing (TDM) and statistical time division multiplexing (STDM). Each lends itself to particular types of applications. TDMs are used when most of the applications must run in real time, including voice, videoconferencing, and multimedia. However, when an input device has nothing to send, its assigned channel is wasted. With STDM, if a device has nothing to send, the channel it would have used is taken by a device that does have something to send. If all channels are busy, input devices wait in queue until a channel becomes available. STDMs are used in situations where efficient bandwidth usage is valued and the applications are not bothered by the queuing delay.

Although used mostly on private networks, both types of multiplexers can interface with the PSTN as well. For example, if a private T1 line degrades or fails, the multiplexer can be configured to automatically switch traffic to a standby ISDN link until the private line is restored to service.

Frame Relay

Frame relay defines a method for routing frames of information across a WAN. Frame relay uses a packet-switching protocol based on X.25 and

ISDN standards. By leaving the error correction and flow control functions to the end points (customer premise equipment), frame relay has relatively low overhead and can move variable-sized packets at high rates. The real benefit of frame relay is its ability to dynamically allocate bandwidth and handle bursts of peak traffic. Today, voice- over-frame relay (VoFR) is receiving growing attention.

Most nonvoice-oriented frame relay access devices (FRADs) use the first in, first out (FIFO) method of handling data traffic. However, in order to achieve the best voice quality, voice frames cannot be allowed to accumulate behind a long queue of data frames. Voice FRADs employ traffic prioritization schemes to minimize delay on voice traffic.

Traffic prioritization schemes ensure that data and voice traffic are interleaved, so voice packets are delivered in a timely manner but data traffic also gets through. During times of network congestion, one of the easiest prioritization methods is to simply discard some of the frames. In such cases, data rather than voice frames are discarded first, giving voice a better chance of making it through the network.

Some service providers offer prioritization of permanent virtual circuits (PVCs) within a frame relay network. Prioritization features on both CPE and frame relay networks can result in better voice application performance. The CPE ensures that higher-priority traffic is sent to the network first, while PVC prioritization within the network ensures that higher-priority traffic is delivered to its destination first.

VoFR equipment compresses a voice signal from 64 to at least 32 Kbps. In most cases, compression to 16 or even 8 Kbps is possible. Some equipment vendors support dynamic compression options. When bandwidth is available, a higher voice quality is achieved using 32 Kbps, but as other calls are placed or other traffic requires bandwidth, a 16- or 8-Kbps compression algorithm is implemented. Most voice FRADs also support fax traffic. A fax can take up as little as 9.6 Kbps of bandwidth for each active line.

VoFR usually allows a company to use its existing phones and numbering plan. In most cases, an internal dialing plan can be set up that allows users to dial fewer digits to connect to internal locations.

A persistent myth about VoFR is that voice calls can be carried free on an existing frame relay network. In fact, VoFR requires special CPE, and entails an increase in the port speed and a boost in the committed information rate (CIR) of the PVCs—all of which have a cost.

SMDS

Switched Multimegabit Data Service (SMDS) is a carrier-provided packet switched data transport service for LAN interconnection and intra-enterprise networking. This is accomplished by providing access to remote sites through the public SMDS "cloud," which greatly reduces the hardware cost associated with conventional leased lines. It also provides a scaled approach when purchasing high-bandwidth services and allows for easy migration to ATM services. Despite the poor marketing efforts of RBOCs and consequent low interest among potential users, SMDS is an important transmission technology because it is less expensive and simpler to operate and manage than frame relay. It also supports multimedia traffic.

As a connectionless service, SMDS eliminates the need for carrier switches to establish a call path between two points before data transmission can begin. The carrier's cell switch reads addresses and forwards cells one by one over any available path to the desired end point. SMDS addresses ensure that cells arrive in the right order. The benefit of this connectionless any-to-any service is that it puts an end to the need for precise traffic-flow predictions and for dedicated connections between locations. With no need for a predefined path between devices, data can travel over the least congested route in an SMDS network, providing faster transmission, increased security, and greater flexibility to add or drop network sites.

The multicast capability of SMDS enhances the throughput of multimedia applications, while conserving bandwidth. A single copy of a multicast transmission is placed on the Subscriber-Network Interface (SNI) and is replicated by the carrier's SMDS switches as necessary for delivery to multiple target SNIs.

The problem with SMDS is that it is not widely available. Some RBOCs that previously offered the service have discontinued it and have migrated their few existing customers to ATM. Such migrations are facilitated by the fact that SMDS and ATM use the same 53-byte cells.

ATM

Asynchronous Transfer Mode is a connection-oriented, cell-switching technology. ATM uses very short, fixed-length packets that can be switched in hardware (rather than software) and therefore have very little transmission delay. This low delay is important when considering interactive and multimedia applications, as well as voice and video, which cannot tolerate delay. The ATM protocols are not tied to a particular transmission rate or physical medium. This gives ATM the ability to operate at whatever rate is appropriate for a given application over whatever facilities are available.

ATM's real value in integrating voice and data is its built-in Quality of Service (QoS) capability, which allows traffic to be prioritized. When the QoS is negotiated with the network, certain performance guarantees go along with it: maximum cell rate, available cell rate, cell transfer delay, and cell loss ratio. The network reserves the resources needed to meet the performance guarantees, and users are required to honor the contract by not exceeding the negotiated parameters. Several methods are available to enforce the contract, among them traffic policing and traffic shaping.

To police traffic, the switches or routers use a buffering technique referred to as a *leaky bucket*. This technique entails traffic flowing (leaking) out of the buffer (bucket) at a constant rate (the negotiated rate), regardless of how fast it flows into the buffer. If the traffic flows into the buffer too fast, the cells are allowed onto the network only if enough capacity is available. If there is not enough capacity, the cells are discarded and must be retransmitted by the sending device.

Traffic shaping is a management function performed at the user network interface (UNI) of the ATM network. It ensures that traffic matches the contract negotiated between the user and network during connection setup. Traffic shaping helps guard against cell loss in the network. If too many cells are sent at once, cell discards can result, which disrupts time-sensitive applications. Because traffic shaping regulates the data transfer rate by evenly spacing the cells, discards are prevented.

ATM was specifically designed to integrate voice and data. The trouble is that the technology is expensive to implement. It requires a whole new infrastructure—from network interface cards, hubs, routers, and switches. In addition, the applications must be upgraded to become "ATM aware."

TCP/IP

TCP/IP nets are capable of converging voice and data to handle information distribution, telephone calls, videoconferencing, and interactive gaming, as well as streaming audio/video and multicast delivery. Using IP makes the global distribution of information and services more efficient and economical than using traditional PSTNs, while permitting their integration with legacy data and applications.

TCP/IP networks are more efficient than circuit-switched PSTNs, which rely on dedicated connections that are set up between end points. These connections are idle much of the time. During a typical voice conversation, as much as 40 percent of circuit-holding time is wasted with silent periods. In an IP network, packets are sent out onto the network only when there is voice to be conveyed. This frees the network to handle much more traffic—both voice and data—over the available bandwidth. The packetization scheme allows the network to handle any mix of voice and data at the same time over the same links.

What TCP/IP has lacked is an effective QoS mechanism. For QoS to work over IP, other protocols must be added. Several protocol standards exist—such as RSVP (Resource Reservation Protocol) and RTP (Rapid Transfer Protocol)—but none are universally implemented, except on private intranets where an organization has more control of the routers and switches within its domain. There also are proprietary techniques for dealing with the latency, but again these are better suited for private nets.

The performance problem of TCP/IP nets has not gone unnoticed by the Internet Engineering Task Force (IETF). New proposals are being considered, such as Differentiated Services (Diff-Serv) and Simple Integrated Media Access (SIMA), which define IP classes of service (CoS), but these are not likely to be agreed upon or deployed any time soon. Currently, and for the foreseeable future, next-generation networks will run IP over ATM on SONET fiber-optic links.

SUMMARY

In the late 1970s, ISDN was touted as the network technology that would provide true voice-data integration. In the early 1990s, attention shifted to ATM as the means of achieving true voice-data integration. In the late 1990s, TCP/IP emerged as the new choice for integrating voice and data. Actually, there are many solutions that are available for the integration of voice and data, including DSL and CATV technologies and frame relay service. This is in addition to TDM over private leased lines. ATM and SMDS are also effective in integrating voice and data, but ATM is the more expensive solution because it requires an entirely new infrastructure, while SMDS is not widely available. Managed TCP/IP nets potentially represent the most economical voice-date integration solution, particularly for companies that control the lines, hardware and applications end-to-end so latency and other impairments that affect real-time applications can be tightly controlled.

See Also **Asynchronous Transfer Mode, Digital Subscriber Line Technologies, Frame Relay, Internet Protocol, Switched Multi-megabit Data Services, Transmission Control Protocol/Internet Protocol**

Voice over Frame Relay

Companies are interested in running voice over frame relay (VoFR) to save money on intracompany and international calls. The major hardware component required for VoFR is a Frame Relay Access Device (FRAD) that supports voice. Low-end FRADs typically create point-to-point links and offer no additional features. The more expensive models offer an array of network routing and management capabilities, and support voice and video as well as LAN traffic.

Voice FRADs create packetized frames of voice data for transmission over frame relay and other lines. Like the packet assemblers and dissasemblers used to package and interpret data frames, voice FRADs are required at each end of the frame relay link. The FRAD's DTE interfaces can operate asynchronously or synchronously. The DCE interface provides a connection to an external DSU/CSU. Some FRADs come with an integrated DSU/CSU. Both EIA-232 and V.35 are supported on the DTE and DCE interfaces.

VoFR is most easily justified for intracompany communications between sites on the corporate enterprise network, using the same facilities that are already in place for the corporate data infrastructure. Similarly, it can be justified by a carrier who wishes to provide voice services for users connecting over its internal frame relay network.

Optimization Techniques

With voice compression algorithms and management of voice and data transmission parameters, voice quality can be maintained in high-traffic networks. Among the techniques vendors use to ensure voice, fax, and data are delivered reliably and that voice signals maintain their original quality are:

COMPRESSION Voice can be compressed as low as 4.8 Kbps and still achieve near toll quality.

PREDICTIVE CONGESTION MANAGEMENT The FRAD can respond to traffic load by varying the queue depths before congestion occurs, in addition to responding to the congestion notification messages issued by the network.

JITTER BUFFERS FRADs with a configurable jitter buffer can funnel the incoming packets to guarantee continuous speech output. This eliminates the variable delays between consecutive packets (jitter), which can impede the ability of the receiving end to smoothly regenerate voice.

FRAGMENTATION This ensures reliable voice packet delivery. The FRAD fragments all the packets so they can traverse the network within acceptable delay parameters. Voice frame sizes delivered to the network can be limited to 83 bytes per frame, asynchronous data to 71 bytes per frame, synchronous data to 72 bytes per frame, and fax to 58 bytes per frame. This fragmentation minimizes end-to-end delay through network switching equipment, ensuring the timely delivery of all traffic. Specifically, the fragmentation of data packets assures that voice and fax packets are not unacceptably delayed behind large data packets. To further assure that voice quality withstands network conditions, a FRAD can be configured to react to predictive congestion management and/or congestion notification messages from the network by varying the voice algorithm digitization rates and/or frame sizes.

PRIORITIZATION Each input signal can be configured into one of several priority queues in the FRAD. Voice and fax signals, which are intolerant of delay, can be placed in the highest-priority queue for most expeditious delivery to the network. Lower-priority data signals are buffered until the higher-priority voice and fax packets are sent.

SUMMARY

Frame relay is rapidly evolving from a single, data-only transmission technology to one with a broad spectrum of uses. VoFR technology consolidates voice and voice-band data (fax and analog modems) with data services over the frame relay network. It has the potential to provide users with greater efficiencies in the use of access bandwidth and to provide cost-effective voice traffic transport for intracompany communications. *See Also* **Frame Relay, Frame Relay Access Devices Video over Frame Relay**

WAN Service-Level Management

There was a time when most organizations relied on their long-distance carrier to maintain acceptable network performance. More often than not, the carriers were not up to the task. This led to the emergence of private networks in the mid-1980s and their tremendous growth, which continued to the mid-1990s. Private networks allowed companies to exercise close control of leased lines with a staff of network managers and technicians. Metrics such as network availability, throughput, and delay were manageable.

In their eagerness to recapture lost market share, carriers have made great strides in improving overall network performance, as well as their response to network congestion and outages. This has gone a long way toward winning back companies that had previously set up and maintained their own networks.

Despite the migration from private to public WANs in recent years, telecommunication managers are still accountable for overall network performance. Consequently, they must ensure that their carrier is equally as concerned about service quality.

An effective approach for ensuring service quality is to implement the relatively new concept of WAN service-level management, a collaborative effort between subscriber and service provider to manage the service quality of public network services. In this arrangement, both parties work together to plan, monitor and troubleshoot WAN service quality.

WAN service-level management offers a number of benefits. Subscribers can increase network availability and performance, reduce the need for recurring support, and ensure that business needs are met at the lowest possible cost. Service providers can reduce operational support costs, prioritize response to alarms, issue trouble tickets, set expectations for service quality, and help justify recommendations to upgrade bandwidth.

System Components

To ensure the success of WAN service-level management requires historical data and the collection of WAN service quality information to arrive at baseline performance metrics. Normally, these are difficult tasks, but a new breed of agent-based system is available that automates the collection, interpretation, and presentation of WAN service-level information. There are three primary components to a service-level management system: data collection, data interpretation, and data presentation.

The data collection component typically relies on intelligent agent software that resides at each WAN access point to perform circuit management—physical and logical layer service monitoring—and application traffic management. These agents may be embedded in CSU/DSUs or other WAN access devices such as multiplexers.

To minimize bandwidth overhead created by management traffic, agents collect service quality information and process the data locally. Telecommunication managers do not have to rely on bandwidth-consuming centralized polling to collect the performance data. Instead, the data is uploaded to the management system only when WAN bandwidth becomes available.

With the data interpretation component, centralized management software on a server automates the uploading and archiving of performance data. This centralized middle manager or server provides partitioned databases and flexible access control mechanisms. This allows service providers to securely offer subscribers access to their own performance information.

Finally, service-level management systems present service data in friendly graphical formats and allow for multiple user access. This information is presented in a way that correlates performance data on a network-wide basis. In addition, a variety of management platforms are supported, notably Web browser access.

SUMMARY

Businesses increasingly rely on wide-area networks to improve productivity, exploit new market opportunities, and secure strategic competitive advantages. So when these networks become severely degraded, corrective measures must be taken as soon as possible to minimize the impact on applications and to head off a possible outage. A proactive approach for ensuring service quality is WAN service-level management, an arrangement where subscribers and service providers collaborate to manage the

service quality of public network services. Agent-based software is becoming available for WAN access devices to facilitate the collection, interpretation, and presentation of WAN service-level information.
See Also **Intelligent Agents, Service-Level Agreements**

Wave Division Multiplexers

Wave division multiplexing (WDM) technology has been in use by long-distance carriers in recent years to expand the capacity of their trunks by allowing a greater number of signals to be carried on a single fiber. Although the technology has been around since the late 1980s, the need among carriers to get more performance and flexibility from their fiber-optic networks arose only in the mid-1990s. WDMs can help carriers ramp up their trunk speeds from 2.5 to over 40 Gbps without having to install additional fiber. The technology is accelerating at such a pace that terabit-per-second data transmission will soon be the norm.

Applications

WDM helps eliminate capacity constraints in carrier networks brought on by the ever-increasing processing power of computers and the need to link multiple users with multiple sites. WDM support applications such as the simultaneous distribution of full-motion video and medical images, and doesn't force carriers to rip out existing fiber backbones and replace them with higher-capacity links. The Department of Defense also sees a role for WDM-based networks for its Advanced Research Projects Agency (ARPA). It intends to deploy WDM to move large amounts of data and images, such as satellite and reconnaissance photos, over long distances in real time.

Cable TV operators are installing WDM-based fiber-optic backbones that connect to their existing cable-based networks, enabling them to provide Internet access to millions of their subscribers. WDM is also useful for large companies whenever high-speed interconnection is needed between widely dispersed sites. The technology is particularly important for corporations that maintain multiple data centers, either for distributed applications, disaster recovery, or CPU redundancy.

WDM works with a variety of existing protocols and technologies, such as Synchronous Optical Network (SONET) services ranging from OC-1 (51.84 Mbps) to OC-256 (13.271 Gbps) and broadband Asynchronous Transfer Mode (ATM) cell switching, which provides seamless connectivity between LANs and WANs.

Operation

Wavelength division multiplexing allows carriers to divide and condense standard fiber optic transmissions into separate wavelengths; each wavelength carries different content. Multiple data channels are transmitted over a single optical fiber using distinct colors of light, or optical wavelengths. This is similar to the way radio stations broadcast at different wavelengths without interfering with each other.

Early WDM systems offered only two or four of these widely spaced channels, which was not particularly cost-effective. But with technological improvements, vendors now offer "dense" WDM solutions that can segment a standard OC-48 (2.488 Gbps) line into as many as 8, 16, or 32 separate channels, each offering a multimegabit-per-second data rate.

Unidirectional and bidirectional fiber amplifiers affect the way WDM is implemented. In a unidirectional system, two fiber lines with one-way amplifiers are needed for two-way communications. For a bidirectional system, one fiber line with bidirectional amplifiers is needed for two-way communications.

With continual improvements, WDM systems will soon reach the terabits-per-second (Tbps) range. SilkRoad, a fiber-technology company, claims it is theoretically possible to achieve up to 10 Tbps using a single laser on a single fiber strand. The company also claims that the technology it is developing will be able to send data in both directions simultaneously over the same strand, with the signal traveling up to 200 miles without repeaters to strengthen the signal. This is four times the distance of present WDM technologies.

SUMMARY ▬ ▬ ▬ ▬ ▬ ▬ ▬

WDM technology provides the solution for economically increasing network capacity, without the expense of installing new fiber, to meet the increasing demand for more bandwidth. Widespread adoption of WDM will be an important step toward the goal of realizing all-optical networks, in which optical-to-electrical conversions are minimized by moving more transport and switching duties into the optical domain. WDM has proved to be the ideal encoding technique for optical switching. Individual bands in a fiber can be tapped with simple add/drop optical switches without disturbing the other channels and without any electronic intermediaries. The resulting network system can emulate any existing network topology, but at throughput that is orders of magnitude higher. Compared to existing technology such systems represent virtually unlimited bandwidth and span capabilities.

See Also **Statistical Division Multiplexing, Time Division Multiplexing**

Web-Based Management Tools

The expanding use of intranets—which are essentially smaller, private versions of the global Internet—has sparked the creation of Web-based tools for monitoring and managing enterprise networks. In fact, a new network management paradigm has emerged that allows administrators to use Web browsers to monitor, configure, and control enterprise networks and their various components. Use of the popular Web browser gives network managers a degree of mobility they did not have before and enables a wider range of users to access network status information. In addition, Web browsers overcome many interoperability issues that are inherent in multiplatform environments.

The convenience and ease of use of Web-based management (WBM) tools is very appealing. A Web-based management tool is easier to use than the command-line interfaces of UNIX systems. Such tools can be used from any PC or workstation with a browser. This also means network planners, designers, and managers do not have to load specialized client software on their laptops. With a Web browser, users can access management information on the corporate intranet at any time—in any building, while traveling, or even at home. There are now WBM tools for virtually any type of network: LANs, frame relay, and ATM data networks; traditional voice networks; and even legacy SNA networks.

In being able to access performance data and implement routine management tasks over the Web, companies with global networks do not have to distribute expensive management platforms to every remote site. The availability of Web-based management tools also has the potential to reduce the cost of managing transmission facilities and services by eliminating the need for proprietary, high-priced equipment offered by carriers. In fact, carriers such as AT&T, MCI WorldCom, and Sprint offer Web-based management tools to their customers, recognizing their need for low-cost alternative management solutions.

Applications

There are three fundamental applications where WBM tools can provide a significant benefit: individual device configuration and management, Web browser access to sophisticated management applications, and corporate access to network status data.

DEVICE CONFIGURATION Web-based configuration and management of individual devices is a capability aimed at administrators of small networks who may not have their own network management system, or branch office locations where technical expertise is not immediately available. Such users need configuration tools that are easy to use and which can be accessed with a Web browser.

This is accomplished by providing an agent with the equipment to be configured. The agent includes a native HTML interface for access over the Web. The manager enters basic configuration parameters for each device by completing a simple online electronic form. Remote monitoring of simple device statistics is also possible via the browser, using graphical displays of basic device information and performance indicators.

NETWORK SUPPORT Access to network management information from any browser-equipped desktop is targeted at enterprise network support staffs who already use such advanced platforms as Hewlett-Packard's OpenView to monitor the network, understand potential faults and alarms, and provide end-users with continuous network availability. The addition of Web browsers provides a low-cost option for easily accessing important information from any location.

For example, a staff member out on the manufacturing floor troubleshooting the network may need to access a particular management application. Through a Web browser running on any convenient PC or laptop computer, the technician can access the necessary application and continue the troubleshooting process. This reduces the time and effort to do the job.

ACCESS TO NETWORK STATUS DATA Web reporting of network status information via the organization's intranet is aimed at Information Systems (IS) group managers who usually do not operate the network or get involved in extremely technical detail. Instead, their goal is to quickly obtain information about the state of the network and view trends over time, so they can identify potential trouble spots. The Web provides an easy, economical way to distribute this type of information to all who need it.

Various people within the organization need different types of information. Members of the finance group, for instance, may need usage information for accounting purposes, while database users may need to determine system status or submit an online trouble ticket and follow it through to resolution. The corporate intranet offers a simple, effective method for distributing this type of information to people who do not normally have ready access to traditional management systems.

Web-Based Management Approaches

There are two basic strategies for implementing WBM: the proxy solution and the embedded approach. While these methodologies can be used in combination, each has advantages.

PROXY SOLUTION   The proxy solution adds a Web-based server to an intermediate workstation called a proxy, which interconnects with the end devices it manages (Figure 130). Users access the proxy through a Web browser, using the HyperText Transfer Protocol (HTTP), while the proxy accesses end devices using the Simple Network Management Protocol (SNMP). Usually, this approach adds a Web server to an existing management product, which optimizes product functionalities, such as database access and SNMP polling. This approach maintains the advantages of the workstation-based management systems while adding flexible access to intranet data. Since the proxy communicates with all the managed devices, the administrator can view all or portions of the company's networks, servers, and desktops, as well as such logical entities as virtual LANs. Since the proxy-to-device protocol remains SNMP, this approach also works with SNMP-only devices.

EMBEDDED METHOD   The embedded method actually installs Web server functionality in each end device. Since each device has its own Web address, the administrator can use a Web browser to visit the managed devices (Figure 131). This approach allows graphical management of those devices.

Enterprise networks can make use of both proxy-based and embedded Web server capabilities. Large organizations can avail themselves of the en-

Figure 130

The proxy solution for Web-based management.

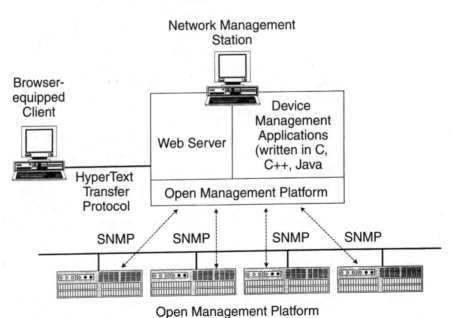

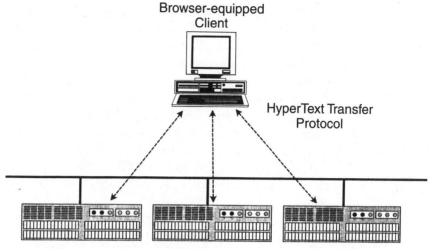

Figure 131
The embedded
approach to Web-
based
management.

Browser-equipped
Client

HyperText Transfer
Protocol

Network Devices with Embedded Web Server Functionality

terprise-wide monitoring and management capabilities that are only provided with the proxy solution. The proxy solution can also manage
SNMP-only devices. In conjunction with proxy-based servers, large corporations could also benefit from embedded Web servers because of their
simple set-up requirements and their ability to manage new devices.

The embedded Web server approach is especially well suited for managing small branch offices. These networks are relatively simple and typically
do not require powerful management systems or need enterprise-wide
views. Users at these remote locations usually are not familiar with device-
control and network procedures. Embedded Web servers allow these users to
become operational quickly with minimal instruction. Although the plug-
and-play nature of Web-based devices simplifies installation and minimizes
troubleshooting, they do not necessarily limit device-level capabilities.

A Web browser can communicate directly with networked devices using HTTP. The most common reasons for connecting directly to individual network devices include configuration and reconfiguration, simple
status monitoring, and implementing specific corrective actions. For
small networks, this approach alone might be a sufficient management
solution; for larger networks, this approach will typically supplement
more global tools in a variety of possible circumstances, including the final stages of problem resolution.

Components of WBM

The components of WBM can be written in any language, including HyperText Markup Language (HTML), Common Gateway Interface (CGI),
and Java.

HTML provides a standardized way to create pages of Web-based information and embed hyperlinks that lead the user to other pages. While HTML pages are usually static and textual, they can be made interactive by embedding graphics and active elements, such as Java applets and CGI applications, within the page. HTML is good for displaying tables of information, such as network inventory details and IP address listings.

CGI is not a language, but a protocol used to access database information. For example, a WBM application may need to display the current number of incomplete work orders. This data could be stored in the database of a proxy workstation. A CGI application can be used to query the database and to format an HTML page on-the-fly to display the information.

Java is an interpretive programming language and is rapidly emerging as an important development tool, especially for management applications. Java code is not compiled before run time but is interpreted by a Java Virtual Machine (JVM) at run time. JVMs are included with such Web browsers as Netscape Navigator and Microsoft Internet Explorer, enabling these browsers to execute Java code.

Java applets, which can be called from either a proxy or embedded Web server, can be used to:

- Display dynamic graphs that interpret network operations

- Illustrate complex situations, such as interactive views of chassis hubs or modules of a stackable hub

- Display real-time data that can be updated from polling and traps

- Add graphics, including animation

Because Java can produce applications that are portable across UNIX, Windows, and other environments, a JVM can be embedded in an end device and the device agent can then execute Java code. Code can be distributed dynamically, ported from a management proxy to devices, and ported between devices or components within a device.

The use of Java within an embedded agent can increase management capabilities by enforcing policy-based management or security rules. For example, assume an administrator wants to enable remote network access between 6:00 A.M. and 6:00 P.M. Traditional methods, such as SNMP sets, have been used to enforce such restrictions but at the cost of increasing network traffic and with the difficulty of scaling to handle thousands of devices. While a device-resident agent could perform this function, agent releases are typically infrequent. An embedded agent with a JVM could independently create the scheduling routine and dynamically distribute the policy to restricted devices, eliminating the need for an agent release.

Standards

The first Web-based management tool was introduced in 1994 by Thomas Conrad Inc., a small networking device vendor that has since been acquired by Compaq. A handful of pioneers followed suit in 1995. By 1996, virtually every major interconnect and network management vendor had either introduced or announced plans for Web-based management products that could be accessed with ordinary Web browsers.

The idea behind Web-based management is simple as it was compelling: it gives managers, administrators, and authorized end users access to management information from dissimilar platforms such as MVS, VMS, Windows NT, UNIX, and NetWare. However, the product development efforts of vendors are unfocused and there is still lack of agreement on standards.

Not only are vendors and carriers approaching Web-based management in different ways, but the tools they offer address different pieces of the management puzzle. Capabilities offered by one vendor are often unaddressed by other vendors. Some vendors' reporting capabilities consist of static displays, while others offer real-time displays that report status changes for as long as the connection is open.

There are several standards efforts underway that attempt to address the functionality, interoperability, security, and performance issues of Web-based management. What makes standards so difficult to achieve is that the current proposals are vendor biased. If the differing views of vendors can be reconciled and a single standard agreed to, the Web browser may emerge as the key building block of tomorrow's network management console.

SUMMARY

The advent of intranets that provide enterprise-wide data has spawned two main Web-based management approaches: the proxy Web server method and the embedded Web server approach. While both methods can be used together, generally the proxy method is more useful for larger enterprises that want to complement their workstation-based management systems. The embedded server method is better suited for small groups that do not need the complex functions offered by the first method.

Web-based management standards would provide network managers with what they need most: a cost-effective means for consolidating topology, fault, and performance data from many management platforms, element managers, and devices. Already, many Web-based tools are in the process of being integrated with one or more of the major management platforms, including Hewlett-Packard's OpenView, SunSoft's Solstice SunNet Manager, and IBM's NetView. The reason for optimism here is that these platforms owe their market dominance to their ability to pull data from vast multi-vendor networks, so a high degree of de facto standardization is possible, regardless of the outcome of formal standardization processes.

See Also **Network Management Systems, Simple Network Management Protocol**

Web Portals

A Web portal is a site on the Internet or a corporate intranet that aggregates content, applications, or commerce services in a single place, making it convenient for users to find what they need without having to engage in time-consuming searches of producer-centric Web sites.

When looking into buying a car, for example, instead of calling up the Web sites of each individual manufacturer, users can get everything they need by visiting Microsoft's CarPoint (Figure 132). This Web site sorts au-

Figure 132
Microsoft's CarPoint is an example of a Web portal that aggregates content of interest to potential car buyers.

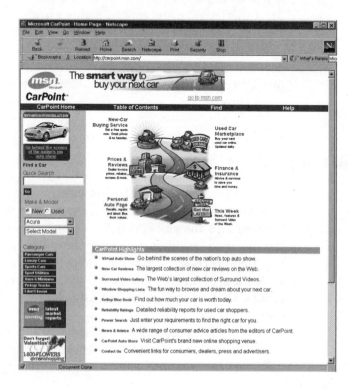

tomobiles into categories that are useful to consumers—new and used, for example—and provides tools for comparison shopping and selection, regardless of vendor.

Users can configure their browsers to call up a portal as their home page. When they log on to the Internet, the portal shows up as the home page, enabling users to quickly go to frequently accessed locations. Users can customize portals and include messaging functions.

Although this type of site is proliferating on the Internet, it is also useful for corporate intranets and extranets. Corporations can set up their Web sites as either the default or a frequently visited destination where employees, partners, and customers can easily find enterprise applications, filtered job-related content, and commerce-based services.

Compared to installing traditional client/server solutions, portals provide a far lower total cost of ownership (TCO) at a fraction of the start-up cost. Unlike many client/server applications, business portals provide immediate and substantial value for an organization, as well as ease of use through the familiar graphical interface of a Web browser.

Organizations can realize immediate top-line revenue and savings by using business portals. The total cost of ownership for a business portal is less than half that of a traditional enterprise application. A business portal can be up and running in days or weeks—not the typical months or years it takes to develop and implement client/server applications.

For mobile professionals, business portals provide a single place to handle virtually every aspect of their job. In addition to crucial application functionality, business portals save time because they quickly deliver information and job-related content and services.

SUMMARY

Web portals are emerging as a seamless integration point for content, applications, and e-commerce from within and outside the enterprise. The impetus for portal development in the corporate environment is the out-of-control proliferation of Web sites that are popping up across organizations. The result is that most employees are not aware of all the resources available online to help them do their jobs. Companies are looking to restore some semblance of order that will enable these resources to become more available and used more effectively in the furtherance of personal productivity and organizational objectives. Web portals provide the aggregation of resources that makes all this possible.

See Also **Client/Server**

Wide-Area Networks

A wide-area network (WAN) interconnects telephone systems, computers, and/or local-area networks over a relatively large geographical area that extends beyond the local service area of a telephone company, cable operator, or Internet service provider (ISP). The WAN may consist of private leased lines, carrier-provided services, or a combination of both. The largest WAN is the Internet, which interconnects over 20 million computer systems in more than 200 countries.

A WAN can be quite complex, consisting of many types of transmission media, including T-carrier, optical fiber, terrestrial wireless, and satellite. These transmission media enable network operators to offer many types of services, including plain old telephone service (POTS), Switched 56K, ISDN, packet data services, and frame relay. Specialized devices are used to interconnect facilities and services on the WAN, including circuit switches, multiplexers, cross-connect systems, routers, and bridges.

SUMMARY

WANs are intended to handle the traffic of many users simultaneously, whereas LANs serve all users by controlling access to the medium so the full bandwidth is available to only one user at a time. Ethernet LANs accomplish this with a contention-based scheme called Carrier Sense Multiple Access with Collision Detection (CSMA/CD), while token-ring LANs rely on a deterministic scheme called token passing. Until a few years ago, WANs were considered a bottleneck when used to interconnect high-speed LANs at different locations. With increasing reliance on optical fiber rather than T-carrier, however, this situation is more the exception than the rule.

See Also **Campus-Area Networks, Enterprise Networks, LAN/WAN Integration, Metropolitan-Area Networks**

Windows NT

The TCP/IP-based Internet is a global collection of servers interconnected by routers over carrier-provided lines and services. Most of the servers are based on the UNIX operating system, but increasingly Windows NT machines from Microsoft Corp. are being used because they are easier to set up and administer. Like UNIX, Windows NT has built-in support for TCP/IP and many of the tools necessary to set up and manage Internet/intranet sites.

Windows NT is a 32-bit operating system with preemptive multitasking, strong networking support, no memory limits, and no dependency on DOS. Windows NT surpasses Windows NT 3.1 and Windows 95 to claim equivalent power to UNIX. It also offers the opportunity to run both technical and business applications on a single desktop platform, which in most cases is far less expensive than a UNIX workstation.

At the most basic level, Windows NT servers include communications protocol support, network utilities, and application programming interfaces (APIs) that allow them to communicate with most UNIX variants.

Windows NT servers has built-in TCP/IP support. This means they can be used for Internet/intranet communications right out of the box. Among the core TCP/IP protocols included with Windows NT are the user datagram protocol (UDP), address resolution protocol (ARP), and the Internet control message protocol (ICMP).

More than a dozen basic network utilities are included with Windows NT, including utilities that provide terminal access to or file transfer capabilities to and from most UNIX-based systems. The basic TCP/IP connectivity applications include finger, gopher, file transfer protocol (FTP), and trivial file transfer protocol (TFTP). TCP/IP diagnostic utilities include hostname, ipconfig, nbstate, netstap, ping, and route.

Windows NT servers provide several facilities for integrating Windows NT-based systems into networks that employ the simple network management protocol (SNMP), a common TCP/IP-based network management facility. This component allows a Windows NT server to be administered remotely using such enterprise-level management platforms as Hewlett-Packard's OpenView, IBM's NetView, and Sun's Solstice SunNet Manager.

Windows NT includes native support for NetBEUI and IPX/SPX, as well as TCP/IP. Regardless of the protocol used, each system hides the details of the underlying network from the applications and end users. Network administrators can choose the protocol that best addresses the company's network requirements.

DHCP Support

With regard to TCP/IP, Windows NT is the only server operating system that provides integral Dynamic Host Control Protocol (DHCP) management. DHCP assigns TCP/IP addresses, eliminating the process of manual address allocation every time users request a connection.

The proliferation of TCP/IP-based networks, coupled with the growing demand for Internet addresses, makes it necessary to conserve IP addresses. Issuing IP addresses on a dynamic basis provides a way to recycle this increasingly scarce resource. Even companies with private intranets are increasingly using dynamic IP addresses, instead of issuing unique IP addresses to every machine. With DHCP, IP addresses can be doled out from a pool of IP addresses as users need them to establish network connections. When they log off the net, the IP addresses are released and become available to other users. Assigning addresses can consume a majority of a network manager's time if done manually. With DHCP support, Windows NT greatly reduces the time spent on this chore.

A related feature provided by Windows NT is the Windows Internet Name Service (WINS), which maps computer names to IP addresses, allowing users to refer to their machines with an easy-to-remember plaintext name rather than by an IP address such as 123.456.789.22, for example.

Remote Access

Portable systems represent a large and growing share of personal computers. They are used by telecommuters, system administrators, and mobile workers. Through Windows NT Server's Remote Access Server (RAS), network administrators can extend the local-area network across a phone line, allowing remote computers to appear as if they are directly attached to a network. Up to 256 sessions are supported and can participate fully in the network, sharing files and printers, accessing databases, connecting to hosts, and communicating with colleagues via e-mail. Windows NT also supports PPP and SLIP, making Internet access a routine task.

PPP is a set of industry-standard framing, authentication, and network configuration protocols that enable remote access solutions to interoperate in a multivendor network. PPP support allows Windows NT to receive calls from and provide network access to other vendors' remote access workstation software.

SLIP is similar to PPP, except it is an older standard that addresses only TCP/IP connections over serial lines. It does not provide automatic negotiation of network configuration without user intervention. It also does not support encrypted authentication. Although SLIP is not recommended, it is supported by Windows NT to accommodate users who already have SLIP servers installed.

RAS supports any combination of NetBEUI, TCP/IP, and IPX protocols to access remote networks. IPX turns Windows NT into a remote access server for NetWare networks. TCP/IP support makes Windows NT an "Internet-ready" operating system, allowing users to access the vast resources of the World Wide Web with any browsing tool.

Windows NT enhances the RAS architecture by adding an IP/IPX router capability. This allows clients to run TCP/IP and IPX locally and run Windows Sockets applications or NetBIOS applications over their local TCP/IP and IPX protocols.

Security

Windows NT implements user authentication via account numbers and passwords. In addition, NT supports account lockout after a set number of incorrect password attempts. Further, even after authentication, users may access system resources based only on users' access permissions. The Windows NT security system guards against access by unregistered users and ensures that registered users cannot delete crucial system files.

Starting with Windows NT 5.0, otherwise known as Windows 2000, there is an integral Certificate Server, which provides more control and flexibility for security management. It centralizes the administration and use of certificates—essentially small, verifiable files with customizable content. ActiveX controls and Java applets usually employ security certificates on the Web.

Another security option available with Windows NT 5.0 is the Encrypting File System (EFS), which lets users encrypt individual files or entire directories that reside on an NTFS volume. This can protect data from being read by potential network intruders. To encrypt files or directories, users just right-click on them in Explorer and choose Encrypt from the Context menu. The system creates random keys used for the Data Encryption Standard (DES) encrypting algorithm. When an encrypted file is accessed, NT 5.0 transparently validates the user's identity and decrypts the file.

System Management

Windows NT provides a centralized management environment allowing distributed systems to be managed and backed up from a single location. A broad range of tools are provided to simplify management tasks, including a Performance Monitor, a Service Control Manager (to start and stop services), a User Manager (to add/delete/modify user information such as access permissions), and a Disk Administrator (to create/delete disk partitions).

To further simplify administration, Windows NT uses a centralized configuration Registry containing operating system, application, and hardware configuration data. The Registry eliminates the need for configuration files such as AUTOEXEC.BAT, CONFIG.SYS, and .INI files. As the Registry contains crucial information, Windows NT ensures that updates to the Registry occur atomically; in other words, either the update completes entirely or not at all.

In addition to DHCP, Windows NT also includes support for DNS (Domain Name System) addresses and the Point-to-Point Tunneling Protocol (PPTP). PPTP enables a dial-up client to access a Windows NT server securely through the Internet. With this protocol, a remote user can access a corporate LAN by dialing a local Internet service provider.

Starting with version 5.0, Windows NT offers a feature called Active Directory, a hierarchical directory service designed to ease network management. Active Directory supports three key capabilities: single network log-on, a single point of administration for all network objects, and the ability to query any attribute of any object, such as a network user's phone number or e-mail address.

The Active Directory management scheme is like White Pages for a network. It holds information about domains, users, groups, organizational units, and printers—all hierarchically stored within the directory. This architecture allows administrators to configure a directory to resemble their organization and network topology. The Active Directory also conforms to industry standards such as the Lightweight Directory Access Protocol (LDAP) and HyperText Transfer Protocol (HTTP), so its capabilities can be extended over the Internet. Potentially, resources across the Internet could be located within the directory—with access that appears seamless to administrators and users.

Openness

Windows NT supports interoperability with other systems using TCP/IP, DCE RPC, and SNA. Further, Windows NT interfaces with LAN Manager, Novell NetWare, DEC Pathworks, and Banyan VINES.

Windows NT offers an environment for client/server computing and supports a number of mechanisms for client applications to communicate with server applications. One of these, DCE (Distributed Computing Environment)-compliant RPC, provides complete interoperability with other DCE/RPC-compliant applications. NT supports the Distributed Component Object Model (DCOM), which integrates client/server applications across multiple computers on a network. Other mechanisms supported include Named Pipes (providing asynchronous messaging support) and Windows Sockets (Winsock).

Internet Information Server

Windows NT comes with Internet Information Server (IIS), a fully functional Internet server. IIS supports Web pages, gopher, anonymous FTP, and Telnet sessions. Also included is support for Active Server Pages, Index Server, and NetShow. In addition, IIS includes support for the Internet Server Applications Programming Interface (ISAPI) Web programming interface, as well as a full complement of server-side application environments such as CGI, Perl, Java VM, and Visual Basic scripting.

Windows NT does not come with a Usenet news server, but Microsoft offers one separately. When implemented for private discussion groups on an intranet, Microsoft's News Server can be integrated with Windows NT security, which means that authenticating users can be relatively painless as long as users have accounts on an NT server or domain. As an Internet server, however, Microsoft News Server falls short in that it does not support streaming-mode NNTP (Network News Transfer Protocol). Although it is not a part of the NNTP specification (RFC 977), it has become a de-facto standard among Usenet sites throughout the Internet, and Microsoft's server is among the few that does not support it.

Peer Web Services

With Peer Web Services, users can create a personal Internet or intranet server. This capability is convenient for development, testing, and peer-to-peer publishing. Peer services are implemented in much the same way as the peer networking services had been in Windows for Workgroups. All that is needed to expose a user's personal Web site is an appropriate Internet or intranet connection. Peer Web Services supports all ISAPI extensions and filters, as well as the Secure Sockets Layer (SSL) for secure transactions.

SUMMARY ▪▪ ▪▪ ▪▪ ▪▪ ▪▪ ▪▪ ▪▪ ▪▪

Windows NT is aimed at all corporate server needs. It offers a robust, reliable kernel and brings the one thing still not provided by UNIX—standardization. Unlike UNIX, which comes in different flavors from different vendors, NT's code base is controlled by Microsoft. This means that advanced multithreaded applications written for NT do not have to contend with differences in operating systems. The software should run as long as the server architecture is the same.

Although Windows NT has the flexibility to provide a solid Internet server platform, it is limited in the number of users it can support. This varies between 200 and 300 users, depending on the type of applications running at any given time. Although Windows NT can support dual processors, this configuration usually does not result in an appreciable performance gain. While Windows NT offers many of the capabilities of UNIX and is easier to set up and manage, it does not match UNIX in terms of scalability.

See Also **UNIX**

Wireless Centrex

Centrex is an outsourced solution that offers businesses the same capabilities and features as an on-site PBX, but delivers them from a local central office. The service lets businesses take advantage of advanced voice and data management features without the cost of on-site switching equipment, maintenance, and administration. The local telephone company takes responsibility for managing the service for its customers.[5]

A relatively recent innovation in the delivery of Centrex service is the use of wireless technology to create a private, wireless environment at a company's business location without incurring expensive cellular airtime charges. Only two RBOCs offer wireless Centrex: Bell Atlantic and Pacific Bell.

Pacific Bell's Wireless Centrex, for example, does this by combining its conventional wireline Centrex offering with Ericsson's Freeset Business Wireless Telephone system. At the corporate facility, multiple, overlapping cells cover assigned areas. The number of cells required at any given company is determined by traffic density.

System Components

Freeset is a low-power, digital, wireless telephone system made up of the following components:

RADIO EXCHANGE All incoming and outgoing calls are routed via the radio exchange and base stations. Functions such as powering and control, and facilities for connection to Centrex, are handled by the radio ex-

---

5. For a more in-depth discussion of conventional wireline Centrex, see my book *Desktop Encyclopedia of Telecommunications,*" published by McGraw-Hill (ISBN: 0-0704-4457-9). This 600-page reference can be conveniently ordered through the Web at Amazon.com.

change. A single Centrex line or extension of a Centrex line is used for each portable telephone.

BASE STATION This relays calls from the radio exchange to portable telephones. Each base station provides eight simultaneous speech channels. The coverage of each base station depends on the character of the environment, but it is typically between 8,000 and 15,000 square feet.

PORTABLE TELEPHONE The handset is a small, lightweight unit containing the intelligence needed to accommodate roaming and cell-to-cell handover. All Centrex features are available on the handset user, plus some unique features like predialing and display editing.

When the radio exchange receives an incoming call, it transmits the identification signal of a portable telephone to all base stations. Because the portable telephone communicates with the nearest base station, even in standby mode, it receives the signal and starts ringing. When the call is answered, the portable telephone selects the channel with the best quality transmission.

Configuration, administration, and maintenance can be handled locally or remotely using software available with the system. An internal test and maintenance module routinely tests the Freeset system, while a set of online diagnostics provides early fault detection and prompts remedial action.

Call Integrity

At the customer premises, multiple, overlapping cells cover assigned areas. The number of cells required is determined by traffic density at a given location. A number of system features help maintain the integrity of the calls:

ROAMING Calls can be made and received with the portable telephone whenever it is within an area covered by a base station. When a user roams between base stations, the Freeset system keeps track of the phone's location.

SEAMLESS CALL HANDOVER The portable telephone continuously seeks the best channel to use, and performs an undetectable handover to that channel. Each base station provides both vertical and horizontal coverage.

SECURITY AND AUTHENTICATION The system uses full speech encryption to ensure complete privacy and protection against eavesdropping. The Freeset authentication process assigns a unique identity to each portable telephone.

Options

Pacific Bell's Wireless Centrex service gives business users full wireless mobility, plus the following options:

- Account codes
- Authorization codes
- Automatic callback
- Automatic recall
- Call diversion/call forwarding
- Call diversion override
- Call hold
- Call transfer
- Call waiting
- Call pickup
- Speed calling
- Call park
- Conference/three-way calling
- Remote access to network services
- Distinctive/priority ringing
- Do not disturb
- External call forwarding
- Executive intrusion/executive busy override
- Individual abbreviated dialing/single digit dialing
- Speed dialing
- Last number redial
- Loudspeaker paging
- Message waiting indication
- Remote access to subscriber features
- Select call forwarding

SUMMARY

Wireless Centrex is an adjunct to conventional wireline Centrex service and is designed to integrate with existing telephone systems to relay calls as users move throughout a building or company location. There are no airtime charges or service fees with wireless Centrex. This communication solution has the added advantage of eliminating paging delays and reducing noise from overhead paging systems within buildings.

***See Also* Wireless LANs, Wireless Local Loop, Wireless PBX**

Wireless Home Networking

As more PCs, peripherals, and intelligent devices are installed in the home, and as network connections proliferate, users are faced with new opportunities for accessing information as well as challenges for sharing resources. For example, many users want to:

- Access information delivered via the Internet from anywhere in the home
- Share files between PCs and share access to peripherals no matter where they are located within the home
- Control electrical systems and appliances whether in, around or away from home
- Effectively manage communications channels for phone, fax, and Internet usage

Each of these capabilities requires a common connection between the various devices and networks found in a home. However, in order to truly be effective, any home network must meet certain criteria:

- It must not require additional home wiring. Most existing homes are not wired for networking and retrofitting them would be too labor-intensive and expensive. A wireless solution is a viable alternative.
- Wireless connections must be immune to interference, especially with the growing number of wireless devices and appliances emitting radio frequency (RF) noise in the home.
- The range of wireless connections must be adequate to allow devices to communicate from anywhere within and around a typical family home.
- The network must be safe and protected from unwanted security breaches.
- It must be easy to install, configure, and operate for nontechnical users. Most home users do not have the expertise to handle complex network installation and configuration procedures.
- The entire system must be easily and spontaneously accessible— anytime and from anywhere in or even away from the home.

These issues have been addressed by a consortium of 70 vendors called the Home Radio Frequency Working Group (HomeRF WG), which has developed a platform for a broad range of interoperable consumer devices. Its specification, called the Shared Wireless Access Protocol (SWAP), is an open standard that allows PCs, peripherals, cordless telephones, and other consumer electronic devices to communicate and interoperate with one another without the complexity and expense associated with installing new wires.

SWAP is designed to carry both voice and data traffic and to interoperate with the Public Switched Telephone Network (PSTN) and the Internet. It operates in the 2.4-GHz ISM (Industrial, Scientific, Medical) band and uses frequency-hopping, spread-spectrum radio for security and reliability. The SWAP technology was derived from extensions of existing cordless telephone (Digital Enhanced Cordless Telephone, or DECT) and wireless LAN technologies to enable a new class of home cordless services. It supports both a TDMA (Time Division Multiple Access) service to provide delivery of interactive voice and other time-critical services, and a CSMA/CA (Carrier Sense Multiple Access/Collision Avoidance) service for delivery of high-speed packet data. The following are the main system parameters of SWAP (from the HomeRF Working Group):

FREQUENCY HOPPING NETWORKS   50 hops/second

FREQUENCY RANGE   2.4 GHz ISM band

TRANSMISSION POWER   100 mW

DATA RATE   1 Mbps using 2FSK modulation and 2 Mbps using 4FSK modulation

RANGE   Covers typical home and yard

SUPPORTED STATIONS   Up to 127 devices per network

VOICE CONNECTIONS   Up to six full duplex conversations

DATA SECURITY   Blowfish encryption algorithm (over one trillion codes)

DATA COMPRESSION   LZRW3-A algorithm

48-BIT NETWORK ID   Enables concurrent operation of multiple colocated networks

Applications

The SWAP specification provides the basis for a broad range of new, home networking applications, including:

- Shared access to the Internet from anywhere in the home, allowing users to browse the Web from a laptop on the deck or have stock quotes delivered to a PC in the den.

- Automatic intelligent routing of incoming telephone calls to one or more cordless handsets, fax machines, or voice mailboxes of individual family members.

- Cordless handset access to an integrated message system to review stored voice mail, faxes, and electronic mail.

- Personal intelligent agents running on the PC for each family member, accessed by speaking into cordless handsets. This new voice interface would allow users to access and control their PCs and all of the resources on the home wireless network spontaneously, from anywhere within the home, using natural language commands.

- Wireless local-area networks (LANs) allowing users to share files and peripherals between one or more PCs, no matter where they are located within the home.

- Spontaneous control of security, electrical, heating, and air conditioning systems from anywhere in or around the home.

- Multiuser computer games playable in the same room or in multiple rooms throughout the home.

Network Topology

The SWAP system can operate either as an ad-hoc network or as a managed network under the control of a Connection Point. In an ad-hoc network, where only data communication is supported, all stations are equal and control of the network is distributed between the stations. For time critical communications such as interactive voice, a Connection Point is required to coordinate the system. The Connection Point, which provides the gateway to the PSTN, can be connected to a PC via a standard interface such as the Universal Serial Bus (USB) that will enable enhanced voice and data services. The SWAP system also can use the Connection Point to support power management for prolonged battery life by scheduling device wakeup and polling. The network can accommodate a maximum of 127 nodes. The nodes are of four basic types:

- Connection point that supports voice and data services
- Voice terminal that uses only the TDMA (Time Division Multiple Access) service to communicate with a base station

- Data node that uses the CSMA/CA (Carrier Sense Multiple Access with Collision Avoidance) service to communicate with a base station and other data nodes
- Integrated node, which can use both TDMA and CSMA/CA services

SUMMARY

Home users need a wireless network that is easy to use, cost-effective, spontaneously accessible, and able to carry voice and data communications. Through SWAP, consumer electronics and small appliances in and around the home will contain technology that enables them to "talk" to each other without being tethered to the existing wiring in the home while distributing the power of the PC throughout the home. For example, a mobile display pad linked to the Internet could access recipe information in the kitchen, be taken into the yard to assist with plant and disease identification in the garden, or go into the garage for the do-it-yourselfer looking for the latest automobile mechanical updates. Other solutions will compete with SWAP, including Digital Power Line, which uses the home's existing electrical system as the network, and Phoneline Networking, which uses the existing telephone extension wiring as the network. Only SWAP, however, offers true mobility through wireless communication.

See Also **Digital Power Line, Phoneline Networking**

Wireless Internet Access

Until recently, retrieving information from a central database required a modem-equipped laptop computer, which enabled users to dial into the Internet from a remote location and from there gain access to a secure corporate Web server, where the information was kept. Alternatively, users could come into the office with their laptops and insert them into a docking station that was already connected to the LAN, and then access the information on a server. In both cases, the files on all machines can be synchronized, ensuring that users are working with the most current information.

These access methods are still used, of course, and they work well for people who are only occasionally mobile. But for workers who are out of the office most of the time and who may be at different locations on a daily basis, the docking station approach is impractical and the dial-in method inconvenient because a phone jack must be nearby. It can also be expensive, especially if access requires a long-distance call.

These problems can be overcome with various wireless access alternatives. Wireless access provides mobile computer users with the ability to connect to various public and private network resources and communicate with each other via e-mail and other messaging techniques. Through gateways, users of these wireless alternatives can communicate with any resource or person connected to a wireline network, and vice versa.

Being able to connect whenever and wherever desired greatly enhances this system's flexibility, and increased productivity results because users do not have to find or wait for an available phone line to make a connection. There is also the convenience of Internet and e-mail accounts being in one location.

Wireless access seems to be affordable for business organizations. A flat monthly rate provides everything users need to operate the system, including wireless modems, Web browsers, and messaging software. Numerous optional services are also available, such as headline news and stock quotes.

Wireless Access Methods

There are several methods to wirelessly access the Internet, including analog cellular networks, Cellular Digital Packet Data networks, packet radio services, and satellite services.

CELLULAR NETWORKS Analog cellular phone subscribers can send files and e-mail wirelessly via the Internet by connecting a modem to their phones. Of course, this method of access requires that users have a cell phone in the first place, as well as an adapter cable to connect the phone to the modem. Cellular modems work anywhere a cell phone does. The problem is that they work only as well as the cell phone does at any given moment. Checking e-mail on the Internet can be slow, and connection quality varies from network to network.

To access Web content on the Internet, however, really requires a digital cellular service (also called PCS, for personal communication service) and a special phone equipped with a liquid crystal display (LCD) screen. AT&T's PocketNet phone, for example, lets users access the Internet in areas served by its Cellular Digital Packet Data (CDPD) network.

With this type of cellular service, standard analog cell phones will not work; a digital CDPD phone is required. An alternative is to use a dual-mode phone, which operates in standard analog cellular mode for voice conversations and CDPD mode for access to the Internet at 19.2 Kbps. AT&T's PocketNet service includes a personal information manager (PIM) that contains an address book, calendar, and to-do list that are maintained on AT&T's Web site and can be accessed through the phone. The personal address book is tied to the PocketNet phone's "easy dialing" feature for fast, convenient calling.

CDPD service is available in only 20 metropolitan areas, so it is not as ubiquitous as standard analog cellular services. However, there is a specification under consideration that expands coverage nationwide by creating a gateway between CDPD and circuit-switched cellular. Developed by AT&T, Ameritech, GTE, and others, the gateway allows customers to access the CDPD network from anywhere in the country—either through the cellular network or via a landline connection.

Linking the two types of cellular networks could prove crucial to the future of CDPD. It effectively gives cellular service providers a way to deliver nationwide CDPD—even though the digital packet transport is now up and running in a small number of cities. Once users access the CDPD system, they can take advantage of all its special features, including security, customized billing, and access to the Internet.

Another feature offered by CDPD is mobility management, which routes messages to users regardless of the location or the technology. With gateways in place, the cellular network will be able to recognize when subscribers move out of the CDPD coverage area and transfer messages to them via circuit-switched cellular.

PACKET RADIO NETWORKS   Metricom, which provides network solutions and wireless data communications for industrial and PC applications, has developed the Ricochet service, which comprises radios, wired access points (WAP), and network interconnection facilities (NIF). The service enable data to be sent across a network of intelligent radio nodes.

The network uses low-power, frequency-hopping, spread-spectrum packet radios, a large number of which are installed throughout a geographical region in a mesh topology, usually placed on top of street lights or utility poles. These radio receivers, about the size of a shoe box, are also referred to as *microcell radios.* Only a small amount of power is required for the radio, which is received by connecting a special adapter to the street light. No special wiring is required. These radios are placed about every quarter to a half mile, and take only about five minutes each to install. Using 162 frequency-hopping channels in a random pattern accommodates many users at the same time and provides a high degree of security.

Distributed among the radios are WAPs, which route the wireless packets to the wired backbone. Gathering and converting the packets so they can be transmitted to the wired backbone is accomplished via a T1-based frame relay connection. The packets can then be sent to another WAP, the Internet, or the appropriate service provider. If the packet is sent to another WAP, it is being used as an alternate route through the mesh. The number of paths a packet can take through the network enhances speed throughput, since network blockage is not as frequent and many possible repeaters exist for the packet.

The Ricochet modem weighs only eight ounces and is thus very portable. It can be plugged into the serial port of a computer and can be connected to online services and networks just like a standard phone modem. The modem works with most communications software, both Intel-based and Macintosh hardware platforms, and Windows 3.1, Windows 95, and Macintosh operating systems.

Ricochet's wired backbone is based on standard Internet Protocol (IP) technology, routing data via a metropolitan service area. If a data packet has to move across the country, Ricochet's NIF system is used. The NIF functions as a router, collecting packets from the WAPs, and using leased lines to connect with NIFs situated in the different metropolitan areas. A Name Server is part of the Ricochet network backbone, providing security by validating all connection requests.

At this writing, Metricom is testing a new wireless data technology that provides data rates equal to that provided by wired ISDN 128K service. This new technology, called Ricochet II, uses two bands of unlicensed spectrum: 900 MHz and 2.4 GHz. The new network will be compatible with the company's existing Ricochet network and modems and will provide current Ricochet subscribers with existing modems increased performance, comparable to a wired 56K modem. Subscribers with the new Ricochet II modems will have performance up to ISDN data rates.

Other radio networks that provide access to the Internet for messaging applications provide more points of access than Ricochet. These include the RAM Mobile Data Network and ARDIS. As the oldest network, RAM Mobile has the best coverage, extending to 93 percent of the U.S. business population. It is also supported in Canada via the Cantel network. But it also has the lowest data rate—just 4.8 Kbps, which is fast enough for text-only e-mail but too slow for binary attachments or Web browsing.

ARDIS has been upgrading its network to 19.2 Kbps, which matches the speed of CDPD. ARDIS reaches the top 400 metropolitan areas in the U.S., serving 80 percent of the population and 90 percent of business locations.

SATELLITE SERVICES Satellites are a powerful solution for the rapidly growing demand to transmit Internet and other networking traffic, because they offer reliable connections to virtually anywhere in the world.

CyberStar, for example, offers a global broadband IP multicasting service that allows business users to send high-bandwidth voice, video, and data files to branch offices. Customers pay for the bandwidth they use rather than signing up for a flat-rate service. Pricing is based on the amount of traffic sent each month and the number of sites that receive the traffic. Customers are charged initial installation costs per site, which includes antennas, satellite receiver cards, and service activation fees.

Instead of launching its own satellites to support its services, CyberStar relies on its sister company, SkyNet, for its geosynchronous earth orbit (GEO) satellites. CyberStar and SkyNet are subsidiaries of Loral Space & Communications. CyberStar's existing services are based on store-and-forward technology, but the company plans to launch a videostreaming service that will let users send traffic over CyberStar's network in real time.

Other satellite service providers also support Internet traffic. DigitalXpress, for example, offers its XpressNet satellite-based Internet access service. The service allows companies to link LANs in far-flung locations to the Internet at 512 Kbps. The service requires users to purchase a receiving dish and an integrated receiver decoder device. In addition to the one-time equipment cost, users need a dial-up landline connection to the Internet. Additional monthly fees include a charge for sending traffic upstream through the satellite and a usage charge for every 100MB of traffic sent through the satellite.

Hughes Network Systems (HNS) has been offering its DirecPC Internet access service for several years, as well as a LAN-based Internet access service called DirecPC Network Edition, which is limited to NetWare LANs. This service supports Internet access speeds of 400 Kbps. Now HNS also offers a Windows NT version of the service. The service requires a receiving dish and an ISA card, as well as software. DirecPC Network Edition charges for every 64MB of traffic sent over the satellite. Like XpressNet, Network Edition requires users to have a dial-up connection to an ISP.

In the near future, low earth orbit (LEO) satellites such as Teledesic will provide Internet access and support other broadband applications, with bandwidth on demand ranging from 64 Kbps to 155 Mbps.

Service Caveats

Service availability may be an issue with some types of wireless services. Metricom's Ricochet network, for example, is established in only three cities—Seattle, San Francisco, and Washington, D.C.—with extensive suburban coverage in each area. Access is also available at 11 major airports, including New York's LaGuardia, Minneapolis-St.Paul, and Los Angeles. However, access is not available at Chicago's O'Hare or in Atlanta, which are among the world's two most visited airports. Metricom's expansion rate elsewhere depends on one-by-one contracts with cities and utility companies to hang its transceivers.

Radio modems are a reliable way to stay in touch from the road, but their range is limited to areas where service is available. Fortunately, radio service is available in all medium to large metropolitan areas and surrounding communities.

With regard to satellite services, the equipment and service charges are much higher than for other wireless services. In addition, broadband satellite data services are a good three to five years away from availability. The other current unknown is pricing. Satellite service providers, including Teledesic, have not said much about service costs, except that rates will be determined by resellers. LEO providers have a tough road ahead striking international deals with carriers and ISPs to provide uniform global coverage and support. It remains to be seen how these relationships will work out.

Wireless Access Protocol

The Wireless Access Protocol (WAP) Forum has published the first version of a wireless communication specification it hopes will accelerate the use of cellular phones and other wireless devices to access the Internet and advanced telephony services.

The WAP Forum—which includes founders Ericsson, Motorola, and Nokia—hopes to establish the specification as an industry standard. Such a standard would offer potential economies of scale, encouraging cellular phone and other device manufacturers to invest in developing compatible products.

WAP Version 1.0 works with most major network types, including GSM (Global Systems for Mobile) communications and CDMA (Code Division Multiple Access). The specification includes the following architectural elements:

- Wireless Application Environment, which defines the wireless markup language, script language, and wireless telephony application interface
- A transport layer that includes support for Datagram Protocol, Transaction Protocol, Control Message Protocol, and Transport Layer Security
- A Wireless Session Protocol specification that enables both connection-mode and connectionless services

Unwired Planet's UP.Browser software already supports WAP. This browser is available for CDPD, GSM, and CDMA handsets and allows users to explore the Web and retrieve content. More vendors have joined the

WAP bandwagon and will support it in their products without waiting for it to become a standard.

SUMMARY

In the future, people will be able to perform any white-collar task at any place in the world over the Internet—even as they travel by air from one location to another. This will be facilitated by new LEO satellite networks, such as Teledesic, which hand off the sessions from one satellite to another to provide users with uninterrupted Internet access.
See Also **Third-Generation Wireless Networks**

Wireless Internetworking

Corporations are making greater use of wireless technologies for extending the reach of LANs where a wired infrastructure is absent, impractical, or too costly to install. Wireless bridges and routers can extend data communication between buildings in a campus environment or between buildings in a metropolitan area. A variety of technologies can extend the reach of LANs to remote locations, including microwave, laser, and spread spectrum. All rely on directional antennas at each end and a clear line of site between locations.

Wireless Bridges

Short-haul microwave bridges, for example, provide an economical alternative to leased lines or underground cabling. Because they operate over very short distances—less than a mile—and are less crowded, they are less stringently regulated and have the additional advantage of not requiring an FCC license.

The range of the bridge is determined by the type of directional antenna. A four-element antenna, for example, provides a wireless connection of up to one mile. A ten-element antenna provides a wireless connection of up to three miles.

Directional antennas require a clear line of sight. To ensure accurate alignment of the directional antennas at each end, menu-driven diagnostic software is used. Once the antennas are aligned and the system ID and channel are selected with the aid of configuration software, the system is operational. Front-panel LEDs provide a visual indication of link status and traffic activity. The bridge unit has a diagnostic port, allowing performance monitoring and troubleshooting through a locally attached terminal or remote computer connected via modem.

Because it is fully compatible with the IEEE 802.3 Ethernet standard, microwave bridges support all Ethernet functionality and applications without the need for special software or network configuration changes. For Ethernet connections, the interface between the microwave equipment and network is virtually identical to that between a LAN and any cable medium, where retiming devices and transceivers at each end of the cable combine to extend the Ethernet cable segments. Typically, microwave bridges support all Ethernet media types via AUI connectors for thick Ethernet (10Base5), 10Base2 connectors for thin Ethernet, and twisted pair connectors for 10BaseT Ethernet. These connections allow microwave bridges to function as an access point to wired LANs.

Like conventional Ethernet bridges, microwave bridges perform packet forwarding and filtering to reduce the amount of traffic over the wireless segment. The microwave bridges contain Ethernet address filter tables, which help to reduce the level of traffic through the system by passing only the Ethernet packets bound for an inter- or intrabuilding destination over the wireless link. Since the bridges are self-learning, the filter tables are automatically filled with Ethernet addresses as the bridge learns which devices reside on its side of the link. That way, Ethernet packets not destined for a remote address remain local. The table is dynamically updated to account for equipment added or deleted from the network. The size of the filter table can be 1,000 entries or more, depending on vendor.

With additional hardware, microwave bridges can pull double-duty as a backup to local T1/E1 facilities. When a facility degrades to a preestablished error-rate threshold or is knocked out of service entirely, the traffic can be switched over to a wireless link to avoid loss of data. When line quality improves or the facility is restored to service, the traffic is switched from the wireless link to the wireline link.

Wireless Routers

Wireless, remote-access routers scale wireless-connect, geographically disbursed LANs by creating a wireless WAN over which network traffic is routed at distances of 30 miles or more via microwave, laser, or spread-spectrum technology.

Applications of wireless routers include remote-site LAN connectivity and network service dissemination. Organizations with remote offices such as banks, health care networks, government agencies, schools, and other service organizations can connect their computing resources. Industrial and manufacturing companies can reliably and cost-effectively connect factories, warehouses, and research facilities. Network service providers can distribute Internet, VSAT, and other network services to their customers. Performance is comparable to commonly used wired WAN connections, approaching T1 speeds with a 1.3-Mbps data rate.

Unlike wireless bridges, which simply connect LAN segments into a single logical network, wireless routers function at the network layer with IP/IPX routing, permitting the network designer to build large, high-performance, manageable networks. Wireless routers are capable of supporting star, mesh, and point-to-point topologies that are implemented with efficient MAC protocols. These topologies can even be combined in an internetwork.

A polled protocol (star topology) provides efficient shared access to the channel even under heavy loading (Figure 133). In the star topology, remote stations interconnect with the central base station and with other remote stations through the base station. Only one location needs a line of site to the remotes. Networks and workstations at each location tie into a common internetwork. The maximum range between the central base station and remote stations is approximately 15 miles.

For small-scale networks, a CSMA/CA protocol supports a mesh topology (Figure 134). In the mesh topology, each site must have a line of sight to every other site. The CSMA/CA protocol ensures efficient sharing of

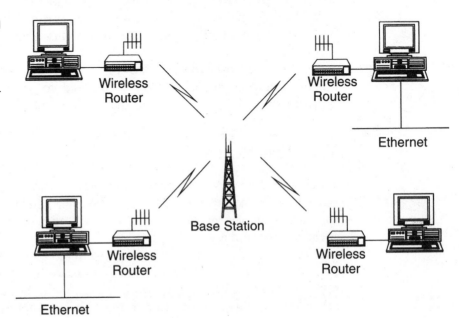

Figure 133
Star topology of a router-based wireless WAN.

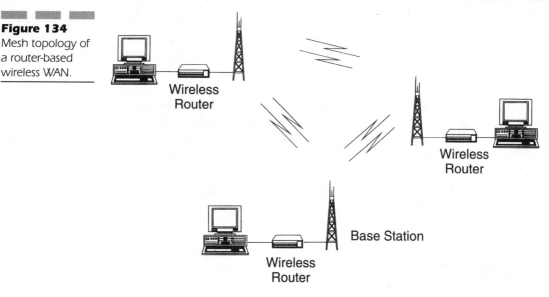

Figure 134

Mesh topology of a router-based wireless WAN.

the radio channel. The range with omnidirectional antennas is up to 3.5 miles.

The point-to-point topology is useful where there are only a few sites (Figure 135). A router can also function as a repeater link between sites. The single-hop, node-to-node range is up to 30 miles, depending on factors such as terrain and antennas, with a multiple hop range extending up to a hundred miles. Clusters of nodes can also be connected using a point-to-point protocol for large-scale internetworks.

Network Management

In addition to the Windows-based tools for setting up and configuring wireless LANs, the Simple Network Management Protocol (SNMP) is available for managing wireless internetwork end to end. The same SNMP

Figure 135

A point-to-point topology of a router-based wireless WAN.

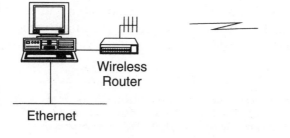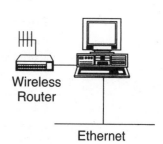

tools for managing the rest of the enterprise network can support the management of the bridges and routers on a wireless WAN. The result is a manageable network with reach extending to metropolitan, suburban, rural, remote, and isolated areas.

SNMP is usually implemented with a vendor-supplied proxy agent. This is an application that continually polls the managed devices for changes in alarm and status information and updates its locally stored Management Information Base (MIB). Events such as major and minor alarms cause the device to generate enterprise-specific traps directed to the network management system (NMS). General alarm and status information stored in the MIB is made available to the NMS in response to SNMP's Get and GetNext requests.

SUMMARY

Interconnecting LANs with bridges and routers that use wireless technologies is an economical alternative to leased lines and carrier-provided services. Installation, setup, and maintenance are fairly easy with the graphical management tools provided by vendors.

See Also **Wireless Centrex, Wireless Home Networking, Wireless Internet Access, Wireless LANs, Wireless Local Loops, Wireless PBXs**

Wireless LANs

A wireless local-area network (LAN) is a data communications system implemented as an extension—or as an alternative—to a wired LAN. Using a variety of technologies including narrowband radio, spread spectrum, and infrared.[6] Wireless LANs transmit and receive data through the air, minimizing the need for wired connections.

Applications

Wireless LANs have gained strong popularity in a number of vertical markets, including health care, retail, manufacturing, warehousing, and academia. These industries have profited from the productivity gains of

---

6. For a more in-depth discussion of these technologies, see my book "*Desktop Encyclopedia of Telecommunications*," published by McGraw-Hill (ISBN: 0-0704-4457-9). This 600-page reference can be conveniently ordered through the Web at Amazon.com.

using hand-held terminals and notebook computers to transmit real-time information to centralized hosts for processing. The following are among the many applications of wireless LANs:

- Hospital staff are more productive because hand-held or notebook computers with wireless LAN capability deliver patient information, regardless of their location.

- Consulting or accounting audit teams, small workgroups, or temporary office staff can increase productivity with quick network setup.

- Network managers in dynamic environments can minimize the overhead caused by moves, extensions to networks, and other changes with wireless LANs.

- Network managers can quickly install networked computers in older buildings with wireless technology, without having to upgrade existing wiring or install new wiring.

- Warehouse workers can use wireless LANs to exchange information with central databases, thereby increasing productivity.

- Branch office workers can minimize setup requirements by installing preconfigured wireless LANs.

Advantages

With wireless LANs, users can access shared information without looking for a place to plug in, and corporate managers can set up or augment networks without installing or moving wires. Wireless LANs offer the following advantages:

MOBILITY Wireless LANs can provide users with access to real-time information anywhere in an organization. This mobility improves productivity.

INSTALLATION SPEED AND SIMPLICITY Installing a wireless LAN can be fast and easy, since it eliminates the need to pull cable through walls and ceilings.

INSTALLATION FLEXIBILITY Wireless technology allows users to go where wires cannot go.

REDUCED COST-OF-OWNERSHIP While the initial investment required for wireless LAN hardware can be higher than the cost of wired LAN hardware, overall installation expenses and life-cycle costs can be significantly lower. Long-term cost benefits are greatest in dynamic environments requiring frequent moves, adds, and changes.

SCALABILITY Wireless LANs can be configured in a variety of topologies to meet the needs of specific applications and installations. They can be expanded with the addition of access points and extension points to accommodate virtually any number of users.

Technologies

There are several technologies to choose from when selecting a wireless LAN solution, each with its own advantages and limitations.

NARROWBAND A narrowband radio system transmits and receives user information on a specific radio frequency. Narrowband radio keeps the radio signal frequency as narrow as possible to pass the information. Undesirable cross-talk between communications channels is avoided by carefully coordinating different users on different channel frequencies.

A radio frequency is much like a private telephone line in that people on one line cannot listen to the calls made on other lines. In a radio system, privacy and noninterference are accomplished by the use of separate radio frequencies. The radio receiver filters out all radio signals except the ones on its designated frequency.

SPREAD SPECTRUM Most wireless LAN systems use spread-spectrum technology, a wideband radio frequency technique developed by the military for use in reliable, secure, mission-critical communications systems. Spread spectrum is designed to trade off bandwidth efficiency for reliability, integrity, and security. In other words, more bandwidth is consumed than in the case of narrowband transmission, but the trade off produces a signal that is, in effect, louder and thus easier to detect, provided that the receiver knows the parameters of the spread-spectrum signal being broadcast. If a receiver is not tuned to the right frequency, a spread-spectrum signal looks like background noise. There are two types of spread-spectrum radio: frequency hopping and direct sequence.

FREQUENCY HOPPING Frequency-hopping spread spectrum (FHSS) uses a narrowband carrier that changes frequency in a pattern known to both transmitter and receiver. Properly synchronized, the net effect is to maintain a single logical channel. To an unintended receiver, FHSS appears to be short-duration impulse noise.

DIRECT SEQUENCE Direct-sequence spread spectrum (DSSS) generates a redundant bit pattern for each bit to be transmitted. This bit pattern is called a chip (or chipping code). The longer the chip, the greater the probability that the original data can be recovered. Of course, this method requires more bandwidth. If one or more bits in the chip are damaged during transmission, statistical techniques embedded in the radio can recover the original data without the need for retransmission. To an unintended receiver, DSSS appears as low-power wideband noise and is ignored.

INFRARED   Infrared (IR) systems use very high frequencies just below visible light in the electromagnetic spectrum. Like light, IR cannot penetrate opaque objects; to reach the target system, the waves carrying data are sent in either directed (line-of-sight) or diffuse (reflected) fashion. Inexpensive directed systems provide a very limited range of not more than three feet. They are typically used for personal-area networks, but are occasionally used in specific wireless LAN applications. High-performance, directed IR is impractical for mobile users and is therefore used only to implement fixed sub-networks. Diffuse IR wireless LAN systems do not require line-of-sight, but cells are limited to individual rooms.

Operation

As noted, wireless LANs use electromagnetic waves (radio or infrared) to communicate information from one point to another without relying on a wired connection. Radio waves are often referred to as radio carriers because they simply perform the function of delivering energy to a remote receiver. The data being transmitted is superimposed on the radio carrier so it can be accurately extracted at the receiving end. This process is generally referred to as carrier modulation. Once data is modulated onto the radio carrier, the radio signal occupies more than a single frequency since the frequency or bit rate of the modulating information adds to the carrier.

Multiple radio carriers can exist in the same space at the same time without interfering with each other if the radio waves are transmitted on different radio frequencies. To extract data, a radio receiver tunes into one radio frequency while rejecting all other frequencies.

In a typical wireless LAN configuration, a transmitter/receiver (transceiver) device, called an access point, connects to the wired network from a fixed location using standard cabling. At a minimum, the access point receives, buffers, and transmits data between the wireless LAN and the wired network infrastructure. A single access point can support a small group of users and can function within a range of less than one hundred to several hundred feet. The access point (or the antenna attached to the access point) is usually mounted high but may be mounted essentially anywhere that is practical as long as the desired radio coverage is obtained.

Users access the wireless LAN through wireless LAN adapters, which provide an interface between the client network operating system (NOS) and the airwaves via an antenna. The nature of the wireless connection is transparent to the NOS.

Configurations

Wireless LANs can be simple or complex. The simplest configuration consists of two PCs equipped with wireless adapter cards that form a network whenever they are within range of one another (Figure 136). This peer-to-peer network requires no administration. In this case, each client would have access to only the resources of the other client and not to a central server.

Installing an access point can extend the operating range of a wireless network, effectively doubling the range at which the devices can communicate. Since the access point is connected to the wired network, each client could access the server's resources as well as to other clients (Figure 137). Each access point can support many clients—the specific number depends on the nature of the transmissions involved. In some cases, a single access point can support up to 50 clients.

Figure 136
A wireless peer-to-peer network.

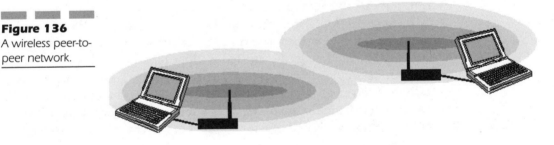

Figure 137
A wireless client connected to the wired LAN via an access point.

Access Point

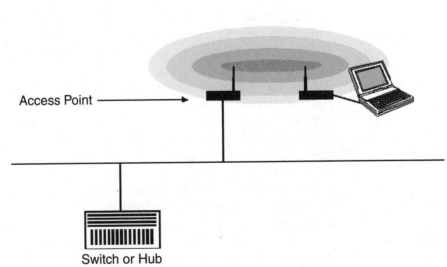

Switch or Hub

Access points have an operating range of about 500 feet indoors and 1000 feet outdoors. In a very large facility such as a warehouse or on a college campus, it will probably be necessary to install more than one access point (Figure 138). Access point positioning is determined by a site survey. The goal is to blanket the coverage area with overlapping coverage cells so clients can roam throughout the area without ever losing network contact. Access points hand the client from one to another in a way that is invisible to the client, ensuring uninterrupted connectivity.

To solve particular problems of topology, network designers might choose to use extension points (EPs) to augment the network of access points (Figure 139). Extension points look and function like access points (APs), but they are not tethered to the wired network as are APs. EPs function just as their name implies; they extend the range of the network by relaying signals from a client to an AP or another EP.

Another component of wireless LANs is a directional antenna. If a wireless LAN in one building must be connected to a wireless LAN in another building a mile away, one solution would be to install a directional antenna on the two buildings—each antenna targeting the other and connected to its own wired network via an access point (Figure 140).

Figure 138

Multiple access points extend coverage and enable roaming.

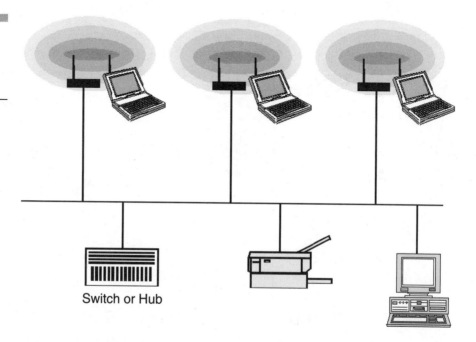

Switch or Hub

Figure 139
Use of an
extension point in
a wireless
network.

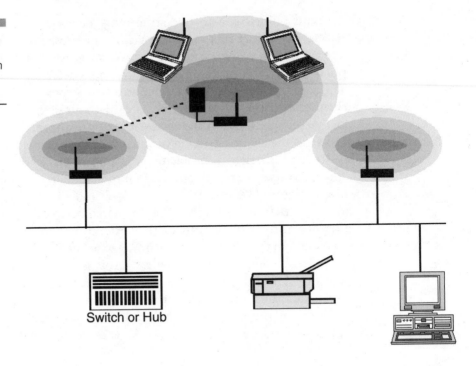

Switch or Hub

Figure 140
A directional
antenna can be
used to
interconnect
wireless LANs in
different buildings.

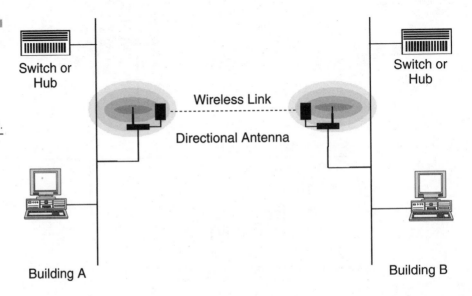

Switch or
Hub

Wireless Link

Directional Antenna

Switch or
Hub

Building A

Building B

Network Management

Wireless LANs are typically set up and managed with Windows-based tools. Lucent Technologies, for example, offers a Windows-based site survey tool to facilitate remote management, configuration, and diagnosis of its spread-spectrum WaveLAN wireless LANs, which include access points and adapters that are available in 900-MHz and 2.4-GHz versions.

WaveManager makes it easy for system administrators to monitor the quality of communications at multiple WaveLAN stations in a wireless network (Figure 141). It can also be used to verify building coverage, identify coverage patterns, select alternate frequencies, locate and tune around RF interference, and customize network access security. Five basic functions are offered by WaveManager:

COMMUNICATIONS INDICATOR This is located on the Windows 95 taskbar and provides mobile users with graphical, real-time information on the level of communication quality between a WaveLAN station and the nearest WavePoint access point.

LINK TEST DIAGNOSTICS Verifies the communications path between neighboring WaveLAN stations, as well as between a WaveLAN station

Figure 141
Lucent's
WaveManager
provides an
administrative
graphical user
interface through
which WaveLAN
wireless LANs can
be configured,
managed, and
maintained.

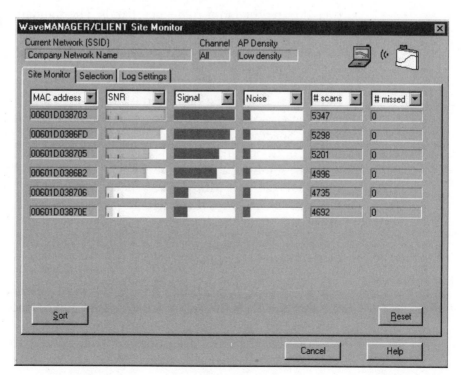

and WavePoint access points within one wireless cell. Link Test Diagnostics measure signal quality, signal-to-noise ratio, and the number of successfully received packets.

SITE MONITOR Ensures optimal placement of WavePoint access points. While carrying a WaveLAN-equipped computer through the facility, Site Monitor graphically displays changing communication quality levels with the various access points installed in the building. This tool makes it easy to locate radio dead spots or sources of interference.

FREQUENCY SELECT Manages RF Channel selection. It enables users to choose from up to eight different channels (in the 2.4-GHz frequency band).

ACCESS CONTROL TABLE MANAGER Enables system administrators to provide extra levels of security by restricting access to individual computers in a facility.

Wireless LAN Standard

A wireless LAN standard has been in the making since 1990. In mid-1997, the IEEE 802.11 committee defined a transmission rate of up to 2 Mbps over infrared or radio frequency bands. The standard also includes the media access control protocol Carrier Sense Multiple Access with Collision Avoidance (CSMA/CA), which allows devices implementing the standard to interoperate with wired Ethernet LANs.

The standard provides for an optical-based, physical-layer implementation that uses infrared light to transmit data. It also provides two physical-layer choices based on the radio frequency (RF): Direct Sequence Spread Spectrum (DSSS) and Frequency Hopping Spread Spectrum (FHSS). Both operate in the 2.4-GHz ISM band.

In the standard, a 2-Mbps peak data rate is specified for DSSS with optional fallback to 1 Mbps in very noisy environments. The standard defines the FHSS implementation to operate at 1 Mbps and allows for optional 2-Mbps operation in very clean environments.

The 802.11 media access control (MAC) can work seamlessly with standard Ethernet via a bridge to ensure that wireless and wired nodes on an enterprise LAN are logistically indistinguishable and can interoperate. The 802.11 MAC is necessarily different from the wired Ethernet MAC— the wireless LAN standard uses a carrier sense multiple access with collision-avoidance scheme, whereas standard Ethernet uses a collision-detection scheme—but the difference is masked by an access point that connects a wireless LAN channel to a wired LAN backbone.

The roaming provisions built into 802.11 also provide several advantages. It includes mechanisms to allow a client to roam among multiple access points that can operate on the same or separate channels. For example, an access point transmits a beacon signal at regular intervals. Roaming clients use the beacon to gauge the strength of their existing connection to an access point. If the connection is weak, the roaming station can attempt to associate itself with a new access point.

Although the 802.11 standard addresses roaming, it is with the understanding that all the access points in an installation are manufactured by the same vendor. The standard does not ensure that clients can roam among access points from different vendors. With the advent of 802.11 products, however, users may want to mix and match access points. For example, some customers might need standard commercial-grade bridges in the office but want ruggedized bridges for the factory floor.

To address multivendor roaming, Aironet Corp., Digital Ocean, Inc. and Lucent have collaborated to develop the Inter Access Point Protocol (IAPP) specification. That will extend the 802.11 multivendor interoperability benefits with comprehensive roaming protocols. Several others, including IBM, have voiced support for IAPP as a necessary step toward true multivendor interoperability.

The 802.11 specification adds features to the MAC that can maximize battery life in portable clients via power-management schemes. Power management causes problems with wireless LAN systems because typical power-management schemes place a system in sleep mode (low or no power) when no activity occurs for a user-definable time period. This can cause a sleeping system to miss critical data transmissions. To support clients that periodically enter sleep mode, the 802.11 standard specifies that access points include buffers to queue messages.

The 802.11 standard also addresses data security. The standard defines a mechanism through which the wireless LANs can achieve Wired Equivalent Privacy (WEP). The optional WEP mechanism is especially important because RF transmissions—even spread-spectrum transmissions—can be intercepted more easily than wired transmissions.

The next step in the evolution of the 802.11 standard will likely be the inclusion of higher data rates, expected in the 10-Mbps and above range in the 5.2-GHz band. The higher speed could encourage development of such applications as streaming video, telephony, and multimedia for wireless networks.

SUMMARY ▉ ▉ ▉ ▉ ▉ ▉ ▉ ▉

Wireless LANs provide all the functionality of wired LANs, without the physical constraints of the wire itself. Wireless LAN configurations range from simple peer-to-peer topologies to complex networks offering distributed data connectivity and roaming.

See Also **Wireless Internetworking**

Wireless Local Loops

Wireless local loop (WLL) is a generic term for an access system that uses wireless links rather than conventional copper wires to connect subscribers to the local telephone company's switch. Also known as fixed wireless access (FWA) or simply fixed radio, this type of system uses analog or digital radio technology to provide telephone, facsimile, and data services to business and residential subscribers.

WLL systems provide rapid deployment of basic phone service in areas where the terrain or telecommunications development makes installation of traditional wireline service too expensive. WLL systems can be easily integrated into a wireline public switched telephone network (PSTN) and can usually be deployed within a month of equipment delivery, far more quickly than traditional wireline installations, which can take several months for initial deployment and years to grow capacity to meet the pent up demand for communication services.

WLL solutions include analog systems for medium- to low-density and rural applications. For high-density, high-growth urban and suburban locations, there are WLL solutions based on Code Division Multiple Access (CDMA). TDMA (Time Division Multiple Access) and GSM (Global System for Mobile) telecommunications systems are also offered. In addition to being able to provide higher voice quality than analog systems, digital WLL systems can support higher-speed fax and data services.

WLL technology is also generally compatible with existing operations support systems (OSS), as well as existing transmission and distribution systems. WLL systems are scalable, enabling operators to leverage their previous infrastructure investments as the system grows.

WLL subscribers receive phone service through a radio unit linked to the PSTN via a local base station.[7] The radio unit consists of a transceiver,

---

7. The International Telecommunication Union (ITU) V5.2 open standard interface enables network operators to mix and match local exchange equipment and local access equipment, irrespective of competing suppliers. Enhancements to V5.2 enable the interconnection of WLLs with existing PSTN switching platforms.

power supply, and antenna. It operates off ac or dc power, may be mounted indoors or outdoors, and usually includes battery backup for use during line power outages. On the customer side, the radio unit connects to the premise's wiring, enabling customers to use existing phones, modems, fax machines, and answering devices (Figure 142).

WLL subscribers can access all the usual voice and data features, such as caller ID, call forwarding, call waiting, three-way calling, and distinctive ringing. Some radio units provide multiple channels, which are equivalent to having multiple lines. The radio unit offers service operators the advantage of over-the-air programming and activation to minimize service calls and network management costs.

Figure 142

The fixed wireless terminal is installed at the customer location. It connects several standard terminal devices (telephone, answering machine, fax, computer) to the nearest cell site Base Transceiver Station (BTS).

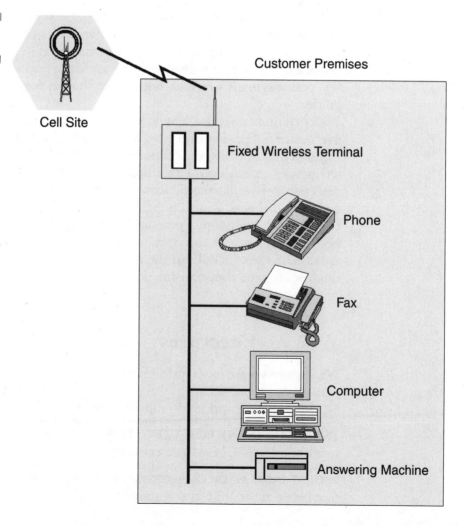

Cell Site

Customer Premises

Fixed Wireless Terminal

Phone

Fax

Computer

Answering Machine

The radio unit contains a coding and decoding unit that converts conventional speech into a digital format during voice transmission and back into a nondigital format for reception. Many TDMA-based WLL systems use the 8-Kbps Enhanced Variable Rate Coder (EVRC), a Telecommunications Industry Association (TIA) standard (IS-127). EVRC provides benefits to both network operators and subscribers.

For operators, the high-quality voice reproduction of EVRC does not sacrifice the capacity of a network nor the coverage area of a cell site. An 8-Kbps EVRC system, using the same number of cell sites, provides network operators with more than 100-percent additional capacity than the 13-Kbps voice coders deployed in CDMA-based WLL systems. In fact, an 8-Kbps EVRC system requires at least 50-percent fewer cell sites than a comparable 13-Kbps system to provide similar coverage and in-building penetration.

For subscribers, 8-Kbps EVRC systems use a state-of-the-art background noise-suppression algorithm to improve the quality of speech in noisy environments, typical of urban streets where there is heavy pedestrian and vehicular traffic. This is also an advantage compared to traditional landline phone systems, which do not have equivalent noise-suppression capabilities.

Depending on vendor, a radio unit may also include special processors to enhance call privacy on analog WLL systems. Voice privacy is enhanced through the use of a speech coder based on Digital Signal Processor (DSP) technology, an echo canceler, a data encryption algorithm, and an error detection/correction mechanism. To prevent eavesdropping, the speech data (encoded with a low bit rate), is encrypted using a private key algorithm, which is randomly generated during a call. The key is used by the DSPs at both ends of the communications link to decrypt the received signal. The use of a DSPs in the radio units of analog WLL systems also improves fax and data transmission.

WLL Architectures

WLL systems come in several architectures: a PSTN-Based Direct Connect network, a Mobile Telephone Switching Office/Mobile Switching Center (MTSO/MSC) network, and proprietary networks.

PSTN-BASED DIRECT CONNECT   There are several key components of a PSTN Direct Connect network:

PSTN-TO-RADIO INTERCONNECT SYSTEM   Provides the concentration interface between the WLL and wireline network

SYSTEM CONTROLLER (SC) Provides radio channel control functions and serves as a performance monitoring concentration point for all cell sites

BASE TRANSCEIVER STATION (BTS) Is the cell site equipment that performs the radio transmit and receive functions

FIXED WIRELESS TERMINAL (FWT) Is a fixed radio telephone unit that interfaces to a standard telephone set acting as the transmitter and receiver between the telephone and the base station

OPERATIONS AND MAINTENANCE CENTER (OMC) Is responsible for the daily management of the radio network and provides the database and statistics for network management and planning

MTSO/MSC An MTSO/MSC-based network contains virtually the same components of the PSTN Direct Connect network, except that the MTSO/MSC replaces the PSTN-to-Radio Interconnect system. The key components of an MTSO/MSC-based network are:

- Mobile Telephone Switching Office/Mobile Switching Center (MTSO/MSC), which performs the billing and database functions and provides a T1/E1 interface to the PSTN
- Cell Site equipment, including the Base Transceiver Station (BTS)
- Fixed Wireless Terminal (FWT)
- Operations and Maintenance Center (OMC)

For digital systems such as GSM and CDMA, the radio control function is performed at the Base Station Controller (BSC) for GSM or the Centralized Base Site Controller (CBSC) for CDMA.

In GSM systems, there is a Base Station System Controller (BSSC), which includes the Base Station Controller (BSC) and the transcoder. The BSC manages a group of BTSs, acts as the digital processing interface between the BTSs and the MTSO/MSC, and performs GSM-defined call processing.

In CDMA systems, there is a Centralized Base Site Controller (CBSC), which consists of the Mobility Manager (MM) and the transcoder subsystems. The MM provides both mobile and fixed call processing control and performance monitoring for all cell sites as well as subscriber data to the switch.

As in PSTN-based networks, the FWT in MTSO/MSC-based networks is a fixed radio telephone unit that interfaces to a standard telephone set acting as the transmitter and receiver between the telephone and the base station.

Operations and maintenance functions are performed at the OMC. As in PSTN-based networks, the OMC in MTSO/MSC-based networks is responsible for the day-to-day management of the radio network and provides the database and statistics for network management and planning.

The PSTN Direct Connect network is appropriate when there is capacity on the existing local or central office switch. In this case, the switch continues to provide the billing and database functions, the numbering plan, and progress tones. The MTSO/MSC architecture is appropriate for adding a fixed subscriber capability to an already existing cellular mobile network or for offering both fixed and mobile services over the same network.

PROPRIETARY NETWORKS While MTSO/MSC-based and PSTN Direct Connect networks are implemented with existing cellular technologies, proprietary WLL solutions are designed specifically as replacements for wireline-based local loops. One of these proprietary solutions is Nortel's Proximity I, which is used in the U.K. to provide wireline-equivalent services in the 3.5-GHz band. The TDMA-based system was designed in conjunction with U.K. public operator Ionica, which is the source of the I designation. The I Series provides telecommunications service from any host network switch, providing toll quality voice, data, and fax services. The system is switch-independent and transparent to DTMF tones and switch features. The Proximity I system architecture consists of the following main elements:

- Residential service system (RSS), which is installed at the customer premises and provides a wireless link to the base station
- Base station, which provides the connection between the customer's RSS and the PSTN
- Operations, administration, and maintenance system, which provides functions such as radio link performance management and billing.

RESIDENTIAL SERVICE SYSTEM (RSS) The RSS offers two lines that can be assigned for both residential and home office use, or for two customers in the same 2-km area. Once an RSS is installed, the performance of the wireless link is virtually indistinguishable from a traditional wired link. The wireless link is able to handle high-speed fax and data via standard modems, as well as voice. The system supports subscriber features such as call transfer, intercom, conference call, and call pick-up.

The RSS has several components: a transceiver unit, residential junction unit (RJU), network interface unit, and power supply. The transceiver unit consists of an integral 30-cm octogonal array antenna with a radio transceiver encased within a weatherproof enclosure. The enclosure is mounted on the customer premises and points toward the local base station.

The RJU goes inside the house, where it interfaces with existing wiring and telephone equipment. The Proximity I system supports two 32-Kbps links for every house, allowing subscribers to have a voice conversation and data connection for fax or Internet access at the same time. At this writing, work is under way to develop systems that can handle ISDN speeds of 64 Kbps and beyond. Further developments will result in RSSs that can handle more lines per unit for medium-sized businesses or apartment blocks.

The network interface unit, mounted internally or externally, is a cable junction box that accepts connections from customer premises wiring. The unit also provides access for service provider diagnostics and contains lightning-protection circuitry.

The power unit is usually mounted internally and connects to the local power supply (110/220 volts ac). The power unit provides the dc supply to the transceiver unit. A rechargeable battery takes over in the event of a power failure and is capable of providing 12 hours of standby and 30 minutes of talk time.

BASE STATION The base station contains the radio frequency equipment for the microwave link between the customer's RSS and the PSTN, along with subsystems for call-signal processing, frequency reference, and network management. This connection is via radio to the RSS by either microwave radio, optical fiber, or wireline to the local exchange. The base station is modular and can be configured to meet a range of subscriber densities and traffic requirements. The base station has several components: transceiver microwave unit, cabinet, power supply, and network management module.

The base station's dual antenna transceiver microwave unit provides frequency conversion and amplification functions. Each unit provides three RF channels, the frequency of which can be set remotely. The unit can be configured for a maximum of 18 RF channels. The antennas are available in omnidirectional or sectored configurations, depending on population densities and geographical coverage. An omnidirectional system can support 600 or more customers, while a trisectored antenna can serve more than 2,000 customers. Base stations in rural areas can be sited up to 20 km from a subscriber's premises.

The base station can be configured with either an internal or external cabinet. The internal cabinet is for location in an equipment room, while the external cabinet is weather-sealed and vandal-proofed for outside locations. Both types of cabinets house the integrated transceiver system, transmission equipment, optional power system and batteries. A separate power cabinet provides dc power to the base station from the local 110/220-volt ac source. This cabinet may include battery backup with battery management capability and power distribution panel that provides power for technicians' test equipment.

The network management module is the base station polls individual RSS units to flag potential service degradation. Reports include link bit error rate (BER), signal-to-noise ratio, power supply failure, and the status of the customer standby battery.

The connection from the base stations to the local exchange on the PSTN is via the V5.2 open standard interface. In addition to facilitating interconnections between multivendor systems, this interface enables operators to take full advantage of Proximity I's ability to maximize spectrum usage through allocation of finite spectrum on a dynamic per-call basis, rather than on a per-customer basis. Concentration allows the same finite spectrum to be shared across a much larger number of customers, producing large savings in infrastructure, installation, and operation costs for the network operator.

OPERATIONS, ADMINISTRATION, AND MAINTENANCE OA&M functions are implemented through an element manager accessed through a field engineering terminal. In Nortel's Proximity I, the element manager is built around Hewlett Packard's OpenView. The network of base stations and customer equipment communicate through the Airside Management Protocol, which is based on the OSI Common Management Information Protocol (CMIP). The field engineering terminal can operate in a remote operations center, but is primarily intended for use by on-site maintenance engineers who are responsible for the proper operation of the base stations.

All the applications software in the customer premises equipment is downloadable from the element manager. This software provides the algorithms that convert analog voice signals into 32-Kbps digital ADPCM, which provides toll-quality voice transmission. Other application software includes algorithms for controlling the draw of battery delivered power, in the event of a 110/220-volt ac power failure.

Via the Air Interface Protocol, customer equipment can provide the element manager with information about its current status and performance, the most useful of which are measurements taken during the transmission of speech. This allows the management system to flag performance degradation for corrective action.

SUMMARY

Wireless local loops eliminate the need for laying cables and hard-wired connections between a local switch and the subscriber's premises, resulting in faster service startup and lower installation and maintenance costs. And because the subscriber locations are fixed, the initial deployment of radio base stations need provide coverage only to areas where immediate demand for service is apparent. Once the WLL system is in place, new customers can be added quickly and easily. Such systems support standard analog as well as digital services, and provide the capability to support the evolution to new and enhanced services as the needs of the market evolve. *See Also* **Wireless Centrex, Wireless LANs, Wireless PBX**

Wireless PBX

With office workers spending increasing amounts of time away from their desks—supervising various projects, working at temporary assignments, attending meetings, and just walking corridors—there is a growing need for wireless technology to help them stay in touch with colleagues, customers, and suppliers. The idea behind wireless PBX is to facilitate communication within the office environment, enabling employees to be as productive with a wireless handset as they would if they were sitting at their desk.

Applications

Typically, a wireless switch connects directly to an existing PBX, key telephone system, or Centrex service,[8] converting an office building into an intracompany microcellular system. This arrangement provides wireless telephone, paging, and e-mail services to mobile employees within the workplace through the use of pocket-sized portable phones, similar to those used for cellular service. Almost any organization can benefit from improved communications offered by a wireless PBX system, including:

MANUFACTURING Roving plant managers or factory supervisors do not have to leave their inspection or supervisory tasks to take important calls.

RETAIL Customers can contact in-store managers directly, eliminating noisy paging systems.

HOSPITALITY Hotel event staff can stay informed of guest's needs and respond immediately.

---

8. For a more in-depth discussion of PBXs, key telephone systems and Centrex, see my book *Desktop Encyclopedia of Telecommunications*, published by McGraw-Hill (ISBN: 0-0704-4457-9). This 600-page reference can be conveniently ordered through the Web at Amazon.com.

SECURITY Guards can relay emergency information quickly and clearly, directly to a control room or police department, without trying to reach a desktop phone.

BUSINESS Visiting vendors or customers have immediate use of preassigned phones without having to use employee offices.

GOVERNMENT In-demand office managers can be available at all times for instant decision-making.

A wireless PBX is especially appropriate in areas such as education and health care, or any operation with multiple buildings in a campus environment. In the health care industry, for example, a typical environment for a wireless PBX would be a hospital with staff members who are typically away from their workstations one third of the workday.

System Components

Many of the wireless office systems on the market today are actually adjunct systems that interface to an existing PBX that provides user features and access to wireline telephones and outside trunk carrier facilities. The advantages of this approach include cost savings in terms of hardware, space requirements, and power. A wireless PBX typically consists of several discrete components (Figure 143).

ADJUNCT SWITCH An adjunct contains the CPU and control logic. Its function is to manage the calls sent and received between the base stations. The adjunct is a stand-alone unit that can be wall-mounted for easy installation and maintenance. It can be collocated with the PBX or connected to the PBX via twisted pair or optical fiber from several thousand feet away. Optional battery backup is usually available, permitting uninterrupted operation should a power failure occur. System control, management, and administration functions are provided through an attached terminal that is password-protected to guard against unauthorized access.

As portable telephones and base stations are added to accommodate growth, line cards are added to the PBX and radio cards are added to the adjunct switch to handle the increasing traffic load. Each adjunct is capable of supporting several hundred portable telephones. Additional adjuncts can be added as necessary to support future growth.

BASE STATIONS Antenna-equipped base stations, about the size of smoke detectors, are typically mounted on the ceiling and connected by twisted pair wiring to the wireless PBX. They send and receive calls between the portable telephones and adjunct unit. As users move from one cell to another, the base station hands off the call to the nearest base sta-

Figure 143

A typical wireless PBX system in the corporate environment. In this case, workers can roam between the office and home using the same handset. When the handset moves within range of the local cellular service provider, the signal is handed off from the wireless PBX to the cellular carrier's nearest base station.

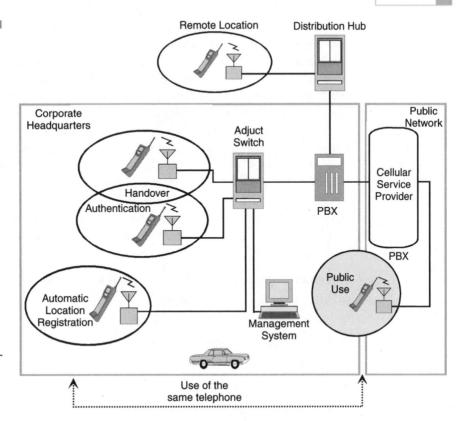

Remote Location   Distribution Hub

Corporate Headquarters

Public Network

Adjuct Switch

Cellular Service Provider

Handover

Authentication

PBX

PBX

Automatic Location Registration

Public Use

Management System

Use of the same telephone

tion with an idle channel. When the next base station grabs the signal, the channel of the former base station becomes idle and is free to handle another call.

To facilitate the handoff process, each base station can be equipped with dual antennas (antenna diversity). This improves signal detection, enabling the handoff to occur in a timely manner. This is accomplished when the base station samples the reception on each of its antennas and switches to the one offering the best reception. This process is continuous, ensuring the best voice quality throughout the duration of the call. Some vendors offer optional external antennas for outdoor coverage or directional coverage indoors.

Wireless PBX systems can be easily expanded—portable telephones and base stations added as needed. Eliminating traditional phone moves, adds, and changes can result in substantial savings over time. There is also significant savings in cabling since there is less need to rewire offices and other locations for desktop telephones.

TELEPHONE HANDSET Each portable phone has a unique identification number that must be registered with the adjunct switch. This allows only authorized users to access the communications system. The portable phone can be configured to have the same number as the user's desk phone, so when a call comes in both phones ring. A user can even start a conversation on one phone and switch to the other. If the portable and desktop phones have different numbers, the user can program each one to forward incoming calls to the other.

Since the adjunct switch becomes an integrated part of the company's existing telephone system, users can access all of its features through their portable phones. They can even set up conference calls, forward calls, and transfer calls. If the handset is equipped with a liquid crystal display (LCD), the unit can also be used to retrieve e-mail messages, faxes, and pages. An alphanumeric display shows the name and number of the person or company calling. The portable phone offers a number of other features, including:

- Private directory of stored phone numbers for quick dialing
- Multilevel last-number redial
- Audio volume, ring volume, and ring tone control
- Visual message-waiting indicator
- Silent vibrating alert
- Electronic lock for preventing outgoing calls
- In-range and out-of-range notifications
- Low battery notification

When the portable phone is not being used, a desktop unit houses the phone and charges both an internal and a spare battery. An LED indicates when the battery is fully charged. Recharging takes only a few hours and varies according to the type of battery used: nickel cadmium (NiCd) takes about two and a half hours, while nickel metal hydride (NiMH) and the newer lithium ion (Li-Ion) batteries take about one and a half hours. Li-Ion batteries offer longer life and are lighter weight than NiCd and NiMH batteries. Some vendors offer an intelligent battery charging capability that protects the battery from overcharging.

Distribution Hub

Distribution hubs are used in large installations to extend and manage communications among base units in remote locations that are ordinarily out of range of the adjunct unit. They also allow high-traffic locations to be divided into smaller cells, called *microcells,* with each cell containing multiple base stations. This arrangement makes more channels available to handle more calls.

The distribution units are connected to the adjunct unit with twisted pair wiring or optical fiber. Optical fiber is an ideal medium for an in-building wireless network because its low attenuation over distance (approximately one decibel per kilometer) allows high-quality coverage even in large buildings and campus environments. Fiber is also immune to electromagnetic interference, allowing it to work effectively alongside other electronic equipment in installations such as factories and warehouses.

Frequency Bands

Wireless PBXs operate in a variety of frequency bands, including the unlicensed 1910- to 1930-MHz Personal Communications Services (PCS) band. The term *unlicensed* refers to the spectrum used with equipment that can be bought and deployed without FCC approval because it is not part of the public radio spectrum. In other words, since wireless PBX operates over a dedicated frequency band for communications within a very narrow geographical area, it has little chance of interfering with other wireless services in the surrounding area. The individual channels supported by the wireless PBX system are spaced far enough apart to prevent interference with one another.

Standards

The Telecommunications Industry Association (TIA) and the American National Standards Institute (ANSI) have defined a North American standard that ensures interoperability between portable phones and wireless PBXs from different vendors. The TIA TR41.6.1 Subcommittee based its development of the Personal Wireless Telecommunications (PWT) standard on the Digital European Cordless Telecommunications (DECT) standard. Portable phones that support PWT, formerly known as the Wireless Customer Premises Equipment standard, will interoperate with PWT-compliant wireless PBXs from any vendor.

For a wireless handset to communicate with any wireless PBX, manufacturers of both devices must agree on how the signal should be handled. As part of the PWT standard, the Customer Premises Access Profile defines the features that each side of the air interface must support to provide full, multivendor interoperability for voice services. As with most standards, vendors can add proprietary extensions to support additional features and differentiate their products.

The air interface is a layered protocol, similar to the Open Systems Interconnection (OSI) architecture of the International Organization for

Standardization (ISO). Accordingly, the air interface is composed of four protocol layers:

PHYSICAL LAYER Specifies radio characteristics such as channel frequencies and widths, modulation scheme, and power and sensitivity levels. This layer also specifies the framing, so each handset can translate the bits it receives.

MEDIA ACCESS CONTROL (MAC) LAYER Specifies the procedures by which the portable phone and the base station, or antenna, negotiate the selection of the radio channels.

DATA LINK CONTROL (DLC) LAYER Specifies how frames are transmitted and sequenced between the handset and the base station.

NETWORK LAYER Specifies messages that identify and authenticate the handset to the wireless PBX.

Call Handoff Scenario

Examining the handoff from one base station to another can provide an illustration for the operation of these protocols. Handoff occurs when a mobile user walks out of the range of one base station and into the zone or cell of another base station. When the handset detects a change of signal strength from strong to weak, it will attempt to get acceptable signal strength from another channel offered by the same base station. If there is a better channel available, an exchange of messages at the MAC level occurs, which allows the conversation to continue without interruption. This channel change takes place without notification to the DLC layer.

If an acceptable channel is not available to the current base station, the handset searches for another base station. An exchange of messages at the DLC and MAC layers secures a data link via a radio channel to the new base station while the call through the original base station continues. When the data link to the second base station is established, the handset drops the old channel and begins processing the frames received through the new one. This process occurs without the network layer being notified. This means the caller and the wireless PBX are not aware that a handoff has happened.

SUMMARY

Businesses everywhere have put a high priority on increasing the productivity of their workforce, even while they continue to cut back on staff. In order to improve profitability, serve customers, and increase market share, organizations must find ways to do more—cheaper, faster, and better. Most companies are focused on increasing efficiency and productivity while reducing the time to market and improving customer service. This puts workers between a rock and a hard place; they must be mobile, away from their desks and offices, but not far from their telephones. Wireless PBX technology meets both demands.

See Also **Wireless Centrex, Wireless LANs**

Workflow Management

A workflow process provides an efficient assembly-line approach to compound document processing whereby operators at each stage develop or revise a portion of the information contained in the document before passing it on. Special workflow management software controls the movement of documents from station to station, typically eliminating redundant tasks and potential confusion concerning which version of a particular document is the most recent. Administrators typically assign new work to available individuals and monitor the quality of work. Workflow management tools are useful in a variety of document processing applications, including insurance claim processing and mortgage loan applications. Such tools are also useful in coordinating the development and maintenance of corporate Web sites.

Today's Web sites can be comprised of a virtually unlimited number of individual pages scattered across hundreds of servers. Each page may have its own set of links to more pages, applications, or databases. In addition to text, each of these pages can contain active image maps, animated graphics, imaged documents, Java applets, ActiveX controls, audio and video clips, and other objects. Because each of these components may require different skills and expertise to develop, many people can be involved in maintaining the content of a Web site. On large distributed Web sites, this greatly complicates the task of content management.

In these types of environments, special workflow management tools are required to aid content revisions, enforce quality standards, maintain a consistent look and feel, and speed production. With such tools, managers can keep track of the work being performed by all staff members and outside contractors to increase efficiency and eliminate redundancy.

Graphical Management Tools

Two representative products amply illustrate the capabilities of today's graphical workflow management tools: Aziza's Enterprise Web Manager and MKS's Web Integrity.

AZIZA'S ENTERPRISE WEB MANAGER   Aziza's Enterprise Web Manager decentralizes content authorship and centralizes management, allowing one or more administrators to manage multiserver sites.

The Aziza Enterprise Web Manager consists of three software components: a Web Object Manager, two content servers, and three management clients that complement a corporation's existing infrastructure of Web servers, browsers, page editors, and firewalls.

The first component, Web Object Manager, is the repository for both Web content and Web metadata. Web content can include text, images, applications, and other documents in their native form. Metadata includes title, author, version history, and creation, modification, and expiration dates.

Embedded in the Web Object Manager is the Verity search engine. Aziza uses Verity to index not only HTML but also native documents such as Lotus 123, Microsoft Word, Microsoft PowerPoint, and Adobe PDF. Verity can search not only full-text content, but also all page metadata including page name, author name, page size, and creation, modification, and expiration dates. These metadata are available for all page types, not just HTML.

The Web Object Manager also guarantees link integrity. Page checkout and check-in facilities guarantee that no other author can simultaneously edit the same page. The check-out operation serves to warn other potential authors that the page is under edit by someone else. If the page is available, an author can edit the content with any file-based editor such as Claris HomePage, Microsoft FrontPage, or Netscape Composer.

When pages change, Aziza does more than simply alert the user to bad links; it prevents them. As destination pages move, the links follow or are reassigned upon page deletion.

Uniform security is also provided by Web Object Manager, even across multiple platforms. Aziza authenticates users on the basis of username, password, and domain name or IP address. Wildcards are supported in both domain name and IP address. User authentication content can be loaded in batch from user and group files. Authentication data is replicated across all Aziza Enterprise Web Managers.

The second component, Content Servers, deliver Web content to browsers and editors, and accept new and edited content. For the HTTP protocol, Aziza supports Netscape's FastTrack and Enterprise Web. For the FTP protocol, Aziza provides its own server.

The third component, the Management Clients, provides access to authors via a Java applet, administrators via a Windows NT application, and browsers via Javascript and HTML forms.

For authoring, Aziza does not require software to be installed on client machines. Any user can access all content managed by Aziza with an ordinary browser. Because the Aziza Author's Console (Figure 144) is implemented as a Java applet, no software needs to be installed on client platforms for authoring purposes. The applet is simply retrieved from the server when needed.

With Aziza's drag-and-drop Administrator's Console (Figure 145), a Web administrator can distribute or replicate content across multiple Web servers, whether on a LAN or across the WAN. In implementing load balancing, Aziza ensures that replicated content is always identical across servers. Aziza plugs into each server, allowing any administrator or user to access content across all servers. Each Web server functions as a gateway to the content on all Aziza servers—disentangling logical and physical views. A page is identified by the same URL, regardless of the machine that serves up the page.

Multiple administrators can manage the same site. Administrator operations are defined in a way that ensures that no two administrators can interfere with each other's operations, but which gives both full access to all contents of the Web site.

Figure 144

Aziza's Java applet Author's Console.

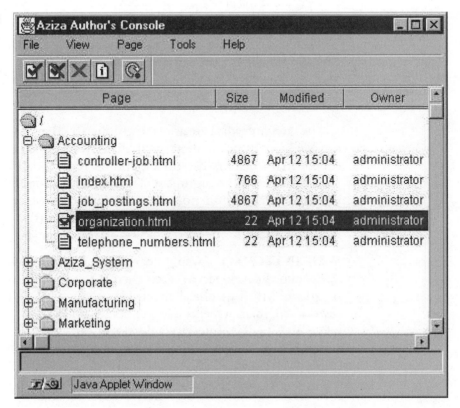

Figure 145
Aziza's drag-and-
drop
Administrator's
Console.

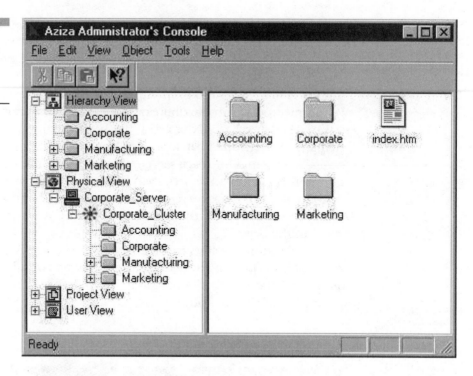

Aziza offers an extensive reporting mechanism. In the Administrator's Console, administrators can access the properties of users, groups, pages, folders, projects, servers, clusters and other objects. These properties include access controls, version histories, replication status, page attributes, etc.

The same metadata are available through a series of a dozen reports accessible by reference to designated URLs. When one of these URLs is accessed, Aziza delivers the corresponding report for the object (e.g., the page) specified as an argument. These individual reports may be linked together into custom reports as frames in an encompassing page. Access to reports can be limited by setting access controls on specific URLs. Finally, SQL tools can be used to access all the metadata.

WEB INTEGRITY   Web Integrity from Mortice Kern Systems Inc. (MKS) allows developers to access their Web sites by using a Web browser or Sidecar, MKS's graphical client-side product. MKS's Web Integrity is a distributed management solution for changing and maintaining Web content. Using Web Integrity, a developer can:

- Freely make changes to a Web site without any assistance from the Webmaster or team leader
- Re-create past documents with complete accuracy

- Safely record new versions of documents for later use
- Easily view a document's full history
- See an audit trail of changes within a document: who, what, and when
- Lock documents for editing, stopping one author from overwriting another

Web Integrity consists of two software components: the Integrity Engine, a server-side product, and Sidecar, a Java client-side interface. The Integrity Engine functions as an executable common gateway interface (CGI), and works with either Netscape's FastTrack and Enterprise servers as well as Microsoft's Internet Information server. Sidecar's Java interface is geared to the nontechnical corporate knowledge worker. In addition to Sidecar, power users can work with a command-line interface through the browser to connect to a project through the Integrity Engine.

Integrity Engine works in the background as the user starts to retrieve and make changes to the site. A browser can be used to retrieve the files by typing a command line similar to the following:

http://strategic.com/nathan/index.htm; version=head:lock

The head revision of the page would appear with a lock on it so others will not be able to make changes on it at the same time. The extra extensions on the URL are Web Integrity's version control syntax. This allows the developer to make changes on the site from anywhere, provided a Web browser is available.

Sidecar offers a much simpler method of communicating with the Web Integrity Engine. Users do not have to remember URL extensions and can revise files using a preferred HTML editor. Sidecar also allows the entire history of an individual page or graphic file to be viewed. The file can be retrieved and locked for editing, then previewed on a browser, uploaded to the Web server, and viewed remotely from the browser to ensure it is correct. The file can then be unlocked so others can work on it.

Web Integrity saves only the files users explicitly check out, saving space for important changes, not simple changes that do not affect the page design or layout. The entire history of the file can be viewed in the browser by clicking on the Changes button of Sidecar. The developer can view or download any previous version, making the re-creation of the site a very simple matter of uploading the file and saving it as the current version.

With Web Integrity's Visual Difference utility, developers can compare text-based files individually, side by side, or in a merged view. And with Track Integrity (Figure 146), managers can view problems by their type,

Figure 146
MKS's Track
Integrity utility for
Web Integrity.

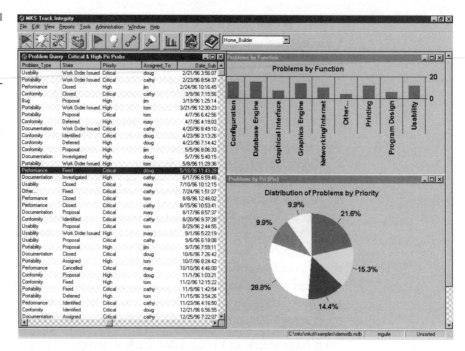

state, priority, who they are assigned to, and the date/time they were submitted. Problem types can be grouped by function for comparison in a bar chart, and problems by priority can be compared in a pie chart.

Editing, approving, and publishing Web content one file at a time has been a tedious and frustrating experience for many managers and developers. Web Integrity includes a feature that eases the content approval and publishing process. Web Lens allows corporate users to group a set of Web files together to travel simultaneously through the automated approval and publishing process. For example, a corporate communications specialist can now submit an entire online newsletter, with all its associated HTML, graphics, and Java files, to a VP for final approval.

SUMMARY

While useful for any document processing application that involves multiple steps, workflow management tools are ideally suited to enhancing the efficiency of Web site development and maintenance. Many corporate Web sites beginning to grow into large, complex databases that are hard to control. While it is easy to add servers and storage to support ever-larger page counts, the long-term concern is how to update and maintain the integrity of all that data. As data amasses and old pages go unchecked, links tend to break down, making the site difficult to navigate and causing users to lose confidence in the validity of the information.

Manual procedures for grooming the data are not sufficient. Consequently, a new generation of tools has become available that manage workflow and automate many database maintenance chores. Their graphical user interfaces make it easy to catalog the pages and links, add content, reassign links, and manipulate entire collections of Web documents that may be distributed among multiple servers. These tools will become even more important as organizations come to rely on their intranets and extranets for mission-critical applications.

See Also **Document Imaging Systems, Groupware**

World Wide Web

The World Wide Web (WWW) is a service available over the Internet. It began as a project at CERN, the European Laboratory for Particle Physics in Geneva, Switzerland. There, Tim Berners-Lee, now director of the World Wide Web Consortium (W3C), developed a vision of the project in 1980. It was finally realized a decade later.

Initially, his project was personal in nature; he was trying to find a way to organize his own activities, but found existing databases and spreadsheets insufficient for his purposes. Furthermore, he noticed that in a place like CERN—where so many people came in with great ideas, did some work, and eventually left—there was virtually no trace of what all these people did and why they did it. Berners-Lee thought there needed to be some place to record and store organizational knowledge so everyone could access it and benefit from it. He finally wrote his so-called "World Wide Web program" at the end of 1990, which provided a browser editor and a full client. By simplifying the existing hypertext language syntax, called Standard Generalized Markup Language (SGML), Berners-Lee developed a language anyone could use to put documents online for public access. Documents are written in the now familiar Hypertext Markup Language (HTML), which includes tags for creating links to other documents, regardless of their location. Through the browser, anyone can read the documents and even search through them by entering keywords.

The first browsers were text-based. Graphics capability was added by Mark Andreeson at the National Center for Supercomputer Applications (NCSA) at the University of Illinois. There, he developed Mosaic, the first graphical Web browser. Andreeson quickly found himself deluged by excitement in Mosaic. Eventually, he left the university to found Netscape. His company's products—Netscape Navigator and Netscape Communicator—now dominate the browser market, along with Microsoft's Internet Explorer.

Since 1990, the Web has grown into one of the most sophisticated and popular services on the Internet. All major companies, educational institutions, associations, libraries, museums, government agencies, and military branches have a presence on the Web, as do millions of individuals—each contributing to the vast storehouse of information that anyone with an Internet connection and browser-equipped computer can explore. Although no specific organization exercises administrative control of the Web, order is imposed by the languages and protocols that constitute worldwide standards, such as the Hypertext Markup Language and HyperText Transfer Protocol (HTTP).

Hypertext Markup Language

HTML is a set of tags that enables documents to be published on the Web. As noted, HTML started as a subset of SGML and has been in general use since 1990. Since then, HTML has branched off into new directions—some of them proprietary.

The purpose of HTML tags is to give Web browsers the information they need to properly render documents so they appear as the author intended. The tags encapsulate or surround various elements of a document, such as headings, paragraphs, lists, forms, tables, and frames. Tags also specify items such as hypertext links, fonts, colors, and backgrounds. In combination with various scripts, tags can also point to Java applets, JavaScript functions, and Common Gateway Interfaces (CGI) for forms processing and database access.

HTML is not a programming language in the normal sense. It is more like the simple notations a magazine editor uses to get an article ready for publication, which is why it is referred to as *markup language*. HTML is essentially ASCII or plain text embedded with special tags that specify how documents are to be rendered by Web browsers for viewing by clients. A feature's end tag is indicated by a slash (/). Together, start and end tags have the following format:

```
<TAG>. . .</TAG>
```

The dots between the tags represent the specific text areas of the document that will be rendered by a browser according to the tags' instructions. The following are some of the most frequently used HTML tags, including those used to build hypertext links:[9]

---

9. There are many more HTML tags, but a complete description of all of them is beyond the scope of this book.

<HTML>. . .</HTML> Delimits the start and end of the HTML document (this is now optional).

<HEAD>. . .</HEAD> Delimits the start and end of the header portion of the HTML document.

<BODY>. . .</BODY> Delimits the start and end of the body portion of an HTML document.

<TITLE>. . .</TITLE> Delimits the title string of the HTML document, which appears above the menu bar of the Web browser's GUI.

<P> Indicates the end of a paragraph and separates two paragraphs with one line of white space.

**
** Equivalent to a hard return, and does not add a line of white space.

<ADDRESS>. . .</ADDRESS> Delimits the address text in an HTML document, which is used to frame such information as the name of the document author, e-mail address, and document modification date.

<TEXTAREA>. . .</TEXTAREA> Delimits more than one line of text in a scrollable area, the dimensions of which are defined by the ROWS and COLS attributes.

<PRE>. . .</PRE> Delimits text to be displayed in a nonproportional font with all spacing intact and without automatic line wrap.

<H1>. . .</H1> Delimits a level-one heading, providing the largest font size. H6 provides the lowest font size.

<B>. . .</B> Delimits the start and end of boldface text.

<I>. . .</I> Delimits the start and end of italic text.

<U>. . . Delimits underlined text. (Not often used because it may be confused with a hypertext link, which is also indicated by underlined text.)

<UL>. . .</UL> Delimits an unordered (bulleted) list.

<OL>. . .</OL> Delimits an ordered (numbered) list.

<LI> Indicates a bulleted item in an unordered list or a numbered item in an ordered list.

<DL>. . .</DL> Delimits a definition list, consisting of a variable number of alternating terms and definitions.

<DT> Indicates the start of a term within a definition list.

<DD> Indicates the start of a definition within a definition list.

<SELECT>. . .</SELECT> Delimits a multiple-choice list, which is typically rendered as a drop-down or pop-up menu. Each item in the menu list starts with an OPTION tag.

<A HREF="goto.html">. . .</A> Anchor tag that indicates a hypertext link. The text within quotation marks refers to the name of a target document or program.

<A NAME="#. . .">. . .</A> Indicates a label within the same or target document that is used as the target of a hypertext link. The text within quotation marks is the label and is preceded by a pound sign (#).

<IMG SRC="graphic.gif"> Indicates the location of an image within the document. Images are usually in GIF or JPG format.

<TABLE>. . .</TABLE> Delimits a table, the rows of which are defined by table row . . . tags and cell contents by table data . . . tags.

<FRAMESET>. . .</FRAMESET> Delimits and specifies two or more HTML files that will be rendered adjacent to each other as separate display areas.

<FORM>. . .</FORM> Delimits the area of the HTML document that provides input fields of a form.

<INPUT> Describes the input field using such attributes as type, name, and size.

<SCRIPT>. . .</SCRIPT> Delimits the area of the HTML document that contains the coding for a scripting language such as JavaScript, as in <SCRIPT language="JavaScript">. . .</SCRIPT>.

<APPLET>. . .</APPLET> Delimits the area of the HTML document that identifies a Java applet.

HTML documents can be created with a simple text editor, a word processor, or a graphical development tool that makes it unnecessary for authors to learn the growing number of HTML tags and their proper usage. With graphical development tools, users simply drag and drop various elements into a workspace or highlight portions of a document for attributes such as font size, font style, paragraph, or list as if using a word processor such as Microsoft Word or Corel WordPerfect.

HyperText Transfer Protocol

HTTP is used to transfer hypertext documents among servers on the Internet and, ultimately, to a client—the end user's browser-equipped computer (Figure 147). Collectively, the tens of thousands of servers distributed world-

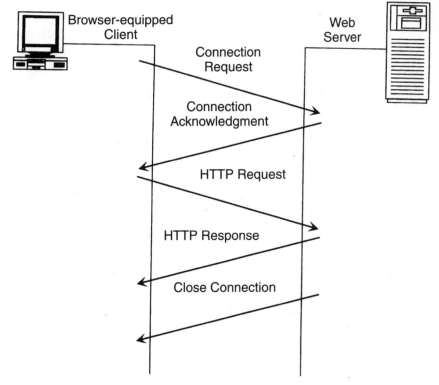

Figure 147
The HyperText Transfer Protocol (HTTP) delivers documents from Web servers to browser-equipped clients in response to specific requests, then closes the connection until a new request is made from the client.

wide that support HTTP are known as the World Wide Web. HTML is used to structure information that resides on the servers in a way that can be readily rendered by browser software, such as Internet Explorer, Netscape, or Mosaic, installed on the clients. HTML makes documents portable from one computer platform to another and is intended as a common medium for tying together information from a multitude of sources.

Business Applications

Many organizations are finding that the Web is an ideal medium for distributing documents and software, while others are pursuing electronic commerce. Some companies use the Web to enhance support of their networks and systems. Among the routine tasks that can be implemented over the Web are:

- LAN managers at distributed locations can access an HTML-coded database stored in an internal Web server to troubleshoot system and network problems. The Web server can make a valuable adjunct to the help desk, especially when corporate locations are spread across multiple time zones.

- Service requests can be dispatched electronically to carriers, vendors, and third-party maintenance firms via standardized HTML forms.

- For remote sites that are too small to be economically tied into the corporate backbone network, the use of HTML forms can convey move, add, and change information to a central management console to facilitate asset management.

The Web is also used for more sophisticated network support. Many interconnect vendors, for example, are using Java to build network management applications that can be accessed through Web browsers. Through hypertext-linked home pages set up by the vendor, network managers can use their Java-enabled Web browsers to launch various network management applications. Routers, switches, hubs, multiplexers, and CSU/DSUs—virtually any network device—can be configured, monitored, and maintained in real time from any location. Applications that provide trend analysis, network reports, access to the vendor technical support, and online documentation are also integrated through a Web browser so configuration changes and network planning can be accomplished using real-time data instead of guesswork.

World Wide Web Consortium

The World Wide Web Consortium (W3C) was created in 1994 to develop common protocols that enhance the interoperability and promote the evolution of the Web. W3C is a vender-neutral industry consortium of over 220 members jointly run by the MIT Laboratory for Computer Science (LCS) in the U.S., the National Institute for Research in Computer Science and Control (INRIA) in France, and Keio University in Japan. Members include hardware and software vendors, telecommunications companies, content providers, corporate users, and government and academic entities. Funding comes from membership dues and public research funds. As noted, Tim Berners-Lee is the organization's director.

SUMMARY

The Web is the fastest growing communications medium of all time. According to the investment firm Morgan Stanley, radio took 38 years to attract 50 million users, television took 13 years, cable took 10 years, and the Web has taken only 5 years. Various research firms peg the number of Web users worldwide at 200 to 300 million. The Web has become a cost-effective medium for information distribution, electronic commerce, and the delivery of support services. Its capabilities are continually being expanded. In addition to text and images, the Web is routinely used for telephony, videoconferencing, group scheduling, faxing and paging, collaborative computing, and electronic commerce.

See Also **Internet Servers, Web Portals**

Year 2000 Compliance

The year 2000 (Y2K) problem stems from the inability of computers and embedded chips to accept the two-digit date for the year 2000 (00) and misread it as 1900, leading to possible system shutdowns among businesses, transportation, banking, electric power, and telecommunications.

The source of this date problem originates from the early days of computers, which had limited memory capacity. Valuable space was saved by abbreviating the year to only two digits. The year 1978, for example, was stored simply as 78. This scheme worked well over the years, except that with the year 2000 approaching, computers will interpret 00 as 1900. This has potentially disruptive consequences for any computer-controlled system that implements date-sensitive functions.

Bellcore (now known as Telcondia Technologies) has determined that 75 percent of voice-networking equipment, up to 35 percent of data networking devices, and almost 100 percent of network-management devices are date-sensitive. Date-sensitive functions include service routing and scheduling, message reporting, clock maintenance, event alarm, and security of log-ins and passwords.

Potential Problems

If left uncorrected, the seriousness of this problem would range from minor nuisances to major disasters. For example, e-mail programs sort messages by date. When a program interprets the date as 1900 rather than 2000, it will put current message at the bottom rather than the top of the pile. Automated inventory control programs are likely to initiate orders too soon or too late. Cable TV subscribers may experience a variety of service disruptions including erratic commercial insertions, inability to order pay-per-view channels, and improper billings.

On corporate LANs, if date incompatibilities in distributed hardware and software are not caught in time, local execution problems will not only disrupt users but also corrupt central data during interaction with servers and mainframes.

With regard to telecommunications, the consequences of the Y2K problem are more serious. Because U.S. and global economies are dependent on continuous communications, a failure could affect telecommunication networks, major utilities, transport, production lines, security systems, and other systems that rely on software and microprocessors that have any date awareness or time dependency. Because telecommunication is based on the seamless interconnection of networks, the international dimensions of the Y2K problem are of particular concern. A problem in Bulgaria, for example, could mean worldwide phone systems would be unable to accept, terminate, or route calls to or through Bulgaria, thus affecting the global system. Countries without the resources to adequately address the Y2K issue can pose problems long into the future.

Of a more serious nature are problems with 911 emergency communication systems used by police, fire, and rescue organizations. There are three components in every 911 call: the public switched telephone network (PSTN), the public safety answering point (PSAP), and the automatic location database (ALI). The PSAP receives 911 calls and logs their date, time, and duration. The PSAP sends the caller telephone numbers to the ALI database and uses them to search the ALI database and retrieve the proper addresses. The Y2K problem could lead to a failure to incorporate ALI database updates, thus resulting in incomplete or incorrect location information; incomplete transmissions between the PSAP and the ALI database; and, particularly if a 911 call originates from behind a private branch exchange (PBX) or similar type of private switch, a failure of the 911 call to reach the PSAP.

The most serious ramifications of the Y2K problem involve disruption to air and sea navigation systems. At airports, for example, any delay in processing information could put lives and property at risk. At a minimum, one airport's problems would delay or ground flights, affecting travel at other airports.

Corrective Action

Resolution of the Y2K problem calls for the intense and comprehensive upgrade of system hardware and software facilities around the world. A rather vexing problem is that of a shortage of engineers and programmers able to quickly tackle the Y2K problem.

There are literally hundreds of millions of program instructions that have to be reviewed and tested in order to bring all systems into compliance. Further, dates can be stored in programs and databases in many different formats. All date references have to be found and checked. The problem has been equated to simultaneously changing all of the light bulbs in a large city. This is not a difficult technical challenge, but it is a logistics nightmare.

To help weed out Y2K problems, vendors are offering tools that can help network administrators find and fix problems in far-flung PCs, servers, and network gear. Tivoli Systems, for example, offers a module for its Tivoli Enterprise, the company's IT management software. The module helps users more effectively review and continually measure Year 2000 readiness on desktop computers, in their network, and in the rest of their distributed enterprise. The module uses information collected by the Tivoli Inventory application about server and desktop hardware and software configurations, and network device information collected by Tivoli NetView. Tivoli Inventory allows the review process to be automated and controlled from a central console, eliminating the intensive manual effort required to perform this task. The collected information is then analyzed by Tivoli Decision Support to aid customers in reviewing the Year 2000 readiness status of different elements in the enterprise.

The Network Testing Committee (NTC), a telecommunications industry forum sponsored by the Alliance for Telecommunications Industry Solutions (ATIS), began internetwork interoperability testing in January 1999 to evaluate the impact of the Year 2000 date change on the nation's PSTN. This was accomplished by rolling forward the dates of interconnected switches in a laboratory environment to stimulate date rollovers for December 31, 1999 to January 1, 2000, plus three additional significant date changes: February 28 to February 29, 2000; February 29 to March 1, 2000; and December 31, 2000 to January 1, 2001. During each simulated rollover, the signaling network was monitored to ensure that the public network responded in a satisfactory manner. Additionally, individual service providers and vendors of telecommunications equipment engaged in the testing of their networks, products and support systems in a stand-alone environment.

SUMMARY

Billions of dollars have been directed at the Y2K problem by most of the industrialized world, and some estimates have put the global cost at more than $1 trillion. There are many different opinions on what might happen as a consequence of ignoring or not adequately addressing the problem. Most experts agree that, when the millenium rolls over, computer code that cannot recognize the new date will cause computers to fail. What is unknown is the scale of those failures, how quickly the systems can be restored, the extent of the damage that may result, and what long-terms effects, if any, the disruptions will have.

INDEX

Note: **Boldface** numbers indicate illustrations.

ABOUT THE AUTHOR

Nathan J. Muller is an independent consultant specializing in advanced technology marketing, research, and education. He wrote the companion *Desktop Encyclopedia of Telecommunications*, as well as *Mobile Telecommunications Factbook*, 14 additional books and more than 1500 articles on a variety of networking and telecommunication topics. He is a regular contributor to Gartner Group's *Datapro Research Reports*.